Ecosystem

TEMPERATE FORESTS

Revised Edition

Ecosystem

TEMPERATE FORESTS

Revised Edition

Michael Allaby

Illustrations by Richard Garratt

®Facts On File
An imprint of Infobase Publishing

TEMPERATE FORESTS, Revised Edition

Copyright © 2008, 1999 by Michael Allaby

Facts On File, Inc.
An Imprint of Infobase Publishing
132 West 31st Street
New York NY 10001

ISBN-10: 0-8160-5930-6
ISBN-13: 978-0-8160-5930-0

Library of Congress Cataloging-in-Publication Data
Allaby, Michael.
    Temperate forests / Michael Allaby; illustrations by Richard Garratt. — Rev. ed.
        p. cm. — (Ecosystem)
    Includes bibliographical references and index.
    ISBN 0-8160-5930-6
    1. Forest ecology—Juvenile literature. 2. Forests and forestry—Juvenile literature. I. Garratt, Richard, ill. II. Title.
    QH541.5.F6A46 2007
    577.3—dc22   2006028859

Text design by Erika K. Arroyo
Illustrations by Richard Garratt
Photo research by Elizabeth H. Oakes

Printed in the United States of America

Bang Hermitage 10 9 8 7 6 5 4 3 2 1

This book is printed on acid-free paper.

Contents

Preface

Increasingly, scientists, environmentalists, engineers, and land-use planners are coming to understand the living planet in a more interdisciplinary way. The boundaries between traditional disciplines have become blurred as ideas, methods, and findings from one discipline inform and influence those in another. This cross-fertilization is vital if professionals are going to evaluate and tackle the environmental challenges the world faces at the beginning of the 21st century.

There is also a need for the new generation of adults, currently students in high schools and colleges, to appreciate the interconnections between human actions and environmental responses if they are going to make informed decisions later, whether as concerned citizens or as interested professionals. Providing this balanced interdisciplinary overview—for students and for general readers as well as professionals requiring an introduction to Earth's major environments—is the main aim of the Ecosystem set of volumes.

The Earth is a patchwork of environments. The equatorial regions have warm seas with rich assemblages of corals and marine life, while the land is covered by tall forests, humid and fecund, and containing perhaps half of all Earth's living species. Beyond are the dry tropical woodlands and grassland, and then the deserts, where plants and animals face the rigors of heat and drought. The grasslands and forests of the temperate zone grow because of the increasing moisture in these higher latitudes but grade into coniferous forests and eventually scrub tundra as the colder conditions of the polar regions become increasingly severe. The complexity of diverse landscapes and seascapes can, nevertheless, be simplified by considering them as the great global ecosystems that make up our patchwork planet. Each global ecosystem, or biome, is an assemblage of plants, animals, and microbes adapted to the prevailing climate and the associated physical, chemical, and biological conditions.

The six volumes in the set—*Deserts, Revised Edition*; *Tundra*; *Oceans, Revised Edition*; *Tropical Forests*; *Temperate Forests, Revised Edition*; and *Wetlands, Revised Edition*—between them span the breadth of land-based and aquatic ecosystems on Earth. Each volume considers a specific global ecosystem from many viewpoints: geographical, geological, climatic, biological, historical, and economic. Such broad coverage is vital if people are to move closer to understanding how the various ecosystems came to be, how they are changing, and, if they are being modified in ways that seem detrimental to humankind and the wider world, what might be done about it.

Many factors are responsible for the creation of Earth's living mosaic. Climate varies greatly between Tropics and poles, depending on the input of solar energy and the movements of atmospheric air masses and ocean currents. The general trend of climate from equator to poles has resulted in a zoned pattern of vegetation types, together with their associated animals. Climate is also strongly affected by the interaction between oceans and landmasses, resulting in ecosystem patterns from east to west across continents. During the course of geological time, even the distribution of the continents has altered, so the patterns of life currently found on Earth are the outcome of dynamic processes and constant change. The Ecosystem set examines the great ecosystems of the world as they have developed during this long history of climatic change, continental wandering, and the recent meteoric growth of human populations.

Each of the great global ecosystems has its own story to tell: its characteristic geographical distribution; its pattern of energy flow and nutrient cycling; its distinctive soils or bottom sediments, vegetation cover, and animal inhabitants; and its own history of interaction with humanity. The books in the Ecosystem set are structured so that the different global ecosystems can be analyzed and compared, and the relevant information relating to any specific topic can be quickly located and extracted.

The study of global ecosystems involves an examination of the conditions that support the planet's diversity. But environmental conditions are currently changing rapidly. Human beings have eroded many of the great global ecosystems as they have reclaimed land for agriculture and urban settlement and built roads that cut ecosystems into ever smaller units. The fragmentation of Earth's ecosystems is proving to be a serious problem, especially during times of rapid climate change, itself the outcome of intensive industrial activities on the surface of the planet. The next generation of ecologists will have to deal with the control of global climate and also the conservation and protection of the residue of Earth's bio-diversity. The starting point in approaching these problems is to understand how the great ecosystems of the world function, and how the species of animals and plants within them interact to form stable and productive assemblages. If these great natural systems are to survive, then humanity needs to develop greater respect and concern for them, and this can best be achieved by understanding better the remarkable properties of our patchwork planet. Such is the aim of the Ecosystem set.

Acknowledgments

All of the maps and diagrams in this book were drawn by my friend and colleague Richard Garratt. Richard and I have been collaborating for more years than I am sure either of us would care to count and, as always, I am deeply grateful to him. I drew up a list of photographs to illustrate the text, but the task of finding them fell to Elizabeth Oakes, with results that are obvious. Without her enthusiasm and skill this would have been a much duller and less informative book. Finally, I wish to thank my friend and editor of many years Frank K. Darmstadt, whose encouragement and guidance helped me greatly.

Introduction

A long time ago, and subsequently for tens of thousands of years, northern Europe, Canada, and most of the northern United States lay buried beneath vast ice sheets, thousands of feet thick. As the world then warmed and the ice began to melt along its southernmost edge, plants began to appear. Seeds and spores, carried on the wind or dropped by birds, germinated and brought color to cold, barren lands. In time the first plants gave way to others, to woody shrubs, and, among the shrubs, a few trees. Little by little, the bare ground freed by the retreat of the ice, turned to forest. This was the forest of the temperate regions, the temperate forest.

In fact, it was not one forest but many. Different forests supported different communities of trees and the plants and animals that live with them, but the forests comprised a number of principal types. In the south, around the Mediterranean and in some of the southern parts of North America, the trees and shrubs were broad-leaved but retained their leaves through the winter. These were broad-leaved evergreen forests. Elsewhere, in regions of mist and very high rainfall, there were temperate rain forests. In other places, and more extensively, the forest trees were broad-leaved but leafless in winter—the broad-leaved deciduous forests. To the north of them, in a wide belt running across Canada, Europe, and Asia, grew a vast coniferous forest, the boreal forest known in Russia as the *taiga*. Together, these different types of forest constitute the temperate forests of the world, forming one of the great vegetation types, or biomes. They occur almost wholly in the Northern Hemisphere because in the Southern Hemisphere there is much less land in temperate latitudes. This book is about the temperate forests. It begins with a few descriptions and definitions. What do we mean by temperate forest? Where is it to be found? How did it come to be the way it is?

■ SETTING THE SCENE

Having framed the questions, the book provides some of the answers. These involve several excursions through time.

They describe how the lands where the forests grow came to lie in the temperate regions and how the way continents have moved explains some otherwise rather curious facts about the distribution of particular trees. Forests grow in certain types of soil. The book explains how soil forms and is classified scientifically, how water flows through it, and what forest soils are like.

Soils develop because of a process called *weathering*. Despite the name, not all weathering is directly caused by the weather, but much of it is, and the book tells some of the story of the climates of temperate regions, why weather happens the way it does, and how it affects the forests. Climates are changing constantly and trees are quite sensitive to those changes. This makes it possible for scientists to use evidence of trees that grew in the distant past to trace the history of those changes.

■ TREES AND FOREST COMMUNITIES

Having set the scene, as it were, with the lands in their proper places and the soils described, the book explains a little about how plants work in general and trees in particular. Trees and shrubs produce wood, and the book explains what wood is and describes how it forms. Part of the description of trees involves explaining the differences between the two main kinds of trees found in temperate forests, those with flowers and broad leaves and those with narrow, sometimes needlelike, leaves and bearing *cones*: the conifers.

A forest contains more than just trees. There are plants of many kinds and animals that live on and among those plants. A forest is a community of living things, and the scientific study of relationships within communities and between them and the nonliving environments they inhabit is called *ecology*. Quite a large section of the book is devoted to an outline of ecological ideas, especially as these apply to forests. Forests support a wide variety of species of plants, fungi, animals, and single-celled organisms. In other words, they are biologically diverse. The expression *biological diversity* is usually contracted to *biodiversity*. The book explains what biodiversity means and why preserving it is important.

The following pages are devoted to descriptions of individual species of trees. Obviously, it cannot hope to be complete, but it does include the most common coniferous and broad-leaved trees of North America and Eurasia. It also tells where each species originated. Most are now cultivated in places thousands of miles from where they grow naturally, so it is worth remembering where they began. What is now the most common tree in Britain, for example, is a native of the coastal regions of northwestern United States and far western Canada.

■ FORESTS AND PEOPLE

The final section of the book, amounting to about one-third, concerns the relationship between people and the temperate forests. There has always been a relationship, of course, for as the forests expanded northward following the end of the last ice age, humans were not far behind, and they very soon began to alter the forests. That was long before the earliest written record, of course, and so the book explains a little of how scientists can reliably reconstruct what happened so long ago.

By the time of the first written records of forests, the changes wrought by human activity were well advanced, and in southeastern Europe some of the adverse consequences were being experienced. In northern Europe steps started to be taken to conserve the forests. Clearly, forests were important, but they were not always loved. The book contrasts the efforts to conserve forests with the way forests were portrayed in folklore and literature, often as a dark, dangerous, inhospitable wilderness.

Whether or not people liked their forests, they certainly depended on them for fuel, building timber, and wood for making furniture and a vast array of everyday articles, each tree yielding wood with characteristics that suit it to particular uses. This section of the book describes some of the products of temperate forests and the ways they have been used. Society still depends on its forests, of course, nowadays for paper as well as wood, so the condition of the temperate forests concerns everyone. The following section of the book describes the effects on forests of pollution by acid rain, of the possible consequences of climatic change, and also of the effect on soils of deforestation. This is followed by accounts of the present state of each of the principal types of forest and of forest conservation.

■ FOREST MANAGEMENT AND PLANTATION FORESTRY

Productive forests must be managed, and in Europe they have been managed for more than one thousand years. The final section of the book describes some of the types of management that have been used for centuries and the kinds of forest they produced. More recently, the growing of trees in plantations has become the most common technique for providing forest products, and in years to come plantations are likely to become still more widespread as ancient and old-growth forests are protected for conservation reasons. Plantation forestry and the modern way of controlling forest weeds, pests, and diseases are described in some detail. This is partly because of the obvious importance of plantations and partly to show that modern forestry is a great deal more complicated than it may seem. There is much more to it than simply planting rows of trees.

Times change and attitudes change with them. When plantation forestry began, the goal was a dependable supply of good quality timber, mainly from straight, relatively fast-growing coniferous trees such as spruces, pines, hemlocks, firs, and larches. Now it is realized that forests are extremely valuable as intricate mosaics of natural habitats that support a wide variety of species. Forest composed of self-sown trees and other plants that have remained relatively undisturbed for a long time provides the best habitat, but plantations support a considerable amount of wildlife and, with some modification, they can support a great deal more. Plantations are now planned with wildlife conservation in mind. Forests within convenient reach of cities also have a very high amenity value. They are places where families can walk, picnic, and relax in safe, interesting surroundings. The more accessible modern plantations provide recreational opportunities as well as timber.

The task has taken several centuries, but at last landowners and managers have tamed the forests. They are no longer the wild, dark, frightening places of legend, to be cleared so that danger could be seen approaching from afar. Farmers now produce as much food as their nations need, so they no longer need to clear away the forests to make fields for growing crops or raising livestock. Perhaps for the first time people are able to appreciate forests for their beauty, for their tranquillity, and for the plants and animals that live in them. We are learning to cherish them by preventing the felling of the more valuable or interesting natural forests, relying on plantations to supply the forest products we need, and seeking to enhance the biological and aesthetic value of those plantations.

That journey, from the primeval forest to the plantation and from the forest seen as the enemy to the forest as a friend, is the subject of this book.

Temperate Forests, Revised Edition contains a certain amount of scientific and technical detail, but I have tried to present it clearly, and it should not be difficult to follow. So far as possible, technical terms are avoided, and where they cannot be avoided, they are defined in simple language. These terms are also defined in a comprehensive glossary.

Scientists use units of measurement that are defined in the Système International d'Unités, known for short as the SI system. These units are listed in an appendix, together with their equivalents in units that may be more familiar. In the text the customary units are given first, with the SI units in parenthesis. Many SI units are also metric units, and in these cases the conversion in the text is between the most obvious comparisons—inches and centimeters, miles and kilometers, pounds and kilograms, for example. There is one exception to this rule: By convention, amounts of rainfall are always expressed in millimeters, not centimeters. This is to avoid confusion, so precipitation in different places and at different times can be compared without the risk of making mistakes.

▪ A REVISED EDITION

Several years have passed since the first edition of *Temperate Forests* appeared. Scientific research never ceases, and much has been learned about trees and forests during those years, and even more about the climatic conditions that favor them and the climate cycles that make them expand and contract. I was very glad, therefore, when my friends at Facts On File invited me to revise the book for a new edition.

I have updated all of the facts and figures contained in the book and have rearranged or rewritten most of the text. This makes the revised edition substantially different from the first edition. I hope that the changes I have made explain important ideas and concepts more clearly.

Some explanations that are necessary to an understanding of deserts would interrupt the flow if they formed part of the main text. Broad-leaved forests are renowned for their splendid fall colors—but why do leaves change color? Each type of cloud has its own name—but how does cloud-naming work? Winds blow around weather systems due to the *Coriolis effect*—but what exactly is that? It would be possible to rely on brief glossary definitions to explain such ideas, but that would be unsatisfactory and would mean having to turn to the back of the book to find them. Instead, I have placed these explanations in sidebars. Sidebars allow me to explain an idea fully without getting in the way of the main text.

I have also used sidebars to provide brief biographical details about individuals who have made important contributions to our knowledge of the Earth and of ecology. The list includes Alfred Wegener, the German meteorologist who first proposed the idea of continental drift, Vasily Dokuchayev, the Russian scientist who was the first to classify soils, and the ecologists Frederic Clements, Arthur Tansley, Charles Elton, and Henry Gleason. In all there are more than 30 sidebars, distributed through the book.

At present the world is growing slowly warmer, and climate change may mean that temperate forests disappear from some places but new forests develop in others. I hope that with all its many revisions *Temperate Forests, Revised Edition* will supply useful information to help readers engage in the ongoing debate about the implications of global warming.

—Michael Allaby
Tighnabruaich, Scotland
www.michaelallaby.com

1

Geography of Temperate Forests

The temperate regions of the world lie between the Tropics and the Arctic and Antarctic Circles, and the forests of the temperate regions—the temperate forests—are adapted to the soils and climates found there. Temperate climates are less extreme than those of the Tropics and the polar regions, but they vary widely and that variation is reflected in the diversity of temperate forest types.

This chapter begins by defining temperate forest and the characteristics of the trees that grow in them. Those trees have particular climatic requirements, and the variety of temperate climates supports quite different kinds of forest. Climates change over long periods, and forests have not always stood on the lands that now support them. They migrate to colonize newly hospitable areas and adverse environmental changes can cause them to disappear.

■ WHAT IS A "TEMPERATE FOREST"?

As the last of the winter snow melts and water, sparkling in the spring sunshine, drips from the boughs, the first green shoots break through the ground surface. Soon a carpet of spring flowers will pattern the forest floor in bright colors and then disappear as the buds, first on one tree then another, burst into leaves, intercepting the sunlight and plunging the floor into deep shade.

Summer is the green time, when the trees are in full leaf. It ends when the days grow shorter and the nights cooler, the green is transformed into the reds, oranges, and browns of fall, and the forest is revealed in what many people consider its true magnificence (see the sidebar on page 2). That, too, is brief. Each in turn, the trees shed all their leaves into the swirling winds that heap them in careless piles through which children wade and kick their way, leaving the trees themselves bare, their twigs silhouetted as delicate traceries against the sky. Rain now gives way to snow, puddles freeze, the ground is frozen hard, and the yearly cycle is complete.

Everyone delights in forests that change their appearance so dramatically with the seasons and, of course, it is not only the trees and herbs that change. Forests support a wide variety of animals and they, too, adjust their lives to the seasons. Spring, which brings the brief flowering of herbs, also brings the first migrant birds. Summer brings out the insects. Mammals, though rarely seen, forage and hunt while food is available.

Not all forests change in this way. The changes are adaptations to a seasonal climate, and they do not occur in the *Tropics,* where the seasons are absent or not so clearly defined. Changing forests, then, are the products of seasonal climates, and these are found in temperate regions of the world. Changing forests are also temperate forests.

Deciduous or Evergreen?

Trees that shed their leaves in fall are said to be *deciduous,* from the Latin *de,* meaning "down," and *cadere,* "fall." Most deciduous trees, but not all of them, also have broad leaves. Oaks, maples, hornbeams, and beeches are broad-leaved deciduous trees. They are *angiosperms,* plants that produce seeds enclosed within an *ovary* (see "Flowering plants" on page 101). Plants producing seeds that are not enclosed in ovaries are called *gymnosperms,* and the most abundant examples are the *coniferous* trees. Their seeds are borne in the female cones and their leaves are often in the shape of needles; *acicular* is the botanical term describing this leaf shape. Needles are shed, but not all at the same time, so most coniferous trees are *evergreen.* Larches are an exception. They are deciduous conifers. Their needles, like the leaves of broad-leaved deciduous trees, turn golden in fall, bringing patches of bright color to the otherwise dark green pine forest.

Just as some coniferous trees are deciduous, there are also broad-leaved trees that are evergreen. Holly is a familiar example, and some oaks are also evergreen. Holm oak, cork oak, and kermes oak of the Old World and the live oaks of the New World are evergreen oaks.

The fall colors of broad-leaved deciduous forests are spectacular and nowhere more so than here, in the Appalachians. *(Fogstock.com)*

All of these trees grow in forests, although they are often to be seen growing by themselves in parks or patches of woodland much too small to be considered forests. They are adapted to seasonal climates, and so they are trees of temperate forests—because it is in temperate regions that climates are seasonal—but the seasons to which they are adapted vary. Broad-leaved evergreen trees require mild, wet winters and can tolerate hot, dry summers. Broad-

Why Leaves Change Color in the Fall

Chlorophyll, the green pigment that absorbs light photons during the first stage in *photosynthesis,* is present in all the cells immediately below the protective outer *cuticle* ("skin") of plant leaves. It is chlorophyll that gives the leaves their green color.

In winter, when water in the soil is likely to be frozen and, therefore, unavailable to the plant roots, the plant faces a problem. If it continues to photosynthesize, it will need to absorb carbon dioxide and rid itself of oxygen, but to do that it will need to open its *stomata,* and while they are open water will evaporate through them. The leaves will lose moisture that cannot be replaced, and the plant is likely to suffer from dehydration that may prove fatal. Many herbs die down in autumn and spend the winter in a dormant state. Others, including evergreen trees and shrubs, have leaves that minimize the loss of moisture.

Deciduous trees deal with the problem differently—by shedding their leaves in autumn. The tree cannot photosynthesize without its leaves so it shuts down for winter.

In preparation for winter, the tree moves essential nutrients from its leaves and back into the stem, where they are stored. In spring they will be available for the new leaves. Autumn leaves cease making chlorophyll, and as their stored chlorophyll breaks down, they lose their green color. This allows other pigments, present in the leaf all the time but masked by the chlorophyll, to become visible. The chemical reactions involved in removing nutrients from the leaves also produce new pigments. The chlorophyll gone, leaves become brown, yellow, orange, or red, depending on the balance of pigments they contain.

Finally, plant hormones start breaking down the very thin walls of cells in the *abscission layer* at the base of each *petiole* (leafstalk). A layer of *cork* forms at the base of the abscission layer, sealing the tree stem against infection. Then the weight of the leaf and the wind combine to break the petiole at the abscission layer, and the leaf falls.

leaved deciduous trees are best suited to warm, wet summers but can tolerate winters cold enough for the ground surface to freeze, provided it does not remain frozen for very long. Coniferous deciduous trees tolerate severe winters and grow well in northern Siberia, but they thrive because of their ability to grow rapidly during the short spring and summer. Coniferous evergreens also tolerate severe winters and short summers.

What Is a Forest?

Clearly, temperate forests are highly variable. Their character and composition change from one region to another. They are also by far the most extensive of all forest types. Tropical forests occur within the Tropics, as their name suggests, but beyond the tropic of Cancer in the North and tropic of Capricorn in the South, the forests are temperate. In high latitudes temperate forests extend to the limit of tree growth, their trees gradually becoming more widely spaced and more stunted until the forest gives way to *tundra*. Over the world as a whole, at present, temperate forests occupy about 6.6 million square miles (17 million km²). This is a large area, amounting to approximately 11.5 percent of the world's total land area, but modern forests cover a smaller area than they did before the trees were felled for timber and to provide land for farming (see "Forest Clearance in Prehistory" on pages 222–224).

This great diversity of forest types is made more comprehensible by a system of classification, and the first word requiring definition is *forest* itself. The word is derived from the Latin *foris,* meaning "out of doors," and in medieval times people talked of the *forestem silvam,* the wood (*silva*) that lay outside the land enclosed by fences. Cultivated fields were enclosed, as were parks in which deer were raised, and the forest lay outside the fences. In England under Norman rule, substantial parts of this forest were set aside for hunting, much of it for the use of the king, and it was subject to special forest laws. The designation of the hunting forest used the word *forest* in its old sense of "outside," regardless of whether or not the area was covered with trees. Much was what would be called forest today because that was the natural vegetation in most lowland areas, but the name is still retained for certain upland areas, heaths, or bogs where trees have never grown in historical times. Dartmoor, for example, is a large area of upland moor in southwest England and a national park. Forest plantations (see "Plantation Forestry" on pages 294–300) occupy some areas, but most of Dartmoor is too exposed and has soils that are too thin for trees to thrive. Nevertheless, the southern part of the moor is still shown on maps as "Dartmoor Forest." The apparently inappropriate name recalls the fact that at one time it was an area set aside for hunting, probably by the king.

Nowadays, the word *forest* means either a community of trees growing together so closely that in summer the leaves of one overlap those of its neighbors to form a continuous *canopy,* or the trees making up such a community. As a verb, *to forest* means "to plant with trees." To qualify for the name, the canopy of a forest should cover at least 60 percent of the total area. If the canopy covers less than 60 percent of the area so that the crowns of most of the trees do not touch, the plant community is called *woodland.* Woodland is more open than closed forest, resembling parkland. Its trees are more widely spaced, and the trees themselves usually spread themselves more widely than the more crowded trees of a forest. Between the trees there are areas of grass or shrubs, contributing to the parklike appearance.

Types of Temperate Forest and Where They Grow

Forests are also named according to their type. At the most general level they may be broad-leaved deciduous (also known as summer deciduous), broad-leaved evergreen, or coniferous (also called *boreal*) forest. In each case at least 80 percent of the trees must be of the type giving the forest its name. A mixed forest contains both broad-leaved and coniferous trees, and there will be more of one than of the other, but the less numerous type must constitute at least 20 percent of the total. These names are of limited use, however, because they describe types of forest that may be typical of entire regions of continents, with many local variations, and it is often the variations that matter.

Forests are also identified by the trees that dominate them. Usually two species are named but sometimes three, and the *dominant* is named first. In ecology, the dominant species is the one having the greatest influence on the composition and form of a community. This is commonly, but not invariably, the biggest or most abundant (see "Forest Structure" on pages 157–164). There are many species of plants in the oak–maple forests of Illinois and the pine–hemlock forests of Maine, but it is the oaks and maples trees that provide the overall impression of those forests and therefore give them their names. Those are the plants the observant visitor remembers. As a rule of thumb, the named trees should amount to at least half of all the trees present and should give the forest its character.

That character is quite likely to change from place to place through a large forest, making the forest more like several distinct types of forest growing side by side. In the pattern of vegetation natural to the southeastern United States, for example, broad-leaved evergreen forest along the coastal plain of the Carolinas gives way to oak–pine forest on the higher ground to the west and then to oak–chestnut forest in the Blue Ridge Mountains. These names refer to the natural communities and are retained even if the communities, or species within them, have long since disappeared. American chestnuts still exist, of course, but almost all of those forming part of natural forest communities were killed in the first half of this century by an outbreak of chestnut blight (see "Tree Predators and Parasites" on pages 173–180).

Temperate forests, then, are the forests that occur naturally in those parts of the world where the climate is temperate. This is a vast area and climates vary considerably within it, although all are strongly seasonal, and there are several quite distinct types of forest reflecting those variations.

■ WHERE ARE TEMPERATE FORESTS TO BE FOUND?

Bordering the Tropics in both hemispheres is a rather vaguely defined subtropical belt. The subtropics are the latitudes in which the biggest of the world's deserts are found. Moving farther from the equator, beyond the arid subtropics and the semiarid regions where droughts are frequent and often prolonged, are the world's grasslands. It is in the

still higher latitudes of the temperate regions that temperate forests occur, bounded on one side by the temperate grasslands—the North American prairie, South American pampa, and Eurasian steppe—and on the other by tundra and the polar deserts.

As a map of the world shows, however, the distribution of temperate forests is far from even. Excluding Antarctica, there is much more land in the Northern Hemisphere than there is in the Southern Hemisphere, and the southern landmasses do not extend into such high latitudes as those in the north. The southernmost tip of Africa lies within the subtropical belt, and although South America extends much farther south, the continent tapers sharply. Even then, Cape Horn, the southernmost tip of South America at 55.78°S, is in the same latitude as Copenhagen and Edinburgh. Dunedin, in the south of South Island, New Zealand, is at 45.87°S, in approximately the same latitude as Portland, Oregon, and Venice, Italy. In the Northern

---

The location of the world's broad-leaved and coniferous temperate forests

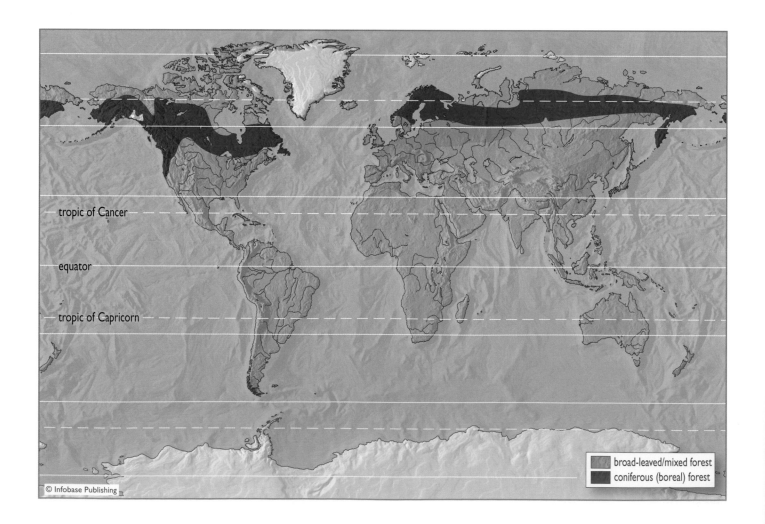

tropic of Cancer

equator

tropic of Capricorn

broad-leaved/mixed forest
coniferous (boreal) forest

© Infobase Publishing

Hemisphere there is very much more land in latitudes higher than these.

It is in these higher latitudes that there are substantial areas within which temperate forest of one kind or another is the natural vegetation. As the map on page 4 shows, apart from the southern tip of South America and New Zealand, temperate forest is found only in the Northern Hemisphere. There it forms a continuous belt across Canada and northern Europe and Asia (together known as Eurasia), and it also occurs in the eastern United States, with southward extensions on the eastern side of North America and Asia.

A traveler through a part of the world shown on vegetation maps as temperate forest should not expect to see forest everywhere. Indeed, over large areas there are rather few forests. Most land in France and Britain, for example, is farmed, and the landscapes are of fields rather than forests. There are large forests in Pennsylvania, covering well over half the land area of the state, but these have been planted to replace the original forest, which was cleared to provide land for agriculture and fuel for industry. Vegetation maps show the type of vegetation that existed prior to human interference and that might be expected to reestablish itself if that interference were to cease.

The Need for a Moist Climate

Forests do not occur in the interior of large continents, even in middle latitudes. The climate there is too strongly continental (see "Continental and Maritime Climates" on pages 85–89), with hot summers, very cold winters, and low rainfall. It is not the winters that limit the extent of forests but the summers and the overall rainfall. Where hot, dry summers last for more than about eight months of the year and the annual rainfall is less than about 16 inches (400 mm), most trees find it difficult to survive. The vegetation becomes scrub, dominated by shrubs with isolated trees scattered among them. As the climate becomes still drier the trees become rarer, eventually giving way to semidesert scrub and *chaparral,* and finally to *desert.* Where winter is the dry season, forests give way to grassland.

Broad-leaved deciduous trees require a minimum of about 120 days a year in which the temperature is higher than 50°F (10°C), and it is short summers that mark the high-latitude boundary of broad-leaved forests. Hardier species can tolerate rather fewer than 120 warm days a year, but the trees become increasingly interspersed with coniferous species, forming mixed forest. Conifers need only about 30 days a year with temperatures about 50°F (10°C), so they can grow in much higher latitudes than can broad-leaved trees. The most northerly forest in the world is in Siberia, at 70.83°N. It is a larch forest, dominated by dahurian larch (*Larix gmelini*), a magnificent tree which grows to a height of almost 100 feet (30 m) and confines its leaf production and growth to the very brief summer.

Forests of the Far North

As summers grow shorter with increasing distance from the equator, only the hardier tree species manage to thrive and the trees become more widely spaced. The forest becomes thinner, with trees occurring in isolated clumps separated by areas of heathlike vegetation and *lichens.* This belt of parklike vegetation separates the closed-canopy conifer forest from the tundra. Some people call it *taiga,* but Russian ecologists generally use the Russian word *taiga* to describe the whole of the coniferous forest that stretches all the way from Norway eastward to the Pacific, and this use has been adopted by most ecologists and geographers of other nationalities. Those who use *taiga* as the name for the entire coniferous forest call the parklike border between it and the tundra *lichen forest.* The coniferous forest is also known as the boreal forest. This simply means "northern," from the name of Boreas, the Greek god of the north wind. There is

The boreal (northern) forest, dominated by coniferous trees and also known by its Russian name as the taiga, extends in a belt across northern North America and Eurasia. This is the vast taiga. *(Vanderbilt University)*

Kermes oak, also called prickly oak and grain tree (*Quercus coccifera*), is a broad-leaved evergreen that grows in the lands around the Mediterranean. It is seldom more than 6.5 feet (2 m) tall and is well adapted to the long, dry summers. These trees are growing in Crete. *(Peter Duinker)*

no paradox over using the name for Southern Hemisphere forests, for no such forests exist in the Southern Hemisphere. No land extends into a sufficiently high southern latitude.

Within these great forest belts, there are areas where trees grow sparsely, if at all. Temperature decreases with altitude, just as it does with latitude, and beyond a certain elevation this restricts tree growth and eventually inhibits it altogether. It is not true that with increasing height up a mountainside, the vegetation patterns change in the same way as they do with increasing distance from the equator, but they do change, and alpine vegetation is somewhat reminiscent of tundra, although composed of different species.

Timberline and Tree Line, Where the Forests End

On some mountains, lower temperatures resulting from increasing elevation cause trees to become more widely spaced and stunted in a belt of what is called *Krummholz,* a German word meaning "crooked wood." Elsewhere this does not happen, and the timberline is abrupt. The *timberline* is the boundary below which there is forest and above which there are no trees at all.

Since temperature decreases with height and the elevations of mountains are invariably measured from sea level, it is the sea-level temperature that determines the location of the timberline. Sea-level temperature varies with latitude, so the lower the latitude the higher the timberline. In the Sierra Nevada, for example, the timberline is at about 11,500 feet (3,500 m). In the Canadian Rockies it is at about 6,500 feet (1,980 m), and

in the central Alps of Europe it is at about 6,800 feet (2,000 m), but in the mountains of New Guinea, where the temperature at sea level is much higher, it is at about 12,600 feet (3,800 m). Within a particular geographic region, the height of the timberline varies from place to place, depending on the amount of shelter from winds and exposure to sunlight.

Forests in Lands Where Summers Are Hot and Dry

Where forests have not been cleared over the centuries so that the present vegetation is natural to the area, there are substantial differences in forest type. Regions with a Mediterranean climate are found in southern California, on a much smaller scale in parts of Chile and the southern tip of South Africa, and, of course, around the shores of the Mediterranean itself.

Here, the summers are dry and hot, the winters wet and warm, and the annual rainfall averages 20 to 40 inches (500–1,000 mm), falling mainly in winter. There are long, dry periods, and the plants are *xeromorphic*—adapted to withstand drought. The forests comprise *sclerophyllous* trees, and so the forest is said to be sclerophyllous.

Sclerophyllous trees are evergreens with small leaves that are often hard, leathery, or stiff, and they sometimes have prickles around the edges, like holly. The bark is thick and the buds are well protected. These are all adaptations to drought. In California and around the Mediterranean, evergreen oaks and pines dominated the original forest. New and Old World species differed, but the overall composition of the natural forests is similar.

Temperate Rain Forests and Giant Redwoods

Farther north the winters are cooler, the annual rainfall is higher, and there are warm temperate rain forests. Most of the trees are broad-leaved evergreens in North America, including evergreen oaks and magnolias in the forests of this type that grow along the Gulf and Atlantic coasts as far north as North Carolina.

In some places the broad-leaved species are mixed with conifers, the most famous of which are the Sierra redwoods (*Sequioadendron giganteum*) and giant redwoods (*Sequoia sempervirens*) of California and, to the north of them, forests of western hemlock (*Tsuga heterophylla*), western red cedar (*Thuja plicata*), and Douglas fir (*Pseudotsuga menziesii*).

Temperate rain forests also occur in New Zealand, where southern beech (*Nothofagus* species) is dominant in the north, growing together with kauri pine (*Agathis australis*). This type of forest is absent from Europe and rare in Asia, although it does occur in southern China and Japan, dominated mainly by evergreen oaks and magnolias.

Where Temperate Forests Once Grew

When most people picture broad-leaved forest, however, what they have in mind is the broad-leaved deciduous type, also known as summer deciduous forest. People think of this for the simple reason that it is by far the most extensive and it is the type of forest that grows naturally in the most densely populated regions of Europe and North America. Originally, forests of this kind covered some 3 million square miles (7.8 million km²), blanketing almost all of Europe and much of the eastern half of the United States.

Unfortunately for the trees, the European climate that favors them also favors farming, and little of the original forest remains. Substantial areas have been allowed to survive in North America, however. Again, their detailed composition changes from place to place, and in the north the broad-leaves are interspersed with conifers to produce a belt of mixed forest between the summer deciduous forest and the boreal forest.

The boreal forest covers a total of nearly 5 million square miles (12.76 million km²), making it the second biggest forest in the world. It is considerably more extensive than the tropical rain forests, although tropical forests of all types cover 7 million square miles (18 million km²). The boreal forest has long been exploited for timber and other wood products, but much of it remains intact. It lies mainly in sparsely populated areas where the climate is harsh, so it is less accessible than the broad-leaved forests to its south and less familiar to most people. Nevertheless, were we to choose two types of forest to represent the forests of Earth,

one of them the tropical forests, the other would have to be the boreal conifer forest.

■ HOW DID TEMPERATE FORESTS DEVELOP?

Three centuries ago, much of North America was blanketed in forest. One thousand years ago, so was most of Europe. To people trying to find food in them, the primeval forests must have seemed timeless as well as endless, as though they had always been there and had always been just as they were then. As the centuries passed, of course, vast areas of forest were cleared, but had they remained undisturbed many people like to think that today they would be able to see them still in that same, pristine condition. It is tempting to think of them as essentially eternal.

The temptation must be resisted, for nothing is eternal on the dynamic Earth. Everything changes, but over time spans that are so long compared with our brief life span that change can be difficult to detect. Were the great-great grandmother of someone living today to walk beside that person through a surviving remnant of the original forest, it is unlikely that she would notice any difference from the forest she remembered from her own youth. It would look just the same. Yet, on an Earth that has supported life for close to 4 billion years, five human generations amounting, perhaps, to 125 years is the merest instant. Even 1,000 years is a very short time indeed. Forests change, but they change slowly, and it usually takes them a great deal longer than that.

At one time, much more than 1,000 years ago, many areas that now support forest lay deep beneath ice sheets (see "Ice Ages" on pages 91–92), and no plants of any kind grew there, far less broad-leaved trees. Delve deeply enough into the past and there was a time before the trees of our modern broad-leaved forests existed at all. They had not yet evolved, and at a still earlier time neither had those of the coniferous forests (see "Evolution of Trees" on pages 91–101).

There was a time, then, before the "timeless" forests existed. There was a period during which they developed. One day, far in the future, perhaps they will vanish. Forests do change.

How Trees Migrate Naturally

Trees also move around, not individually, of course, but they do so by distributing their seeds over a wide area. This allows them to colonize available sites and so to extend their range as opportunity affords. They may encounter barriers that they cannot cross, such as oceans, deserts, and mountain ranges, but they will establish themselves wherever they can.

That in itself is a mechanism for change. When seeds from an invading species germinate successfully, the young plants occupy space and take up from the soil nutrients that would otherwise have been available to other species, and as they grow taller they shade any smaller plants nearby. By exploiting resources more efficiently, or just differently, new arrivals can increase in number, but their increase implies a decrease in the numbers of species that are rivals for those resources. In time, the invaders may replace their rivals completely, and the plant community will have changed its composition. It will still be forest, but forest of a subtly different type.

What Happens When the Environment Changes

Even without human intervention, forests change their composition over time. *Rhododendron ponticum,* native to southern Europe and Turkey and now an invasive introduction and troublesome weed in Britain, grew naturally in Britain about 120,000 years ago when the average temperature was about 4°F (2.2°C) warmer than it is today.

Like all plants, trees have particular nutrient requirements, and there are limits to the range of temperature and moisture they can tolerate and much narrower limits within which they grow most vigorously. Each species has its own preferred environmental conditions, although those of many species overlap. As the example of *Rhododendron ponticum* demonstrates, small differences can have disproportionately large effects.

Nutrients and moisture are supplied by the soil, and soils develop through the combined influences of climate and biological activity (see "Soil Formation and Development" on pages 15–22). Temperature, of course, is a feature of climate. Climate, therefore, is by far the most important factor determining the type of vegetation that establishes itself and the global distribution of vegetation types closely follows that of climate types (see "Climatic Regions" on pages 81–85). The term *temperate forest* reflects this affinity. It follows that when the climate changes, after a delay the type of vegetation also changes by a kind of ecological opportunism.

Suppose, for example, that an area of sclerophyllous forest is adjacent to broad-leaved deciduous forest and, farther north, that the broad-leaved forest gives way to coniferous forest. Tree seeds regularly travel across these boundaries, but the boundaries remain. Sclerophyllous trees grow slowly, and whenever their seeds germinate in an area supporting deciduous species the sclerophyllous seedlings are crowded out by their neighbors, which grow much faster. In the north, seedlings of broad-leaved trees that germinate among the conifers are killed by the long winter to which the conifers but not the broad-leaves are adapted. Trees also determine many of the characteristics of the soils in which they grow. The needles of conifers and the leaves of broad-leaved deciduous trees decompose differently and at different rates, for example, and the soils of broad-leaved and coniferous forests are dissimilar. This difference may be enough to inhibit invaders.

What would happen if the climate were to change, becoming warmer overall and in places drier? Broad-leaved deciduous trees require moderate rainfall. If the climate becomes drier, they will suffer, but drought causes no difficulties for the sclerophyllous species. Their seedlings will now survive better than those of the deciduous species around them, and the sclerophyllous forest will spread. At the boundary between broad-leaved and coniferous forest, broad-leaved species will also be able to expand into regions that have become warm enough for them. Such migrations happen slowly, of course. Many ecologists fear the response may be too slow for the rapid climatic warming of which most climatologists are now warning (see "Forests and the Greenhouse Effect" on pages 268–272), but in the past there have been many climatic changes with consequent changes in vegetation patterns.

How People Alter the Forest

People also alter the character of forests. Change of this kind is happening all the time and it is particular noticeable in Britain.

People have been importing trees to Britain and planting them at least since Roman times. At first the introduced species are usually difficult to grow, but eventually some become naturalized to the British soils and climate. Then they can—and usually do—escape from cultivation by seeding themselves, and in the case of trees, this allows them to invade forests and establish themselves there. Were land to be abandoned to the plants that have established themselves naturally, inevitably the resulting community would include naturalized species. Ecologists distinguish native species from naturalized species, defining native species as *natural,* when classifying forests by age, for example as *primeval forest* or *old-growth forest* (see "Primeval Forest, Ancient Woodland, Old-Growth Forest, and Plantations" on pages 144–147).

One of the most invasive trees in Britain is *Acer pseudoplatanus,* the sycamore or great maple (note that the American sycamore is any one of several species of plane trees, *Platanus,* and is quite distinct from the European sycamore, which is a maple). The European sycamore was introduced to Britain some time before 1500, but it was not planted widely until the 18th century. It now produces seed more prolifically and regularly than most native trees, colonizing disturbed ground and invading old, neglected stands of trees, and it is advancing rapidly. Most forests now contain some sycamore, and many are dominated by it.

Rhododendron ponticum, known in Britain, for good reason, as the common rhododendron and closely related to the North American *R. catawbiense,* was introduced in the 18th century as an ornamental (see "Rhododendron" on page 217). It escaped from cultivation and naturalized itself in the 19th century, and it is now highly invasive. It spreads rapidly, shading the ground and preventing the growth of tree seedlings.

Today, the most common tree in Britain is Sitka spruce (*Picea sitchensis*), although it grows naturally only in areas adjacent to where it has been planted deliberately. It was imported for its timber from western coastal regions of North America and has become naturalized. If the countryside of Britain were to be abandoned, allowing the natural vegetation to return, the forests would include Sitka spruce. People have also moved trees around within their native regions, planting them in places where they do not occur naturally.

If introducing trees can alter vegetation patterns, so can their selective removal. The first use people made of the primeval European and American forests was as a source of timber, but not just any timber. Some trees yield wood that is good for building houses or ships; some grew very straight and provided masts and spars for sailing ships; the wood from others makes durable railroad ties, strong and attractive furniture, or the handles of tools. Some trees provided fuel, as wood or charcoal. Individual trees were sought, as they are sought today in the tropical forests, and forest areas containing few trees of economic value were ignored. In time, the valuable forests shrank and the less valuable ones expanded. In North America, many species of oak and maple were of little value, which is why those trees are now more common than they once were.

Can Change Be Reversed?

The causes of change may be reversed. People may cease the *selective logging* of economically valuable species. Land that was turned to other uses by the clearance of forest may be abandoned. Climates that have warmed may grow cooler.

Ecological change will follow, but the emerging pattern is unlikely to reproduce that which existed prior to the first change and may not even approximate to it because the starting point has shifted. New communities will include naturalized species, for example, and the abandonment of selective logging does not mean that the logged species will return because the resources they need were commandeered long ago by other species, and winning them back may be impossible.

Clearing forest and putting the land to other uses also alters the soil. There are substantial differences between forest soils and farm soils, and the differences are even greater if the ground has been used for buildings or roads. Trees may colonize the area, but it is unlikely that the original forest will return.

The forests that people see and enjoy today are the products of a continuing process of change. Human activities have strongly influenced that process over the centuries, and they continue to do so, but those influences were not imposed on an essentially static environment, changing forests that would otherwise have remained unaltered for hundreds or even thousands of years. Forests have been changing since they first appeared. Forests that humans have not disturbed are in the state they have reached thus far and from which they are already moving. They are changing even as we look, but at a rate so slow our senses cannot detect it.

2

Geology of Temperate Forests

The rocks on which the forest stands form large blocks that move around each other, sometimes drifting apart, sometimes colliding. This constant motion carries the continents across the face of the Earth, and the moving continents carry with them the plants and animals living on their surfaces. This chapter begins by describing the moving continents and how the distribution of plants and animals provided clues that helped scientists discover what is now called plate tectonics.

As the continents move into different latitudes, their climates change and climatic forces—wind, rain, ice, and chemicals dissolved from the rocks by rainwater—shatter rocks and grind them into soil. Soils have life histories. They form from bare rock, mature, age, and finally become infertile and to all intensive purposes they die. The wide variation in soil types has led many scientists to attempt to classify them, and this chapter outlines the history of soil classification. Today there are many classification schemes, but two are especially important, one developed by soil scientists at the United States Department of Agriculture and the other for international use.

All plants need water and they obtain it from the soil. Water moves through the soil both horizontally and vertically. The ease with which it moves is determined by the permeability of the material from which the soil is made, and the ability of a soil to retain water depends on its porosity. When water draining downward meets a layer of impermeable material, it accumulates as groundwater.

■ PLATE TECTONICS AND TREE DISTRIBUTION

Plants travel. They migrate from one part of the world to another, much as animals do, but they move a great deal more slowly. Seeds are carried on the wind or by birds, float on rivers or the sea, cling to the coats of animals, and sooner or later they fall to the ground. If they land on soil that is moist enough for them and contains the nutrients the young plants need and if the temperature suits them, the seeds germinate and the plants colonize the new area into which chance has carried them.

This is how plants appear on bare or disturbed ground, and it is the way the land is recolonized after the retreating ice sheets of a *glaciation,* or ice age, has left it bare (see "Holocene Recolonization" on pages 91–95). Eventually, most of Europe and North America came to be forested after the most recent glaciation, and it was in this gradual way that the trees arrived. Obviously, there are some barriers that plant seeds are unable to cross. They cannot cross a wide ocean, for example, or a huge desert. Nor may they be able to cross a range of high mountains. Consequently, although the temperate forests of North America and Europe are of similar types, the trees themselves are of different species, although closely related ones.

Unfortunately, though, there is a difficulty. A plant species must start its journey from somewhere. There must be a region of the world in which it first appeared and from where it spread. As it spread, it must have occurred at some time in all the regions through which it passed. Even if it is not present now, there should be traces of its passage, the plant equivalent of the footprints an animal leaves and by which its progress can be traced. It will have left remains, isolated individuals perhaps, or at the very least its *pollen* in the soil.

Pollen grains are sealed within coats (called *exines*) so tough they can survive millions of years without decaying, especially in peaty soils. These coats bear patterns of markings, called colpi (singular *colpus*). Colpi patterns vary in such a way that they can be used to identify the family, and sometimes the genus or even the species, of the plants that produced the pollen. Palynologists, the scientists who study ancient pollen grains, provide information with which paleobotanists and paleoecologists reconstruct past vegetation patterns and communities. Plant migrations can usually be traced, but there are exceptions.

Trees That Grow on Either Side of a Vast Ocean

Magnolias are attractive trees that have been planted in parks and gardens all over the world, wherever the climate is mild and moist enough for them. Like all plants, there are certain places where they occur naturally, places where humans did not plant them. These are the places to which they are native, one of which is the eastern side of the United States, through Central America and the Caribbean islands, with two smaller regions in tropical South America and a larger area in southeastern South America. In the temperate parts of these regions, magnolias are components of natural forest, and it is not too hard to imagine them having migrated overland to the places where they are found today. They are not confined to the New World, however. Magnolias also occur naturally in the southern tip of India and throughout Southeast Asia as far as New Guinea, yet they are not native to Australia, just a short distance from New Guinea.

Plane trees (*Platanus*), including the North American sycamores, have a similarly odd distribution. They are natives of the southeastern United States and southeastern Europe. Southern beeches (*Nothofagus*) occur naturally, and there are forests dominated by them in New Guinea, southeastern Australia, Tasmania, and New Zealand. They are also found down the southwestern coastal strip of South America. *Araucaria,* the genus which includes the Chile pine or monkey puzzle tree (*A. araucana*), occurs naturally in South America, New Guinea and northeastern Australia, and islands of the South Pacific.

These examples and others have what botanists call a discontinuous, or *disjunct,* distribution. They turn up in widely separated places, and there are no traces of the route they followed to reach them. Admittedly, *Araucaria* might have been carried from island to island across the Pacific, but the journey would have involved crossing large expanses of open sea. It is not impossible, but the distributions of the other plants cannot be explained in this way. Some animals, such as the marsupials of Australia, New Guinea, and North and South America, also have a disjunct distribution. One of the most curious botanical examples is *Glossopteris,* a genus of trees known only from fossils. These have been found in Antarctica, South America, South Africa, and Australia but nowhere farther north.

Did the Continents Once Fit Together?

If this is curious, so is something else. Look at a map of the world, and it seems as though the coastlines of South America and Africa ought to fit together. If it were possible to push them toward one another, they would join, and the fit is even better in a map that shows the edges of the *continental shelves,* where the land is covered by water.

Over the centuries, many people have noted this curiosity and, as time passed, scientists noted other strange coincidences. The rocks on the western coast of Africa are identical to those in eastern South America and also to rocks in Antarctica, Australia, and India. High mountains contain fossils of seashells. How did they get there? It was not until 1967 that all the pieces of the puzzle were brought together in what is now known as the theory of *plate tectonics* (*tectonic*, an adjective from the Greek *tectonikos,* "carpenter," refers to deformations of the Earth's crust and changes caused by such deformations).

Many scientists contributed partial explanations. In 1879, Sir George Howard Darwin (1845–1912), a son of Charles Darwin, suggested that soon after the Earth formed, what is now the Moon broke away from it. This led Osmond Fisher (1817–1914) to propose, in 1882 and 1889, that the Pacific basin is what remains of the scar left when the Moon separated. America and Asia, he suggested, are moving together to close the scar, and this has split the New World from the Old, causing the Atlantic to fill the gap. An Austrian geologist, Eduard Suess (1831–1914) concluded that at one time all the continents of the Southern Hemisphere had been joined as a single landmass. He called this supercontinent Gondwanaland (nowadays usually shortened to Gondwana), the name of a region of India once inhabited by a people called the Gonds.

Then, in 1912, a German physicist and meteorologist gathered together a vast amount of evidence, including that from the disjunct distribution of certain plants, to propose a still more comprehensive explanation. Alfred Lothar Wegener (see the sidebar on page 12) produced measurements, which later turned out to have been wrong, showing that in the course of the 19th century Greenland moved one mile (1.6 km) farther from Europe. He also asserted that Washington, D.C., and Paris were moving apart by 15 feet (4.5 m) a year and that San Diego and Shanghai were approaching one another by six feet (1.8 m) a year. Wegener proposed that at one time all the continents had been joined together to form a single supercontinent he called *Pangaea*, surrounded by an ocean he called *Panthalassa.* The names are Greek and mean "all-world" and "all-ocean" respectively. He published his ideas in 1915 in a book called *Die Entstehung der Kontinente und Ozeane*; in 1924 the book appeared in the English language as *The Origin of Continents and Oceans*. Wegener calculated that Pangaea began to break apart about 200 million years ago, forming smaller continents that continued to drift apart. At times they moved into very high latitudes and were covered in ice sheets, and at other times they lay close to the equator and supported tropical plants and animals.

Alfred Lothar Wegener (1880–1930) and Continents That Move

Alfred Wegener (1880–1930) is remembered today mainly as the scientist who first suggested that the continents and oceans have not always occupied their present positions. He proposed that the continents were once joined, forming a *supercontinent* he called Pangaea, and that since Pangaea broke up the continents have continued to move across the surface of the Earth, constantly but extremely slowly, reconfiguring the map of the world.

This was his passion, but Wegener was first and foremost a meteorologist who devoted considerable time, much effort, and eventually his life to studying the climate of Greenland.

Alfred Lothar Wegener was born in Berlin on November 1, 1880, the son of a minister of religion who was also director of an orphanage. Wegener was educated at the Universities of Heidelberg, Innsbruck, and Berlin. In 1905 the University of Berlin awarded him a Ph.D. in planetary astronomy. Wegener immediately changed his field of interest to meteorology and took a job at the Royal Prussian Aeronautical Observatory, not far from Berlin. (Berlin, now the capital of Germany, was then the capital of the kingdom of Prussia.) He used balloons and kites to carry instruments into the air high above the surface, and he and his brother Kurt also flew hot air balloons. In 1906 the two men remained airborne for 52 hours, setting a new world endurance record for balloonists.

Later in 1906 Wegener joined a Danish two-year expedition to Greenland as official meteorologist. Wegener used kites and tethered balloons to study conditions in the polar air, returning to Germany in 1909. He was then appointed lecturer in meteorology and astronomy at the University of Marburg.

In 1912 Wegener married Else Köppen, the daughter of a famous climate scientist, Wladimir Köppen. Later that same year, Wegener returned to Greenland as a member of a four-man team that crossed the ice cap. Theirs was the first expedition to spend the winter on the ice.

War broke out in 1914, and Wegener was drafted into the army. He was wounded almost at once, however, and spent a long time recuperating in hospital, passing the time by developing his theory of *continental drift*. When he was well enough, he joined the military meteorological service. After the war Wegener returned to Marburg, and in 1924 he took up a post created for him, as professor of meteorology and geophysics at the University of Graz, in Austria.

In 1930 he embarked on a third expedition to Greenland as leader of a 21-strong team of scientists and technicians. They planned to study the weather over the ice cap from three bases, all at 71°N, one on each coast and one called Eismitte 250 miles (450 km) inland, but bad weather delayed them. A party set out on July 15 to establish the Eismitte base, but bad weather prevented necessary supplies from reaching them. These included their radio transmitter and the hut in which they were to live. On September 21, Wegener and 14 companions headed for Eismitte with 15 sleds loaded with supplies. The weather was so bad that all but Wegener, Fritz Lowe, and Rasmus Villumsen were forced to turn back. The three finally reached Eismitte on October 30. They stayed long enough to celebrate Wegener's 50th birthday on November 1; then Wegener and Villumsen departed for the base camp. Lowe was too exhausted and too badly frostbitten to accompany them.

Wegener and Villumsen never reached the base camp. At first the people at the base camp assumed they must have decided to overwinter at Eismitte, but by April they had still not appeared, and a party went in search of them. They found Wegener's body on May 12, 1931. He appeared to have suffered a heart attack. Villumsen had carefully buried him. The would-be rescuers marked the grave with ice blocks, and later a huge iron cross was erected there. Despite a long search, Villumsen was never found.

The Alfred Wegener Institute for Polar and Marine Research at Bremerhaven, Germany, commemorates the life, work, and bravery of Alfred Wegener.

Continental Drift, Seafloor Spreading, and Plate Tectonics

Wegener's idea won some converts, but it was not generally accepted. At the time, the *crust* of the Earth and the *mantle* beneath it were both believed to be solid, and it seemed impossible that continents could move across the crust of which they were part. In the 1920s, however, the German-born American geophysicist Beno Gutenberg (1889–1960) discovered that, although the mantle is made from extremely dense, tough material, *convection currents* could flow through it. If so, this convective motion might be enough to move the solid blocks forming the continents. Arthur Holmes (1890–1965), a British geophysicist, suggested in 1928 that the decay of radioactive elements in the mantle might be a sufficient source of heat to drive these convection currents, but his idea attracted little attention.

Evidence was accumulating slowly, and in 1937 a South African geologist, Alexander Logie Du Toit (1878–1948), published his discoveries in a book called *Our Wandering Continents: An Hypothesis of Continental Drift*. His studies had confirmed Suess's idea of a single southern continent, but he maintained that the northern continents had also once been joined. He named the northern continent *Laurasia,* combining the names Laurentia, which is the ancient rock mass forming most of Canada, and Eurasia.

At around that time the oil companies were conducting extensive research into the structure of the crust, part of which involved studies of the ocean floor. They employed *seismic* techniques, detonating explosions that send shock waves through the rocks. Analysis of those waves at receiving stations reveals the density and structure of the material through which the waves have traveled. One scientist engaged in this work was the American geophysicist and oceanographer William Maurice Ewing (1906–74). He made two important discoveries: One was that the crust beneath the oceans is very much thinner than the crust beneath continents. The other, which he made in 1957, was that near the center of oceans all over the world there is a ridge, like a submarine mountain chain, similar to the one already known to exist in the Atlantic.

It was in 1961 that the discoveries made so far were brought together independently by two American scientists, the naval oceanographer Robert Sinclair Dietz (1914–95) and the geophysicist Harry Hammond Hess (1906–69). Both of them proposed that material is extruded from the rifts at the centers of all mid-ocean ridges, like lava seeping from a volcano. This material spreads to either side of the rift and solidifies to form ocean crust. As it emerges, the pressure it exerts holds the rift open and, in doing so, pushes away the crust to the sides. This makes the oceans grow wider. Dietz called this *seafloor spreading,* and Hess went further. He suggested that the continents rest on blocks of solid crustal rock that are carried around the world.

Obviously, if new crust is being formed, old crust must be disappearing at the same rate, otherwise the world would grow continually larger and its mantle would be depleted of material. Hess proposed that as the ocean crust spreads from the center, at the edges it plunges beneath continents and back into the mantle. This process is called *subduction.* Hess summarized all the discoveries made so far in his book *Essay in Geopoetry,* published in 1960 with a second edition in 1962. In 1963, the Canadian geophysicist John Tuzo Wilson (1908–93) confirmed these ideas with his discovery that the age of ocean crust increases the farther it is from the mid-ocean ridge, proving that young rock is being added constantly at the centers of oceans. He was the first person to use the word *plates* to describe the large blocks of crust Hess had identified. Wilson also discovered *transform faults,* in 1965: These are places where two plates are moving past one another.

There are submarine mountains to either side of each rift, and they, too, are carried sideways. As they move away from the central ridge, they are gradually eroded. Some disappear entirely. Others survive, far from the ridges, as isolated, flat-topped mountains on the seafloor, usually known as *seamounts.* Hess, who discovered several, called them *guyots,* after Arnold Guyot (1807–84), another American geologist.

By now the arguments and the evidence supporting them were very persuasive. It really did look as though continental drift and seafloor spreading were genuine phenomena, but an important ingredient was still missing, and despite the arguments and evidence many scientists remained unconvinced. No one had yet suggested a plausible mechanism by which it could happen. If huge blocks of crustal rock were wandering over the surface of the Earth, what was the force driving them? That explanation was found in 1967, and all the other pieces of the puzzle fell into place. Dan McKenzie (born in 1942), a British geologist at Cambridge University, showed how the movement of crustal plates could be driven by the convective flow of material in the mantle. Plates move apart at mid-ocean ridges, now called *constructive plate margins,* where new crust is being formed.

Plates slide against one another, along transform faults at *conservative plate margins,* often causing earthquakes and volcanoes as they do so, and where two adjacent plates move in opposite directions one is subducted beneath the other, at a *destructive plate margin,* or crumpled to form mountains. *Sedimentary rock,* forming a surface layer over the harder crustal rock, can also be scraped from a subducting plate and raised as a mountain chain. That is how high mountains can contain the fossils of marine animals. With everything brought together, the theory of plate tectonics was complete and was quickly accepted.

The crust is composed of many plates of varying sizes. The Pacific Plate is about 7,500 miles (12,000 km) across, for example, and the Nazca Plate about 1,800 miles (2,900 km). Sometimes plates cease to move in relation to one another, and their margins become sealed as a permanent join called a *suture.* At present, the eight major plates are the African, Eurasian, Pacific, Indian, North American, South American, Antarctic, and Nazca (between the eastern edge of the Pacific Plate and the western edge of the South American Plate). The more important minor plates include the Cocos (to the north of the Nazca Plate, close to western Central America), Caribbean, Somali (in the western Indian Ocean), Arabian, Philippine, and Anatolian (beneath Turkey). The map on page

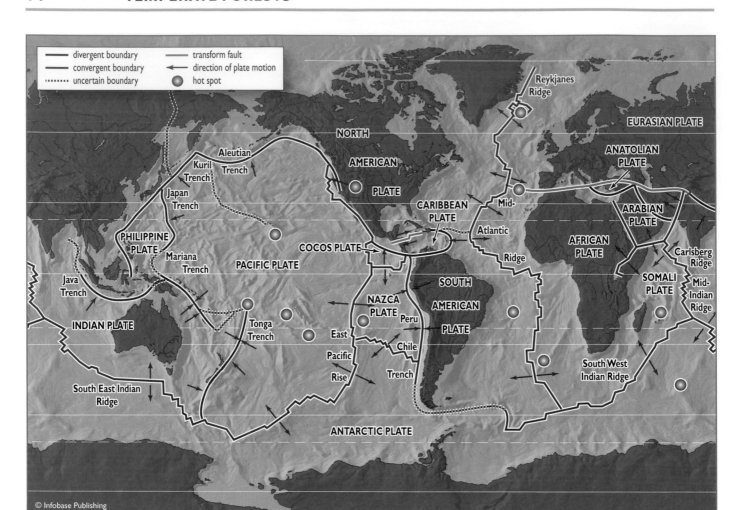

The principal tectonic plates and mantle hot spots, showing the types of plate boundaries, transform faults, and directions of plate motion

14 shows their location. Continental plates are about 75 miles (120 km) thick, but thicker beneath mountain ranges, and those beneath the oceans are about 40 miles (64 km) thick.

Their movement is continuous, and scientists have been able to trace the journeys of the continents into the distant past. The small map shows what the world might have looked like 360 million years ago when most of the continents were joined in a single supercontinent. This was not Pangaea, which had not yet formed, but it shows that what are now North America and Europe straddled the equator. It was then, around the beginning of the Mississippian epoch of the Carboniferous period (see the sidebar "Geologic Timescale" on pages 16–17), that tropical coasts were lined by forests growing in swamps. Falling vegetation was trapped in the airless mud of the swamps, eventually forming coal. It explains why, all these millions of years later, coal is abundant in these continents.

Continents are still moving. At present, for example, the Atlantic is growing wider by about 0.8 inch (20 mm) a year as the North and South American Plates move westward. The Himalayas are still being formed as a result of a collision that began 40 million years ago when the Indian Plate, moving northward, reached the Eurasian Plate and kept going. The

A map of the world as it was 360 million years ago, early in the Carboniferous period

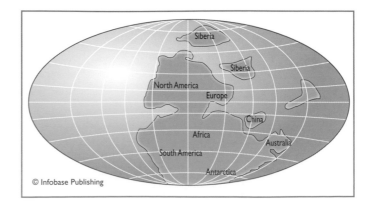

Red Sea is a rift that is opening. Eventually, as the Eurasian Plate moves away from the African Plate, it will become a vast ocean. The lowest of the three maps shows the present map of the world, with the directions in which the continents are being carried. The middle map shows the situation about 65 million years ago (Ma). Then, the Atlantic was much narrower than it is today, North and South America were not joined, and India had not yet reached Asia. As the top map shows, in the still more remote past, about 135 million years ago, North America was joined to Europe and South America to Africa.

It was the shapes produced by the separation of South America and Africa that first encouraged people to suppose that the two might once have been joined. These timescales seem very long, but compared with the age of the Earth itself, which is about 4,600 million years, the crustal plates are moving fairly quickly.

Reconstructing the past movements of continents solves the riddle of the disjunct distribution of plants and animals. These groups lived on a continent that split, and they were carried thousands of miles in opposite directions on the drifting fragments. This mechanism also shows that more than the climate is involved in determining the composition of forests. Plants spread by the scattering of their seed, but over a much longer period they are also taken wherever wanderings of the land on which they grow happen to carry them.

Since the continents are still drifting, it is possible to speculate about where they are heading. One day, millions of years from now, North America will reach Asia. Probably the two continents will collide, perhaps raising a new range of high mountains, like the Himalayas. Then plants that cannot cross the wide Pacific will be able to move from one continent to the other, provided the newly formed mountains do not form a barrier high and continuous enough to prevent them. Asian and American species will mix, there will be winners and losers in the resulting competition, and the composition of temperate forests will change yet again. Meanwhile, as Europe continues its northward journey, conifer forests will replace its broad-leaved deciduous forests, and the tree line, marking the high-latitude limit for tree growth, will extend deeper into the continent.

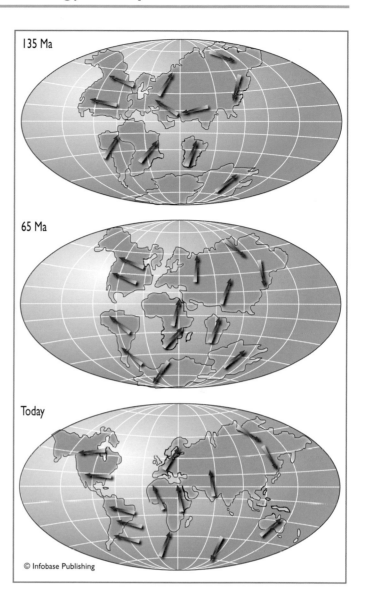

135 Ma

65 Ma

Today

© Infobase Publishing

Continental drift has carried the continents across the surface of the Earth, and it continues to do so. The maps show the world as it looked 135 million years ago, 65 million years ago, and the direction the continents are moving at present.

■ SOIL FORMATION AND DEVELOPMENT

Walk through a forest and it is the trees that hold the attention. A forest is its trees, after all. Away from well-trodden paths, however, walkers need to take more care. The ground on which they walk may be soft, springy, and comfortable or hard and uneven. After rain it may be muddy, or water may soak into it quickly, leaving it dry underfoot. Here and there a bank may expose some of the roots of a tree.

What the walker's feet feel beneath the layer of dead leaves or pine needles is the soil. Forests consist of trees, trees are plants, and plants grow in soil. Soils are not all the same, and each type is best suited for a particular kind of vegetation. The soil found in a forest differs from the soil in a cornfield, and even forest soils vary from one type of forest to another (see "Soil classification" on pages 22–25). As any farmer will attest, soils can vary considerably from one field to the next and even from one part of a field to another part of the same field. What is more, soils are changing constantly in response to changes in the conditions to

Geologic Timescale

EON/ EONOTHEM	ERA/ERATHEM	SUBERA	PERIOD/ SYSTEM	EPOCH/SERIES	BEGAN MA
Phanerozic	Cenozoic	*Quaternary*	Pleistogene	Holocene	0.11
				Pleistocene	1.81
		Tertiary	Neogene	Pliocene	5.3
				Miocene	23.03
			Palaeogene	Oligocene	33.9
				Eocene	55.8
				Paleocene	65.5
	Mesozoic		Cretaceous	Upper	99.6
				Lower	145.5
			Jurassic	Upper	161.2
				Middle	175.6
				Lower	199.6
			Triassic	Upper	228
				Middle	245
				Lower	251
	Palaeozoic	Upper	Permian	Lopingian	260.4
				Guadalupian	270.6
				Cisuralian	299
			Carboniferous	Pennsylvanian	318.1
				Mississipian	359.2
			Devonian	Upper	385.3
				Middle	397.5
				Lower	416
		Lower	Silurian	Pridoli	422.9
				Ludlow	443.7
				Wenlock	428.2
				Llandovery	443.7

which they are subjected. Like plant communities, soils are dynamic systems.

The Life Story of the Soil

Anything that changes over time has a history, a record of past times when it was different from the way it is now, and histories usually have beginnings. If soils have histories, this suggests there was some time, perhaps long ago, before those histories began and, therefore, before the soils themselves existed. Put this way, we can think of soils being

"born," passing through a period when they are "young," becoming "mature," then "aging," and eventually "dying." *Pedologists,* the scientists who study soils, use most of these words, although they do not usually think of soils dying.

Trees grow from seeds produced by older trees, which are their parents. Being born implies the existence of parents. It follows that if soils can be born, they must have parents, and so they do. A soil is born from *parent material,* which is rock, and the process by which it is born and ages is called *weathering* (see the sidebar on page 18). Weathering will reduce rock to fine grains. Even without the help of

EON/ EONOTHEM	ERA/ERATHEM	SUBERA	PERIOD/ SYSTEM	EPOCH/SERIES	BEGAN MA
			Ordovician	Upper	460.9
				Middle	471.8
				Lower	488.3
			Cambrian	Furongian	501
				Middle	513
				Lower	542
Proterozoic	Neoproterozoic	Ediacaran			600
		Cryogenian			850
		Tonian			1000
	Mesoproterozoic	Stenian			1200
		Ectasian			1400
		Calymmian			1600
	Paleoproterozoic	Statherian			1800
		Orosirian			2050
		Rhyacian			2300
		Siderian			2500
Archaean	Neoarchaean				2800
	Mesoarchaean				3200
	Palaeoarchaean				3600
	Eoarchaean				3800
Hadean	Swazian				3900
	Basin Groups				4000
	Cryptic				4567.17

(Source: International Union of Geological Sciences, 2004.

Note: *Hadean* is an informal name. The Hadean, Archaean, and Proterozoic Eons cover the time formerly known as the Precambrian. *Quaternary* is now an informal name, and *Tertiary* is likely to become informal in the future, although both continue to be widely used.)

water, at least for hundreds of millions, or perhaps billions, of years, it has produced the desert sands of Mars and, even without wind, the dust of the Moon. On Earth, though, the weathering is also biological.

Birth of a Soil

Certain bacteria can thrive inside the cracks in rocks, where they are sheltered from the wind and direct sunlight and can obtain nutrients from the substances dissolved from the rock into rainwater. To protect themselves further, the bacterial colonies secrete a tough, jellylike covering and lie sealed beneath it.

As the colonies penetrate deeper into the rock, along microscopically tiny fissures, their own waste products also react with ingredients of the rock. The bacteria contribute to the weathering process until there is enough purchase for lichens, the next arrivals.

A lichen is a partnership between two kinds of living organisms: a *fungus* and an *alga* or *cyanobacterium*. Fungi comprise one of the kingdoms into which living organisms are divided, alongside plants (kingdom Plantae) and

Weathering

Soil is a mixture of rock particles and organic material. All of the physical and chemical processes that transform a mass of rock into boulders, pebbles, gravel, and eventually to fine particles are described as weathering.

Physical weathering begins with movements within the Earth's crust that bend and fracture rock, so its surface bears many cracks and fissures. Whenever rock breaks, small fragments are detached. These can be blown by the wind or carried by water, which throw them against one another and against larger rocks, breaking them into still smaller fragments.

Away from the Tropics, in winter the water may freeze. When water freezes, it expands with great force, widening the cracks and causing more tiny fractures. In spring the ice melts and more fragments are freed.

In summer, the Sun beats down on the rock, heating it. As it is heated, the rock expands, but it is only the surface that expands because that is where the heat is trapped, insulating the rock below the surface. This causes yet more fractures and detaches yet more fragments.

The movement of water rolls stones across each other. The rounding of pebbles on a riverbed or seashore results from this action.

Rainwater is naturally slightly acid. It will react with certain compounds exposed on the surface of rocks to produce soluble reaction products that flow away in solution leaving pits in the rock surface, thereby increasing the area of surface that is exposed to other erosive forces. Water moving belowground carries compounds dissolved from the rocks it has crossed. These compounds react with others. All of these reactions remove constituents of the rock, weakening the mass. These reactions constitute *chemical weathering*.

animals (kingdom Animalia). Fungi obtain mineral nutrients through a network, in some cases a very extensive network, of fine filaments called *hyphae*. An alga (plural algae) is a simple plantlike organism, individuals of many species consisting of only one cell. Cyanobacteria (which used to be known as blue-green algae) are a type of bacteria. Both algae and cyanobacteria can use sunlight to convert carbon dioxide and water into sugars. Combined in a lichen, the fungus provides secure anchorage and supplies minerals, and the alga or cyanobacterium supplies sugars. Lichens can grow on apparently bare rock surfaces.

As material continues to accumulate from the wastes and dead remains of the bacteria and lichens, in time the first true plants can gain a hold. These are often mosses, and

These patches of yellow lichen are growing on bare rock, from which they can extract nutrients. This is a first stage in the formation of soil. *(Mike T. Friggens)*

within deeper cracks, where more organic material has collected, there may be a blade or two of grass or a small flowering herb.

Mosses, grasses, and herbs add more organic material and slowly, as this mixes with the tiny mineral fragments from the weathered rock, soil starts to develop. On Mars and the Moon, where there is no contribution from organic material, the surface material is properly called *regolith,* not soil.

Pedogenesis, which is the scientific term for the formation and development of soils, is often a slow process, but not always. Not all soil develops from solid rock, and if the parent material has already been broken into small fragments, it can become a true soil much more quickly. Very fine mineral particles, deposited as silt on river floodplains, can be blown by the wind to places where it accumulates as *loess.* Glaciers scour the rocks over which they flow, grinding stones into small grains that are left behind when the climate warms and the glaciers retreat. Some types of volcanoes eject vast amounts of ash, which is rock reduced to a powder. These materials are rich in plant nutrients. If they occur in places where there is enough warmth and rain for plants to grow, they can be converted into soil in less than a century. The careful reclamation of mineral waste tips produced by mining can produce several inches of soil in just a few years.

Digging Deep to Reveal the Soil Layers

Once soil has started to form, bigger plants can grow in it. Their roots penetrate more deeply, and when the roots die, they leave channels through which air and water can move, as well as vegetable matter on which a variety of bacteria and small animals can feed. Under favorable conditions, little by little the maturing soil grows deeper, and the community of organisms living in it becomes more complex until the soil is a fascinating and bewilderingly complicated ecosystem in itself, while all the time the weathering processes continue.

As it matures, the soil also acquires a definite structure. It forms layers, which can be exposed by cutting a trench through the soil, deep enough to reach the underlying parent material. The exposed cross section of the soil is called a *soil profile,* and the layers it reveals are called *horizons.*

Horizons are highly variable and often incomplete, but the illustration shows them as they might appear in a profile through the soil of a broad-leaved deciduous forest. As the diagram shows, there are many horizons, grouped into four principal categories: O, A, B, and C. Pedologists often label soil horizons in much more detail than is shown here, with subscripted abbreviations to indicate chemical composition, but this simplified diagram illustrates the general principle.

At the surface there is a thin layer, O1, of leaves, twigs, and other plant and animal material. Fungi and small animals are busily feeding on this. Beneath the O1 horizon, the

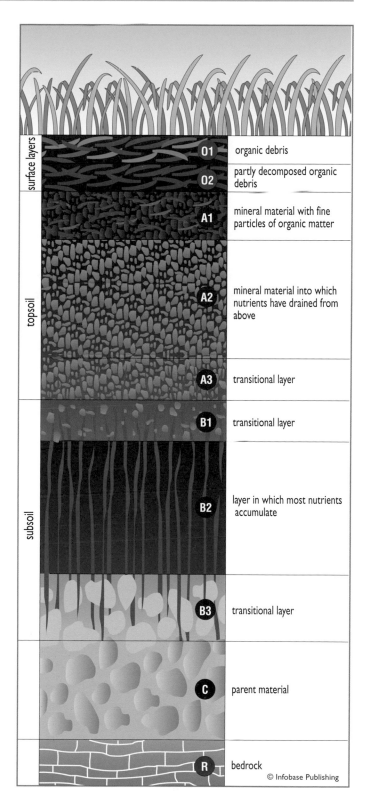

A soil profile, showing all the horizons

O2 layer consists of plant and animal material that has been partly decomposed by the activities of the organisms feeding on it. The two O horizons contain more organic material than the deeper horizons.

Decomposition breaks down complex organic molecules into smaller molecules. Many of these are soluble and as rain drains downward through the soil (see "Water Flow and Drainage" on pages 28–32) they dissolve and are carried into lower horizons. This transport of soluble compounds is called *leaching* and it occurs mainly in the A horizons.

The A1 horizon consists mainly of mineral particles, but these are mixed with fine organic particles from above, which usually give the A1 horizon a dark color. Compounds leaching from A1 flow down into A2, which is much paler. This horizon is composed of mineral particles through which soluble compounds are leaching. Beneath A2, the A3 horizon is similar, but mixed with material from the underlying B horizon, so it is transitional between A and B. The A horizons are what farmers and gardeners call the topsoil.

B horizons, or subsoil, are where the leached compounds accumulate. At the top of the layer, B1 is another transitional horizon, generally similar to B2, but mixed with some material from A3. The B2 horizon, which in many soils is the thickest, comprises mineral particles and the accumulation of dissolved compounds. Beneath it, B3 is transitional to C, and the C horizon is the partly weathered material from which the overlying soil has developed. It lies above the bedrock, which some scientists call the R horizon.

How Weather Makes the Soil

The rate at which a soil forms and develops depends largely on the climate, proceeding fastest where rain falls reliably throughout the year and the ground is frozen for only a few weeks each year or not at all. This describes temperate climates, and these are the climates in which soils develop most quickly. In a dry desert there is too little rain to support the vigorous plant growth needed to provide organic material and to supply the water that dissolves and leaches compounds. Soils may start to develop, but horizons fail to form, and so the soils remain at an early stage of development for as long as the climate stays arid. Much the same happens in very high latitudes, where the ground remains frozen most of the year, and there are regions of *permafrost* where a layer below the soil surface remains frozen all year round. Again, there is little opportunity for soils to mature. They fail to develop and lack clearly marked horizons.

Climates also exert a more subtle influence. Gradations of temperature and moisture, which in themselves have little effect on soil development, nevertheless may have a large effect on vegetation. The difference between the climatic conditions suitable for sclerophyllous, summer deciduous, and boreal forest are not great. All of these grow in temperate climates, but differences in vegetation are reflected in chemical differences. Nutrient requirements vary from one tree species to another, especially between coniferous evergreens and broad-leaved deciduous species, and fallen leaves decay at different rates. Comparing the floor of a broad-leaved deciduous forest with that of a conifer forest demonstrates this difference. Needles lie a long time on the ground, forming a mat, and decay slowly. Leaves of sclerophylls, which are tough and usually have a waxy outer coat, also decay slowly. Holly leaves remain on the ground much longer than the leaves of summer deciduous trees, which rot quickly and are mixed into the surface soil horizon.

Grasses grow quite differently from trees. Some, including many prairie grasses, root deeply, while others form dense but shallow mats of root fibers. Again, this affects soil development and grassland soils are different from forest soils.

How Soils Grow Old

The level of organization and complexity exposed in a complete profile of the kind shown in the illustration on page 19 represents the mature phase in the life cycle of a soil. At first, the young soil is not so strongly stratified and may consist of nothing more than a thin layer of organic material overlying the main mass of undifferentiated mineral particles. A young soil may have only two horizons, A and C, and its properties are derived mainly from the parent material, with little contribution from living organisms. Nor will the mature soil remain mature. As compounds dissolve from the upper horizons and leach deeper and deeper, gradually they will be lost. Rainwater, which carries them into the lower horizons, eventually joins the groundwater (see "Groundwater and the water table" on page 32) and flows with it away from the area. The lost compounds may be replaced by those leaching into the area, but this is not always the case, especially on high ground where there is little soil further up the slope to supply them, and eventually the parent material will become depleted.

As leaching continues, eventually the soil ages and its character changes. Differences between the A and B horizons become extreme, as the A horizon is depleted of substances that accumulate in the B horizon. Fertility declines in the old soil, with consequences for the vegetation it supports. This may become more diverse, as the more aggressive, dominant species start to fail. These are plants that compete vigorously for nutrients but require large amounts of them. As the supply dwindles, the dominants lose vigor, and a wider range of species with more modest requirements is able to flourish. Aging leads to senility. Only the least soluble compounds survive in a senile soil, and the vegetation comprises shallow-rooted plants exploiting nutrients released by decomposition near the surface. Finally, the soil is fully weathered. Depleted of even its relatively insoluble constituents, its fertility is extremely low.

Soil Types

Climate and vegetation may combine to produce and develop soils, but they have no influence at all over the mineral composition of the original parent material. This is rock, and rocks vary widely. Granite is quite different from limestone, for example, and limestone is different from sandstone. Rock is made from minerals, which are crystalline substances with definite chemical compositions, but there are many minerals and countless combinations in which they can occur. As a rock weathers, its mineral composition will determine the kind of chemical reactions that take place and the size of the resulting particles. Soils weathered from different parent materials will differ chemically and physically, and this may affect their subsequent development.

A scientific examination of the mineral composition of a soil often begins in the field where the soil occurs, by taking a handful and feeling it. If the soil is dry it should be moistened until it has the consistency of modeling clay. Then a piece is pinched off and rubbed between finger and thumb. If the soil feels gritty it is mainly sand; if it feels rough but not gritty it is mainly silt, and if it feels smooth and rather greasy it is mainly clay. If the soil dries quickly and does not stick to the fingers when squeezed, it is mainly silt. If it dries slowly and is sticky, it is clay. Further tests involve squeezing a lump of soil to see how easily it breaks apart, rolling it into a thread to see how thin it can be made without breaking, and attempting to press it into the longest possible ribbon shape. These tests reveal the relative proportions of sand, silt, and clay.

Alternatively, the scientist may prefer to take a soil sample back to the laboratory to perform the same kind of test, but more precisely. Mechanical analysis allows for the fact that all the soil minerals are of similar density (1.5 ounces per cubic inch or 2.65 g/cm³, which is the density of silica). Mixed in water to form a suspension, the order in which they settle depends only on their size—sand first, then silt, and finally clay. The pedologist mixes the sample in a large volume of water and stirs it thoroughly, then takes samples of the suspension, all of the same volume, immediately after five minutes, and after eight hours. The first sample will contain sand, silt, and clay, the second silt and clay, and the last only clay. The investigator places each sample in a dish and waits for all the water to evaporate and then weighs the particles left in each dish. This reveals the proportion of each type present in the soil. It is then possible to name the type of soil by locating these proportions on a standard diagram, called a soil textural triangle, shown in the illustration. (The triangle shown here is the one used by the U.S. Department of Agriculture; the one used in Britain is slightly different.) Loam is a soil texture intermediate between the main types.

To a pedologist, the words *pebble*, *sand*, *silt*, and *clay* have quite precise meanings. There are several classification sys-

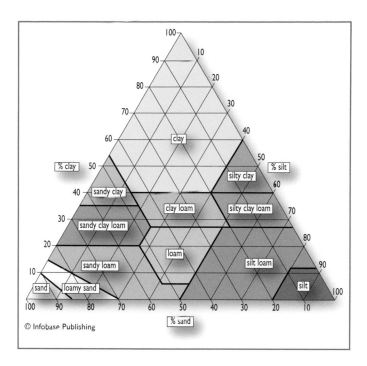

A soil textural triangle. When a soil scientist has measured the proportions of clay, silt, and sand in a soil sample, the textural triangle allows the soil type to be named.

tems for particle sizes, but the table below gives the sizes in the Udden-Wentworth Scale, which is widely used. The measurements are in millimeters (mm); 1 mm = 0.0394 inch.

Soil texture strongly influences many of the soil's important characteristics, especially its capacity for retaining moisture and the ease with which water moves through it (see "Porosity and Permeability" on pages 28–29). The differences arise because the difference in particle sizes is

Udden–Wentworth Scale of Grain Sizes (mm)	
boulder	more than 256
cobble	64–256
pebble	2–64
granule	2–4
very coarse sand	1–2
coarse sand	0.5–1
medium sand	0.25–0.5
fine sand	0.125–0.25
very fine sand	0.062–0.125
silt	0.031–0.062
clay	less than 0.031

huge. A grain of sand is 1,000 times bigger than a particle of clay.

Texture also affects the rate at which weathering affects the soil. This is a matter of geometry. Weathering occurs at surfaces, and the smaller the particle the bigger its surface area is in relation to its volume. Suppose the particles to be spherical (which they are not, but the argument still holds). The volume of a sphere with a diameter of, say, 2 will be 4.2 and its surface area 12.6. Dividing the surface area by the volume gives a ratio of 3:1. Double the diameter to 4, and the volume is 33.5, the surface area 50.3, and the ratio 1.5:1. A given volume of soil will contain more particles if they are small than if they are large, and those particles will have a greater combined surface area on which weathering processes can act. Consequently, that soil will weather faster if the two soils experience similar weathering conditions.

The fertility of the two soils will also differ for the same reason. It is weathering processes that release soluble compounds into the water moving through the soil, and it is from the resulting soil solution that plant roots obtain their mineral nutrients. Clay soils, therefore, are inherently more fertile and productive than sandy soils, but because their tiny particles pack together more closely, water does not drain easily through them. They are heavy and difficult to work, tending to form a hard block when dry and sticky mud when wet.

■ SOIL CLASSIFICATION

At one time, people believed soil was simply a mixture of mineral particles, organic matter, and air. It was pretty much the same everywhere, had always existed, and did not change over the years and centuries. To farmers it was the medium that supported plant growth; to engineers it was the base on which buildings, bridges, and roads were constructed.

About the middle of the last century, however, scientists were beginning to realize that this was a very inadequate description. Indeed, it was misleading to talk about "soil" at all. The soil in one place might be so different from the soil somewhere else that it would be more sensible to talk of "soils" in the plural, and of each individual soil as occupying a particular area to a particular depth and surrounded by other, different soils. This concept was strengthened by the publication in 1840 of a book called *Die organische Chemie in ihrer Anwendung auf Agrikulturchemie und Physiologie* (Organic chemistry in its application to agricultural chemistry and physiology) by the German chemist Justus Liebig (1803–73). Liebig showed that plants take specific compounds from the air and soil, rather than being nourished by directly consuming humus, which was the prevailing idea, and that soils can be improved by the addition of fertilizers in the form of simple, inorganic, chemical compounds.

Clearly, soils can change and can be changed, and they are not all the same.

A professor at the University of Giessen, Liebig was already famous for his discoveries and for having revolutionized the teaching of chemistry by being the first teacher to establish a laboratory in which his students could gain "hands-on" experience. In later years he was recognized as the greatest chemist in Germany and probably in the world. He was made a baron in 1845 and became Baron Justus von Liebig. Not surprisingly, his findings about soil were widely read and led to the founding of the fertilizer industry.

A Russian Beginning

Scientists began to study soils in detail. From those studies they hoped to be able to describe each type of soil, classify them all, and map their distribution. Many European and American scientists took part in this enterprise, but one of the first and most influential was a Russian geologist and geographer, Vasily Vasilyevich Dokuchayev (see the sidebar on page 23).

Climate was held to be the main factor in the formation of what Dokuchayev called "normal" soils, and in the first of the 14 volumes of his report on the survey he made of soils around the city, now called Gork'iy, he proposed a method for classifying them. His "normal" soils, typical of the climatic region in which they occurred, came to be known as *zonal* soils. Soils that were altered because of local circumstances, such as waterlogged wetland soils and soils with excessive amounts of salt, were called intrazonal. Soils that had barely started to form, such as desert sands, or that had been transported by wind or water away from the region where they formed were called azonal.

A system of classification had begun in which each of these main soil orders (zonal, intrazonal, and azonal) was subdivided into suborders and "great soil groups." Reflecting the large volume of work done by the Russian scientists, many of the great soil groups had Russian names. The Russian word for "soil" is *zemlya,* and many of the names end in *zem.* Chernozem is the black soil of steppe grassland; brunizems are prairie soils; podzols are fairly old, weathered, ash-colored soils; solonchaks are saline soils; solonetz is a poorly drained soil; sierozem is a pale, desert soil, and *rendzinas* are grassland soils developed over chalk or limestone.

Russians led the field for many years. *Pochvovedeniye* (which means "soil science"), the world's first journal on soil science, was Russian and was launched in 1899, and the first soil science textbooks were also Russian. Language isolated the Russian scientists from colleagues in other countries. In 1908 Konstantin Dimitrievich Glinka published a book called *Pochvovedeniye.* Few scientists outside Russia heard of it, but in 1914 a German translation was published with the title *Die Typen der Bodenbildung, ihre Klassifikation und*

Vasily Vasilyevich Dokuchayev (1840–1903)

Vasily Dokuchayev, widely recognized as the founder of soil science, was born at Milyukovo, a city in western Russia, on February 17, 1840 (since then, the Russian calendar has been changed; what was February 17 is now March 1). Dokuchayev originally intended to become a priest, but while a student at St. Petersburg University, he switched to a course in natural sciences. He was appointed curator of the geological museum at the university in 1872, and in 1879 he joined the geology faculty and established the university's first course in Quaternary geology (the geology of the last two million years).

In 1877 a group called the True Society of Economics commissioned him to undertake a four-year survey of the soils of the steppes. The society wished to boost farm production in the grasslands and realized that they needed to know more about grassland soils. Dokuchayev published the results of his survey in 1882, as a book called *Russian Chernozem. Chernozem,* which is Russian for "black soil," is a term he introduced to describe the black soil that is rich in carbonates and *humus* that occurs in the temperate latitudes of Russia. In his book, Dokuchayev suggested that soils evolve through the combined action of physical weathering and biological activity and that they have histories, which begin with their formation.

Between 1892 and 1895 Dokuchayev reorganized and directed the Novo-Aleksandr Institute of Agriculture and Forestry at St. Petersburg University, adding departments of soil science and plant physiology. He later became director of the Kharkov Institute of Agriculture and Forestry in Ukraine, and the year his work on steppe soils was published, he embarked on another four-year survey, this time of soils in the province of Nizhny-Novgorod (now Gork'iy) to the east of Moscow.

In the following years Dokuchayev organized soil surveys throughout most of Russia. He believed soils result from the interaction between climate, bedrock, and living organisms. In 1886 he introduced a classification of Russian soils based on the principle that similar bedrocks give rise to different soils, depending on climate.

Later, the work he had started was continued by one of his former students, Konstantin Dimitrievich Glinka (1867–1927). Dokuchayev died at St. Petersburg on October 26 (November 8 in the revised calendar), 1903.

geographische Verbreitung (The types of soil formation, their classification and geographical distribution). Curtis Fletcher Marbut (1863–1935) translated the German edition into the English language and it appeared in 1927 as *The Great Soil Groups of the World and Their Development.* The first American journal on the subject, *Soil Science,* began publication in 1916 at Rutgers College, and the first British journal, *Journal of Soil Science,* appeared in 1949.

At first only the soil surface was examined. It was a Danish scientist, P. E. Müller (1840–1926), who showed that the lower layers of a soil need to be taken into account in the description and classification. Glinka developed this and worked out a way to label the subsurface layers, but it took a long time for the idea to gain acceptance, and it was not fully incorporated into soil classification until the Second International Congress of Soil Science, held in Russia in 1932.

much to promote pedology in the United States and whose work achieved worldwide recognition. Russian publications were often translated into German, and Marbut translated them from German into English—as he had with Glinka's influential book.

The Americans conducted many soil surveys, and the U.S. Soil Survey was established in 1898 with Milton Whitney (1860–1927) as its first director. Marbut was appointed consultant scientist to the U.S. Soil Survey in 1910, and in 1913 he became director, or "scientist in charge," as the director was known.

Surveying and attempts at classification continued and by the late 1940s American scientists had made much more progress than their colleagues elsewhere. Gradually the old Russian names were replaced, although not by familiar American words.

The United States Soil Survey

Meanwhile, American soil scientists were as active as their Russian colleagues but were less concerned with investigating the processes involved in the formation and development of soils. They were kept informed of the Russian work by Curtis Fletcher Marbut, a geologist and geographer who did

Soil as a Base for Buildings and Civil Engineering

Most people think of soil as a medium for plant growth. That is the way farmers and horticulturists regard it, but theirs is not the only point of view. To a geologist, soil is a type of rock. It is not a solid mass, like the rocks that crop out through the thin turf on hillsides, but neither is gravel or the scree covering a mountain slope and they are clearly rock. Soil can be studied and analyzed just like any other form of rock. Ask a builder or a mining or civil engineer for a definition and the reply will be even simpler. Soil is the loose material on which buildings and other structures are erected, the source of raw materials used in building, and the material miners must remove to gain access to near-surface deposits.

The properties of the soil are no less important to the builder or engineer than they are to the farmer or gardener, but they are different properties. What matters is whether the soil will provide a firm base for whatever is to be constructed upon it and also the ease with which machines can excavate it to install pipes for services. It must not be subject to erosion, and it must not contain large amounts of particular clays, such as montmorillonite, that swell when they are wet and shrink when they dry.

Modern Soil Taxonomy

Compiling a complete classification proved very difficult. By 1975, when the U.S. Department of Agriculture published a manual describing the scheme which its soil survey staff had devised, more than 10,000 soil types had been identified in the United States alone. The work had been thorough, however, and what is now known as the U.S. Soil Taxonomy was widely adopted outside the United States as well as nationally.

Soil scientists in other countries had developed their own classification systems that suited local soils and climate, and it was not always easy to merge these with the U.S. Soil Taxonomy. Eventually scientists from many countries collaborated in compiling an international classification that was published under the auspices of the Food and Agriculture Organization (FAO) of the United Nations (see the sidebar).

The U.S. Soil Taxonomy begins by dividing soils into 21 main groups, called orders. These are then divided further into 64 suborders, and below these are great groups, subgroups, families, and soil series, with six phases in each soil series. The entire system generates tens of thousands of names, all of which look very strange to anyone meeting them for the first time. The 12 orders (usually written with an initial capital) are: Entisols, Vertisols, Inceptisols, Aridisols, Mollisols, Spodosols, Alfisols, Ultisols, Oxisols, Gelisols, Adisols, and Histosols (see the sidebar on page 25). It is possible for anyone who understands the origin of their prefixes to work out the meaning of the suborder names. *Aqu-*, for example, suggests water, so aquents are wet surface deposits. *Psamm-* means sandy. *Xer-* means arid. *Plagg-* is derived from *plaggen,* for a farmed soil produced by many years of plow-

National and International Soil Classifications

Farmers have always known that soils vary. There are good soils and poor soils, heavy soils containing a large proportion of clay, sandy soils that dry out rapidly, and light soils rich in loam that retain moisture and nutrients. In the latter part of the 19th century, Russian scientists were the first to attempt to classify soils. They thought that the differences between soils were due to the nature of the parent material—the underlying rock—and the climate. They divided soils into three broad classes and gave soils Russian names that are still widely used, such as *chernozem, solonchak,* and *podzol.*

American soil scientists were also working on the problem, and by the 1940s their work was more advanced than that of their Russian colleagues. By 1975 scientists at the United States Department of Agriculture had devised a classification they called Soil Taxonomy. It divides soils into 12 main groups, called orders. The orders are divided into 64 suborders, and the suborders are divided into groups, subgroups, families, and soil series, with six "phases" in each series. The classification is based on the physical and chemical properties of the various levels, or horizons, that make up a vertical cross section, or profile, through a soil. These were called *diagnostic horizons.*

National classifications are often very effective in describing the soils within their boundaries, but there was a need for an international classification. In 1961 representatives from the Food and Agriculture Organization (FAO) of the United Nations, the United Nations Educational, Scientific, and Cultural Organization (UNESCO), and the International Society of Soil Science (ISS) met to discuss preparing one. The project was completed in 1974. Like the Soil Taxonomy, it was based on diagnostic horizons. It divided soils into 26 major groups, subdivided into 106 soil units. The classification was updated in 1988 and has been amended several times since. It now comprises 30 reference soil groups and 170 possible subunits.

ing and manuring. *Ud-* describes a soil that is moist for most of the summer and *ust-* one that holds a limited amount of moisture during the period of maximum plant growth. *Fluv-,* from the Latin *fluvius,* "river," refers to stream deposits.

Soil Orders and Reference Groups: Orders of the U.S. Soil Taxonomy

Entisols   Soils with weakly developed horizons, such as disturbed soils and soils developed over alluvial (river) deposits. Entisols, a made-up word, are soils found only locally and are formerly known as azonal. The suborders, all ending with -ents, are aquents, arents, fluvents, orthents, and psamments.

Vertisols   Soils with more than 30 percent clay that crack when dry. Vertisols, from the Latin *verto*, "turn," are soils that have been inverted, so what was once below the surface now lies on top. Vertisol suborders, all ending in -erts, are torrerts, uderts, usterts, and xererts.

Inceptisols   Soils with a composition that changes little with depth, such as young soils. Inceptisols, from the Latin *inceptum*, "beginning," are very young soils. Inceptisol suborders, all ending in -epts, are andepts, aquedepts, ochrepts, plaggepts, tropepts, and umbrepts.

Aridisols   Soils with large amounts of salt, such as desert soils. Aridisols, from the Latin *aridus*, "dry," are desert soils. There are only two suborders, both ending in -ids, the argids and orthids.

Mollisols   Soils with some horizons rich in organic matter. Mollisols, from the Latin *mollis*, "soft," are the soft soils of grasslands and forests, the most fertile and agriculturally productive of soils. The suborders, all ending in -olls, are albolls, aquolls, borolls, rendolls, udolls, ustolls, and xerolls.

Spodosols   Soils rich in organic matter, iron, and aluminum; in older classifications known as a podzol. Spodosols, from the Greek *spodos*, "wood ash," are acid, ash-colored, weathered soils from which most soluble compounds have leached. They used to be called podzols, and the process by which iron and aluminum oxides and hydroxides leach from the upper soil horizons is called podzolization. The suborders, all ending in -ods, are aquods, ferrods, humods, and orthods.

Alfisols   Basic soils in which surface constituents have moved to a lower level. Alfisols, a made-up word, are similar to Spodosols. They develop in moist, temperate climates, often over a clay parent material. The suborders, all ending in -alfs, are aqualfs, boralfs, udalfs, ustalfs, and xeralfs.

Ultisols   Acid soils in which surface constituents have moved to a lower level. Ultisols, from the Latin *ultimus*, "last," are red or yellow podzolized soils, more heavily weathered than Alfisols. They are ancient soils, near the end of soil development. The suborders, all ending in -ults, are aquults, humults, udults, ustults, and xerults.

Oxisols   Soils rich in iron and aluminum oxides that have lost most of their nutrients through weathering; old soils often found in the humid Tropics. Oxisols, from the French *oxide*, "oxide," are even more weathered than Ultisols and well into soil senility. They contain lumps or layers of solid iron and aluminum oxides and hydroxides called laterite. Oxisols occur mainly, but not exclusively, in the tropics. The suborders, all ending in -ox, are aquox, humox, orthox, torrox, and ustox.

Histosols   Soils rich in organic matter. Histosols, from the Greek *histos*, "tissue," are the soils of bogs, consisting mainly of organic material. The suborders, all ending in -ists, are fibrists, folists, hemists, and saprists.

Gelisols   Soils with permafrost within 6.5 feet (2 m) of the surface.

Andisols   Soils formed on a basis of volcanic ash.

The FAO classification uses some of the same names as are used for orders in the U.S. Soil Taxonomy, but most are different (see the sidebar on page 26).

■ FOREST SOILS

Forests are mainly associated with the Soil Taxonomy orders Inceptisols, Alfisols, and Spodosols. Inceptisols were formerly known as brown earths. These are young soils in the sense that the leaching of soluble compounds from the A horizons is at an early stage. There is a considerable variety of Inceptisols, but because they are young, all of them lack a clearly defined layer in which leached compounds are accumulating. The lack of an obvious B horizon means the soil is often brown all the way to the C horizon, and hence its old name (from the German *Braunerde*, first used in 1905). The color depends on the parent material, however, and can be distinctly reddish if this includes sandstone or red marl, a very fine clay derived from seabed or lakebed deposits. Inceptisols drain well. This characteristic, combined with the high concentration of plant nutrients that have not yet leached from the upper layers and are within reach of plant roots, makes soils of this type fertile and easy to cultivate.

Under certain exceptional conditions Inceptisols can be found in the Tropics, but most occur in middle and high latitudes, where they have formed since the retreat of the ice sheets of the last glaciation. They are the predominant soils in a belt down the eastern United States from the Adirondack Mountains, through Pennsylvania and the Appalachian Mountains as far south as northern Alabama. They also occur in the northwest, in parts of Washington. Where the

FAO Reference Soil Groups

Histosols Soils with a peat layer more than 15.75 inches (40 cm) deep.

Cryosols Soils with a permanently frozen layer within 39 inches (100 cm) of the surface.

Anthrosols Soils that have been strongly affected by human activity.

Leptosols Soils with hard rock within 10 inches (25 cm) of the surface, or more than 40 percent calcium carbonate within 10 inches (25 cm) of the surface, or less than 10 percent of fine earth to a depth of 30 inches (75 cm) or more.

Vertisols Soils with a layer more than 20 inches (50 cm) deep containing more than 30 percent clay within 39 inches (100 cm) of the surface.

Fluvisols Soils formed on river (alluvial) deposits with volcanic deposits within 10 inches (25 cm) of the surface and extending to a depth of more than 20 inches (50 cm).

Solonchaks Soils with a salt-rich layer more than 6 inches (15 cm) thick at or just below the surface.

Gleysols Soils with a sticky, bluish-gray layer (gley) within 20 inches (50 cm) of the surface.

Andosols Volcanic soils having a layer more than 12 inches (30 cm) deep containing more than 10 percent volcanic glass or other volcanic material, or weather volcanic material within 10 inches (25 cm) of the surface.

Podzols Pale soils with a layer containing organic material and/or iron and aluminum that has washed down from above.

Plinthosols Soils with a layer more than 6 inches (15 cm) deep containing more than 25 percent iron and aluminum sesquioxides (oxides comprising two parts of the metal to three of oxygen) within 20 inches (50 cm) of the surface that hardens when exposed.

Ferralsols Soils with a subsurface layer more than 6 inches (15 cm) deep with red mottling due to iron and aluminum.

Solonetz Soils with a sodium- and clay-rich subsurface layer more than 3 inches (7.5 cm) deep.

Planosols Soils that have had stagnant water within 40 inches (100 cm) of the surface for prolonged periods.

Chernozems Soils with a dark-colored, well structured, basic surface layer at least 8 inches (20 cm) deep.

Kastanozems Soils resembling chernozems, but with concentrations of calcium compounds within 40 inches (100 cm) of the surface.

Phaeozems All other soils with a dark-colored, well structured, basic surface layer.

Gypsisols Soils with a layer rich in gypsum (calcium sulfate) within 40 inches (100 cm) of the surface, or more than 15 percent gypsum in a layer more than 40 inches (100 cm) deep.

Durisols Soils with a layer of cemented silica within 40 inches (100 cm) of the surface.

Calcisols Soils with concentrations of calcium carbonate within 50 inches (125 cm) of the surface.

Albeluvisols Soils with a subsurface layer rich in clay that has an irregular upper surface.

Alisols Slightly acid soils containing high concentrations of aluminum and with a clay-rich layer within 40 inches (100 cm) of the surface.

Nitisols Soils with a layer containing more than 30 percent clay more than 12 inches (30 cm) deep and no evidence of clay particles moving to lower levels within 40 inches (100 cm) of the surface.

Acrisols Acid soils with a clay-rich subsurface layer.

Luvisols Soils with a clay-rich subsurface layer containing clay particles that have moved down from above.

Lixisols All other soils with a clay-rich layer within 40–80 inches (100–200 cm) of the surface.

Umbrisols Soils with a thick, dark-colored, acid surface layer.

Cambisols Soils with an altered surface layer or one that is thick and dark-colored, above a subsoil that is acid in the upper 40 inches (100 cm) and with a clay-rich or volcanic layer beginning 10–40 inches (25–100 cm) below the surface.

Arenosols Weakly developed soils with a coarse texture.

Regosols All other soils.

climate is suitable, they support broad-leaved deciduous forest. The mild climate and fertility of their soils place such forests at risk from agricultural development. Historically, whenever there has been competition for land between primeval forest and farming, the farmers have won.

As they mature and become more acid, Inceptisols grade into Alfisols and then Spodosols. Alfisols occur to the south of the Great Lakes, from Ohio to southern Wisconsin, and also in parts of Texas, Colorado, Montana, North Dakota, and southern California. These soils also support broad-leaved deciduous forest. Spodosols, more suited to coniferous forest, occur in New England, on the shores of Lakes Huron, Michigan, and Superior, and in parts of Florida.

Soil Acidity

Acidity is measured on the *pH* scale, where pH 7.0 is neutral. Values lower than pH 7.0 are *acid,* and values higher than pH 7.0 are *alkaline.* A *base* is a substance that will react with an acid to form a *salt.* Soils rich in base elements are able to absorb acidic materials without becoming acidic themselves. This protective effect is called *buffering.*

Many forest soils tend to be acid, and many forest trees, especially coniferous species, prefer a slightly acid soil. European larch (*Larix decidua*), for example, grows best in a pH of 5.0–6.5, white spruce (*Picea glauca*) 5.0–6.0, and black spruce (*P. mariana*) prefers even more acid conditions, with a pH of 4.0–5.0. Black and white spruces are important sources of American woodpulp for making paper.

Some broad-leaved trees also like acid conditions, but not so acid as those best suited for conifers. American beech (*Fagus grandifolia*) prefers a pH of 5.0–6.7, and pin oak (*Quercus palustris*) 5.0–6.5, but black oak (*Q. velutina*) and sugar maple (*Acer saccharum*) are best suited to a more neutral soil, with a pH of 6.0–7.0 and 6.0–7.5 respectively. There are many exceptions, but in general the soils beneath a broad-leaved deciduous forest are less acid than the soils beneath a coniferous forest.

Increasing acidity is a common and natural consequence of the development of soils. Rainwater is naturally acid because of the carbon dioxide, sulfur dioxide, and nitrogen oxides that dissolve into it from the air, and the ordinary metabolism of plants and animals releases acid products into the soil. Base elements present in the soil, such as calcium, potassium, or magnesium, will react with the acids, in the case of carbonic acid (H_2CO_3) forming bicarbonate (HCO_3). Bicarbonate dissociates to release hydroxyl (OH) ions, which react with the acids and so reduce the acidity. The effectiveness of this buffering depends on the amount of base elements present. This varies, soils derived from such rocks as limestone, chalk, and dolostone (or dolomite) containing relatively large amounts, and soils derived from granites and sandstone usually containing little. Buffering in

soils derived from different parent materials is what makes some soils more susceptible than others to acidification due to acid precipitation (see "Acid Rain" on pages 255–264).

Natural drainage tends to leach out the base elements, however, so soils become progressively more acid as they age. The base elements are also plant nutrients that are taken up by plant roots and returned to the surface in dead plant material which decomposes, allowing plants to absorb the elements once more. Constant recycling makes the acidification of soils a slow process. Forests prefer slightly acid soil, but continued weathering and leaching may eventually make them too acid. When that happens, land that now supports temperate forests will support either a different type of vegetation or forests of a markedly different composition. Clearing natural forests can greatly accelerate leaching and the consequent acidification of forest soils, called podzolization.

How Trees Alter Soils

Plants help to make the soil, and forests have a much stronger influence on soil development than most types of vegetation. Tree roots are large and spread far to the side, although, contrary to what people used to believe, they do not penetrate very deeply into the soil (see "Roots" on pages 103–104). The roots are also long-lived, because trees themselves live for a long time. Nonwoody plants, such as grasses, may root to a considerable depth, but individual roots do not live long and when they die their decomposition returns nutrients directly to the subsurface horizons. Trees, in contrast, move nutrients from belowground to the surface, where they are deposited as leaves, twigs, and deadwood.

Accumulated organic matter alters the profile of forest soils by adding to the O and A horizons (see "Digging Deep to Reveal the Soil Layers" on pages 19–20). In forests around the Great Lakes, these horizons contain nearly 40 tons of organic matter per acre (82.8 t/ha). In coniferous forests, this organic material decays very slowly. Usually it is cleared from time to time by fires, leaving ash, rich in plant nutrients, that is quickly incorporated into the soil. On sloping ground, however, the ash is often washed away by rain, increasing the erosion of the bare, exposed soil. If there is no fire to remove the surface litter, it continues to accumulate until the ground feels very spongy underfoot. At its base the litter is compressed, wet, and airless, and will form peat.

Trees do die, of course, and the standing remains of dead trees (called *snags*) and fallen trees are common. On sloping ground, soil that moves downhill will be caught by timber lying on the ground. The accumulation of soil on the upslope side reduces soil erosion.

The woody remains of broad-leaved trees decay much more quickly than those of coniferous species, partly because broad-leaved forests grow in warmer climates, where decomposition is faster. In old-growth forests (see "Primeval Forest, Ancient Woodland, Old-Growth Forest, and Plantations" on pages 144–147) of western hemlock (*Tsuga heterophylla*) and Douglas fir (*Pseudotsuga* species), up to 20 percent of the ground may be covered by dead timber. In broad-leaved forests, dead logs often cover no more than 3 percent of the ground surface.

How Trees "Plow" the Forest Soil

Dead trees often remain standing for many years. Stripped of their leaves, they offer much less wind resistance than living trees. They fall at last, of course, and living trees can also be blown down.

As it falls, a tree dislodges the soil around its own base. On sloping ground, some of this loose soil is likely to be moved downhill, so tree falls can accelerate erosion. Falling trees also produce pits in the surface where the base of the tree and some of its roots have been wrenched from the ground to be exposed as a *root plate* as well as mounds of earth that accumulate on the upslope side. This soil is protected from erosion by the fallen tree itself. The ground surface in natural woodland is very uneven for this reason. Changes in the configuration of the surface that were caused by tree falls remain long after the dead trees have decomposed and vanished. Water and litter accumulate in the pits, where there is also generally a smaller variation in temperature than is found in the surrounding surface soil.

Falling trees scatter topsoil from the O and A horizons and expose subsoil from the B horizon (or C horizon on Inceptisols). This is rather like the effect of plowing, and in time a natural forest will turn over a considerable proportion of its soil. The effect is patchy, however, because the wind will not blow down trees at the same rate everywhere, so a natural forest has undisturbed soil in some places and in others soil that has been turned over many times. There is a reason for this: The falling tree leaves a gap in the leaf canopy, so a small area receives direct sunlight. The warm sunshine favors seed germination, and the light favors the growth of the seedling. Consequently, it is easier for young trees to establish themselves on the mounds of earth produced by tree falls than it is for them to grow elsewhere. Added height also gives the young tree an advantage because it starts life closer to the open sky. As it grows to its full height, however, the new tree is likely to project above the canopy, exposing itself to the full force of the wind. This increases the likelihood that it will also be blown down, so trees often fall in the same place repeatedly.

No other type of vegetation alters its soil to this extent. The overall effect is that, although temperate forests are associated with particular soil orders, the trees themselves produce substantial modifications. The soil beneath an old-growth forest comprises a complex mosaic of types, all derived from the same original soil.

◼ WATER FLOW AND DRAINAGE

When it rains, in most places water does not accumulate on the ground. Puddles may appear on city streets and in parking lots, but on a forest floor the rain usually disappears almost at once, except in depressions along well-trodden and worn paths. The soil may be damp but it is not waterlogged, and if the weather turns fine and warm, before very long the soil will look and feel dry. It dries because water evaporates from it, but this affects only the upper layer. There is moist soil a few inches beneath the surface. Even when it has not rained for some time, the soil will be moist belowground, although during a prolonged drought the moist layer may be fairly deep.

Water moves through soil laterally, vertically downward, and vertically upward. Plants, which must absorb water at least occasionally, obtain it from the soil, not directly from the falling rain. All the precipitation that falls from clouds comes originally from the sea, of course, and all of it returns to the sea, but it moves slowly, allowing plants to capture a proportion of it (see "Groundwater and the Water Table on page 32).

Porosity and Permeability

Apart from the peaty soil found in bogs or on land that was once boggy and has since been drained, soil consists mainly of mineral particles. Even in a soil rich in organic matter most of the volume comprises solid particles weathered from rock. Particles pack together, but their shapes are irregular and they do not fit closely, like bricks. There are spaces, called *pores,* between them and it is through these spaces that water can travel. The total amount of space between soil pores determines the *porosity* of the soil.

The smaller the particles are, the more of them will be contained in any given volume. Surprisingly, however, the size of the particles does not necessarily affect the total amount of *pore space,* the porosity, of the soil. The diagram on page 29 illustrates this with three equal boxes of particles, all of the particles spherical in shape but differing greatly in size. It looks as though there is more empty space in the box containing the bigger particles, but in fact the amount of pore space is the same in all three. If these were soil particles, all three soils would be equally porous. As a simple check on this, a jar filled in turn with large, medium-sized, and small particles and then with as much water as will fill the jar to the brim will give a measure of the volume of pore space. It will be the same, regardless of the size of the particles.

Porosity is an indication of the amount of water the soil will hold. In this case, all three soils will absorb the same amount, but they will not retain equal amounts. That is because water can flow more easily through some soils than through others. Although the total amount of pore space is the same in each, water can pass through a few large spaces more easily than it can through many small ones.

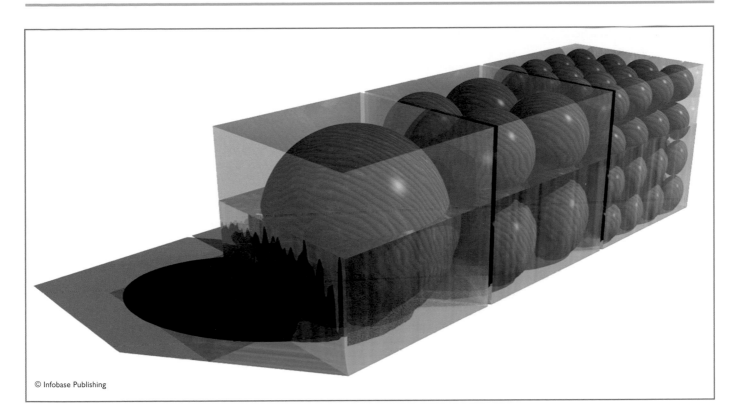

Porosity and permeability. The particles occupy the same volume in all three compartments.

Permeability is a measure of the ease with which water flows through a soil (or any other material). The soil in the box on the left, containing one very large particle, is more permeable than the one in the center, with four particles, which in turn is more permeable than the soil on the right, with many small particles. People often confuse porosity and permeability, but they mean quite different things.

Drainage

Rain falling onto the box on the left in the diagram will drain downward very rapidly, and it will not be long before the surface dries thoroughly. It will take rather longer for water to drain through soil made of smaller particles like those in the middle box, so its surface will remain moist for longer. When rain falls onto very fine soil, however, like that in the right box, it will soak into the ground but then drain away only very slowly. This soil will remain moist much longer than the other two. Permeability and water retention, then, depend not on the total amount of pore space but on the size of the pores and the tiny channels that, in a real soil, link them.

The shape of the particles also affects permeability. Sand grains, for example, are usually very angular. This increases the size of pore spaces without increasing overall porosity and greatly increases permeability. Clay particles, on the other hand, are minute and flat-sided, so they stack together leaving extremely small pore spaces. Clay soils are very porous but relatively impermeable. They hold water, and after prolonged rain, clay can turn into a heavy, sticky, cold, structureless mass.

If the rain is very heavy and the ground is bare of vegetation, the impact of raindrops can batter fine soil particles closer together until they form a thin but impermeable *cap* over the surface. Water then flows horizontally across the surface of sloping ground or lies in pools on level ground. Surface flow on hillslopes is a major cause of soil erosion.

Vegetation slows the fall of water drops. Rain falls much more gently in a forest than it does on an open field. Rain in a forest drips from leaves and branches and trickles down trunks and the stems of smaller plants. Unless the rain is very prolonged, ground sheltered by the leaf canopy often remains dry. Less of it reaches the ground because of the amount that remains in small depressions on branches and the much greater amount that forms thin films over leaves and branches. This water evaporates rather than falling farther.

How Drainage Can Produce Different Soils from the Same Material

Water and the compounds dissolved in it play an important part in the weathering processes by which soil is produced (see the sidebar "Weathering" on page 18). Variations in the way water drains through a soil and moves belowground can lead to the development of markedly different soils all from the same parent material. A group of neighboring and closely related soils that develop from the same parent material but differ because of drainage or some other factor is called a *catena* or toposequence. Catenas sometimes form on hillsides. The diagram shows how this can happen. Near the top of the hill, the ground is fairly level, and there the soil is well drained. As water drains downward through the soil, however, gravity causes it also to move some distance down the slope and a few small particles tend to move with it. Where the slope is steepest, the gravitational, downslope movement accelerates and the drainage is excessive. At the bottom of the slope, the soil receives the water and all the sediment from higher levels. Consequently, there is a progressive change in the type of soil from the top of the hill to the bottom, with the soils at the top and bottom being substantially different from one another. At the top of the slope, the soil is deep, well drained, and has a well developed profile (see "Digging Deep to Reveal the Soil Layers" on pages 19–20). Farther down the slope, water drains faster, indicated by the dark tint, and water accumulates at the foot of the slope.

Frost action and even burrowing animals can alter drainage patterns and lead to the development of a catena. It is sometimes possible to hear water moving underground through burrows excavated by small animals.

A catena of the kind that forms on a slope where the drainage varies

Capillarity—How Water Can Move Upward

Watch water soaking into the ground and you might suppose that it sinks under its own weight, that it is continuing its fall from the sky. Indeed, water draining vertically is called *gravitational water,* but the image is misleading. Water does have weight, causing it to move downward, but it is more accurate to think not of water sinking into the ground, but of the water being sucked into the ground from below, at a speed determined by the permeability of the soil (see "Porosity and Permeability" on pages 28–29). The "sucking" is caused by *capillarity* or capillary attraction, and it is due to a peculiar property of water itself.

A molecule of pure water consists of two atoms of hydrogen and one of oxygen. Hydrogen atoms carry a positive electrical charge, oxygen atoms a negative charge, and they are bonded by the attraction of opposite charges. The molecule is not symmetrical, however, both hydrogen atoms being bonded to the same side of the much larger oxygen atom and separated by an angle of 104.5° measured from the center of the oxygen atom. The oxygen and hydrogen atoms share electrons (carrying negative charge), so on the hydrogen side of the molecule the hydrogen nuclei—protons with positive charge—project from the oxygen atom. This leaves the molecule with a very small positive charge on its hydrogen side and an equal negative charge on its oxygen side. Such molecules are said to be *polar*. Their polarity allows water molecules to attach to one another, by hydrogen bonding of the hydrogen side of one molecule to the oxygen side of another. *Hydrogen bonds* are weak, but they also allow water molecules to attach to other substances.

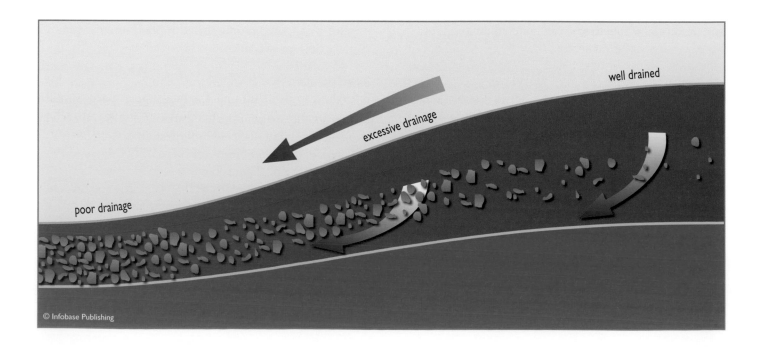

© Infobase Publishing

When water is held in a container, water molecules, because they are polar, will cling to the sides (this is called *adherence*) as well as to one another (called *coherence*). Adherence makes the water climb up the sides of the container, and coherence makes it draw more water behind it. Eventually the water will reach a height where the force making it rise precisely balances the weight of the water column itself. After that it can rise no further. Consequently, the smaller the container, the less the water column inside it weighs, and the higher water will climb.

Soil Moisture Tension

There is another way to think of this. Picture a beaker filled with water. At the bottom, water molecules are being subjected to a pressure by the weight of water above them. This is the *water pressure*. Obviously, the water pressure decreases with height in the beaker, because there is less water above, and at the surface the water pressure is zero.

Now imagine that a narrow tube is inserted vertically into the water, as shown in the drawing. Inside the tube, above the surface, the water pressure will be still lower. If the water pressure at the surface is zero, then above the surface it will be less than zero and it will continue to decrease with increasing height up the tube. It is that difference in water pressure that causes water to rise. Water pressure above the column is negative (less than zero), and working with negative quantities is inconvenient, so soil scientists simply eliminate the minus sign, giving the pressure a positive value, and call it *soil moisture tension* (SMT). It then represents a pulling force (tension) rather than a pushing force (pressure).

SMT varies inversely with the size of the pore spaces. The smaller the pore spaces, the greater is the SMT. Values for water pressure and SMT are given in the same units as those used to report atmospheric pressure—pounds per square inch, inches or millimeters of mercury, millibars, or pascals. This can cause confusion because these forces are due wholly to the water and not at all to atmospheric pressure.

Capillarity can exert a force in any direction, however. As water drains down through a soil, the pore spaces beneath it are dry and there is an SMT drawing it down faster. At the same time it is also being drawn to the sides. You can see this effect by spilling a colored liquid onto a thick stack of kitchen towels or blotting paper because they also absorb liquids by capillarity. The color soaks sideways as well as downward.

How Soils Retain Water

When water soaks downward into the soil, almost all of it does so because of capillary attraction. Some may drain directly through the larger pore spaces, but only if these are open at the surface. In the boxes of particles shown in the

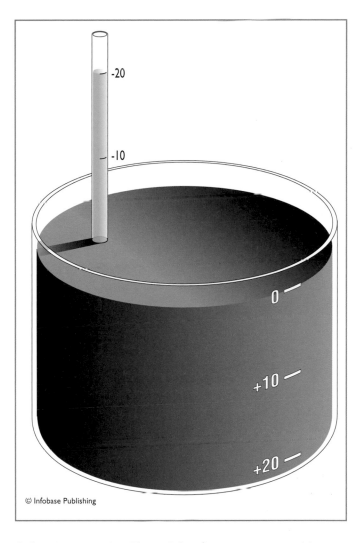

© Infobase Publishing

Soil moisture tension. The weight of water exerts a positive pressure inside the vessel. At the surface the pressure is zero, and above the surface the water exerts a negative pressure. Removing the minus sign converts a pressure pushing downward into a tension pulling upward.

illustration of porosity and permeability (see pages 28–29), water drains readily through the bigger pore spaces simply because they are so large; in reality they are unreasonably so. If the large pore spaces are closed, atmospheric pressure will prevent water flowing out of them in the same way that it prevents water flowing from a tube that is inserted it into water with both ends open, then one end is sealed and the tube withdrawn. It is rare for soil pore spaces to be open. Moving water tends to wash small particles into them, thus sealing them.

Suppose, though, that the water is draining through a layer of fine-grained soil that overlies a layer of gravel or coarse sand. Common sense seems to suggest that when the water reaches the more permeable material it will drain even faster. In fact, though, this is not what happens. Its movement ceases, at least temporarily, because the more permeable material contains much bigger pore spaces and, therefore, exerts a smaller SMT. The SMT pulling the water sideways in the fine-grained soil is greater, so it will sink no farther until the fine-grained layer holds so much water that almost all the pores are filled, meaning the soil is close to saturation. This reduces the SMT and when it falls to a value lower than that in the coarse-grained layer, the water will resume its downward movement. Some of the best farm soils in the northwestern United States work like this. In the basin of the Columbia River there are fine soils overlying sand and gravel which are renowned for their capacity to retain water.

An ordinary shower following a long period of dry weather will wet the ground, but the moisture penetrates only a certain distance and below it there is dry soil. Water soaks only so far, but the wetted soil may then remain moist for quite some time. The upper, moist soil, from which water has drained as far as it can, is said to be at *field capacity.* The amount of water that remains, held by the soil, varies widely depending on the size of the soil particles. The water is held there by SMT. This also varies, but water will move out of this region only if it is drawn by an SMT greater than the SMT that retains it.

Groundwater and Water Table

Finally, if enough water arrives at the surface, the downward movement will continue until a layer of impermeable material checks it. The impermeable material may be rock or tightly compacted clay. Water can drain no farther vertically, so it accumulates, completely saturating the soil above the impermeable layer. It is then called groundwater, and the upper boundary of the saturated soil, above which some pore spaces are still free of water, is called the *water table.* Just above the water table, in the unsaturated soil, capillarity produces a layer, called the *capillary fringe,* where groundwater is being drawn upward.

The height of the water table varies from time to time, depending on the amount of water draining into the groundwater. During a drought the water table falls and after prolonged rain, or when a thick layer of snow melts in spring, it rises. Where rainfall is distributed fairly evenly through the year and droughts are rare, the height of the water table will change little. A well is a hole cut from ground level to a depth below the water table. Groundwater will fill the bottom of the hole to the height of the water table.

Groundwater also flows, but very slowly, because it must move through soil pore spaces. Speeds vary widely, depending on the permeability of the material through which it flows, permeability being much more important than the gradient. Commonly, groundwater moves from a few feet a day to a few feet a month, but through limestone with many

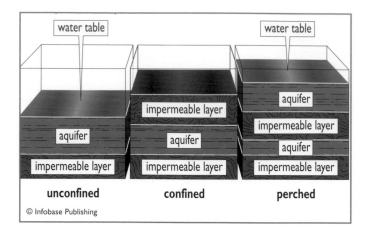

An aquifer is a layer of permeable material that lies above an impermeable layer. These are the three possible types of aquifer.

underground caverns it can travel at more than one mile per hour and through clay or shale, with low permeability, it may move less than one foot in a century.

Aquifers

Flowing groundwater is called an *aquifer.* It is through aquifers that chemical compounds leached from the surface soil horizons are removed. Aquifers eventually discharge their water. Where the soil is shallow or drains poorly, the water bubbles through the surface, as a spring or as a seep, evident as an area of very wet ground. Beneath deeper soil it emerges only on low ground, where the water table, as measured by its level in the higher ground to either side, is above the surface level. There the water in the aquifer flows as a river and the low ground becomes a river valley.

An aquifer into which water can drain from directly above, so there is no impermeable layer between the water table and the ground surface, is said to be unconfined. In its journey, eventually to a river, lake, or the sea, the aquifer may flow through a layer of permeable material that lies beneath an impermeable layer. It is then moving between two impermeable layers, one above and one below, and is said to be confined. Above the upper impermeable layer, often formed by densely compacted clay forming a lens, water draining from the surface will accumulate to form a second, perched aquifer, with its own water table. The diagram illustrates the three types of aquifer.

Because a perched aquifer lies fairly close to the surface, the soil above it will be shallow and poorly drained. This, too, may lead to the development of a catena (see "How Drainage Can Produce Different Soils from the Same Material" on pages 29–30).

3

Atmosphere and Temperate Forests

Temperate forests grow in climates that are cooler and drier than those of the humid Tropics, cooler and moister than those of the arid subtropics, and warmer and moister than those of the Arctic and Antarctic. The climates of the world vary from place to place, and they also vary through the year as the seasons progress. This chapter describes what climate is and how the warmth of the Sun drives the movements of the air and oceans that produce the weather. It explains what happens when water freezes and melts, evaporates and condenses, how clouds form, and why the air temperature decreases with increasing altitude. It describes the fronts and jet streams that generate weather systems, and it describes the extreme conditions that produce hurricanes, tornadoes, blizzards, and floods. Scientists have devised many schemes for classifying the different kinds of climate. This chapter explains the two that are most widely used.

The climate determines the type of plants found in any particular area. This also means that the plants present in that area indicate the type of climate. Certain trees are especially useful as climate indicators, and traces of them found in ancient soils have been used to reconstruct the climates prevailing at the time they were alive.

■ CIRCULATION OF THE ATMOSPHERE

Temperate forests grow between latitudes 25° and 50° in both hemispheres. Latitude 25° is close to the Tropics at 23.5°N and S, and 50° is the approximate latitude at which the average sea-surface temperature is 50°F (10°C). The region between latitudes 25° and 50° is known as the temperate belt, middle latitudes, or midlatitudes. Boreal forest—the predominantly coniferous forest of northern North America and Eurasia—extends approximately to the Arctic Circle (66.5°N).

People living in the temperate belt enjoy a temperate climate. The climate in lower latitudes bordering the temperate belt is subtropical, and in higher latitudes it is subarctic. Near the equator, the climate is hot and humid; near the poles it is very cold and, surprisingly perhaps, extremely dry. The Arctic and Antarctic regions are cold deserts, despite all their ice.

This gradation of temperature from the equator to the poles seems so obvious that it is easy to take it for granted, yet the reason for it is not quite so obvious as it seems. No other planet in the solar system has a climate even remotely like our own. On the Moon, for example, which is a satellite, but one big enough to be treated as though it were a small planet, the climate is quite unlike any climate on Earth, even though the Earth and the Moon are both the same distance from the Sun. At noon, the temperature on the Moon is about 260°F (127°C), and just before dawn, which is always the coldest time, it falls to -280°F (-173°C). The temperature is much the same at all latitudes and there are no seasons.

Why Earth and Moon Have Very Different Climates

The difference, of course, is that the Earth has a very much denser atmosphere than the Moon. It is not quite true that the Moon has no atmosphere; it does, but in daytime the lunar atmosphere contains only about one-ten trillionth (10^{-13}) the amount of gas that is in the Earth's atmosphere. At night, when some of the constituents condense onto the surface, it contains only one-tenth of the daytime amount. If Earth had an atmosphere as thin as that on the Moon, temperatures here would range from about 260°F (127°C) by day to about -280°F (-173°C) by night, just as they do on the Moon.

When a body is heated, it radiates the heat it has absorbed at a wavelength that is inversely proportional to its temperature. In other words, the hotter the body is, the shorter the wavelength at which it radiates. When the Moon is heated by the Sun, its surface temperature rises and it radiates its heat back into space. During the hours of daylight, the difference between the heat absorbed and the heat reradiated produces the high daytime temperature—heat is being absorbed much faster than it is being radiated into space. At night, when there is no input of heat, the surface continues to radiate the heat it absorbed during the day and so its temperature falls to the predawn minimum, after which it starts to rise again.

On Earth, the surface of land and sea is also warmed by sunshine, and it also radiates its absorbed heat but not all of it. The warmed surface is in contact with the atmosphere, and air at the surface is warmed by that contact. Air can and does move, but it cannot escape from the planet, so it acts like a blanket, keeping the surface warmer than it would be otherwise.

The Greenhouse Effect

If that were all, the average temperature over the surface of the Earth would be about -9°F (-23°C). Scientists know this because they can measure the amount of radiation the Earth receives from the Sun and can calculate the rate at which it should radiate heat to space. All the oceans would be frozen to a considerable depth and the air would be very dry indeed. There would be no forests and no people. The frozen Earth could support no living things more complex than bacteria. In fact, of course, the average surface temperature is a comfortable 59°F (15°C). Simply having an atmosphere is not enough. What matters is the composition of that atmosphere and, in particular, the presence in it of very small concentrations of certain gases.

The Sun radiates at all wavelengths, but very short-wave gamma, X, and most ultraviolet radiation is absorbed at the top of the atmosphere. Radiation reaches the surface most intensely at wavelengths between 0.2 μm and 4.0 μm (1 μm = 1 micrometer = one-millionth of a meter = about 0.00004 inch). This is the waveband of visible and long-wave ultraviolet light and of infrared radiation (heat). Nitrogen and oxygen, the gases comprising nearly all of the atmosphere (it is 78.1 percent nitrogen and 20.9 percent oxygen) are almost completely transparent to radiation at these wavelengths. The radiation passes through them as though they were not there and is absorbed at the surface.

The warmed surface then radiates its absorbed heat. The surface is still quite cool, so it emits radiation at the longer, infrared wavelengths between about 4.0 μm and 100 μm. Some of this radiation is absorbed by minor gaseous constituents of the atmosphere, principally water vapor, carbon dioxide, and methane. Water vapor absorbs radiation at wavelengths between 5 μm and 8 μm and also in certain longer wave bands. Carbon dioxide absorbs at 4 μm and between 13 μm and 17 μm. Methane, ozone, nitrous oxide, and some other gases also absorb long-wave radiation, each at particular wavelengths. This leaves an atmospheric window, between about 8.5 μm and 13.0 μm, where radiation is not absorbed and escapes into space.

Atmospheric molecules that absorb radiation then reradiate it in all directions, some of it outward into space, some sideways, but about two-thirds of it down toward the surface. This radiation warms the air and the surface. At night, when the surface is not being warmed by the Sun, the blanket of warmed air slows the rate at which the surface cools, but radiation continues to escape so that overall there is a balance between the amount of incoming and outgoing radiant energy. The air and surface continue cooling until about an hour before dawn. Then, as the sky begins to grow lighter, the temperature ceases to fall and as the Sun appears over the horizon the temperature begins to rise to a peak in the middle of the afternoon.

This delayed cooling is known as the *greenhouse effect*. The name is slightly misleading because in a greenhouse the air is warmed by the sunlight entering through the glass and then trapped by the glass so it cannot mix with cooler air from outside that would reduce its temperature. The atmospheric mechanism is quite different, but the popular name is now so firmly established that it is fruitless to try substituting a more accurate one. Increasing the thickness of the panes of a greenhouse would have no effect on the temperature inside (actually, it might lower it slightly by blocking some of the incoming sunlight), but increasing the atmospheric concentration of gases that absorb in the infrared wave band would raise the air temperature. Such an induced warming is also known as the greenhouse effect, but is more correctly called the enhanced greenhouse effect (see "Forests and the Greenhouse Effect" on pages 268–274). The gases that absorb long-wave radiation (except for water vapor, which is usually excluded) are known as *greenhouse gases*.

Why Surface Color Matters

When light and heat from the Sun passes through the atmosphere, some of it is reflected by clouds. More is reflected by

Albedo	
SURFACE	**ALBEDO (%)**
fresh snow	75–95
cumulus cloud	70–90
stratus cloud	59–84
dry sand	35–45
desert	25–30
concrete	17–27
broad-leaved deciduous forest	10–20
plowed field	5–27
asphalt	5–17
coniferous forest	5–15
field crops	3–15
black road	5–10

the surface. Reflected sunlight—including heat—does not warm the surface at all. How much is reflected depends on the color of the surface. People make use of the same principle when they wear pale clothes in summer to reflect heat and dark clothes in winter to absorb heat. Scientists call the reflectivity of a surface its *albedo*. Albedo can be measured and its value is expressed as a percentage of the incoming radiation that is reflected, either directly (e.g., 80 percent) or as a decimal fraction (e.g., 0.8). Obviously, radiation that is not reflected is absorbed, warming the surface, and the warmed surface warms the air in contact with it. The albedo has a direct effect on local climates.

As the table shows, the albedos of different surfaces vary widely. Some are self-evident. Cloud and freshly fallen snow reflect so much bright sunlight that it is dazzling. Pilots and anyone outdoors on a sunny day when snow covers the ground wear dark glasses, without which the brilliant light could actually be painful. Other surfaces may be less obviously reflective. Broad-leaved deciduous forest reflects more light and heat than coniferous forest, for example. This is

another way of saying that coniferous forest is a much darker color than broad-leaved forest. This also means that coniferous forest, being darker, absorbs more radiant heat than does broad-leaved deciduous forest. Their lower albedo may help coniferous trees to keep warm in the cool regions where most of them grow.

The Difference Latitude Makes

Regardless of albedo, the Sun warms some parts of the Earth more intensely than others. Earth differs from the Moon in this respect because it is much bigger. It is a matter of geometry and the illustration shows why. Picture a place on the surface where the Sun is directly overhead, as shown in

The Sun is almost directly overhead in the Tropics (A), so it illuminates a relatively small area but shines very intensely. It is lower in the sky in higher latitudes (B), so it illuminates a larger area but less intensely.

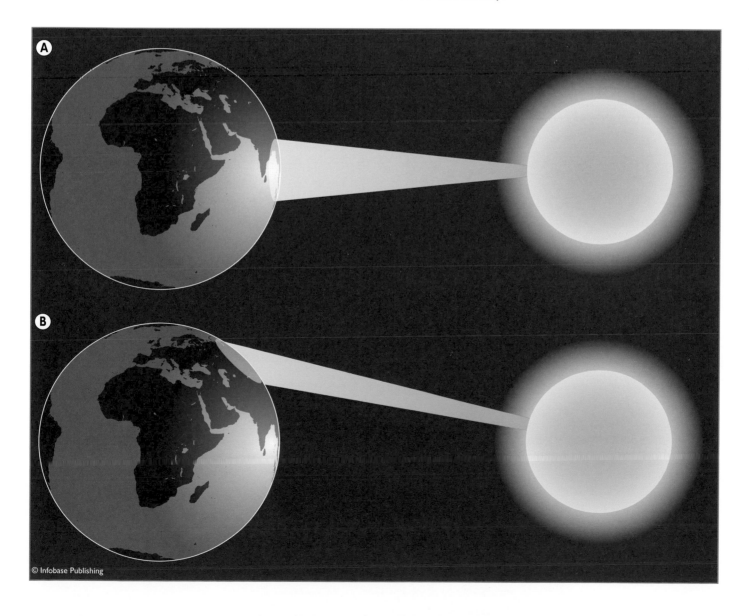

Seasons

The Earth turns on its own axis, taking 24 hours to complete one revolution. Earth also orbits the Sun, taking one year to do so. However, the axis around which the Earth turns is not at right angles to the track of its solar orbit. The axis is tilted approximately 23.5° from the vertical.

Because of the tilt, as Earth orbits the Sun, first one hemisphere and then the other faces the Sun more directly, as shown in the diagram. In December the Southern Hemisphere and in June the Northern Hemisphere faces the Sun. The *solstices* are the two days in the year when this is most extreme—midsummer and midwinter days, on June 21–22 and December 22–23. At the *equinoxes*, on March 20–21 and September 22–23, the Earth is in an intermediate position. Both hemispheres are illuminated equally, and day and night are equal in length everywhere in the world.

The angle of axial tilt, of 23.5°, determines the location of the Tropics and the Arctic and Antarctic Circles. As the diagram illustrates, the solstices occur when the Sun is directly overhead at noon over the latitude equal to the angle of the axial tilt—23.5°N and 23.5°S. These latitudes define the tropics of Cancer and Capricorn. On the midwinter solstice, the Sun does not rise above the horizon in latitudes higher than 90 - 23.5 = 66.5°. On the midsummer solstice, it does not sink below the horizon in these latitudes. Latitudes 66.5°N and S mark the Arctic and Antarctic Circles. At the equinoxes, the noonday Sun is directly overhead at the equator.

The seasons occur because the Earth's tilted axis results in first one hemisphere and then the other facing the Sun.

drawing A, and suppose there is a narrow beam of sunlight. Where the Sun is overhead, it illuminates a relatively small surface area. A beam of similar width illuminates a larger area in B, which is located in a higher latitude where the Sun is lower in the sky. The beam transmits the same amount of energy in both cases, but in A the energy is concentrated into a smaller area than it is in B. Consequently, A will be warmer than B.

The Tropic of Cancer is at latitude 23.5°N, and the Tropic of Capricorn is at 23.5°S. The Tropics are defined as the region, between these latitudes, where the Sun appears directly overhead at noon on at least one day in the year. The Arctic and Antarctic Circles, at latitudes 66.5°N and S respectively, define the regions to the north and south of which there is at least one day in the year when the Sun never sinks below the horizon and one day in the year when

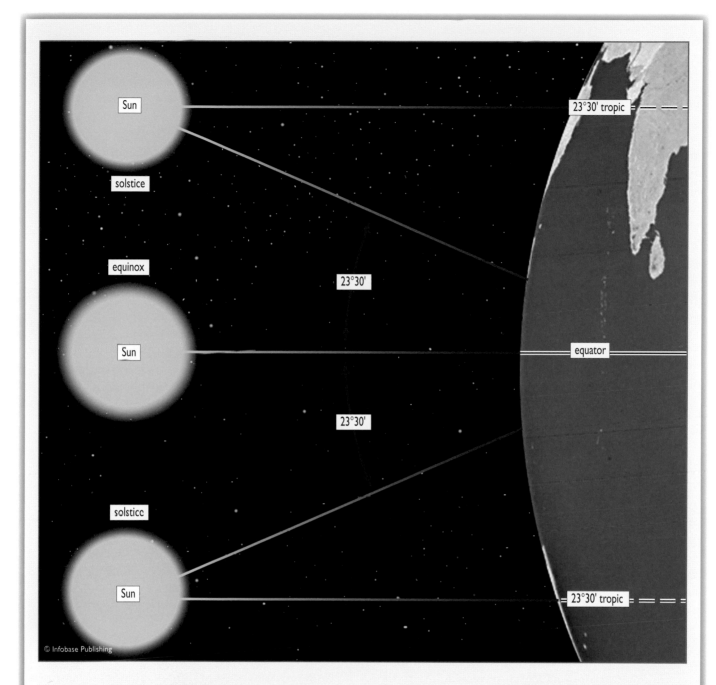

The Earth's axis is tilted 23.5° from the vertical, and the two Tropics are at latitudes 23.5°N and 23.5°S. The Tropics are where the Sun is directly overhead at noon on one solstice. At the equinoxes the noonday sun is directly overhead at the equator.

it never rises above the horizon. These boundaries exist because of the tilt in the Earth's axis of rotation (see the sidebar "Seasons" on page 36).

This shows why land at the poles is colder than land at the equator, but when the general picture is considered in much more detail, and actual amounts of energy are calculated, a curious fact emerges. In a belt between 40°N and 35°S, the surface receives more energy than it reradiates, and in latitudes higher than about 75°N and S, the surface radiates more energy than it receives. Consequently, the average temperature at the equator should be about 25°F (14°C) warmer than it is and it should be about 45°F (25°C) colder at the poles. There can be only one explanation of the more moderate temperatures that actually exist. Heat is being transferred from low to high latitudes. Heat transfer of this kind requires a fluid medium, and

Earth has two. The heat is carried by the atmosphere and the oceans.

Ocean Currents and the Trade Winds

Currents circulating in all the oceans form *gyres*—approximately circular flows of surface water. The water is warmed near the equator, and as it moves into higher latitudes, it warms the air in contact with it. Then it curves back toward the equator as the cold-current part of the system to be warmed again. Air in high latitudes is warmed and air in low latitudes is cooled. There is also a much bigger movement of water that carries cold water from the edge of the sea ice in the North Atlantic and replaces it with warm water from the equator. This circulation is known as the Great Conveyor (see the sidebar), and in the course of its circulation it transports water through all of the oceans.

Air also moves, both vertically and horizontally, and it, too, transports heat away from the equator. Like many scientific discoveries, our understanding of how it does so came as a by-product of the search to explain an apparently quite different puzzle.

Centuries ago, when European sailing ships took to plying the world, sailors found that winds in the tropics were very reliable. North of the equator they nearly always blew from the northeast and south of the equator from the southeast. Indeed, these winds were so dependable that they were called *trade winds*. Originally, the word *trade* meant a track or path, suggesting movement in a constant direction. Very close to the equator itself, however, there were regions where winds were often light and variable or the air was quite still, and ships were becalmed. This is where the trade winds converge in a belt called the Intertropical Convergence Zone (ITCZ). These windless areas came to be called the doldrums, and they were encountered in the horse latitudes. Ships often carried cargoes of horses, and when becalmed for so long that supplies of drinking water were exhausted, some of the horses might die from thirst. Being becalmed was not merely inconvenient, it was also extremely dangerous. Scientists then began to wonder why the trade winds are so reliable.

The first person to attempt an explanation was Edmund Halley (1656–1742), the English astronomer (and Astronomer Royal) for whom the famous comet is named. He suggested in 1686 that air over the equator is heated strongly. It expands, rises, and is replaced near the surface by cooler air, as the trade winds. He was nearly correct but not quite because the circulation he described would make the trade winds blow from due north and due south. He had not explained the easterly component. It was not until 1735 that an explanation was offered for that, by the English meteorologist George Hadley (1685–1768). Hadley accepted much of the Halley explanation. His idea

The Great Conveyor

The oceans influence climates in several important ways. One of these is by transporting heat from the Tropics into polar latitudes by a system of currents known as the Great Conveyor, or just simply as the Conveyor.

Seawater with an average salinity of 35 percent is at its densest at a temperature of 32°F (0°C), and it freezes at 28.56°F (-1.91°C). When seawater freezes, the salt dissolved in it separates from the ice crystals and enters the surrounding water, making that water denser. Water close to the edge of the sea ice is denser than the adjacent water because its temperature is close to freezing and because of the salt it contains. To the north of Iceland, this dense water sinks all the way to the ocean floor, forming the North Atlantic Deep Water (NADW).

The cold, dense water flows southward along the ocean floor all the way to the edge of the Antarctic Circle, where it rises over the Antarctic Bottom Water, which is formed in the same way as the NADW. The water then joins the Circumpolar Current flowing from west to east. The Circumpolar Current divides, with one branch entering the Indian Ocean where it follows a clockwise circular path. The other branch enters the South Pacific to the east of New Zealand, flows northward to about latitude 50°N, then turns in a clockwise direction to head westward, between the islands of Indonesia, and into the Indian Ocean, where it joins the other branch. The current then flows around the southern tip of Africa, through the South Atlantic and North Atlantic, and finally to the area around Iceland where it replaces the water sinking to form the NADW.

The cold water that flows southward as the NADW remains cold until it rises in the South Pacific and Indian Oceans. It warms during its passage through the Tropics, and it returns to the North Atlantic as a warm current. The system transports cold water toward the equator and warm water away from the equator.

was that the warm air rises to a great height, flows away from the equator in both directions, and then sinks and flows back toward the equator. In fact, the air forms a vast convection cell. He also pointed out that while the air is moving, the Earth is also turning in an easterly direction.

It is this motion which swings the returning air so that it approaches the equator from the northeast and southeast. Like Halley before him, he was almost correct. Hadley had described the convection cell, but he was wrong about the reason for the change in direction of the flowing air. That was not explained until 1856 by the American meteorologist William Ferrel (1817–91), who said that the swing is due to the tendency of moving air to turn about its own vertical axis, like water flowing down a drain.

The Three-Cell Model of Atmospheric Circulation

What Hadley had identified was the convection cell driving the circulation. He believed that the equatorial air travels all the way to the poles, sinks there, and returns at low level, so there is one cell covering the entire Earth. We know now that the real situation is much more complex and that the rotation of the Earth prevents the formation of a single, planetwide cell. What really happens is that in several equatorial regions warm air rises to a height of about 10 miles (16 km), moves away from the equator, cools, sinks between latitudes 25° and 30°N and S, and flows back to the equator as the surface trade winds. There are several of these convection cells, and they are known as *Hadley cells,* in recognition of the man who came so close to describing them accurately in the course of explaining the trade winds.

Air over the poles is also sinking and flowing away at low level, forming a second system of *polar cells.* The Hadley-cell circulation in the tropics and the polar-cell circulation drive a third, midlatitude set of cells, discovered by William Ferrel and known as the *Ferrel cells.* This is the central feature of the general circulation of the atmosphere. It is called the three-cell model, and the diagram shows how the system works, the arrows indicating the direction in which the air moves. It is the mechanism by which air movements transport heat from the equator to the poles, although the meridional (north or south) movement of air is fairly weak, except in the Hadley cells.

Why the Equator Is Wet and the Sahara Is Dry

The three-cell model also indicates some of the consequences of this circulation. When air is warmed from below, it expands and rises (see "Evaporation, Condensation, and Precipitation" on pages 43–48). This produces a region of low atmospheric pressure at the surface. As the air rises it cools, and as it cools, its water vapor condenses, producing cloud and precipitation. In equatorial regions, where air is rising into the Hadley cell circulation, there is generally low

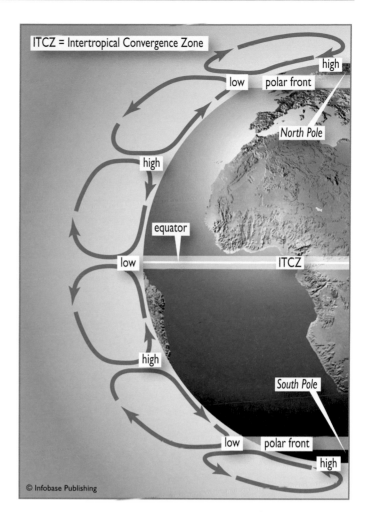

The three-cell model of the circulation of the atmosphere. The three cells are the Hadley, Ferrel, and polar cells.

surface pressure and a great deal of rain. That is why the equatorial climate is humid. By the time it reaches its maximum height, the air is very dry and very cold. It then sinks on the descending side of the Hadley cells. As it descends it is compressed, and when air is compressed it grows warmer. The subsiding air reaches the surface as warm but still extremely dry air and produces a region of generally high surface pressure in the subtropics. That is why a belt of hot, dry deserts circles the Earth in the subtropics.

When air is cooled, it contracts and becomes denser. Denser air should sink beneath less dense air, so it seems paradoxical that extremely cold air can move away from the equator above warmer air that should be less dense. The concept of *potential temperature* explains the apparent paradox (see the sidebar on page 45).

Not all the air on the descending side of the Hadley cells flows back toward the equator. Some spills the other way, toward the poles. In middle latitudes, this warm air flowing away from the subtropics at low level meets cold air flowing away from the poles. Where the two types of air meet, there is a boundary, called the polar front. It is shown as a straight line in the illustration, but in fact it is wavy and moves with the seasons. It also produces the polar front jet stream (see "The Polar Front and Jet Streams" on pages 57–63), which is a very fast, high-level, westerly wind. Local areas of low pressure called depressions form beneath the jet stream and are dragged eastward by it.

The middle latitudes are where warm and cold air meet and where complex weather systems move in a generally easterly direction in both hemispheres, losing moisture as they cross continents and gaining moisture as they cross oceans. Combine this with the overall weakness of the Ferrel-cell circulation and the seasonal migration of the polar front and its jet stream toward the equator in winter and toward the poles in summer, and it is easy to see why the temperate regions have such variable weather.

In summer, the polar front jet stream over North America usually lies across Washington, Montana, then slightly south across the center of Lake Michigan, and north again just to the south of the Canadian border. In winter it forms a wavy line from about Los Angeles through northern Texas and to South Carolina. To the north of the front there is cold, polar air, and to its south there is warm, tropical air.

Wind Belts

Air moving toward the equator is deflected to the west, producing generally easterly winds, and air flowing away from the equator is deflected to the east, producing westerly winds. Over the Earth as a whole, the easterly and westerly winds balance.

It is as well that they do. Air moving over the surface experiences friction. This slows the wind, especially over land (which is why wind speeds are lower over land than over the sea), but it also slows the Earth in its rotation. Obviously, the effect is much greater on the winds than on the solid Earth, but friction works both ways and, though small, the effect is real. If there were more westerly than easterly winds, gradually they would accelerate the Earth's rotation, and the days would grow steadily shorter. If there were more easterlies than westerlies they would slow it, making the days longer. There are occasions when the winds blow more strongly in one direction than in the other, but these are short-lived, and the extremely small effects on the Earth's rate of rotation cancel out, so we can be confident that the latitudinal winds balance. In fact, though, the rate of rotation is slowing, making the days longer by about 0.002 second per century, but this is due to friction from the tides.

The easterly and westerly winds form latitudinal wind belts, shown in the illustration, where the arrows indicate the direction of prevailing winds. This diagram on page 41 completes the picture. It shows the vertical circulation as actually being more complex than the diagram of the three-cell

model suggests, and it also marks the position of the permanent areas of high pressure. Low-pressure areas lie between the high-pressure areas. The ITCZ moves with the seasons, but its average position is slightly north of the equator, with the easterly trade winds to either side of it. In midlatitudes, between the subtropical high-pressure region and the polar front, the prevailing winds are westerlies, and in polar regions they are easterlies. The *tropopause* is the upper boundary to the lowest layer of the atmosphere, the *troposphere,* which is where weather occurs (see "Fronts" on pages 52–53).

The atmosphere works like a vast machine transporting heat and insulating the surface. It is this machine that produces our weather and, in the midlatitudes, the climates in which temperate forests thrive.

Cycles of Climate

The weather changes from day to day, especially in the middle latitudes. Sometimes a particular type of weather will continue for longer than usual, as a spell of wet, dry, cool, or warm weather. The weather also changes with the seasons. Changes in the weather are not the same thing as changes in the climate. When scientists talk of climate, they mean the average weather in a particular place measured over many decades and averaging all of the seasonal and short-term changes in the weather.

Climates do change, but over a much longer period. The most dramatic changes are those associated with ice ages (see "Holocene Recolonization" on pages 91–95). Ice ages alternate with interglacial periods in a cycle related to changes in the Earth's orbit and rotation. The most recent ice age ended approximately 10,000 years ago, and a new ice age may well commence some time in the next few thousand years. This is a cycle with a very long period. Other cycles have shorter periods, however. From about 600 C.E. until 1300 C.E., climates all over the world were warmer than in the preceding or succeeding centuries. Known as the Medieval Warm Period, this was when Norsemen established colonies in Iceland and Greenland, and Britain was a major wine producer. The Medieval Warm Period was followed by the Little Ice Age, lasting from the second half of the 16th century until the early 20th century. The rise in average global temperatures during the first half of the 20th century was probably the final stage of recovery from the Little Ice Age.

On a much shorter timescale, droughts afflict the Great Plains of the United States at intervals of 20–23 years. The most severe drought of modern times lasted from 1933 until the winter of 1940–41, producing the dust bowl. There were also droughts in the 1950s, 1970s, and 1990s.

Climate cycles affecting the northwestern United States influence the salmon populations on which the fishery depends. Salmon are relatively plentiful when the

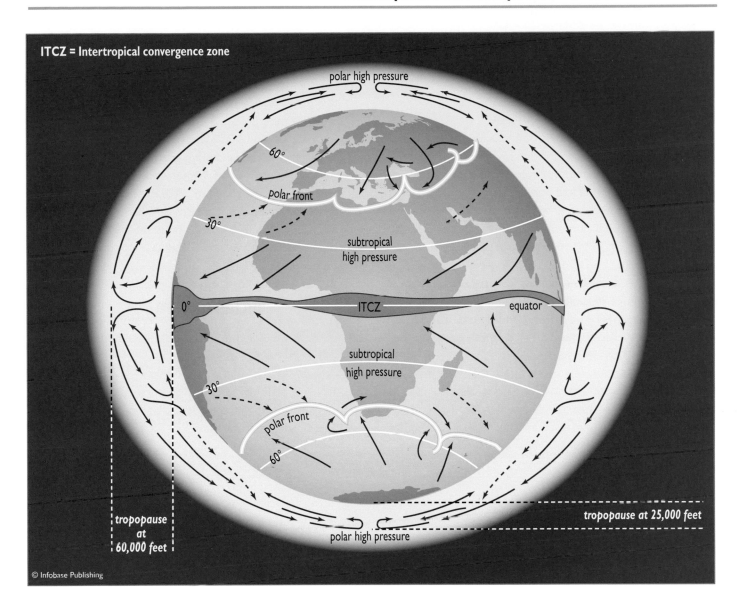

ITCZ = Intertropical convergence zone

The global wind belts are produced by the circulation of the atmosphere influenced by the Earth's rotation.

weather is cool and wet, and relatively scarce when the weather is warm and dry. These conditions alternate, and in 1996 Steven Hare, a fisheries scientist investigating this cycle, gave it a name: the Pacific Decadal Oscillation (PDO). It has a period varying from about 15 years to about 70 years. The PDO is a cycle of changes in the temperature of surface water in the North Pacific Ocean, and its effects extend into the Tropics. The cycle was in a cool phase from 1890 to 1924, warm from 1925 to 1946, cool from 1947 to 1976, and warm from 1977 until the middle 1990s. El Niño is the best known short-period climate cycle. It is similar to the PDO but centered in the Tropics rather than in the North Pacific, and it comes and goes over a much shorter period.

No one knows what causes the PDO, but it is variations in the distribution of pressure that drive a climate cycle centered on the North Atlantic region. This cycle used to be called the North Atlantic Oscillation (NAO), but it is now known to affect the weather over a very much larger area, so it has been renamed the Northern Annular Mode (see the sidebar on page 42).

El Niño

On either side of the equator, the easterly trade winds drive ocean currents from east to west. In the equatorial South Pacific Ocean, the South Equatorial Current transports warm water from the eastern side of the ocean basin to the western side. On the eastern side of the ocean, near the South American coast, the layer of surface water heated by the Sun is relatively shallow. The cool Peru Current, flowing northward parallel to the coast, has many upwellings that bring cold water from near the ocean floor all the way to

The Northern Annular Mode

The prevailing westerly winds that blow across the middle latitudes of the Northern Hemisphere propel the weather systems that cross the North Atlantic from west to east. The strength of those winds depends on the distribution of atmospheric pressure. Ordinarily there is a region of low pressure permanently centered near to Iceland and known as the Iceland low. There is also a region of permanently high pressure centered near the Azores and known as the Azores high. Sometimes this anticyclone occurs farther to the west and is known as the Bermuda high.

In the Northern Hemisphere, air circulates counterclockwise around centers of low pressure and clockwise around centers of high pressure. Consequently, the counterclockwise circulation around the Iceland low and the clockwise circulation around the Azores (or Bermuda) high combine to produce westerly winds across the middle latitudes between them. The strength of the wind is proportional to the pressure gradient—the rate at which pressure changes over a horizontal distance—and the pressure gradient varies according to the difference in pressure between the high and low centers.

The pressure distribution over the North Atlantic varies on a scale measured in decades. This variation used to be called the North Atlantic Oscillation. It was then found to influence weather conditions over a much larger area than the North Atlantic and was renamed the Arctic Oscillation. It is now known to affect all of the Northern Hemisphere and is called the Northern Annular Mode (NAM). *Annular* means "ringlike." The NAM is measured according to an index of values called the NAM index.

When pressure is lower than average over Iceland and higher than average over the Azores, the NAM index is high. A high index brings cold winters in the northwestern Atlantic, mild, wet winters in Europe, and dry weather to the Mediterranean region. A low index, when pressure is higher than average over Iceland and lower than average over the Azores, brings the opposite conditions.

There is a similar pattern of pressure distribution in the Southern Hemisphere. It also varies, so there is a Southern Annular Mode (SAM).

---

the surface. These rise through the thin layer of warm water. They are important because they are rich in nutrients that sustain fish, birds, and seals.

On the western side, around Indonesia, the accumulation of warm water driven westward by the ocean current produces a deep pool of warm water—the warm pool. Evaporation from the warm pool gives Indonesia and the surrounding part of southern Asia a very wet climate. The eastern coastal belt of South America has a very dry climate because the prevailing easterly winds must cross the continent and then the Andes Mountains before reaching the coast, losing their moisture along the way.

The strength of the trade winds, and therefore of the South Equatorial Current, is linked to the pattern of pressure distribution over the South Pacific. Ordinarily, warm air rises over Indonesia, travels eastward at high level, and subsides near to the South American coast, producing low surface pressure on the western side of the ocean and high pressure on the eastern side. At intervals of one to five years, the pattern changes. Pressure rises in the west and falls in the east. This is known as the Southern Oscillation, and the magnitude of the effect is measured as the Southern Oscillation Index (SOI).

A strongly negative SOI indicates a rise in temperature in the eastern South Pacific. When this happens, the trade winds weaken and in extreme cases cease or even reverse direction. Water from the warm pool around Indonesia floods eastward, deepening the layer of warm water off the South American coast and suppressing the upwellings in the Peru Current. With the loss of the warm pool, the weather in the west becomes much drier. Indonesia often experiences severe drought at these times. On the eastern side of the ocean, warm, moist air moves over the coast, bringing heavy rain to the South American coast.

This change in the weather usually happens in late December—the middle of summer in the Southern Hemisphere. Its coincidence with the Christmas season led people to call it El Niño, the (boy) child. It is a gift because it allows South American farmers farther inland to grow bountiful crops, assuring communities of abundant food for the coming year. Unfortunately, by suppressing the upwelling, nutrient-rich cold water, El Niño also drives away the fish on which coastal communities depend.

At other times the SOI is strongly positive. The warm pool is deeper and warmer than usual, and the water on the eastern side of the ocean is cooler than usual. The trade winds intensify and the ocean current strengthens. This is the opposite of an El Niño, called La Niña. The complete cycle of El Niño and La Niña associated with the Southern Oscillation is known as an El Niño–Southern Oscillation (ENSO) event.

■ EVAPORATION, CONDENSATION, AND PRECIPITATION

Someone gazing across a frozen lake in the middle of winter is looking upon what is a beautiful scene and also a very remarkable phenomenon. Beneath the ice, the water is liquid and it will remain unfrozen all winter. Above the lake the air contains water vapor, invisible because it is a colorless gas. In the scene that entrances our observer, ice, liquid water, and water vapor are all present in the same place at the same time.

This is commonplace, of course. Whenever someone takes ice out of the freezer and uses it to cool a drink, water is present as solid, liquid, and gas simultaneously. Everyone may take it for granted, but nevertheless this is really highly unusual. There is no other substance that can exist in all three states at the same time at the temperatures and atmospheric pressures found at the surface of the Earth. What is more, if water did not possess this unusual property our climates would be very different.

The Water Molecule and Water Density

Water is dihydrogen oxide, H_2O, its molecule comprising one atom of oxygen bonded to two atoms of hydrogen arranged at an angle of 104.5°, so there is hydrogen, carrying a positive electric charge, at one side of the molecule and negatively charged oxygen at the other. The hydrogen end of the molecule can then form a further bond with the oxygen of another molecule, as is shown in the illustration. This is a hydrogen bond, and it occurs only between molecules in which hydrogen is bonded to the strongly negative elements oxygen, nitrogen, or fluorine. In its liquid phase, hydrogen bonds link water molecules into short strings. In the solid phase, more hydrogen bonds form, linking each molecule to four others. In the gaseous phase, all the hydrogen bonds are broken and molecules exist as individuals.

Heating makes water molecules move and vibrate faster. Hydrogen bonds break and reform repeatedly, molecules move farther apart, and the liquid expands. Cooling makes the molecules move closer together, and the liquid contracts. As water freezes, however, the angle between the hydrogen atoms on the molecules widens to 109.28°, and the resulting structure is rather open. That is why water expands as it crystallizes into ice. Pure water reaches its maximum density at 39.2°F (4°C), and as the temperature falls below that to 32°F (0°C) and the water freezes, its density decreases. (Impurities in the water alter these temperatures.) Consequently, ice floats on top of water. Floating ice insulates the water below by restricting the loss of heat into the air. Fish and other aquatic organisms are able to live in water that remains liquid because it is protected by a layer of

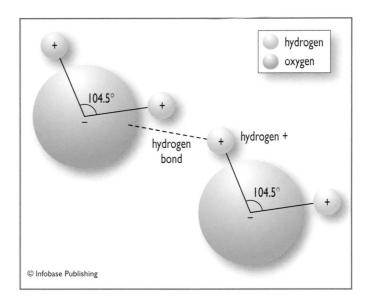

A water molecule consists of two hydrogen atoms bonded at an angle of 104.5° to one oxygen atom. Water molecules link to one another by hydrogen bonds.

ice. If ice were denser than liquid water so that water froze at the surface and the ice sank, the lake would fill with ice from the bottom up and the plants and animals would be killed. Bacteria and algae might survive in lakes that froze in winter, but otherwise such lakes would be lifeless.

Latent Heat

Heat is a form of energy. When a substance is heated, its molecules absorb some of that energy, and it makes them move and vibrate faster. When ice is heated, the molecules in its crystals vibrate faster, and if the ice absorbs enough heat, some of the hydrogen bonds linking molecules together start to break. Eventually, with enough heat absorbed, the molecules will possess sufficient energy to break free from the crystal structures and move about as short strings of molecules. The ice has then melted, and the strings of molecules, constantly breaking and reforming, constitute liquid water. Melting occurs only when ice has absorbed enough energy to break some of its hydrogen bonds.

The absorbed energy does not alter the temperature of the ice or water; all of it is used in breaking hydrogen bonds. In about 1760, the Scottish chemist Joseph Black (1728–1799) found that when he warmed ice gently it slowly melted, but while it was melting its temperature did not change. This led him to conclude that heat has two forms: intensity and quantity. Thermometers measure the intensity of heat but not the quantity. As he warmed the ice, it must have been absorbing a quantity of heat that combined with the ice particles and became latent, or hidden, in the substance of the ice. Black called this *latent heat,* and by the end of 1761 he had proved its existence experimentally. He also measured the amount of latent heat needed to vaporize liquid water. Black never published his work on the topic, although he lectured on it.

Melting one gram of ice absorbs 80 calories (0.335×10^6 joules per kilogram) of latent heat. Much more latent heat is needed to vaporize water because all the hydrogen bonds must break, allowing individual molecules to escape into the air. For pure water, it takes 597 calories of energy to change one gram (1 g = 0.035 ounce) from liquid to gas (2.5×10^6 J/kg). This is more than the amount of latent heat that is required to vaporize most substances. Water can also change directly from solid to gas, without passing through the liquid phase. The direct change from ice to water vapor is called *sublimation*, and it requires 676 calories for each gram (2.83×10^6 J/kg), which is the sum of the latent heats of vaporization and melting.

A sufficient fall in temperature causes water vapor to condense into liquid and liquid water to freeze. Water vapor can also change directly into ice—this process is called deposition, although some people apply the term *sublimation* to the change in either direction. Hydrogen bonds reform during condensation, freezing, and deposition, and exactly the same amount of latent heat is released as was absorbed in breaking them. That heat must come from somewhere, and it is taken from or transmitted to the surrounding medium. In hot weather people cool off by bathing in cold water and allowing the water to evaporate from their skin. This works because the body supplies the latent heat for its evaporation.

Saturation and Humidity

Whether or not water will evaporate or sublime depends partly on its temperature (though it will always evaporate if it is boiled) and partly on the air adjacent to it. Air is a material substance and, therefore, it has mass, and the gravitational attraction between the masses of the air and Earth gives the atmosphere weight. That weight is felt as atmospheric pressure. In effect, air pressure is the weight of the air, all the way to the top of the atmosphere, pressing down on the instrument measuring it.

Air is a mixture of gases, however, each of which contributes to the overall pressure, and these can be considered separately, as the *partial pressure* exerted by a particular gaseous constituent of the atmosphere. For example, oxygen constitutes approximately 21 percent of the atmosphere, so if the atmospheric pressure is *x,* the partial pressure of oxygen is $(21 \div 100)x$, or $0.21x$. Water vapor is also one of the atmospheric gases, but its concentration varies widely. Its partial pressure is known as the vapor pressure.

At the surface of liquid water or ice, detached molecules are constantly leaving and entering the air. They exert a vapor pressure of their own by pushing against the air, and water vapor already present in the air exerts a countervailing pressure, holding them down. The higher the temperature of the water or ice, the more energy its molecules possess, and the greater the vapor pressure from its surface. The drier the air, the lower is its vapor pressure. When the vapor pressure in the boundary layer of air immediately above the water or ice surface is equal to the vapor pressure exerted from the surface, water cannot pass through this layer to the air beyond, and as many molecules return to the surface as leave it. The air above the surface is then saturated. If the vapor pressure above the boundary layer is greater than the vapor pressure from the surface, more molecules will enter the surface than leave it and the water vapor will condense. If the vapor pressure from the surface is greater than the vapor pressure in the surface layer of air, that layer is unsaturated. Molecules will pass through it and the water will evaporate or, if it is frozen, sublime.

The word *surface* suggests a boundary between the air and a lake or the sea. That is a surface, of course, but the tiniest of droplets and ice crystals also have surfaces, and they, too, are surrounded by boundary layers, just a few molecules thick, with vapor pressures that determine whether or not the water or ice will evaporate or sublime. This becomes very important in the formation and dissolution of clouds.

How much water vapor air can hold varies according to the air temperature. Warm air can hold more than cold air, and as the temperature rises the amount of water vapor the air can hold increases rapidly. Air at 104°F (40°C) can hold six times more water vapor than air at 50°F (10°C). The amount of water vapor present in the air is the humidity of the air, and there are several ways to measure it. The mass of water vapor present in a given volume of air is the absolute humidity. This is not a very useful measure and is seldom used because it takes no account of the fact that changes in temperature and pressure can alter the volume of air without affecting the amount of water vapor it contains. Specific humidity is the ratio of the mass of water vapor in the air to a unit mass of air including the water vapor—the number of grams of water vapor in one kilogram of air including the water vapor. The mixing ratio is the ratio of the mass of water vapor to a unit mass of air with the water vapor excluded. The most widely used measure is relative humidity (RH). This is the one given in weather reports and forecasts and the one measured by the type of hygrometer designed for use in the home. Relative humidity is the amount of water vapor present in the air expressed as a percentage of the amount needed to saturate the air at that temperature; the RH of saturated air is 100 percent. As the air temperature rises or falls, the RH changes even though the amount of water vapor in the air remains the same.

Why Temperature Decreases with Altitude

Chill an inflated balloon and it will shrink and its surface will wrinkle. Warm it and it will expand again. Warm it enough and it will burst. The air inside the balloon expands and contracts as its temperature changes, but the quantity of air in the balloon remains constant. It simply takes up more or less space. This alters its density because the same volume of air contains more or fewer molecules.

Consider now what happens when air is warmed by contact with the surface of land or water. It expands, occu-

pying a larger volume and, therefore, decreasing its density. The warm air then weighs less than the air lying immediately above it. Denser, heavier air sinks beneath it and the warm air rises. As the air rises, its molecules must push surrounding molecules of denser air out of the way, and in doing so they expend some of their energy. They move more slowly, which is another way of saying that the rising air becomes cooler. This cooling is called *adiabatic*, from the Greek *adiabatos*, meaning impassable, because no heat is exchanged between the cooling air and its surroundings. When air sinks into a region of denser air, it is compressed. This makes its temperature rise adiabatically as its molecules move closer together and gain energy from the surrounding molecules that compress them.

Rising air grows cooler with increasing distance from the surface, but the temperature of static air that is neither rising nor sinking also decreases with increasing altitude. When the temperature at ground level is 80°F (27°C), the air temperature at 30,000 feet (9,000 m) is usually about -28°F (-32°C). This decrease is due to the fact that air is heated from below, not from above. Sunshine warms the air, but not directly. Air is transparent to sunlight and heat, so sunshine passes through the air without affecting it. Sunshine warms the Earth's surface, and it is contact with the surface that warms the air. Air at high altitude is that much farther from the source of heat—despite being (very slightly) closer to the Sun.

If that sounds paradoxical, there is another paradox: If air contracts and becomes denser as its temperature falls, why does the cold air high in the atmosphere not sink to the surface? The answer to that conundrum is revealed by the concept of potential temperature (see the sidebar).

Lapse Rates

When air is warmed at the surface, it rises and cools adiabatically, and the temperature of static air also decreases with increasing height. The rate at which the temperature decreases, or lapses, is known as the *lapse rate*. There are several different lapse rates.

In air that is unsaturated, the lapse rate is always 5.4°F with every 1,000 feet of altitude (9.8°C/km). This is called the dry adiabatic lapse rate (DALR). All air contains some water vapor, but how much the air can hold depends on its temperature. There will be a height, called the condensation level, at which its temperature falls so low that the air can no longer contain all the water vapor it has been carrying. This is its dew point temperature at which water vapor will start to condense into liquid droplets and cloud will start to form. The condensation level will mark the cloud base, and above the cloud top, the relative humidity (see "Saturation and Humidity" on page 44) of the air will be low enough for droplets to evaporate. If the condensation level is at the

---

Potential Temperature

Air temperature decreases with increasing altitude at an average *lapse rate* of 3.6°F per 1,000 feet (6.5°C/km). As the temperature decreases, air molecules move closer together. A given mass of air occupies a smaller volume and the density of the air increases. High above the ground the air is much colder than air near the surface, and presumably it is denser. Yet it does not subside to the surface.

In order to understand this apparent paradox, it is necessary to consider what would happen if the air were to sink to sea level. Suppose the air temperature at sea level is 60°F (15°C) and at 36,000 feet (11 km) it is -65°F (-54°C). (These are quite realistic temperatures.)

If the air were to subside, as it did so its temperature would rise. This air is extremely dry, and in dry air the lapse rate is 5.4°F per 1,000 feet (9.8°C/km). The temperature would therefore increase by (5.4 × 36)°F, to which its initial temperature (-65°F) must be added. The increase in temperature is, therefore, (5.4 × 36) - 65 = 129.4°F ([9.8 × 11] -54 = 54°C).

The subsiding air would therefore be much warmer, and therefore less dense, than the air near the surface. Consequently, the high-level air is unable to subside.

The temperature a volume of air would have if the pressure under which it is held were adjusted to sea-level atmospheric pressure and its temperature were to change due to compression, is called the potential temperature of the air. In this example, the actual temperature of the high-level air is -65°F (-54°C) and its potential temperature is 129.4°F (54°C).

surface, the condensing water vapor will form mist or fog. Condensation releases latent heat, warming the air, and once the relative humidity of the air reaches 100 percent, as it rises higher the air will cool at the saturated adiabatic lapse rate (SALR). The SALR varies according to the air temperature because it is air temperature that determines the rate of condensation and, therefore, the amount of latent heat being released. If the air is very warm, the SALR may be as low as 2.7°F per 1,000 feet (5°C/km), but at an air temperature of -40°F (-40°C), the SALR is about 5°F per 1,000 feet (9°C/km). An average value for the SALR is about 3°F per 1000 feet (6°C/km). The actual lapse rate, measured at a particular time in a particular place, is called the environmental lapse rate (ELR). It is calculated by measuring the temperature at the surface and at the tropopause and dividing the difference by the height of the tropopause. The ELR is highly variable, often changing by the hour depending on the intensity of sunshine warming the surface.

How Cloud Forms

Water vapor will condense at a relative humidity (RH) as low as 78 percent if the air contains minute particles of a substance that readily dissolves in water. Salt crystals and sulfate particles are common examples. If the air contains insoluble particles, such as dust or fine soil particles, the vapor will condense at an RH of about 100 percent. If there are no particles at all, the RH may exceed 100 percent and the air will be supersaturated, although the relative humidity in clouds rarely exceeds 101 percent.

The particles onto which water vapor condenses are called cloud condensation nuclei (CCN). Over land, each cubic inch of air contains an average of 80,000 to 100,000 CCN (1.3–1.6 million/cm³); over the open ocean there are about 16,000 per cubic inch (0.26 million/cm³). CCN range in size from 0.001 µm to more than 10 µm diameter, but water will condense onto the smallest particles only if the air is strongly supersaturated, and the largest particles are so heavy that they do not remain airborne very long. Condensation is most efficient on CCN averaging 0.2 µm diameter.

At first, water droplets vary in size according to the size of the nuclei onto which they condensed, but an average cloud droplet has a diameter of about 20 µm. Inside the cloud, these droplets start to fall. The buoyancy of the air, updrafts, and friction combine to slow their descent, so their rate of descent, called their terminal fall velocity, is determined by subtracting the effect of retarding forces from the gravitational acceleration. Small droplets fall so slowly that a gentle updraft will keep them airborne, and should they fall into the drier air below the cloud, they will evaporate long before they reach the ground. In order to reach the ground, the droplet must become much larger. Drizzle consists of droplets about 300 µm in diameter, and an average raindrop

has a diameter of about 2,000 µm. To become drizzle, a cloud droplet must increase its volume more than 3,000 times, and to become a raindrop, it must grow 10 million times bigger. Complicating matters further, as droplets grow they also lose water by evaporation because they are warmed by the latent heat of condensation that formed them.

Cloud forms when water vapor condenses into droplets. Clouds vary greatly in appearance, and the type of cloud that forms depends on the amount of vertical air movement, the wind speed and direction at different heights, and the height of the condensation level. Describing clouds is surprisingly difficult. One of the first persons to attempt it was Theophrastus, a Greek philosopher who lived from about 370 B.C.E. to about 288 B.C.E. He wrote about "clouds like fleeces of wool" and "streaks of cloud." It was not until 1803 that Luke Howard (1772–1864), an English chemist and amateur meteorologist, worked out a successful scheme for naming clouds. His method of classification has been modified, but it still provides the basis for the modern international classification (see the sidebar on page 47).

How Snowflakes and Raindrops Grow

In middle and high latitudes, most clouds contain some ice crystals, even in summer. When the temperature at ground level is 60°F (15.5°C), at about 6,000 feet (1,830 m) the temperature of dry air will be 28°F (-2.4°C). Even when the ground-level temperature is 90°F (32°C), the air at 10,500 feet (3,200 m) will be close to freezing temperature. In 1911 Alfred Wegener, who first proposed continental drift (see the sidebar "Alfred Lothar Wegener and Continents That Move" on page 12), suggested that these ice crystals might play a crucial role in forming raindrops. This idea formed part of a theory devised by the Swedish meteorologist Tor Bergeron (1891–1977) and confirmed in 1935 by the German meteorologist Walter Findeisen (1909–45). It is known as the Bergeron–Findeisen process, forgetting all about the contribution made by Alfred Wegener.

Ice always melts when its temperature exceeds 32°F (0°C), but water does not always freeze at that temperature. Just as CCN are needed for water vapor to condense, ice nuclei are needed for ice crystals to form. Without them, water may remain liquid at much lower temperatures as supercooled water. In pure air, water droplets can be supercooled to -40°F (–40°C) before they freeze spontaneously. Where both are present together, there will be many more supercooled droplets than ice crystals because CCN are much more abundant than ice nuclei, of which there may be no more than one in every six cubic inches (3/cm³) of air at -20°F (-29°C). The saturation vapor pressure (see "Saturation and Humidity" on page 45) is greater over ice than it is over water, so water will freeze more readily onto ice than it will condense onto water droplets. Ice crystals will grow, but that removes water vapor

The Naming of Clouds

Each type of cloud has a name within an international system of cloud classification. The definitions of each name, together with photographs of the clouds, are contained in the *International Cloud Atlas,* published by the World Meteorological Organization, which is a United Nations agency with its headquarters in Geneva, Switzerland.

The basic cloud types are defined by their appearance, as layered or heaped, and by the usual height of their base. The defining height of the cloud base varies with latitude.

Layered clouds have names beginning *strat-* from the Latin *stratum,* which is the past participle of the verb *sternere,* to strew. Heaped clouds have names beginning *cum-,* from the Latin *cumulus,* meaning heap. The names of high-level clouds begin *cirr-,* from the Latin *cirrus,*

meaning curl, and those of two of the three middle-level clouds begin *alt-,* from the Latin *altus,* meaning high. Low-level clouds have no identifying prefix. This classification produces 10 basic cloud types, or *genera,* listed in the table below.

The genera are further divided into 14 cloud species: calvus; capillatus; castellanus; congestus; fibratus; floccus; fractus; humilis; lenticularis; mediocris; nebulosus; spissatus; stratiformis; and uncinus.

There are also nine varieties: duplicatus; intortus; lacunosus; opacus; perlucidus; radiatus; translucidus; undulatus; and vertebratus.

In addition, accessory clouds, such as pileus, tuba, and velum, may be attached to the main cloud. Main clouds may also possess such supplementary features as arcus, incus, mamma, praecipitatio, and virga.

Height of cloud base

	POLAR REGIONS		TEMPERATE LATITUDES		TROPICS	
	'000 feet	'000 meters	'000 feet	'000 meters	'000 feet	'000 meters
High cloud	10–26	3–8	16–43	5–13	16–59	5–18
Genera: cirrus, cirrostratus, cirrocumulus						
Middle cloud	6.5 13	2–4	6.5–23	2–7	6.5–26	2–8
Genera: altostratus, altocumulus, nimbostratus						
Low cloud	0–6.5	0–2	0–6.5	0–2	0–6.5	0–2
Genera: stratus, stratocumulus, cumulus, cumulonimbus						

from the air, lowering the relative humidity and causing the supercooled water droplets to evaporate. Their evaporation releases water vapor that is then deposited onto the ice crystals. In other words, in clouds that contain both supercooled water droplets and ice crystals, the ice crystals will grow at the expense of the liquid droplets.

When an ice crystal is big enough it will fall, often gathering other crystals as it does so, to form snowflakes. Snowflakes form best at 23°–32°F (-5°–0°C) because at these temperatures thin films of water coating the crystals freeze on contact, fixing the crystals together. The falling snowflakes will melt if they enter warmer air lower in the cloud or below it and fall as rain. Even in the middle of summer, therefore, most of the rain falling in middle latitudes is in fact melted snow.

That is not the whole story, however, because there are also warm clouds that contain no ice because their tops do

not reach the height at which water freezes. Raindrops form in warm clouds by a process of collision and coalescence.

Cloud droplets are not all the same size. They may average 20 μm in diameter, but some are smaller, and others can grow by further condensation until they are up to 100 μm across. Large droplets fall faster than small ones, so a big droplet passes through a mass of small droplets, and as it falls it will collide with those directly in its path. Colliding droplets may simply ricochet away from each other or, if they are about the same size, may merge temporarily and then break apart into smaller droplets. But the two droplets will coalesce if one droplet is much bigger than the other and they collide more or less head-on. Droplets must reach a minimum diameter of more than about 40 μm before they will sweep a path through the smaller droplets and coalesce with them. Droplets smaller than 20 μm are swept aside by the airflow surrounding the larger droplet.

Once coalescence starts, however, it accelerates, and a drop can grow to a diameter of 400 μm within about 50 minutes. In most clouds, droplets grow big enough to fall as precipitation within an hour.

Collisions are far from inevitable. Clouds give the appearance of being densely packed with water droplets. Anyone walking through dense fog will become very wet, and dense fog is really stratus cloud with its base at ground level. In reality, however, the droplets are widely separated in relation to their size. If droplets were more closely crowded and collisions between them were frequent, almost any cloud would inevitably produce rain, regardless of its size. In fact, it is only fairly deep clouds—clouds of considerable vertical extent—that produce rain because it is only in these clouds that falling droplets can grow large and heavy enough to fall from the base of the cloud.

Where the cloud base is high, trails or streaks often extend from beneath it. Called virga (see the sidebar "The Naming of Clouds" on page 47), this is rain or snow falling from the cloud and then evaporating before reaching the ground. The raindrops or snowflakes evaporate because below the cloud they enter unsaturated air. They also enter warmer air in which ice crystals and snowflakes may start to melt. Water that survives long enough to reach the ground is called drizzle if the droplets are less than 0.02 inch (0.5 mm) in diameter and rain if they are larger. Snow will fall only if the freezing level, the height below which the air temperature is above freezing, is lower than about 1,000 feet (300 m). Snow rarely falls and settles when the air temperature near the ground is higher than 39°F (4°C). In warmer air the snowflakes melt before they reach the ground, and at 34°–35°F (1°–1.5°C) precipitation usually consists of a mixture of snowflakes and raindrops, known as sleet in Britain (in North America, sleet consists of ice pellets the size of drizzle droplets).

Any moisture passing from the air to the surface is called precipitation. Fog and mist are forms of precipitation; it is fog if it reduces horizontal visibility to less than 900 yards (1 km) and mist if the visibility is greater than this. Dew and frost are also types of precipitation, frost or rime ice forming when water vapor is deposited onto surfaces that are below freezing temperature.

Rain and drizzle may fall as supercooled droplets that freeze on contact with a surface. This is freezing rain or freezing drizzle, and it covers everything it touches with a coating of clear ice. This can cause serious damage to trees (see "Avalanches and Ice Storms" on pages 80–81).

Snow, snow pellets, and snow grains are also distinguished by size. Snow pellets are usually spherical, bouncing and breaking when they strike hard ground, and 0.02–0.08 inch (0.5–2 mm) in diameter. If their diameter is less than 0.04 inch (1 mm), they are snow grains.

Hail comprises more or less spherical lumps of ice. These form by accretion in tall clouds, often storm clouds. Ice pellets, formed high in the cloud, fall through the cloud. Water condenses onto them as they reach lower levels where the temperature is above freezing, and the water freezes as the pellets are carried upward by strong updrafts. This continues until the hailstones are too heavy to be carried aloft by the updrafts, so they fall from the cloud. The bigger the hailstones, the stronger the updrafts, and the more violent the conditions inside the cloud. The manner of their growth gives hailstones a layered structure, rather like the layers of an onion. Hailstones that are made partly from snow and are therefore soft; they are called graupel.

Sometimes small ice crystals fall from clouds or water vapor freezes into ice crystals in clear air outside clouds. The resulting minute particles are called ice prisms or diamond dust. They hang suspended in the air and glitter as they catch the sunlight. Water vapor will turn directly into ice, without the presence of ice nuclei, if the temperature falls to -40°F (-40°C), and ice prisms form only when the air temperature is extremely low.

■ TRANSPIRATION

Enter a forest on a bright summer day, and the air will feel cool. There is no mystery about this. The trees cast shade and a person walking beside them is no longer exposed to direct sunshine. If the sky is cloudy, however, the air will still feel cooler than the air outside the forest. The difference will be less marked and will vary from one type of forest to another, but it is real. Scotch pine (*Pinus sylvestris*) forest in Italy is almost 6°F (3°C) cooler in August than the surrounding open countryside. Norway spruce (*Picea abies*) and beech (*Fagus sylvatica*) forests are almost 8°F (4°C) cooler. It is true, of course, that even on an overcast day, diffuse sunshine can deliver a considerable amount of warmth, but this cannot be the full explanation. Provided there is no wind so that shelter from the trees does not reduce windchill, forest air is cooler even in winter, although the difference between temperatures inside and outside the forest is only about half of what it is in summer.

Furthermore, there are exceptions to the rule, which, far from contradicting this curious phenomenon, support it. *Maquis* is a dense type of sclerophyllous vegetation, with shrubs and forests that originally were dominated by holm oak (*Quercus ilex*). In maquis, the summer temperature inside the forest may be as much as 4°F (2°C) *higher* than the temperature outside, although in winter it is about 1°F (0.5°C) cooler. Maquis occurs in the Mediterranean region, and there are similar vegetation types in southern California and South Africa.

Why the Air Inside a Forest Is Cool

These marked temperature differences are due to the evaporation of water, which takes latent heat from the air (see "Latent Heat" on pages 43–44). When it rains over a broad-leaved deciduous forest, on average the trees intercept nearly half of the precipitation in summer when the trees are in leaf and nearly one-quarter of it in winter when they are bare. Water coats leaves in the forest canopy and runs down the branches and trunks. Pine trees intercept about one-third of the precipitation, but this varies according its intensity. When the rain is light, they intercept almost all of it, but when it is heavy, four-fifths flows directly to the ground. All of the water wetting the trees evaporates, and this cools the air.

This explanation is less convincing than it seems. The same amount of rain falls outside the forest as inside, and outside the forest, it also coats the leaves of grasses and herbs. Obviously, being much smaller, they intercept less rain than trees do, so more of the water reaches the ground, but much of it evaporates, just as it does inside the forest. In any case, rain cools the air even before it reaches the ground because the raindrops are cool (most are melted snow, see "How Snowflakes and Raindrops Grow" on pages 46–48), and much evaporation occurs between the cloud base and the ground. What is more, the cooling effect in a forest persists undiminished even when it is not raining and all surfaces are dry.

Forests and the Climate

Some other factor must be at work, and it is. The plants themselves are removing water from the ground and releasing it into the air from their leaves. The process is called *transpiration*. All green plants transpire water, but the amount is proportional to their size, and trees transpire a great deal. In summer an Old World silver (or common) birch tree (*Betula pendula*), bearing about 250,000 leaves, may transpire 95 gallons (360 l) of water a day. An area of birch forest, about 250 acres (100 ha) in area, might contain 80,000 birch trees. They would transpire approximately 7.6 million gallons (34.5 million l) of water a day. A maple transpires about 53 gallons (200 l) an hour. A German forest of Norway spruce has been found to return the equivalent of more than 13 inches (330 mm) of rain a year to the air through the combined effects of transpiration and evaporation from plant surfaces. The rate at which trees transpire water greatly exceeds the rate at which water would evaporate from an open surface, such as a lake, under similar conditions of temperature and relative humidity.

Not only do transpiration and evaporation from surfaces cool the air by adding water to the air, but they also increase the relative humidity. Many years ago this effect was measured outside and inside a North Michigan forest of birch (*Betula*), beech (*Fagus*), and maple (*Acer*), the readings being taken at the same time every day for several months. From mid-June until early September, the relative humidity inside the forest was consistently higher than it was outside. In European forests, the average relative humidity during the year ranges, according to tree species, from about 4 percent to more than 9 percent higher than levels measured outside.

Transpiration moves water from the ground and releases it into the air, so transpiration can occur only when water is available below ground. This explains the apparent anomaly of the maquis forest. There, the rain falls mainly in winter, and the summer is very dry. Sclerophyllous plants adapt to the lack of soil moisture by greatly reducing their rate of transpiration. Consequently, air inside a sclerophyllous forest is warmer and drier than air is outside the forest, where the wind lowers the temperature by sweeping away warm air and replacing it with cooler air drawn down from a higher level.

Although evaporation from surfaces and transpiration are distinct processes, in practice it is impossible to measure them separately in the open. Measurements of transpiration are made under controlled laboratory conditions. Outdoors they must be measured together. Combined, they are called *evapotranspiration*. Because evapotranspiration moves substantial amounts of water from the ground to the air, more cloud may form in the moister air above the forest than there would be otherwise, and some of the cloud will produce precipitation. If the forest is cleared over a large area, say for providing land for building or growing crops, the reduction in evapotranspiration can make the climate generally drier. Trees slow the wind by absorbing the energy from it, so forest clearance usually makes the climate windier. This will tend to dry the soil because the rate of evaporation increases in proportion to the wind speed (see "Climatic Effect of Forest Clearance" on pages 264–268).

How Plant Roots Gather Water from the Ground

All living organisms need water. Green plants use water for photosynthesis (see "Photosynthesis" on pages 116–122), but this requires very little. They also need water to fill out their cells and give them rigidity. A nonwoody plant that is deprived of water wilts as water drains from its tissues, growing limp until eventually it may fall. Plants also need water for the transport to all of their tissues of mineral nutrients from the soil and of sugars made in the leaves. Plants do not retain water, however. More than 90 percent of the water entering a plant leaves it by transpiration. During summer, a broad-leaved tree leaf must replace all the water it contains every hour.

Water containing dissolved minerals—the soil solution—enters plants by passing through the permeable membranes that coat the cells near the tips of root hairs. Inside the root, the soil solution passes freely along the cell walls without entering the cells. Water is then in direct contact with a very large surface area. It is drawn into the plant by a tension that is transferred to the root hairs and from them to the film of water coating soil particles. Roots of some plants exert a strong pulling force of up to 220 pounds per square inch (15 kg/cm²). Others exert much less or virtually none. The root system is able to shut down when necessary, the plant sealing itself to prevent water flowing back into the soil. As the soil solution moves through the root, some of the water and mineral nutrients useful to the plant are allowed to pass through cell walls. This movement, called *osmosis,* allows certain molecules to pass through the membrane from a strong to a weak solution, and it sets up a tension, drawing molecules through the membrane.

Once inside the plant, water is drawn upward, against gravity, all the way to the topmost leaves. If the tree is a Sierra redwood (*Sequoiadendron giganteum*), this may be more than 300 feet (90 m) above the ground. Roots can draw water into the plant, but they cannot raise it to the topmost leaves. Even the most efficient mechanical vacuum pump can raise water no higher than 33 feet (10 m), so water must rise through a tall plant by some other means.

How a Tree Can Raise Water to Its Topmost Leaves

The water is not pushed by a pump but pulled from above, traveling through very narrow vessels that together make up the *xylem.* This water is the *sap* that leaks when a stem is cut.

Leaves are covered with a layer of skin—the epidermis—that has a waxy outer layer called the cuticle. There are microscopically small pores, called stomata (singular stoma), in the epidermis that can open and close by means of guard cells, which shrink or swell in response to changes in light intensity. The stomata open during the day and close at night, although a shortage of water can also make the guard cells close them during daylight. The density of stomata varies widely from one species to another, but many have about 200,000 per square inch (31,000/cm²) of leaf surface, most located on the underside of the leaves.

Stomata open onto a network of air spaces in the *mesophyll* of the leaf—the central layer of the leaf. Photosynthesis takes place in mesophyll cells. It is through the stomata that the plant absorbs carbon dioxide for photosynthesis and excretes oxygen, the by-product. Photosynthesis is driven by sunlight, which is why stomata open during the day. They close at night to prevent further water losses when photosynthesis is impossible. The spaces in the mesophyll are coated with a film of water, and the air in contact with them is saturated with water vapor. Outside the leaf, the vapor pressure is lower, so water is constantly evaporating from the mesophyll, through the stomata, and into the air.

Xylem and Stomata— How the "Pump" Works

Xylem (see "Xylem" on pages 104–105) is plant tissue comprising a system of vessels that transports water from the roots to every part of the plant. As it enters the leaves, the xylem divides into ever-finer vessels. The larger vessels are visible as the *veins* in a leaf. The smallest vessels supply water to the mesophyll layer of the leaf. Consequently, moisture in the mesophyll spaces is directly linked to the xylem. As water evaporates through the stomata, molecules are lost from the film coating the mesophyll spaces. The remaining water is drawn by adhesion into small indentations in the cell walls, but attraction between water molecules resists any increase in the surface area of the water. This pulls the water into a bulge (a meniscus). As the meniscus shrinks and becomes more spherical, its pull increases on molecules in the remaining film of water. These molecules are attached by hydrogen bonds to others, forming a link to the water in the xylem vessels and, through them, all the way to the roots as an unbroken chain of water molecules stretching from the leaf surfaces to the soil. Water molecules also adhere, but less strongly, to the walls of the xylem vessels. In this way evaporation from leaf surface draws water through the plant at up to 30 inches (76 cm) a minute. The tension is so strong that it draws the walls of the xylem vessels inward, sometimes making a measurable difference to the thickness of a tree trunk.

Transpiration is an inevitable consequence of the need all cells have for water and the need plants have for stomata through which to exchange gases, but it is useful to the plant. The flow of water transports mineral nutrients and evaporation cools leaf surfaces. In very hot weather it can cool them by 20–30°F (11–17°C), keeping their temperature below that at which the enzymes needed in photosynthesis are inactivated.

◼ AIR MASSES AND FRONTS

Return for a moment to the person standing by the lakeside, but suppose now that it is summer and a blazingly hot day. She can feel the Sun beating down on her body. If she takes off her shoes, the sandy shore of the lake will burn her feet. When she plunges into the water, however, it feels cold, perhaps very cold.

How can that be? After all, the lake is exposed to just as much sunshine as the dry land beside it. Somehow it just fails to warm up. Anyone who has visited the coast knows that the ocean behaves in precisely the same way. At least, it is like that in summer. In winter the sea is often warmer than the land.

Land heats up in summer much faster than the sea does, and in winter it cools much faster. This greatly affects air passing over land and sea, which is warmed or cooled by contact with the surface. It is why the interiors of continents have much hotter summers and colder winters than coastal regions and islands, which receive air that has crossed an ocean and been cooled by it in summer and warmed by it in winter (see "Continental and Maritime Climates" on pages

Specific Heat Capacity

SUBSTANCE	TEMP.		HEAT CAPACITY	
	(°F)	(°C)	(cal/g/°C)	(J/g/K)
dry air	68	20	0.24	1.006
ice	-6	-21	0.48	2.0
ice	30	-1	0.50	2.1
pure water	60	15	1.00	4.186
seawater	63	17	0.94	3.93
granite	68	20	0.19	0.80
granite	212	100	0.20	0.84
white marble	64	18	0.21	0.90
sand	68–212	20–100	0.19	0.80

85–89). The effect is due to the relatively high *specific heat capacity* of water (see the sidebar).

The amount of heat that a surface absorbs is determined partly by its albedo (see "Why Surface Color Matters" on pages 34–35), but when the Sun is high in the sky and its albedo is low, water absorbs more than any solid because it is transparent. About one-fifth of the solar radiation falling on the ocean penetrates to a depth of 30 feet (9 m), and turbulence in the water carries the warmed water much deeper. The oceans act as heat sinks partly because of the specific high heat capacity of seawater and partly because of their huge volume. The table lists the specific heat capacities of a range of substances.

How Moving Air Transports Heat

Air is constantly on the move. In middle latitudes it is usually traveling from west to east. The air crosses continents, oceans, and islands, but its journey takes time. Air leaving Asia takes days, or even weeks, to cross the Pacific before it reaches America. Then it takes more days to cross the American continent and several days more to cross the Atlantic before reaching Europe. During its travels, the air is in contact with the land or sea surface, and this changes certain of its characteristics.

During the First World War, Scandinavia was no longer able to receive reports from weather stations in other European countries because information about the weather could be of use to an enemy (the broadcasting of weather forecasts also ceased in the Second World War for the same reason). The difficulty for Scandinavians was solved by one of the most famous of all meteorologists, Vilhelm Friman Koren Bjerknes (1862–1951). Born in Oslo, where he also died, Bjerknes was the son of a mathematics professor. He became a professor at Stockholm and Leipzig Universities, and in 1904 he published *Weather*

Specific Heat Capacity

When a substance is heated, it absorbs heat energy and its temperature rises. The amount of heat it must absorb in order to raise its temperature by one degree varies from one substance to another, however. The ratio of the heat applied to a substance to the extent of the rise in its temperature is called the specific heat capacity for that substance. It is also known as the heat capacity and thermal capacity.

Specific heat capacity is measured in calories per gram per degree Celsius (cal/g/°C) or in the scientific units of joules per gram per kelvin (J/g/K; 1K = 1°C = 1.8°F). Specific heat capacity varies slightly according to the temperature, so when quoting the specific heat capacity of a substance, it is customary to specify the temperature or temperature range to which this refers.

Pure water has a specific heat capacity of 1 cal/g/°C (4,186 J/g/K) at 59°F (15°C). This means that at 59°F (15°C) one gram of water must absorb 1 calorie of heat in order for its temperature to rise by one degree Celsius (or 0.56 cal to raise its temperature by 1°F). Seawater at 17°C (62.6°F) has a specific heat capacity of 0.94 cal/g/°C (3,930 J/g/K).

At temperatures between 68°F (20°C) and 212°F (100°C), the specific heat capacity of granite is 0.19–0.20 cal/g/°C (800–840 J/g/K). Within the same temperature range, the specific heat capacity of sand is 0.20 cal/g/°C (800 J/g/K). These values are typical for most types of rock.

It requires almost five times more heat to raise the temperature of water by 1K than to heat granite or sand by the same amount.

Forecasting as a Problem in Mechanics and Physics. This was one of the first scientific studies of weather forecasting. In 1917 he returned to Norway to found the Bergen Geophysical Institute. From there, he and his colleagues established a network of weather stations throughout Norway. These stations fed measurements and observations of weather conditions to the Institute. Studying these reports, from many scattered locations, led Bjerknes and his colleagues, who included his son, Jacob Aall Bonnevie Bjerknes (1897–1975), and Tor Harold Percival Bergeron (1891–1977), to consider in more detail how the atmosphere works to produce weather. In 1921 Vilhelm Bjerknes published *On the dynamics of the circular vortex with applications to the atmosphere and to the atmospheric vortex and wave motion,* a somewhat unromantic title for his description of the conclusions they had reached.

These now form the basis of the scientific understanding of climate and weather, and they begin with the concept of the air mass. While it remains over a large landmass or ocean in a high or a low latitude, the air over a wide area acquires characteristics that are much the same throughout. Everywhere, the pressure, temperature, and humidity will vary little at the surface or when measured anywhere at any level above it.

Continental Air and Maritime Air

Air masses are classified according to the source regions where they acquired their defining characteristics. Those forming over the ocean are called maritime (m), those forming over continents continental (c), and the latitude in which they originate makes them tropical (T), polar (P), arctic (A), or equatorial.

The southwestern United States, for example, lies beneath a continental tropical (cT) air mass, but in winter the continental polar (cP) air mass that is over northern Canada in summer spreads to cover most of the land east of the Rockies and as far south as Texas. Northwestern Europe, ordinarily beneath maritime polar (mP) air, experiences warm weather in summer whenever maritime tropical (mT) air extends northward and bitterly cold weather in winter when cP air spreads westward from northern Siberia. There are no maritime arctic or continental equatorial air masses, because arctic air masses form over surfaces that are permanently covered in ice and oceans cover most of the equatorial region.

As an air mass moves away from the region where it formed, its characteristics change. Contact with the surface may warm or cool it. If it is heated from below, the warmed air rises, rapidly spreading the warming throughout the air mass and, at the same time, making the air unstable (see "Stable and Unstable Air" on pages 64–65). When surface air is cooled, it sinks, becoming stable, but with a layer of warmer air some distance above the surface, which cools much more slowly by radiating its heat into space. This affects the type of clouds that form and the weather they bring. Sheets of cloud (called stratiform cloud), bringing steady drizzle or rain, form in stable air, and heaped clouds (called cumuliform cloud), bringing showers and storms, form in unstable air. When air is warmed, it is able to hold more water vapor, and if the warmed air crosses the ocean, water will evaporate into it. Warm, moist air will lose much of its water if it is cooled, and by the time it has crossed a continent it will be dry.

Why There Are No Temperate Air Masses

There are arctic, polar, tropical, and equatorial air masses, but no temperate air masses. The middle latitudes lack the features necessary for the continents and oceans of middle latitudes to qualify as source regions. The polar front and its associated jet stream (see "The Polar Front and Jet Streams" on pages 57–63) cross the middle latitudes, and fluctuations in the polar front jet stream generate low-pressure weather systems that move beneath the jet stream. Low-pressure systems draw in surrounding air from adjacent air masses. Consequently, the weather systems of middle latitudes pull in air masses from either side but without generating air masses of their own.

Temperate forests experience weather that is warm when tropical air produces it and cold when polar air produces it. Continental arctic air masses strongly influence weather conditions over northern Canada. Polar air masses produce the weather over southern Canada and the northern United States. Continental polar air masses drift southeastward from the Canadian Prairie Provinces, and maritime polar air masses dominate the western and eastern coasts. Farther south, maritime tropical air masses drift northward to the southwestern and southeastern United States, and continental tropical air masses drift northward from Mexico.

Fronts

Since an air mass occupies a defined area, it must have boundaries where it is adjacent to another air mass possessing quite different characteristics. It might seem that where two air masses border one another the air would mix until there is a wide belt of air that is intermediate in character between the two. There is some mixing, but this is not really what happens. The border is up to 120 miles (190 km) wide, but compared with the size of an air mass this is narrow and it is quite clearly defined.

That is how it would remain, with air masses moving all together, were it not for the fact that they move at different speeds. Cold air generally travels at up to twice the speed of warm air, so as a cold air mass advances, it pushes beneath the warm air ahead of it. There is conflict between the two, because cold air is denser than warm air, and when Vilhelm Bjerknes and his colleagues were considering this, the stories in the newspapers were dominated by accounts of First World War battles—conflicts of a different kind. That is what led them to call the boundary between two air masses a front, like the front between opposing armies. Fronts are designated warm or cold depending on the air behind them. The passing of a warm front brings a change to warmer air, and a cold front brings cooler air. "Warm" and "cool" are relative terms, however, meaning only that the air behind the front is warmer or cooler than the air ahead of the front.

Air masses extend from the surface all the way to the tropopause. This is another boundary, between two layers of the atmosphere, the troposphere and the *stratosphere.* In the lower layer, the troposphere, temperature decreases with height (see "Why Temperature Decreases with Altitude" on page 45). The rate of decrease, called the lapse rate, averages about 3.6°F per 1,000 feet (6.5°C/km). At the tropopause, the temperature reaches a minimum. This varies with the season, but usually it averages about -60°F (-50°C) in summer and -90°F (-68°C) in winter.

Above the tropopause, in the lower stratosphere, the temperature either remains constant with height or, more usually, increases. This is a temperature inversion, where warm, less dense air overlies cooler, denser air. It prevents air from rising higher by convection because the rising air encounters air of similar density. Consequently, there is little exchange of air across the boundary, although huge storm clouds may penetrate it.

Surface temperatures are higher at the equator than at the poles and there is more convection there. Given the fairly constant lapse rate, this means that the height of the tropopause varies from an average of 10 miles (16 km) at the equator to five miles (8 km) at the poles, with local variations raising the tropopause higher above warm air and lowering it over cold air.

As air rises and cools adiabatically, its capacity for holding water vapor decreases, and at the tropopause the air is very dry. Almost no water vapor enters the stratosphere.

Highs and Lows

Cyclones, also called depressions and lows, and anticyclones, or highs, are regions in which the atmospheric pressure changes over a horizontal distance to reach a minimum or maximum at the center. Where the pressure is low, air will tend to flow toward the center to equalize it, and where it is high, air will tend to flow outward. Cyclones and anticyclones occupy fairly large areas, up to 1,200 miles (2,000 km) across. Places on a surface (not always the Earth's surface on the charts meteorologists use) where the atmospheric pressure is the same are joined on weather maps by lines called

isobars. Isobars surround low- and high-pressure centers, like the contours around hills and hollows on a topographic map, making the centers easy to identify.

Cyclone is from the Greek *kuklos,* meaning "wheel" (from which we also get our words *cycle* and *bicycle*). The wind flows approximately parallel to the isobars around cyclones and anticyclones, rather than across them, and the strength of the wind is proportional to the distance between the isobars. Whenever air (or water) flows for a long distance over the surface of the Earth, it follows a curved path. The reason for this was discovered in 1835 by the French engineer and mathematician Gaspard-Gustave de Coriolis (1792–1843), and it is known as the *Coriolis effect,* abbreviated to CorF (see the sidebar on page 54).

Pressure Differences and Wind Direction

Air is drawn toward a center of low pressure by the difference in pressure and with a force that is proportional to the rate of pressure change, or pressure gradient. This is called the pressure-gradient force (PGF). As soon as the air begins to move in response to the PGF, however, it is affected by the Coriolis effect (CorF), which deflects it. A component of the CorF is then acting in the same direction as the wind. This accelerates the wind, but because the magnitude of the CorF is proportional to the wind speed, the moving air is deflected still more. Finally, the PGF and CorF balance, and the wind blows parallel to the isobars. The diagram on page 54 illustrates how this happens.

In the Northern Hemisphere the wind blows counterclockwise around centers of low pressure and clockwise around centers of high pressure. In the Southern Hemisphere these directions are reversed. Any increase in either the PGF or CorF causes a balancing increase in the other, and the wind continues to blow parallel to the isobars. The resulting movement is called the *geostrophic wind* (from the Greek *geo,* "earth," and *strepho,* "to turn"). The geostrophic wind occurs only well clear of the ground, however, because near the ground friction slows the air movement, causing the wind to blow slightly across the isobars.

How Fronts Begin and End

The air flows in opposite directions on either side of a front separating cold (high pressure) and warm (low pressure) air, and the advancing cold air moves beneath the less dense warm air. In middle latitudes, where weather systems are strongly influenced by the polar front jet stream (see "How Waves in the Jet Stream Affect Weather at the Surface" on page 60), fronts tend to develop into frontal systems centering on a cyclone, or depression. Frontal systems have a

The Coriolis Effect

Any object moving over the surface of the Earth, but not firmly attached it, does not travel in a straight line. It is deflected to the right in the Northern Hemisphere and to the left in the Southern Hemisphere. As a consequence of this, moving air and water tend to follow a clockwise path in the Northern Hemisphere and a counterclockwise path in the Southern Hemisphere.

The French physicist Gaspard-Gustave de Coriolis (1792–1843) discovered the reason for this in 1835, and it is called the Coriolis effect. It happens because the Earth is a rotating sphere, and as an object moves above the surface, the Earth below it is also moving. The effect used to be called the Coriolis "force" and it is still abbreviated as CorF, but it is not a force—there is nothing pushing or pulling the moving object sideways.

The Earth makes one complete turn on its axis every 24 hours. This means every point on the surface is constantly moving and returns to its original position (relative to the Sun) every 24 hours, but because the Earth is a sphere, different points on the surface travel different distances to do so.

Consider two points on the surface, one at the equator and the other at 40°N, which is the approximate latitude of New York City and Madrid. The equa-

tor, latitude 0°, is about 24,881 miles (40,033 km) long. That is how far a point on the equator must travel in 24 hours, which means it moves at about 1,037 MPH (1,668 km/h). At 40°N, the circumference parallel to the equator is about 19,057 miles (30,663 km). The point there has a shorter distance to travel, so it moves at about 794 MPH (1,277 km/h).

A body of air moving away from the equator will be traveling at the speed of the equator, but the farther it travels, the slower the surface beneath it will be traveling. This will cause it to overtake the surface in an easterly direction. Consequently, its path is deflected to the east. A body moving toward the equator will experience the opposite deflection, as the surface overtakes it. In the Northern Hemisphere the deflection is to the right and in the Southern Hemisphere it is to the left; it is an easterly deflection in both hemispheres.

The magnitude of the Coriolis effect is directly proportional to the speed at which the body moves and to the sine of its latitude. The effect on a body moving at 100 MPH (160 km/h) is 10 times greater than that on one moving at 10 MPH (16 km/h). Since sin 0° = 0 (the equator) and sin 90° = 1 (the poles), the Coriolis effect is greatest at the poles and zero at the equator.

The geostrophic wind blows high enough above the surface for it not to be influenced by friction with the surface. The balance between the pressure gradient force and the Coriolis effect causes the wind to blow parallel to the isobars.

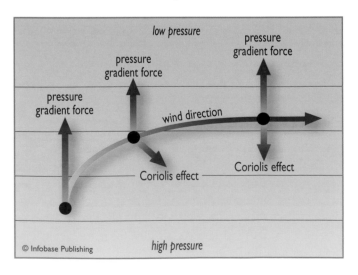

life cycle starting with their formation, called frontogenesis, and ending with their dissolution, called frontolysis. The diagram on page 55 illustrates the steps in the life cycle. This is the sequence that produces the weather typical of middle latitudes, in which temperate forests thrive.

At first the front is straight, with air flowing in opposite directions on either side. Then (1) a tongue of warm air at the surface pushes into the cold air, forming a small wave in the front. Cold air moves around the crest of the wave (2), making it more pronounced and trapping a wedge of warm air. At the same time, warm air rides up the slope of the cold front. Advancing cold air, on the left in the diagram, produces a cold front, and warm air still pushing into the cold air, on the right in the diagram, produces a warm front. It is now a frontal system. At the crest of the wave, where warm air is being lifted most strongly, a center of low pressure forms. This is the depression. Cold air continues to undercut and lift warm air until a substantial part of the warm air has been raised clear of the ground (3). Once this happens, the fronts are said to be occluding.

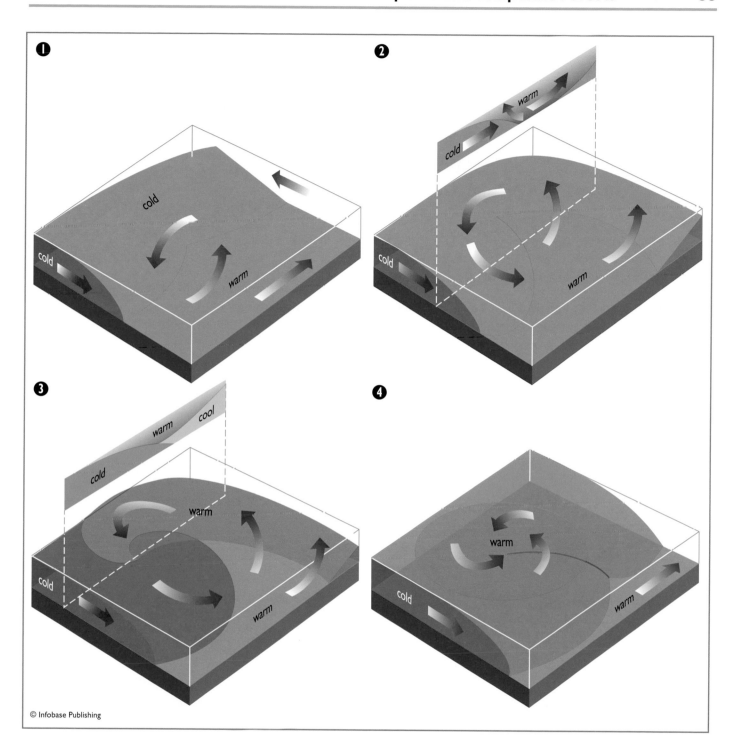

Life cycle of a frontal system. (1) The open stage, when a wave has developed along the front separating warm and cold air. (2) Cold air is now pushing beneath the warm air, and the cold front is overriding the warm front. (3) The cold air is lifting the warm air clear of the surface, and the fronts are occluding. (4) The warm air is completely clear of the surface, and the system is dissipating.

On weather maps, representing the situation at surface level, a cold front is conventionally depicted as a line with triangles along it and a warm front as a line with semicircles along it. If the map is colored, the triangles are always blue and the semicircles are red. Where the fronts are occluding, the triangles and semicircles alternate along the line marking the front. Finally (4), all the trapped warm air has been lifted well clear of the surface, above the cold air. The depression dissipates as the lofted air cools and reaches heights where its density is equal to that of the surrounding air. The frontal wave then shrinks, and the front is more or less straight again. From the first appearance of the frontal wave to the final dissipation of the occluded front, the process usually takes from four to seven days.

Frontal systems develop along undulations in the polar front and beneath waves in the jet stream. There is not usually just one wave in the jet stream, however, but several. Consequently, frontal systems tend to occur as groups, called families, which is why in middle latitudes soon after one depression clears, another takes its place.

Warm Fronts, Cold Fronts, and the Weather They Bring

Frontal systems are three dimensional, and the drawing does not describe what people on the ground experience as one passes overhead. As the system approaches, everyone is in a region of (relatively) cold air. First the warm front arrives, with the region of warm air behind it and then the cold front, with cold air behind it. Weather maps depict fronts as lines across the surface. These mark the surface position of the front and may give the impression that fronts are either confined to the surface or that they rise vertically from it. In fact, fronts rise from the surface along a gentle incline, and the weather associated with a front begins to arrive sometime before the front itself crosses a point on the surface. The diagram shows a vertical cross section through the entire system, which is traveling from left to right (from west to east). The drawing distorts the true situation because it grossly exaggerates the slope of both fronts, although it does show the cold front sloping more steeply than the warm front. This distortion is unavoidable because a diagram of the system which portrayed the frontal slopes to scale, with the tropopause three inches above ground level, would be about 40 feet (12 m) long!

Warm fronts slope very gently indeed, at an angle of about 1/2–1°. Cold fronts have a steeper slope, of about 2°. When the point where the upper edge of a warm front meets the tropopause is directly overhead, the point where its lower edge meets the surface is more than 600 miles (950 km) away, and it will be many hours before it arrives. Similarly, when the cold front passes at ground level, it will be several hours before its upper edge, probably 200 miles (320 km) away, passes overhead and the front clears completely. Since cloud forms as warm air is squeezed between the two fronts, if the warm-sector air is moist, cloud may persist for quite a long time.

Depending on conditions in the warm sector behind the warm front, the first indication of its approach may be wispy, high-level cloud. This thickens steadily to cover more and more of the sky, and its base becomes lower. Precipitation begins when the warm front is quite near, but there is still some time before it arrives. Low cloud usually persists throughout

A cross section through a frontal system, but with the angle of the frontal slopes greatly exaggerated

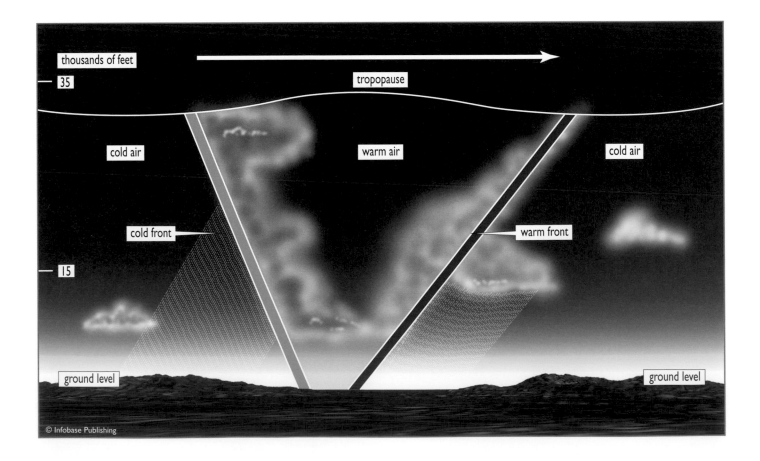

the warm sector, but its type changes with the approach of the cold front, which passes more quickly. The cold front brings more broken cloud, with showers and sunny intervals.

Red Sky at Night— Reliable Forecast or Myth?

Water vapor condenses onto cloud condensation nuclei (see "How Cloud Forms" on page 46), minute particles that float in the air. In dry air, the particles float freely, most densely in the lower atmosphere, and behind a cold front the air is usually dry. Gas molecules and microscopic particles scatter sunlight in all directions, but short wavelengths are scattered more than longer wavelengths. When the Sun is low in the sky so that radiation passes at a shallow angle through a great depth of air holding many particles, the shorter wavelengths are scattered and absorbed, leaving only the longer orange and red wavelengths. That is why the sky is often red at about dawn and sunset, and it is also why a red sunset often heralds a fine day. It means that the air to the west is dry. This is the direction of the sunset and also the direction from which weather is approaching, and by the following morning the dry air will usually have arrived. It is one of the few items of weather folklore that is reasonably reliable. A red sky at dawn, on the other hand, is a less dependable sign. It means that the dry air is to the east and moving away, but it tells you nothing about conditions to the west, from where a frontal system may or may not be advancing.

The first weather forecast was issued from Cincinnati Observatory on September 22, 1869, and it was not until early in the 20th century that regular forecasts were issued for the entire country. Even then it was some time before they were very reliable. For most of history, therefore, farmers, sailors, and others whose lives or livelihoods depended on the weather had to rely on their observations of natural phenomena to predict the weather. The predictions were then encapsulated in sayings such as "red sky at night, shepherd's delight; red sky in the morning, shepherd's warning." Some of this weather lore was accurate, but most was not.

■ THE POLAR FRONT AND JET STREAMS

During World War II, aircraft began to routinely fly at high altitude for the first time. There were flights in both directions across the Atlantic and, in particular, there were many American flights over the Pacific, in the vicinity of Japan. It was not long before aircrews flying near Japan started to report something very strange. Aircraft navigators use two speeds. Airspeed is the speed at which the airplane flies

through the air. That is the speed registered on the airspeed indicator in the cockpit. Groundspeed is the speed with which the airplane crosses the surface. It takes account of the effect of the wind. Navigating an airplane involves using the forecast wind speed and direction to calculate a heading for the pilot to steer in order to follow a desired track over the ground and the groundspeed. Once the groundspeed is known, the journey time can be measured. Crews found, however, that some journeys took much less time than the navigator had calculated before takeoff, some took much longer, and sometimes navigators had to make major inflight corrections to the direction they were flying to avoid their aircraft being carried miles out of their way. Similar peculiarities were reported in trans-Atlantic flights, but it was near Japan that the effect was strongest.

It was the Norwegian meteorologist Jakob Bjerknes, son of the famous Vilhelm (see "How Moving Air Transports Heat" on pages 51–52), who solved the puzzle. Close to the tropopause in certain places, there is often a narrow belt of wind, like a tube, blowing at great speed from west to east. The first jet airplanes were just entering service, and this wind came to be known as the *jet stream*. The map on page 58 shows its approximate location, farther north in summer than in winter. The map also shows how it was that the jet stream affected airplanes flying near Japan, where the jet stream blows in both summer and winter and is especially strong, and how it sometimes affected flights across the North Atlantic but had no effect on airplanes flying anywhere else.

The jet stream made a big difference to flight times. It often reaches speeds of up to 150 miles per hour (240 km/h), and in winter, when it is stronger, it sometimes blows at double that speed, but it is variable. A headwind with that force not only slows the aircraft, it may cause it to run short of fuel. Aircrews cannot rely on it always being there or be sure it will carry them in the direction they wish to fly— and there is more than one jet stream. Apart from the one shown in the illustration, which is known as the subtropical jet stream, there is a more variable polar front jet stream and others that appear locally from time to time and then disappear. There are also jet streams in the Southern Hemisphere, and in summer an easterly jet stream, blowing in the opposite direction to other jet streams, lies over southern India. Modern civil airliners try to avoid the jet stream altogether by flying above it in the lower stratosphere.

The Polar Front

The track of the subtropical jet stream lies close to where equatorial air, on the descending side of the Hadley cells (see "The Three-Cell Model of Atmospheric Circulation" on page 39), meets cooler air along a line that moves northward in summer and southward in winter. It is also close to the polar front marking the boundary between the Ferrel and

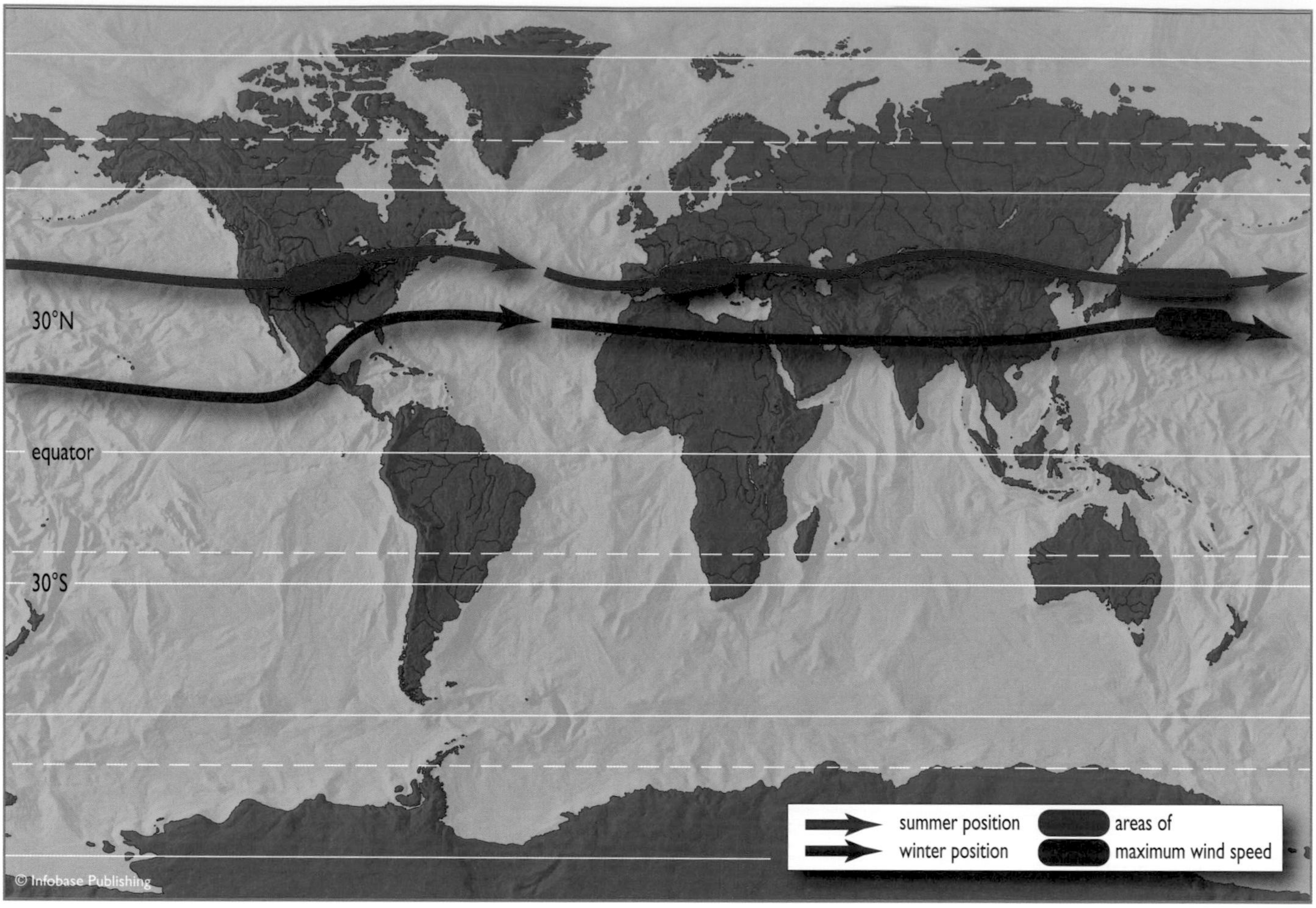

30°N

equator

30°S

| summer position | | areas of |
| winter position | | maximum wind speed |

© Infobase Publishing

The approximate position of the jet stream in summer and winter, showing the regions where the wind speed is greatest

the polar cells. In fact, the polar front jet stream produces the strongest winds, but the subtropical jet stream is more constant, so the average position of the jet stream (its most likely location on any day), which is what the map shows, is close to the position of the subtropical jet stream.

Temperature decreases with increasing latitude between the equator and the poles. In summer, high latitudes grow warmer and the temperature difference decreases. This allows the front between polar and tropical air to move northward. In winter, as the polar regions cool, the front migrates toward the equator. Clearly, the jet stream is related to differences in air temperature.

Why a Difference in Temperature Makes the Wind Blow

Both the direction and the strength of the wind are determined by the balance between the pressure-gradient force and the Coriolis effect (see "Pressure Differences and Wind Direction" on page 53), and the steeper the pressure gradient, the stronger the wind is. On a weather map the distance between isobars indicates wind speed; the closer together the isobars are, the stronger the wind. Well clear of the surface, where friction with the surface has no effect, wind speed increases and the wind blows parallel to the isobars as the geostrophic wind.

Isobars indicate atmospheric pressure across a horizontal surface, but there is another way to think of them. On an ordinary map, contour lines are also drawn on a horizontal surface (the map itself) but they represent heights above the surface. Walkers use contour lines to locate hills and find the easiest (or hardest) gradients up which to climb them. An artist could also use the map to draw a picture of the landscape showing the hills and valleys. Isobars can be used in the same way. The resulting hills and valleys in the sky are invisible, but it is possible to draw them. This makes it easier to see that pressure changes across gradients and that some gradients are steeper than others, and it allows weather forecasters to use the words ridge and trough to describe pressure changes that resemble the ridges and valleys of a landscape in a drawing of this kind.

It is also possible to draw a vertical cross section through the atmosphere that relates pressure to altitude.

This reveals that above one place on the surface the pressure may be, say, 500 millibars at a certain height and nearby it is 500 millibars at a different height. Again, gradients appear because the cross section is through an imaginary surface (called an isobaric surface) across which the pressure is the same everywhere. Diagrams of this type reveal something that would not be immediately obvious without them: Atmospheric pressure decreases with height faster above an area of high surface pressure than it does over an area of low surface pressure. In other words, the 500-millibar level is at a lower altitude in cold than in warm air. This happens because cold, dense air presses down with greater weight than warm, less dense air, and it compresses the lower air more so that a larger proportion of the total amount of air is held at a low level. With increasing height, therefore, pressure decreases more rapidly than it does in warm air, which is less compressed.

The geostrophic wind blows with a force proportional to the pressure gradient, so if that gradient changes with height, so must the geostrophic wind. A cross-sectional diagram showing isobaric surfaces at various heights clearly reveals such changes in gradient. As the drawing on the left in the illustration shows, this gradient increases most across a front where cold air lies to one side and warm air to the other. At some altitudes this can produce a horizontal temperature gradient of up to 1.5°F over five miles (2.7°C over 10 km), leading to a marked difference in the rate at which pressure decreases with height. In the figure, the 500-millibar surface slopes less steeply than the 300-millibar surface, and at still higher levels the gradient would be even steeper. The circles on each surface indicate the wind, their size increasing as wind speed increases.

What the diagram cannot do is identify air on one side as high pressure and on the other as low pressure because

the lines are cross sections through isobaric surfaces, across which the pressure remains constant. Instead, one side is marked "cold" and the other "warm." The thickness of each layer of air—the vertical distance between isobaric surfaces—is proportional to the mean temperature within the layer, and wind blows at right angles to the pressure gradient associated with changing temperature. For this reason it is called the thermal wind. In the Northern Hemisphere the thermal wind blows with the cold air on the left, and in the Southern Hemisphere it blows with the cold air on the right. In both hemispheres, therefore, the jet streams blow from west to east. In the diagram the wind is blowing away from you into the paper.

Average temperatures decrease with increasing distance from the Tropics, and the thermal wind generates high-level westerly winds in both hemispheres. It is at the subtropical and polar fronts, however, that the temperature gradient is most pronounced. The polar front, especially, marks a sharp boundary between polar and tropical air. Not surprisingly, that is where the thermal wind is strongest, and its strength increases with altitude to reach a maximum just below the tropopause. These are the polar front jet streams, blowing from west to east in both hemispheres.

Waves in the Jet Stream

Except at the equator, air flowing over the surface tends to rotate about a vertical axis due to the rotation of the Earth. This is called its *vorticity* (see the sidebar on page 60), with two components, relative vorticity and planetary vorticity; the sum of these two is the absolute vorticity. Because of the conservation of angular momentum (see the sidebar "Conservation of Angular Momentum" on page 70), absolute vorticity remains constant.

Major topographic features aligned across the path of moving air, such as the Rocky Mountains and the Tibetan plateau, deflect the flow of high-level air—the jet stream. If the jet stream is deflected poleward, its planetary vorticity increases but its relative vorticity decreases to maintain a constant absolute vorticity. The decrease in relative vorticity swings the jet stream toward the equator, but this increases its relative vorticity, decreases its planetary vorticity, and swings it back again. This action and reaction set up a series of very long waves along the jet stream. Between three and six complete waves (crest to crest) encircle the Earth, each of them up to 3,700 miles (6,000 km) long.

In 1940 the Swedish-born American meteorologist Carl-Gustaf Arvid Rossby (1898–1957) showed that when there is a certain relationship between the wavelength and the wind speed within these waves, the waves become stationary. They are then standing waves, like the waves that appear in a rope which is fastened at one end while the other

The thermal wind develops because pressure decreases with altitude faster in cold air than it does in warm air, increasing the pressure gradient between the warm and cold air and accelerating the wind. This is what causes the jet stream.

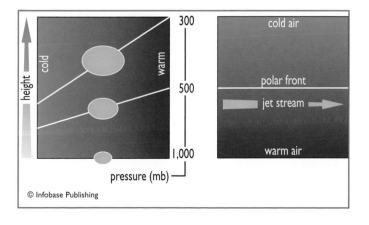

end is moved up and down to produce waves that stay in the same places. These are called *Rossby waves* (and they also occur in ocean currents). By convention, where the wave swings poleward it is said to form a ridge, and where it swings toward the equator it forms a trough.

How Waves in the Jet Stream Affect Weather at the Surface

As the jet stream approaches a trough, it starts to turn in a counterclockwise direction, just like the wind around a cyclone. The cyclonic motion accelerates the wind, producing a region of low pressure and causing air to be drawn into the jet stream to the west of the trough. This is called convergence. The converging air subsides to the Earth's surface, producing a region of high pressure with air moving outward from it. The outward movement is called divergence.

Approaching a ridge, the jet stream starts turning clockwise, and this causes air to flow out of it, or diverge. Divergence produces a region of low pressure, air is drawn up from below to replace it, and a region of low pressure is produced at the surface. As the illustration shows, this leads to the formation of frontal weather systems. In fact, it is the main cause of the formation and movement of such systems in middle latitudes, with cyclones (or depressions) forming beneath jet stream ridges, cold fronts ahead of troughs, and warm fronts behind ridges. In the diagram, the thick, wavy line represents the jet stream, flowing from left to right, and cold and warm fronts are marked with triangles and semicircles respectively.

There are exceptions to this pattern, but it is often accurate and the map on page 61 shows how it affects conditions at the surface. The figure is a typical weather map, but with the jet stream added (and dividing into two in the east). There are cyclones (marked *L*) close to the jet stream ridges and anticyclones (marked *H*) close to the troughs. The map, with the jet stream in its summer position, suggests the Canadian prairie provinces and the central northern states are experiencing wet, windy weather, but there is fine weather over eastern coastal regions, and fine, warm, settled weather is approaching from the Pacific.

Weather systems move. That is how forecasters can be sure that within a day or two the bad weather over central North America will have given way to a spell of cloudy but probably dry and calm weather that is approaching as a weak trough. They can also be confident that it will not be long before the arrival of more settled fine weather presently over the Pacific. It is this constant movement of weather systems that makes midlatitude weather so changeable and difficult to forecast. Frequent change also ensures that precipitation is distributed fairly through the year. It is what produces the climate in which temperate broad-leaved deciduous and coniferous forests flourish.

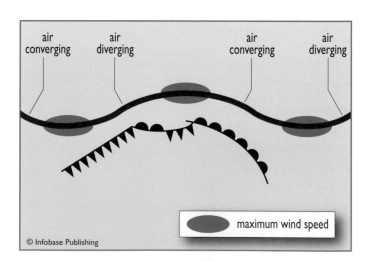

Waves in the jet stream are associated with regions where air is entering (converging) or leaving (diverging) the jet stream, producing high or low pressure that generates frontal systems in the air below.

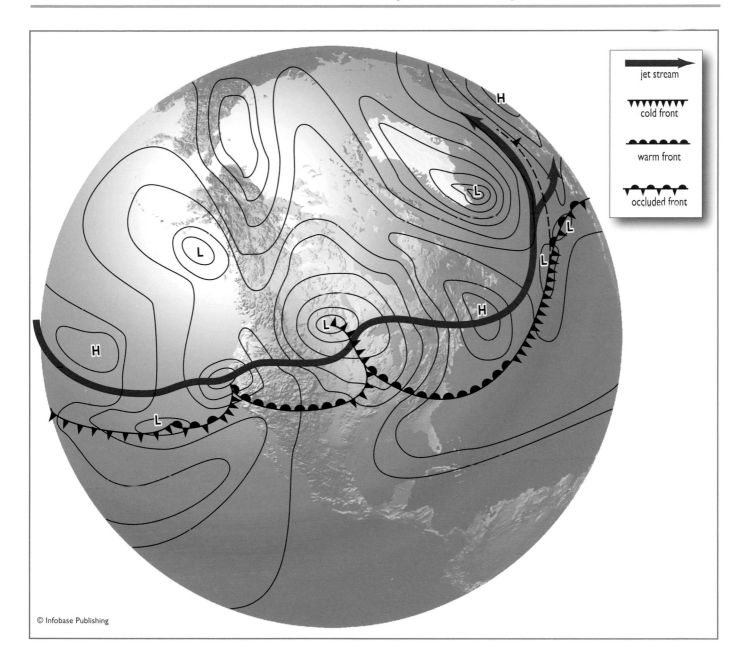

This map shows most of the Northern Hemisphere with the position of the jet stream and the frontal systems associated with it.

Blocking, When the Weather Stands Still

The generally westward movement of midlatitude weather systems is also driven by the jet stream that produces those systems, at a speed roughly equal to 70 percent of the speed of the geostrophic wind in the warm air (behind the warm fronts). Commonly, weather systems travel eastward at about 20 MPH (30 km/h) in summer. In winter, when the north–south temperature gradient is stronger, they move at about 30 MPH (50 km/h). There are times, however, when the weather seems to stick. Warm, dry conditions may persist long enough in summer to produce drought, which can cause severe damage to trees (see "Drought, Water Stress, and Wilting" on page 81). Clear, cold, windless days can follow one another for weeks in winter. Alternatively, rain in summer and snow in winter can continue day after day and seem interminable. When this happens, and a particular type of weather is prolonged, it is often because of changes in the jet stream.

The Hadley cells move air by convection, and this accounts for much of the transfer of heat from low to high latitudes, but it fails to account for all of the transfer in middle latitudes. There, the meridional (south–north) flow of air is weak, and although the predominant winds are from the west, from day to day winds vary greatly in direction and speed. Inefficient heat transfer allows warm air to accumulate in the Tropics. When that happens, high-latitude air becomes relatively cooler, and the north–south temperature gradient grows steadily steeper and the jet stream stronger. Eddies then start to develop along the polar front, some cyclonic with wind circulating counterclockwise around cyclones, others anticyclonic with wind circulating clockwise around anticyclones. This leads to major changes in the jet stream, culminating in its temporary breakdown. West-to-east airflow is called zonal flow, and the state of the upper-level westerly winds is measured by an index cycle related to an index of zonal flow, called the *zonal index*. The sequence of drawings shows, in a very simplified form, the main stages in the index cycle.

At first (1) the flow is strongly zonal, with Rossby waves of small amplitude (distance between trough and crest at right angles to the direction of flow). The amplitude of the waves increases (2) and eventually becomes extreme (3). At this stage, the overall airflow is still from west to east but along a wavy path that in places flows from the northwest and in others from the southwest. Finally (4) the flow breaks down into isolated cyclones (the two lower circular patterns, where the wind flows counterclockwise) and anticyclones (the upper pattern, with a clockwise flow). After that, the

---

The index cycle involves changes in the westerly (zonal) flow of air in middle latitudes. (1) Waves with a long wavelength (Rossby waves) develop in the jet stream. (2) The waves grow bigger. (3) The waves become very extreme. (4) The waves are so large that the flow pattern breaks up and the air circulates in cells, with low-pressure cells to the south and high-pressure cells to the north. After this the smooth westerly flow establishes itself once more.

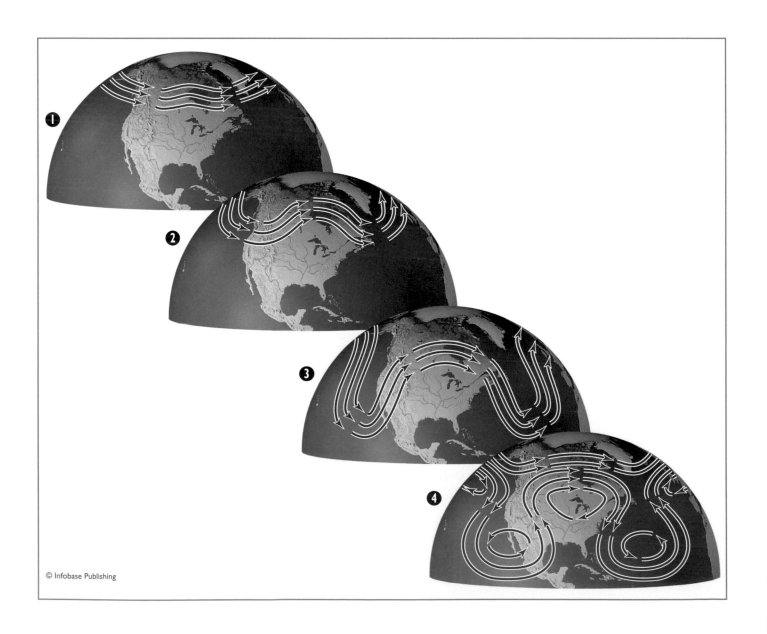

energy of motion (kinetic energy) of the eddies is transferred to the overall flow, and the strongly zonal flow reestablishes itself. The cycle usually starts in a particular place and then moves westward (in the opposite direction to the zonal flow) at about 60° of longitude a week, the entire cycle taking from three to eight weeks to complete. Its effects are less marked in summer than in winter because in winter the temperature gradient is steepest and jet-stream wind speeds are greatest.

Remember that the polar front separates cold, arctic air from warm, tropical air, and this separation produces typical weather. Regions to the north of the front lie beneath arctic air, bringing cold weather at all times of year, and regions to the south of the front lie beneath tropical air. As the index cycle develops, cold air is drawn southward and warm air northward.

These movements of the polar front compensate for the deficiency in heat transfer of the Hadley cell circulation, but they can cause dramatic changes in weather. A trough can carry bitterly cold, arctic air far to the south, and a ridge can take mild, tropical air far to the north. It is not unknown for temperatures to change from record lows to record highs in the space of a week, and the sudden arrival of arctic air can produce a *cold wave*. A cold wave is a sudden drop in temperature over 24 hours that may necessitate emergency measures to protect people and farm animals. It can also cause great distress to wildlife.

Stage 4 of the index cycle isolates huge eddies of air, usually about 900 miles (1,450 km) in diameter, which can remain motionless and persist for days or even weeks. Deep, cold cyclones form to the south, producing wet, windy conditions. To the north, anticyclones bring generally dry weather. As the zonal flow resumes, these eddies survive for a time, and the anticyclones, because they are stationary with respect to the surface, block the passage of weather systems, which are diverted around them. Blocking produces prolonged spells of anticyclonic weather and can contribute to the development of drought, while at the same time exposing areas to the north or south of the block to frontal systems that might otherwise miss them. Each individual blocking anticyclone may last for only a short time, but repetitions of the index cycle can occupy a substantial part of a summer or winter. In a year when this happens, their cumulative effect is considerable.

At present it is impossible to forecast the index cycle. It may be that the zonal flow becomes unstable as its speed increases, and this triggers the breakdown, but no one really knows. Nor do climatologists understand what drives weather cycles that operate over longer periods of years and decades (see "Cycles of Climate" on pages 40–41). What is clear is that the climates of temperate zones, marked by the eastward progress of frontal systems and a westerly airflow, are largely produced by the jet stream. It is also clear that the jet stream is a consequence of the large-scale convective mechanism by which heat is transferred from low to high latitudes. This explains the great changeability of midlatitude weather, one result of which is the distribution of precipitation fairly evenly through the year. It also explains some of the prolonged spells of weather that occur periodically when the westerly flow temporarily breaks down, due to a weakness in the convective mechanism. It is the jet stream that causes the conditions to which temperate forests are adapted, and the index cycle that can, temporarily but radically, alter those conditions.

■ WEATHER SYSTEMS

When air is warmed it expands, its density decreases, and it exerts less pressure (weight) at the surface. When it is cooled it contracts, becomes denser, and exerts a greater pressure. It is the difference in their densities that prevents air masses from mixing. Warm air rides above cold air and cold air, undercuts warm air and lifts it clear of the ground. This is how the "battle of the air masses" is "fought" along the fronts that separate them. The "battle" usually produces cloud and precipitation along fronts and around a depression or, more technically, a cyclone. Cyclone is the meteorological name for an area of low pressure around which the air circulates cyclonically. Air circulates cyclonically when it moves in the same direction as that of the Earth's rotation. This is counterclockwise in the Northern Hemisphere, viewed looking down from a position above the North Pole.

Air movements at the surface are linked to events in the upper atmosphere (see "How Waves in the Jet Stream Affect Weather at the Surface" on page 60). Air converges at a cyclone where the atmospheric pressure is lower than it is in the surrounding air. Convergence occurs because directly above the cyclone there is a region of high pressure in the upper atmosphere from which air is diverging. Divergence in the upper air draws air upward to replace it, and if air is diverging faster than rising air can replace it, pressure at the center of the low-level cyclone will fall sharply and the winds will strengthen.

Cyclone is also the name used in parts of the Indian Ocean for the fierce storms that are an extreme version of a cyclone that occurs only in the Tropics, so storms of this type are known as tropical cyclones (see "Extreme Weather: Hurricanes, Tornadoes, Drought" on pages 67–75). There are also extratropical cyclones. These, too, are ferocious storms that form around areas of low pressure, but in the Arctic, along the edge of the sea ice. An area of high pressure is known as an anticyclone.

Sometimes the flow of air across a north–south mountain range produces a lee depression. As the air rises it adds to the air that is already present. This increases the air pressure at the crest, causing divergence and anticyclonic circulation. As the air descends on the lee side of the range it expands, establishing a local area of low pressure—the lee depression—with air converging on it and turning cycloni-cally. Strong heating of the surface in summer also causes air to rise and clouds to form, and if the rising air starts turning cyclonically it will produce a thermal depression. Intense heating is needed to produce a thermal depression, however. They form in Arizona but rarely in forested regions.

Temperate forests are sometimes affected by polar depressions. These develop in the North Pacific or North Atlantic when unstable maritime air moves south from polar regions along the eastern side of a long ridge of high pressure aligned north–south. No more than about 600 miles (1,000 km) across, these are much smaller than frontal depressions, but they can be very intense and bring severe weather.

Cloud forms when air is cooled below its dew point temperature (see "Lapse Rates" on pages 45–46), and it is cooled by being made to rise. Cold fronts force warm air to rise. Air also rises as it crosses high ground, often pro-ducing a dry region in the rain shadow on the downwind side of a mountain range. In North America, the climate to the east of the Rockies is drier than that to the west for this reason. On a smaller scale, the climate on the western side of the Pennines, a range of hills running in a north–south line down the center of much of England, is wetter than the climate on the eastern side.

Stable and Unstable Air

Clouds are of different types and are classified by their appearance and also as high, medium, and low by the height of their bases (see the sidebar "The Naming of Clouds" on page 47). The height at which cloud starts to form depends on the humidity of the air. Relatively dry air must be lifted higher than relatively moist air before its water vapor starts to condense. Which type of cloud develops depends on the stability of the moist air.

If air is forced to rise, for whatever reason, it will cool adiabatically (see "Why Temperature Decreases with Altitude" on page 45). The temperature of the air surround-ing it also decreases with height, but not necessarily at the same rate. Measure the air temperature at the surface and at the tropopause, divide the difference by the height of the tropopause, and the result is the lapse rate in that place at that time. It is known as the environmental lapse rate (ELR). The pale blue line on the graph represents an ELR of 3.5°F per 1,000 feet (6.4°C/km), with an air temperature at ground level of 50°F (10°C).

Air moving across the ground may be warmer than this, however, say at 60°F (15.5°C), and as it rises it will cool at the dry adiabatic lapse rate (DALR) of 5.5°F per 1,000 feet (6.5°C/km). At 5,000 feet its temperature will be 32.5°F (5.75°C at 1 km), which in the example shown in the graph is the same temperature as the surrounding air. Consequently, the air will rise no higher. Air that rises a little way and then no farther is said to be stable. The dark blue line on the graph represents the DALR.

Lapse rates and the stability of air. Air is stable if the SALR is greater than the ELR. It is unstable if the ELR is greater than the DALR. If the ELR is greater than the SALR but less than the DALR, the air is conditionally unstable. This is the most common type of instability.

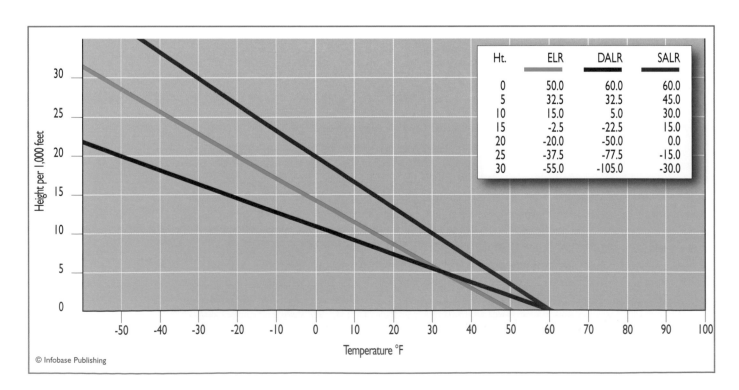

Ht.	ELR	DALR	SALR
0	50.0	60.0	60.0
5	32.5	32.5	45.0
10	15.0	5.0	30.0
15	-2.5	-22.5	15.0
20	-20.0	-50.0	0.0
25	-37.5	-77.5	-15.0
30	-55.0	-105.0	-30.0

© Infobase Publishing

If water vapor is condensing in moist air, the air will cool more slowly at the saturated adiabatic lapse rate (SALR) of 3°F per 1,000 feet (6°C/km), represented by the red line on the graph. It will still be warmer than the surrounding air by the time it reaches the tropopause. Air that continues to rise, because at every height it remains warmer than the surrounding air, is said to be unstable.

In the real world, of course, rising warmed air often starts to cool at the DALR, then water vapor begins to condense, and its rate of cooling slows to the SALR. If dry, stable air is forced to rise, by crossing a mountain for instance, ordinarily it will subside to its former height on the lee side of the mountain. It may happen, however, that cooling at the DALR as the air rises reduces its temperature to the dew point, and that it then cools more slowly than the surrounding air, so it continues to rise. Stable air that becomes unstable in this way is said to be conditionally unstable. Conditional instability occurs when the ELR is greater than the SALR but less than the DALR.

In the graph on page 64, lines above the ELR line represent air that is warmer than surrounding air at that height and lines below the ELR represent air that is cooler.

Frontal Cloud

In a frontal depression of the kind that produces most midlatitude weather, a wedge of warm air is trapped inside colder air. This air may be stable or unstable. If it is stable, it will rise ahead of a cold front to only a limited extent and then sink again to its former level behind the front. A front of this type, called a kata-front, produces very little weather, although cloud may form along it. Unstable warm air will rise vigorously along a cold front, in this case called an ana-front (the Greek *ana* means "up" and *kata* means "down"), and the frontal system will produce a great deal of precipitation. It is possible to have a frontal system in which one front is an ana-front and the other is a kata-front.

As the cold front advances, warm air is lifted. This reduces the surface atmospheric pressure in the warm-air sector. Air converges to fill it and starts turning cyclonically, and a cyclone (depression) forms. The system is now a frontal depression, comprising a warm front, a warm sector, and a cold front. The illustration on page 64 shows such a depression as it is represented on weather charts. The shape is familiar from newspaper and television weather forecasts, but here the isobars are labeled with their pressures and the direction of movement, wind directions, and areas of cloud and precipitation are shown. These indicate that both fronts are ana-fronts. If either was a kata-front, it would produce less high cloud and cloud would not extend so far ahead of the warm front.

A warm front has a very gentle slope of about 1/2–1°, which is a gradient between approximately 1:130 and 1:60.

Air is forced to rise up it by the advancing cold front, but the rise is very gentle. The first air to rise along a warm ana-front is raised all the way to the tropopause. Its water vapor forms minute ice crystals spread by the wind into thin wispy cloud called cirrus (Ci in the illustration on page 66). The appearance of cirrus is often the first indication of an approaching warm front.

Behind it, the cloud spreads to cover most of the sky, but still as a thin sheet of ice crystals through which the Sun remains clearly visible if slightly pale. This is cirrostratus (Cs). The advancing front brings the cloud base lower, but the type of cloud remains the same. Cirrostratus gives way to medium-level altostratus (As), through which the Sun is barely visible as a pale disk with indistinct edges, and this is followed by nimbostratus (Ns), obscuring the Sun and with a base at a still lower level. Precipitation begins as the nimbostratus arrives.

Along a warm kata-front, the cloud is of a more heaped, cumulus type, but the cloud top is much lower, often reaching no higher than 10,000 feet (3,000 m). By the time stratocumulus (Sc) appears blanking out the Sun completely, the front is already quite close, and precipitation usually begins at once. Behind the warm front lies the warm sector, containing the wedge of warm air. There the sky is completely covered by low stratocumulus and stratus (St) cloud, but the layer of cloud is not deep, and it produces little or no precipitation.

The cold front moves at about double the speed of the warm front and is about twice as steep. It shovels warm air upward, and if it is an ana-front, this produces heaped (cumulus-type) cloud mixed with the layered cloud. The cloud extends all the way to the tropopause, and it can include cumulonimbus (Cb), which produces very heavy showers, often with thunder and lightning.

As the front passes, which it does fairly quickly because of its speed, breaks start to appear in the cloud. These grow bigger until what had been fairly continuous precipitation gives way to increasingly isolated showers interspersed with fine intervals.

A cold kata-front produces stratocumulus, which is dense but not deep, and there is no cumulonimbus, a cloud that can form only in unstable air. The precipitation is continuous for a short time but is not unduly heavy, and there are no thunderstorms.

When the depression has passed, there is usually a spell of fine weather associated with the cold air mass, but it does not last. Except when blocking breaks the pattern, midlatitude frontal depressions tend to occur in families of three or four along the ridges in the jet stream (see "How Waves in the Jet Stream Affect Weather at the Surface" on page 60). Each cold front extends to the southwest and is joined at its end to the next warm front, all of them being carried eastward by the general westerly airflow.

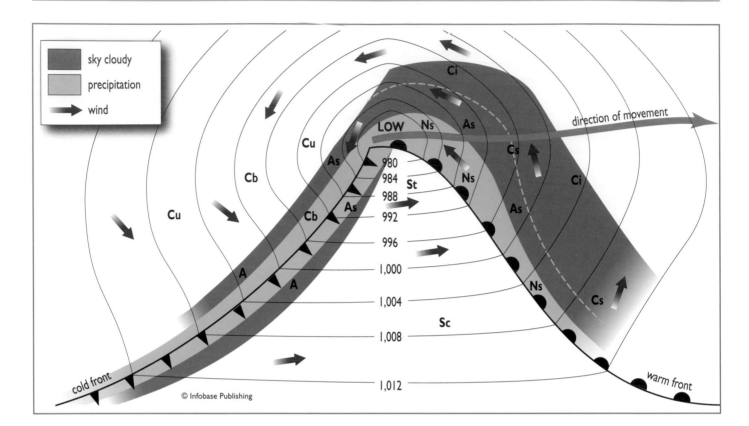

A frontal depression like this might appear on a weather map, but with the pressures (in millibars) added to the isobars and with the areas of cloud and precipitation marked. The cloud types are also shown: As = altostratus; Cb = cumulonimbus; Ci = cirrus; Cs = cirrostratus; Cu = cumulus; Ns = nimbostratus.

Thunderstorms and Hailstorms

Thunderstorms develop in very unstable air (see "Stable and Unstable Air" on pages 64–65). Storm clouds are huge, dark, and threatening, which may seem curious because clouds are made from droplets of water and ice crystals. Water and ice are colorless, so why are storm clouds so dark? The answer is that the clouds are dark because they are huge.

Although they are colorless, water droplets and ice crystals reflect light. As sunlight passes through any cloud, a proportion of the light strikes cloud droplets or crystals and is reflected by them. The reflected light fails to reach the ground, so the area beneath the cloud is shaded. Despite clouds appearing so solid, however, the droplets and crystals inside them are widely scattered. Most of the cloud is empty air. Consequently, most of the sunlight passes unimpeded through the cloud, failing to meet any reflective particle. This changes as the cloud grows vertically. The droplets and crystals are no closer together, but light travels a greater distance through the cloud, so the chance increases that it will be reflected. A full-grown storm cloud

is so tall—it can extend from about 1,000 feet (300 m) to 30,000 feet (9,000 m)—that a large proportion of the sunlight is reflected. The ground beneath it is in deep shade, and the cloud itself appears dark gray or black.

The cloud grows by convection, and the storm derives its energy from the release of latent heat. Warm, moist air rises, its water vapor condenses releasing latent heat, and this raises the temperature of the air so that it continues rising. The resulting cumulonimbus cloud—the storm cloud—towers to a great height. Inside the cloud there are violent vertical air currents, moving at 20–70 MPH (30–112 km/h) and in both directions. An airplane flying through a major storm is alternately lifted and dropped with enough force to cause structural damage.

Water droplets freeze into ice pellets when rising air currents carry them near the top of the cloud where the temperature is below freezing. The pellets fall, and as they do so, they encounter water droplets that are slightly below freezing temperature. These supercooled droplets coat the pellets with liquid water that freezes almost instantly, forming a layer of clear ice. Other supercooled droplets are much colder and freeze on contact with the pellet as tiny ice crystals with air spaces between them. These produce a layer of opaque, white ice with a spongy texture. As vertical air currents continually raise them to the top of the cloud and drop them almost to the bottom, the ice pellets grow into hailstones made of alternating layers of clear and opaque ice. Eventually, hailstones either enter a region of the cloud where down-currents predominate, and they are swept from the base of the cloud or they grow too heavy for the up-currents to raise them. The size that hailstones attain reflects the number of journeys they have made between the top and the bottom of the cloud, which in turn depends on the size of the cloud. The biggest hailstones fall from the biggest clouds.

Within the cloud, ice crystals near the top acquire a positive electric charge, and hailstones and larger water droplets at a lower level acquire a negative charge. The low-level negative charge can then induce a positive charge at the ground surface. Lightning occurs with the sudden and violent heating of air to temperatures of 50,000°F (28,000°C) or more that causes the explosive expansion we hear as thunder when sparks flash between these separated charges.

Lightning is an electric spark, but scientists are not sure what causes it because air is an excellent electrical insulator, and the difference between the positive and negative charges in even the biggest cloud is insufficient to overcome air resistance. One theory is that cosmic radiation—electrically charged particles arriving from space—may trigger a shower of particles moving downward through the cloud. These produce a runaway flow of electrons—the electric current that causes the lightning flash.

Storm clouds die after a time. The circulation of air inside the cloud forms a convection cell in which warm air rises, cools, and descends. When precipitation starts, snow, hail, or rain close to freezing temperature falls from the cloud base. The falling snowflakes, hailstones, and raindrops drag air with them, cooling the air at the same time and causing a strong downdraft of cold air beneath the cloud. As the cold air strikes the ground, it spreads and interferes with the inflow of air feeding the updrafts. The downdrafts come to dominate, cooling the updrafts adjacent to them, and precipitation deprives the cloud of the moisture that supplies its energy. Then the convection cell breaks down, the storm dies away, and the cloud dissipates, sometimes releasing all of its remaining moisture in a cloudburst. Ordinarily this entire process takes no more than an hour or two, which is the lifetime of a typical storm cloud.

Hot Lightning and Cold Lightning

A direct lightning strike can destroy a tree. Lightning can also ignite forest fires—but not always. Even after a long period of dry weather, lightning does not always ignite the tinder-dry plant material covering the forest floor. Lightning that will start a forest fire is known as *hot lightning. Cold lightning* will not start a forest fire. They look identical, but there is a small but importance difference between these two types of lightning.

Lightning can flash inside a cloud, between clouds, or between a cloud and the ground, and thunderstorms usually begin with lightning flashing inside the cloud and illuminating it for a fraction of a second. Flashes between the cloud and the ground follow a little later.

A flash of lightning to the ground begins with a stepped leader. This is a faintly luminous discharge carrying negative charge—a stream of electrons—at a speed of about 225,000 MPH (360,000 km/h). It moves through the electric field beneath the cloud, following the zigzag path of least resistance and imparting an electric charge to the air. Shortly before reaching the ground, the stepped leader is met by a much more powerful return stroke, moving upward. This is the visible lightning flash. Often, a first lightning stroke fails to neutralize the charge inside the cloud, and it is followed immediately by two or three more, the entire set of strokes lasting for about one-fifth of a second. This is the most common type of lightning, but there are others. Sometimes lightning flashes from the ground to the cloud, rather than cloud to ground, producing what looks like upside-down lightning strokes that branch upward instead of downward. Ordinarily, lightning lowers the negative charge, but some lightning lowers the positive charge.

There are also differences in the return stroke. If the current ceases to flow between one return stroke and the dart leader that initiates the next return stroke, the object being struck is not subjected to sustained heating. This is cold lightning. When cold lightning strikes a tree, it may heat the water flowing through the tree's vascular system so rapidly and so intensely that the water vaporizes instantly and the tree explodes, but if it strikes dry material on the ground, it will not raise the temperature high enough to start a fire.

If, on the other hand, the current is sustained for several cycles of stepped leader–return stroke–dart leader–return stroke, the struck object will be heated to a much higher temperature. If it is dry enough, it is likely to catch fire. This is hot lightning.

■ EXTREME WEATHER: HURRICANES, TORNADOES, DROUGHT

Ordinary weather systems bring rain, snow, wind, and all the familiar ingredients of day-to-day weather. They can also generate storms that are sometimes severe. Beyond this ordinary weather, however, the Earth's atmosphere produces conditions that are much more violent. A hurricane—technically known as a tropical cyclone—is a storm that produces sustained winds of more than 75 MPH (120 km/h) together with torrential rain. Tornadoes produce even more violent winds, though affecting much smaller areas. Droughts are less dramatic, but although they develop slowly their effect can be catastrophic.

Hurricanes

Not all depressions are associated with fronts, and there is one type of nonfrontal depression that brings the most savage weather known. Meteorologists call it a tropical cyclone. More familiarly, tropical cyclones that form in the North Atlantic and Caribbean are called hurricanes, those that form in the Pacific and China Seas are typhoons, those that form in the northern Indian Ocean are cyclones, and they go by several other local names. Tropical cyclones can be several hundred miles in diameter, with a central eye up to 40 miles (65 km) across, and they pack the energy of thousands of atomic bombs. All of them form in the same way and bring the same kind of devastation. Hurricanes can demolish large areas of forests, and sometimes they do.

Most wind speeds can be reported using the Beaufort Scale (see the sidebar "The Beaufort Wind Scale" on page 78), devised for the British Admiralty in 1806 by Admiral Sir Francis Beaufort (1774–1857). It allots wind forces from 0 to 12, a force 12 wind being a hurricane with sustained winds of more than 75 MPH (120 km/h). The Beaufort Scale is quite inadequate for categorizing real hurricanes, which generate much stronger winds, and the U.S. Weather Bureau has added a further five categories, known as the Saffir-

Simpson scale, shown in the table. All sustained winds can and do gust to much greater speeds.

As their name suggests, tropical cyclones occur only in the Tropics, but once formed they often move out of the Tropics. In North America they can travel as far as New England and occasionally even to eastern Canada. From time to time one will cross the Atlantic as far as northwestern Europe. Once outside the Tropics, the cyclones weaken, and technically they cease to be tropical cyclones. Nevertheless, their original energy is so great that even when weakened, they remain capable of causing severe damage, and they can regain some of their lost strength if they meet and combine with frontal depressions during their passage across the ocean.

Tropical cyclones are driven by strong convection in very moist air. This constrains the region in which they can develop. To feed them the amount of moist air they need, they can form only over the ocean where the temperature at the sea surface is not less than 80°F (27°C). The sea reaches this temperature only in the Tropics and even then only in late summer, after it has had several months to warm. That is why the season for such storms runs from late summer to late fall. To make them rotate, they need the Coriolis effect (CorF). Its magnitude is zero at the equator, so tropi-

Saffir/Simpson Hurricane Scale

CATEGORY	PRESSURE AT CENTER	WIND SPEED	STORM SURGE	DAMAGE
	mb in. of mercury cm. of mercury	mph kmh	feet meters	
1	980 28.94 73.5	74–95 119–153	4–5 1.2–1.5	Trees and shrubs lost leaves and twigs. Mobile homes destroyed.
2	965–979 28.5–28.91 72.39–73.43	96–110 154.4–177	6–8 1.8–2.4	Small trees blown down. Exposed mobile homes severely damaged. Chimneys and tiles blown from roofs.
3	945–964 27.91–28.47 70.9–72.31	111–130 178.5–209	9–12 2.7–3.6	Leaves stripped from trees. Large trees blown down. Mobile homes demolished. Small buildings damaged structurally.
4	920–944 27.17–27.88 69.01–70.82	131–155 210.8–249.4	13–18 3.9–5.4	Extensive damage to windows, roofs, and doors. Mobile homes destroyed completely. Flooding to 6 miles (10 km) inland. Severe damage to lower parts of buildings near exposed coasts.
5	920 or lower below 17.17 below 69	more than 155 more than 250	more than 18 more than 5.4	Catastrophic. All buildings severely damaged; small buildings destroyed. Major damage to lower parts of buildings less than 15 feet (4.6 m) above sea level to 0.3 mile (0.5 km) inland.

cal cyclones can form only some distance from the equator. They form, then, in late summer and fall over tropical oceans between latitudes 5° and 20° north and south of the equator.

In the Tropics, the prevailing winds are easterlies. These carry the storms westward. As they move, they accelerate, and the CorF swings them away from the equator along a curved path and into the westerly airflow of middle latitudes. It is this that produces the typical Atlantic hurricane track, carrying hurricanes across the Caribbean, northward up the east coast of the United States, then out into the Atlantic again. Some remain over the sea, out of harm's way, but others move farther west, through the Gulf and northward overland through the Carolinas.

They travel at 10–15 MPH (16–25 km/h), and because they are moving, wind speeds around them vary. As the diagram shows, on one side of the eye the wind blows in the same direction as the storm is moving, and on the other side it blows in the opposite direction. Consequently, the wind strength differs on either side of the storm. The lowest wind speeds are always found on the side of the storm nearest to the equator.

Huge cumulonimbus storm clouds surround the eye, fed by moist air that spirals upward. The converging air accelerates as it approaches the eye. The strongest winds are those closest to the eye and, because of the conservation of

angular momentum (see the sidebar on page 70), the bigger the radius of the storm the stronger its winds are.

The extent of its acceleration depends on the radius of the hurricane core, and this depends on latitude because the magnitude of the vorticity (see the sidebar "Vorticity" on page 60) and CorF which cause the air to rotate varies with latitude. Less than 5° from the equator, where vorticity and CorF are close to zero, the distance over which air would need to converge to generate hurricane winds is so large that there is simply not enough air available. At latitude 20°, on the other hand, about at the limit of adequate sea-surface temperatures, air converging from 90 miles (145 km) to a spiraling radius of 20 miles (30 km) would be accelerated from initially being stationary to a speed of about 110 miles per hour (175 km/h).

Around the Eye

A fully developed tropical cyclone covers an area of up to 20,000 square miles (52,000 km²), and its clouds tower to more than 40,000 feet (12,000 m). Pressure is high above the center, and air sinks from there into the eye, warming adiabatically as it descends and adding to the warming due to the release of latent heat from clouds surrounding the eye.

In the eye, winds are light and there are few clouds. Surrounding the eye is a wall of towering cumulonimbus cloud that produces torrential rain, usually accompanied by thunder and lightning. Air is rising rapidly inside the cloud and being swept out and away at the top. It is here, in the wall surrounding the eye, that wind speeds reach their maximum. Beyond this wall of cloud, there is a region free from precipitation where the cloud base is high, often at about 20,000 feet (6,000 m). Still farther from the eye, the

---

Winds are cyclonic around a tropical cyclone, and the storm itself is moving. On one side of the storm, the wind speed and the storm speed produce winds that are stronger than those on the opposite side, where the wind speed and the storm speed moderate them.

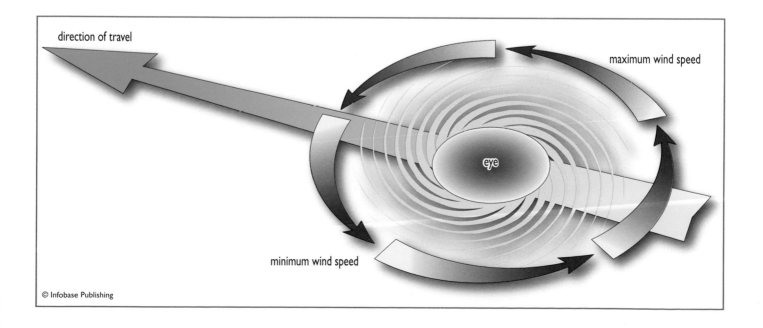

direction of travel

maximum wind speed

eye

minimum wind speed

Conservation of Angular Momentum

Angular momentum is a function of the mass of a rotating body, its radius of rotation, and its angular velocity, which is its speed of rotation measured as the number of degrees through which it turns in a given time (usually one second). Earth, for example, completes one rotation every 24 hours, so it has an angular velocity of 15° per hour (360° ÷ 24) or 15 seconds of arc per second.

Angular momentum is conserved. This means that if any component of angular momentum changes, one or both of the other components will change to compensate, provided the rotating body experiences no friction to slow it down or external force to make it turn faster. In other words, $r \times v \times m$ = a constant, where r is the radius of rotation, v is the angular velocity, and m is the mass.

Air spirals inward toward the core of a hurricane. This decreases the radius of the rotating air in the same way that a dancer might perform a pirouette by commencing with his or her arms outstretched, then drawing them toward his or her body. Call the initial radius r_1 and the final, smaller radius r_2, and let $r_1 = 100$ and $r_2 = 10$. Let $m = 1$ and the initial angular velocity $v = 20$. Then:

$$r_1 \times v_1 \times m = 100 \times 20 \times 1 = 2000 \text{ (the constant).}$$

Reduce r and

$$r_2 \times v_2 \times m = 2000 \text{ (the constant)}$$

therefore

$$v_2 = 2000 \div (10 \times 1) = 200$$

By reducing the radius from 100 to 10, angular velocity increases from 20 to 200. In the case of a hurricane or tornado, angular velocity is the wind speed. The pirouetting dancer spins faster as she or he draws in her or his arms.

dominantly cirrostratus and the lower cloud is cumulus, but there may be up to 200 towering cumulonimbus storm clouds closer to the center.

Once established, the storm sustains itself. Moist, very warm, surface air feeds convection within the cumulonimbus bands and also divergence from the high-level anticyclone that carries away the rising air, drawing in more at the surface. Converging air at the surface produces the winds, but the cyclone can survive only while it has access to an unlimited supply of warm, moist air at the surface. Once it crosses a coast, the air feeding it is much drier. If it remains at sea, as it moves out of the Tropics it crosses colder water. Dry air and a cool sea both weaken it, and few tropical cyclones last longer than a few days before weakening to tropical storms. The air in the core is then cool rather than warm, and wind speeds decrease, but a tropical storm is still capable of producing large amounts of rain and winds of gale force that can do considerable damage.

Extratropical Cyclones

As well as tropical cyclones there are also extratropical cyclones. These are also circular systems, with bands of cloud containing cumulonimbus and a more or less cloud-free eye. Only a few hundred miles in diameter, they are smaller than tropical cyclones, and their winds usually reach no more than about 45 MPH (72 km/h), with gusts to 70 MPH (112 km/h).

Extratropical cyclones form near Antarctica and in the North Pacific and Atlantic Oceans, at fronts between very cold and much warmer air. Fronts of this type can occur where continental polar air from North America meets maritime tropical air warmed by the Gulf Stream and extending unusually far north. They also occur where low pressure over the sea draws in extremely cold air, often at -40°F (-40°C), from above adjacent sea ice, producing a temperature difference that can reach 70°F (39°C).

These cyclones develop rapidly and travel at up to 35 MPH (55 km/h). Like tropical cyclones, they dissipate rapidly when they cross land, but they last long enough to bring heavy falls of snow driven by fierce winds that are quite strong enough to bring down trees.

Squall Lines

Isolated thunderstorms occur in summer, usually late in the day when the weather is hot and humid. They also occur along cold fronts when the air in the warm sector ahead of the front is conditionally unstable (see "Stable and Unstable Air" on pages 64–65). Thunderstorms are not always isolated, however, when they form a little way ahead of a cold kata-front. The cold air advancing into warmer air causes sufficient lifting to trigger instability and produce a line of

cloud base descends again and the rain resumes. Seen in cross section, as in the illustration on page 70, the storm appears as distinct towers of cloud on either side of the eye. From above, these can be seen as spiral bands. Air rises by convection in the clouds, and some sinks from the high-level anticyclone in the regions between bands. At the edges of the tropical cyclone, the upper cloud is pre-

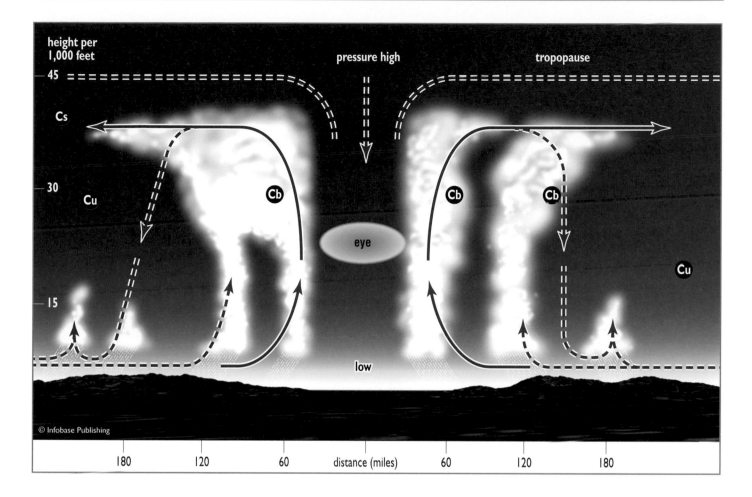

This cross section through a tropical cyclone shows the structure of the storm, with cumulonimbus (Cb) clouds towering to about 40,000 feet (12,200 m) around the calm eye.

thunderstorms. These storms produce gusts of wind that can reach hurricane force. Storms of this type are called squalls, and sometimes they merge to form a continuous line, called a squall line. Squall lines can extend for up to 600 miles (965 km) and, driven by a strong high-level wind, they move across the ground at up to 45 MPH (70 km/h).

It is the winds that drive the storm clouds forward. Wind speed increases with height, so the upper part of each cloud overtakes its own base and the cloud overhangs its base. The overhang captures warm, moist air ahead of the cloud, drawing it into the up-currents inside the cloud. These vertical currents move at up to 100 MPH (160 km/h). The moist air increases the energy available to the cloud and the wind speed in its upper part. The wind blows the top of the cloud into an anvil shape, and where the base of the anvil meets the main body of the cloud, eddies in the air currents produce a characteristic roll of cloud, called a squall cloud. The squall cloud often has many hemispherical protrusions, called mammatus, on the underside.

Downdrafts emerge as gusts that can reach almost hurricane force, forming a gust front that pushes beneath the warm air and shovels it into the cloud. Ahead of the cloud, warm air is drawn in so strongly that it, too, produces a strong wind that can also gust to hurricane force. The line of clouds is then advancing faster than the cold front that triggered its formation. The squall line detaches from the front and advances into the warm sector.

The life of an individual storm cloud is brief, but along a squall line, as one cloud dies, the cold downdraft that spreads beneath it and chokes the inflow of air sustaining it also lifts warm air ahead and to the right. This air also becomes highly unstable and forms a new storm cloud by taking the warm air from its parent so that the death of one cloud causes the birth of another beside it.

Tornadoes

Tornadoes consist of rapidly rotating air, but these twisting are are produced by a mechanism quite different from the one that generates tropical cyclones. Most, but not all, are associated with violent thunderstorms, and especially with squall lines. They occur in all parts of the world outside the Tropics, but the Great Plains of North America experience far more of them than any other region. When a tornado sweeps through a town, it leaves a trail of destruction. When one sweeps through a forest, there may be no one to observe it, but a winding swathe of dead and injured trees, some uprooted or snapped in two, others stripped of their branches and leaves, marks its passing. This trail of destruction lies within a wider area of forest damaged by the storm that spawned the tornado.

Ordinarily, a storm cloud chokes to death when its downdrafts overpower its updrafts, depriving it of the warm, moist air it needs to sustain it. Sometimes, though, a cloud escapes this fate to become a supercell cloud (see the sidebar). Then it can last for several hours, rather than

Supercells and Mesocyclones

In most clouds, the updrafts and downdrafts form several convective cells, but in very large cumulonimbus clouds these sometimes merge into a single supercell that occupies the entire interior of the cloud. Updrafts in a supercell rise at an angle, rather than vertically, so instead of falling directly into the updraft, precipitation falls to one side. The updrafts and downdrafts are then separate, flowing in different parts of the cloud, and the cool downdraft is unable to choke the warm updraft. Ordinary storm clouds dissipate because the downdrafts cool and suppress the updrafts, depriving the cloud of the rising current of warm air that sustains it. This does not happen in a supercell cloud, so the cloud survives much longer than an ordinary cumulonimbus. While it survives, the cloud continues to grow, and the biggest supercell clouds can extend to a height of 60,000 feet (18,300 m). Supercell clouds rarely last for longer than three hours, but the downdraft can lift conditionally unstable air to produce a new one even as the old cloud dissipates.

Inflowing air is converging strongly, and this sets it rotating counterclockwise (in the Northern Hemisphere) and accelerates it because of the conservation of angular momentum. This can start an entire section of the cloud rotating, especially if the wind above the cloud blows at an angle to the air leaving the cloud, giving it an extra twist.

The rotating section, up to 6 miles (10 km) across, is called a mesocyclone; *meso-* is from the Greek *mesos* meaning middle. The mesocyclone begins in the middle of the cloud, and the rotation extends downward.

two hours at most for an ordinary cumulonimbus, and it can grow into a truly immense, terrifying storm.

Inside the supercell cloud, a core of air starts to rotate in the Northern Hemisphere, almost always in a counterclockwise direction. The rotating air is a mesocyclone, and once it has formed the storm is tornadic, which means that it is capable of triggering tornadoes. The cross section through a tornadic storm gives an idea of its structure. The cloud top may be at a height of 60,000 feet (18,300 m) or more, inside the lower stratosphere.

Below the cloud, at the center of the vortex of inflowing air, the atmospheric pressure is extremely low because of the rate at which air is rising. At ground level inside the vortex, the pressure may be equal to that outside the cloud at a height of several thousand feet. This draws down the base of the cloud. The first warning of a tornado is usually this extension below the main cloud, turning fairly quickly and with fragments of cloud moving vertically up and down its exterior. This is the wall cloud.

Inside the cloud, the mesocyclone is also extending downward. This stretches the mass of rotating air vertically, making it narrower as it approaches the base of the cloud, so it is shaped like a funnel. As the funnel grows narrower, the wind speed around it increases due to the conservation of angular momentum (see the sidebar "Conservation of Angular Momentum" on page 70). The vortex is likely to become visible as it emerges from the base of the wall cloud. It appears as a funnel, narrower at the bottom than at the top, snaking this way and that below the cloud. It is dangerous only when it touches the ground, which is when the funnel cloud becomes a tornado.

A storm that can produce one tornado will often produce several, so tornadoes can occur as families. When the tornadic storm is one of many similar storms, strung out along a squall line, tornadoes can appear almost simultaneously in places many miles apart.

In a tornado, all the air is moving upward. The funnel is visible because the extreme low pressure near the core makes inflowing air expand, and as it expands its water vapor condenses. The funnel cloud forms in the vortex itself and is not drawn down from the cloud above. Dust and loose debris are swept up by the inflow and carried into the vortex. They darken it, and when it touches the ground, there is a small cloud of flying debris around its base. Over snow, however, a tornado can be blazingly white. Most tornadoes are no more than half a mile (800 m) across at the base of the funnel, but a vigorous one can produce several other tornadoes, called suction vortices, wandering erratically around its edge, smaller but often even more violent than their parent.

Measuring conditions inside a tornado is very difficult because even the most robust instrument packages are liable to be destroyed and tornadoes are ephemeral. They appear suddenly, and most last for only a few minutes, although

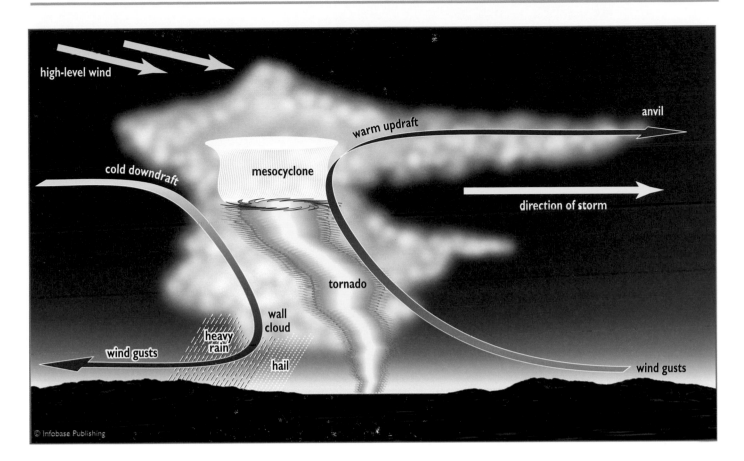

This cross section through a tornadic storm shows how the cold downdraft and warm updraft are separated and the rotating center of the cloud becomes narrower as it stretches downward.

there are exceptions. While they last, they may remain stationary, but most move over the ground at up to about 40 MPH (65 km/h) along an erratic, unpredictable track. Today, tornadoes are studied mainly by radar, from a safe distance. Modern radar can measure the wind speed around the funnel vortex. In a really fierce tornado, this can reach around 300 MPH (480 km/h). No other weather system on Earth can generate winds of this speed.

The Fujita tornado-intensity scale categorizes tornadoes as weak, strong, and violent by their wind speeds and the damage they do. In most cases the force is estimated from the damage that is examined after the tornado has passed. There are two classes in each category, so the scale runs from 0 to 5. A weak tornado with winds of 40–112 MPH (65–180 km/h) can break branches from trees or snap entire trees. Strong tornadoes, with winds of 113–206 MPH (181–330 km/h), will uproot whole trees and leave them flattened. A violent tornado generates winds from 206 to more than 300 miles per hour (331 to more than 480 km/h) and will toss mature trees around as though they were small sticks.

Floods and Flash Floods

A *floodplain* is an area of level ground on either side of a major river. Because the slope is very shallow, many rivers meander across their floodplains. Meander systems advance across the plain. The river erodes the bank along the outside edge of each meander where the water accelerates around the bend. At the inside edge, where the water slows because it has a shorter distance to travel, the river deposits sediment. As the illustration on page 74 shows, this has two effects. The first is to increase the size of the meanders so their winding becomes more extreme. The second is slowly to shift the entire meander system in a downstream direction.

From time to time, the river overflows its banks, inundating the floodplain. The frequency with which this happens varies from one river to another, but all floodplains naturally lie under water for some of the time, the floodplain absorbing the surplus water. That is why homes built on floodplains are especially vulnerable to flooding, and protecting floodplains, for example by building levees, simply transfers the flooding downstream.

Occasional floods and the steady advance of the meander system combine to cover the surface of the floodplain with a layer of silt transported by the river. The silt is rich in plant nutrients, making the soil of the floodplain highly fertile. At one time most North American and European floodplains were covered in forest. The occasional floods did

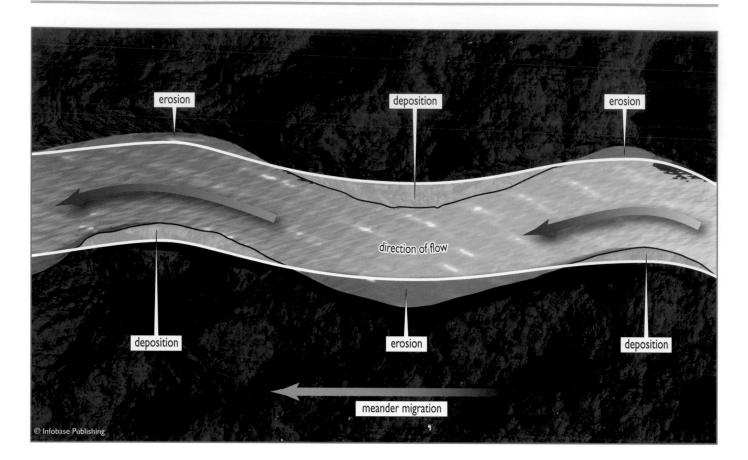

As a river meanders across the plain, the deposition of sediment on one side of each bend and its erosion from the opposite side makes the entire meander system advance slowly.

not harm the trees, because although the ground became waterlogged so that plant roots were unable to absorb oxygen for respiration, the floods receded after a short time. The fertility of the floodplains made them attractive, however, and most floodplain forests were cleared long ago to provide farmland or, more recently, housing.

A flash flood is quite different. It follows a prolonged period of rain that saturates high ground so that the soil can absorb no more water. A very heavy rainstorm then releases a huge amount of water that completely overwhelms the rivers draining the hills. The rivers overflow, sometimes alter their courses and merge with one another, and a wall of water descends onto land near the foot of the slope and the plain beyond. The damage this causes is due to the force with which the water moves. A flash flood sweeps away obstacles in its path, so before long the floodwater carries uprooted trees, boulders, and other debris, all of which it hurls at obstructions farther down the slope. Flash floods are never wide enough to affect a large area of forest, but they can demolish trees along a swath through it.

Why Droughts Happen

Trees grow where soil and climate supply the resources they need. Water is one of those resources, but unlike mineral nutrients in the soil, in temperate regions it is not always available. Ordinarily, this causes no difficulties because trees are well adapted to seasonal variations produced by the climates in which they grow (see "Tree Adaptations to Climate" on pages 138–140). Sometimes, however, the usual precipitation pattern can break down, leading to prolonged drought. This can cause great harm.

A drought is a period of low rainfall or no rain at all at a time of year when rain is usually expected, leading to a shortage of water that is serious enough to cause harm to plants and animals and inconvenience to humans. How long it takes for a dry spell to turn into a drought varies greatly from place to place. In the Sahara several years must pass without rain before a drought is declared. In the United States a drought is defined as a period of 21 days during which the amount of precipitation is no more than 30 percent of the average for that place and that time of year. A drought is declared officially in Britain after 15 consecutive days without more than one millimeter (0.04 inch) of rain. Where it rains every day, even a few days without rain will be sufficient to cause drought. Nowhere is immune from drought, not even the humid Tropics.

There are different types of drought. A desert, where a shower of rain is an exceedingly rare event, experiences permanent drought. Regions where almost all the rain falls in one season experience seasonal drought. An accidental or contingent drought is unpredictable, it can occur anywhere, and its end is no more predictable than its start. If it is especially severe, it is known as a devastating drought. When the rain returns, people often conclude that the drought has ended, but it may not have done. During the drought, the water table fell and the flow of groundwater greatly decreased (see "Groundwater and the Water Table" on page 32). It takes some time and a substantial amount of rainwater to replenish the groundwater and raise the water table to within reach of plant roots. The drought that continues after the rain returns is known as an invisible drought. Droughts are also classified as agricultural if they reduce crop yields and hydrological if the groundwater decreases sufficiently to reduce the flow of water in rivers substantially.

Droughts are often measured on the Palmer Drought Severity Index, a scale devised in 1965 by W. C. Palmer, a meteorologist working for the U.S. Weather Bureau. This scale, shown in the table, measures the extent to which the water supply deviates from the supply that is considered normal in a particular place.

Blocking (see "Blocking, When the Weather Stands Still" on pages 61–63) is the most frequent cause of droughts affecting the forests that grow in middle latitudes. A large anticyclone, or a ridge of high pressure extending from one, becomes stationary and remains so for weeks or even months, blocking the west-to-east path of the frontal depressions that bring rain. The rain does not vanish; it is simply diverted somewhere else, leading to the apparent paradox that a drought in one place can mean floods in another.

It is in summer that blocking anticyclones causes drought in middle latitudes, usually accompanied by high temperatures, which make the soil dry out even faster. Winter blocking also brings dry weather, but that causes fewer problems

than it does in summer because temperate forest trees require little water in winter, and drought then does not harm them. Low temperatures also reduce greatly the rate of evaporation, so soil dries much more slowly. Trees may be damaged by the cold, however, especially in Europe where the anticyclonic circulation draws air westward from Siberia.

What is certain is that forests, like all plant communities, are well able to tolerate the climates in which they grow, with a margin of tolerance for extremes. Fierce winds or drought may cause serious damage, but it is damage from which forests can recover, provided it is inflicted only occasionally. It is the too-frequent repetition of such assaults that causes real, lasting harm.

■ EFFECT OF SEVERE WEATHER ON FORESTS

Trees are living organisms, and like all living organisms, they can be ill and they grow old and die. When a forest tree dies it falls, creating a gap in the canopy and space on the ground. Sunlight penetrates and other plants are able to grow. Then a tree seedling will grow up to replace the fallen tree. The seedling was already present, but the mature tree had shaded it, preventing it from growing. As the young tree grows, the space on the ground fills, and the gap in the canopy closes once more. Forests are dynamic, changing constantly in countless small ways, yet always remaining forest.

Often, it is the wind that brings down old and diseased trees. Most tree diseases are caused by fungi. The word *fungus* suggests the visible mushrooms, toadstools, and brackets found in forests, especially in the fall, but these are just the fruiting bodies from which fungi release their spores, and most disease-causing fungi produce fruiting bodies that are far too small to see. The main part of a forest fungus lies below ground or inside the tissues of trees. A fungus consists of a *mycelium*, which is a network of very fine filaments, called hyphae, often extending over a large area. The fungus absorbs nutrients through its hyphae. If the mycelium is in the soil, it gathers nutrients that help the tree (see "Roots" on pages 103–104). If it extends through the tissues of the tree, however, it weakens the tree and may eventually kill it.

Before a fungus can invade a tree, however, there must be a point through which it can enter. This is usually a wound, left when a branch falls or is broken, caused by a blow from another tree as it falls or a stone catching it a glancing blow as it rolls down a hillside, or, very often, by an animal. Deer and squirrels strip bark from tree trunks, and this can allow fungi to enter. Certain insects bore through tree bark, and they can carry fungal spores with them, transmitting disease in the same way that some biting insects transmit diseases to humans. That is how Dutch elm disease spreads (see "Dutch

Palmer Drought Severity Index Classification	
4.00 or more	Extremely wet
3.00–3.99	Very wet
2.00–2.99	Moderately wet
1.00–1.99	Slightly wet
0.50–0.99	Incipient wet spell
0.49–0.49	Near normal
-0.50–0.99	Incipient dry spell
-1.00–1.99	Mild drought
-2.00–2.99	Moderate drought
-3.00–3.99	Severe drought
-4.00 or less	Extreme drought

Elm Disease" on page 174). A tree that is weakened in this way is less able to withstand the wind than a healthy tree, so often it is the diseased trees which fall first.

Even among healthy trees, some are more likely than others to be brought down by a strong wind. As well as obtaining nutrients and water, roots anchor trees, but beeches and spruces are particularly vulnerable to wind because their main roots snap.

Trees Shaped by the Wind

On exposed sites there are often trees that have been shaped by the wind. Most of their branches and leaves are on the more sheltered side, and the entire tree may lean away from the prevailing wind. The tree appears to be permanently bending in the wind—even when the air is still. This sculpting is due to wind desiccation. Wind has a strong drying effect, because it constantly sweeps away the boundary layer of air adjacent to surfaces. This removes the boundary layer before it can become saturated and replaces it with dry air. That is why people hang laundry outdoors to dry, of course, and what the wind does to the laundry it also does to leaves.

The effect is to evaporate water vapor faster than the tree can replace it by evapotranspiration (see "Forests and the Climate" on page 49), desiccating and killing the leaves. Wind-sculpting is very common along coasts, where the wind blowing off the sea carries salt spray, increasing the drying effect. The sculpted shape also protects the tree from being blown down by the wind. Branches, twigs, and leaves

that might be caught by the wind are on the downwind side and largely sheltered by the trunk.

Why Few Trees Rise Above the Roof of the Forest

Seen from the air, a forest looks like a green carpet. Its surface is rippled, but very few trees rise above the canopy. This may seem curious because trees compete for light, which they need for photosynthesis. The brighter the light, the more strongly a tree can grow and the healthier it will be. It is not so simple, however. A tree that towers above its neighbors projects into a very hostile environment where it loses moisture rapidly and its leaves may find it difficult to survive.

Trees that protrude above the forest canopy are exposed to the wind and its powerful drying effect on the leaves. This effect is especially strong above a forest canopy, partly because wind speed increases with height, so the wind above the forest is always stronger than the wind at ground level. It is also strong because of the contrast between the wind above and below the canopy. The crowns of forest trees touch: That is what makes the closed canopy. They also shelter each other from the wind. It may be blowing a gale above the canopy, but among the crowns just a few feet lower down, the wind is little more than a gentle breeze.

A tree that stands taller than the canopy is called an *emergent*. Its height exposes it to more sunlight, aiding photosynthesis, but also to the wind, which increases the rate of

The drying effect of a strong prevailing wind sculpts trees. *(U.S. Forest Service)*

Seen from above, a closed forest canopy resembles a green carpet. Its surface is rippled, but almost all the trees are the same height. *(Vanderbilt University)*

transpiration and damages the leaves, and so inhibits photosynthesis. Only those trees that have tough, wind-resistant leaves and an abundant supply of water to allow for the high rate of transpiration can survive. It is the wind that produces the fairly level roof of the forest.

Blowdown and Why Some Trees Are More Vulnerable Than Others

The flat, open leaves of a broad-leaved tree catch the wind, and during the growing season, before the tree detaches them in the fall, they are firmly attached. The wind on a single leaf has only a small effect, of course, but a full-sized broad-leaved tree has hundreds of thousands of leaves. This makes broad-leaved trees more vulnerable to the wind in late spring and summer than in fall and winter.

Leaves of coniferous trees, reduced to needles or scales, offer little resistance to the wind, but this advantage is offset by the fact that many conifers grow taller than broad-leaved trees. Where the two occur together in a mixed forest, the crowns of the conifers often stand clear of the main canopy. Their increased exposure means that in a mixed forest, the conifers are more likely to be blown down than the broad-leaved trees.

Wind Force

A strong wind exerts considerable force, and when it blows across a forest, the trees absorb almost all of that force. This has been measured. The wind strength 100 feet (30 m) inside a temperate forest is less than 80 percent of that outside. Move 200 feet (60 m) inside the forest, and the wind speed is halved, and at 400 feet (120 m), it is less than 10 percent of the speed outside.

For broad-leaved deciduous forests, the extent of wind reduction depends on the season. At a particular point inside a mixed forest in Tennessee, the wind speed was reduced to 12 percent of that outside the forest in January when the trees were bare. In summer, in contrast, when the trees were in full leaf, it was only 2 percent of the wind speed outside the forest. This shows how much of the wind force leaves absorb and also that below the level of the canopy, the force is borne by trees growing near to the forest edge. It might seem, therefore, that trees near the edge are most at risk of being blown down, but this is not necessarily the case. Perhaps because they are more exposed, the trees near the edge are also strong. It can even happen that trees inside the forest fall, but those at the edge remain standing.

There are two ways to report the strength of the wind. Meteorological instruments measure it in familiar units of speed—miles per hour, knots (nautical miles per hour; 1 knot = 1.15 MPH), kilometers per hour, or meters per second. The alternative is to report the strength of the wind in terms of its visible effects. This is the method that was devised for the British Admiralty in 1806 by Commander Francis Beaufort (later Admiral Sir Francis Beaufort, 1774–1857). Beaufort was trying to improve seamanship and improve safety at sea. In its original version, his scale described the correct amount of sail a warship should carry in order to

Beaufort Wind Scale

FORCE	SPEED IN MPH (KMH)	NAME	DESCRIPTION
0.	0.1 (1.6) or less	Calm	Air feels still. Smoke rises vertically.
1.	1–3 (1.6–4.8)	Light air	Wind vanes and flags do not move, but rising smoke drifts.
2.	4–7 (6.4–11.2)	Light breeze	Drifting smoke indicates the wind direction.
3.	8–12 (12.8–19.3)	Gentle breeze	Leaves rustle, small twigs move, and flags made from lightweight material stir gently.
4.	13–18 (20.9–28.9)	Moderate breeze	Loose leaves and pieces of paper blow about.
5.	19–24 (30.5–38.6)	Fresh breeze	Small trees that are in full leaf sway in the wind.
6.	25–31 (40.2–49.8)	Strong breeze	It becomes difficult to use an open umbrella.
7.	32–38 (51.4–61.1)	Moderate gale	The wind exerts strong pressure on people walking into it.
8.	39–46 (62.7–74)	Fresh gale	Small twigs are torn from trees.
9.	47–54 (75.6–86.8)	Strong gale	Chimneys are blown down. Slates and tiles are torn from roofs.
10.	55–63 (88.4–101.3)	Whole gale	Trees are broken or uprooted.
11.	64–75 (102.9–120.6)	Storm	Trees are uprooted and blown some distance. Cars are overturned.
12.	more than 75 (120.6)	Hurricane	Devastation is widespread. Buildings are destroyed and many trees are uprooted.

make a specified speed under particular wind conditions. It did not mention the actual speed of the wind. The scale was later adapted to use on land (see the sidebar), with examples of the wind's effect on rising smoke, flags, trees, and similarly familiar sights, and speeds were added.

In the boreal forest, however, windstorms are not the most serious cause of harm. There, fire is much more destructive. Broad-leaved forests usually grow in moister habitats, where it is more difficult to start and sustain a fire and the trees themselves are not especially flammable. Conifers, on the other hand, grow on drier sites, often with a thick layer of dead, dry needles carpeting the forest floor, and the trees themselves contain resin, which burns readily and fiercely. Fires that start naturally rather than being ignited accidentally by people are due to lightning. Although forest fires are often spectacular, frightening, and, of course, very dangerous to homes, they are not necessarily harmful to the forest, despite destroying so many mature trees (see "Fire Climax" on pages 150–152).

Wind, even strong wind, is normal, of course. It can break branches and blow down trees, but although these are events that change the forest locally, they have no effect on the forest as a whole. Occasionally, however, the change may be more drastic.

Tornadoes and Hurricanes

Violent tornadoes demolish everything they touch, and they can cut a swathe of total devastation through a forest. Most do not travel far, but they have been known to leave tracks several miles long (see "Tornadoes" on pages 71–73). Tornadoes can and do happen anywhere, but forests are especially vulnerable across that part of the Great Plains known as "Tornado Alley," which experiences more of them than any other region in the world. Outside Tornado Alley, a 1989 tornado in Cornwall, Connecticut, brought down almost all of a stand of white pines (*Pinus strobus*).

The aftermath of a tornado is easily recognized. Apart from its narrow width, the tornado track is filled and lined with trees lying in all directions, felled by a twisting wind that hit them from several sides. On level ground, windblown trees usually fall in the direction of the wind, and on steeply sloping ground, fallen trees tend to lie with their tops pointing downhill. Trees felled by a tornado lie in a great jumble.

Tropical cyclones (see "Hurricanes" on pages 68–70) are also rotating winds but on a very much larger scale. In the southeastern United States where hurricanes moving in from the Caribbean sometimes cross the coast, conifer forests have been repeatedly devastated.

Windstorm damage can occur outside regions that are usually prone to tropical cyclones, and it has done so throughout history. On December 21, 1694, for example, a gale uprooted entire forests in Scotland. A few years later, in 1703, one of the worst storms in British history, lasting three days, November 24–27, brought winds of up to 80 MPH (129 km/h), equal to category 1 on the Saffir–Simpson hurricane scale. That storm blew down more than 17,000 trees in the southeastern county of Kent (at that time with an area of about 1,600 square miles; 4,100 km²).

Much more recently, a storm in 1987 caused extensive damage in southeastern and eastern England. It struck on October 15 and destroyed 19 million trees. The 1987 storm began as a fairly mild Atlantic hurricane that crossed the ocean and moved up the English Channel. Weak by hurricane standards, its winds were still strong enough to wreak havoc.

Hurricanes that reach New England have already weakened but still retain a substantial proportion of their original force. In 1938, for example, one struck with winds of more than 100 MPH (160 km/h). It killed 600 people, but in Massachusetts it also destroyed a large area of forest that was growing on level ground and on slopes that were directly exposed to the wind.

Predicting wind damage can be difficult. Obviously, trees growing on thin or soft soil are at risk because they are insecurely anchored, and trees growing on high ground are at risk because they are very exposed. Beyond that, the wind is almost whimsical in its effects. The trees most severely affected by the 1938 New England storm were those growing on level ground and wind-facing slopes, but the 1987 English storm was most severe on slopes sheltered from the wind. Deep valleys usually afford shelter to trees, but if the wind direction is approximately aligned with the valley, the wind will be funneled through the valley and accelerated by the shape of the land surface. Tornadoes travel farthest on level ground, but over short distances their tracks are little influenced by the lay of the land. Indeed, a high ridge that tends to trigger severe and sometimes tornadic thunderstorms in warm, moist air crossing it may increase the likelihood of tornadoes on what is otherwise the more sheltered side.

Rotating winds strike trees from first one direction and then another, but to a lesser degree and more erratically all winds strike in this way, especially inside a forest. Wind moving near ground level is deflected by obstructions producing eddies, and the trees produce countless eddies in a wind blowing through a forest. These eddies, which can include very strong gusts, push the trees this way and that.

Wind Damage Leading to Disease

Wind rips branches from trees and hurls them to the ground. This does not kill the trees directly; they can survive the loss of a few branches. It does leave a wound, however, and wounds can become infected in trees just as in animals. Fungal disease spreads slowly into the main body of the tree, weakening the tree and sometimes hollowing it until, some years later, another strong wind topples it. Meanwhile, the disease afflicting one tree may spread to its neighbors, also weakening them.

Not all trees that fall are uprooted. Weakened ones, especially, may simply snap in the wind, leaving behind a very jagged stump, called a *snag*, several feet tall. Breaking the trunk does not kill a tree immediately, but in time it will die, leaving an upright skeleton that will probably fall later.

Falling trees, parts of trees, and branches injure other trees. Forest trees grow close together. When one falls it is more than likely to strike one or two others and as these fall, they may topple still more. Falling neighbors may damage even those trees that remain standing. Branches may be torn from them and bark stripped from their trunks. Falling trees crush smaller saplings and seedlings, along with woody shrubs growing on the forest floor. Indirect wind damage of this kind can be extensive, and, of course, it increases the likelihood of fungal infection.

Snow and Ice

Winter snow can turn a conifer forest into a scene of fairy-tale beauty. Conifers are adapted to cold, snowy winters, of course, but snow can also damage them. Only loose, wet snow will cling to tree branches, but it is heavy. One cubic foot of wet snow weighs about six pounds (100 kg/m³), and over the length of a main branch the weight can be considerable.

Snow transforms the look of the forest. It is very beautiful, but its weight can damage trees. These trees are in Wisconsin. *(Langlade County)*

© Infobase Publishing

Scotch pine   Sitka spruce   Pignut hickory   Pin oak

Silhouettes of two coniferous trees and two broad-leaved trees. From left to right Scotch pine, Sitka spruce, pignut hickory, and pin oak.

Coniferous trees have somewhat flexible branches, and in some species the branches incline downward, helping them to shed snow, but, as the illustration shows, this is not true for all species. Scotch pine (*Pinus sylvestris* var. *scotica*) has branches that extend almost horizontally, and those of Sitka spruce (*Picea sitchensis*) are horizontal or incline upward. A heavy load of snow may break them, although it is usually only small branches or those already weakened by disease that break. As the picture also shows, their greater number of small branches and twigs makes broad-leaved trees much more vulnerable. They can be severely damaged if snow breaks their branches in spring, after leaves have started to open.

Avalanches and Ice Storms

Avalanches move with enough force to uproot trees, and they will sweep away whole areas of forest. Where avalanches are common events, they tend to flow along certain tracks, identifiable by the gaps they leave in forests. One way skiers can recognize areas prone to avalanches is from the paths they have cut through forests.

An avalanche occurs when a layer of snow on sloping ground becomes unstable. This can happen if the layer accumulates more snow, if a warm wind melts the snow at a higher level sending a stream of water beneath the snow layer, or if the bonds holding the snow grains together weaken of their own accord. Once the snow is unstable, the slightest disturbance may set it moving.

There are two principal types of avalanche. A point-release avalanche has the shape of an inverted V and involves only the surface layer of snow. A slab avalanche is much more dangerous. It happens when an approximately rectangular slab of snow, up to half a mile (800 m) wide, moves downhill, accelerating as it goes and often reaching a speed of 100 MPH (160 km/h). A small avalanche can strike objects with as force of up to 0.5 ton per square foot (4.5 t/m²), and a large avalanche can exert a force of more than nine tons per cubic foot (8 t/m²). As it moves, a slab avalanche also pushes air ahead of itself, generating an avalanche wind of up to 185 MPH (300 km/h).

Freezing rain can also break branches. It occurs when supercooled rain (the droplets are below freezing temperature) strikes a surface that is below freezing temperature and freezes instantly. An *ice storm*, caused by heavy freezing rain, can quickly coat surfaces with a thick layer of clear ice—and ice is about 10 times heavier than snow. Ice storms are fairly common. There was a severe one in late November 1996, for example, affecting Texas, Oklahoma,

Arkansas, and Missouri and another on January 6, 1998, in southern Quebec, eastern Ontario, and northern New England. The 1998 ice storm stripped many trees of their branches.

It is usually the smaller branches that snap, stripping trees of their crowns, sometimes over a wide area. The crowns grow again, but in areas especially prone to ice storms, repeated loss of crowns may stunt the trees.

Drought, Water Stress, and Wilting

Drought is more insidious. Trees are affected when the rate of evapotranspiration is high and a prolonged period of low rainfall has lowered the Water Table (see "Groundwater and the Water Table" on page 32), or when water belowground is frozen. Forests growing in Mediterranean climates where summers are ordinarily dry and in high latitudes where winters are cold enough to freeze water all the way from ground level to the top of the water table are adapted to periods when water is not available. Even there, though, summer drought and winter cold can persist long enough to cause harm. Damage occurs much sooner in regions where rainfall is usually distributed evenly through the year.

Some tree species are more tolerant of drought than others. Conifers tend to survive better than broad-leaved species. Birch and beech are especially sensitive, but many species of oak, ash, and chestnut are fairly resistant.

While its roots remain within reach of soil moisture, a tree (or any plant) will take in water and transport it as sap through its stem and branches, and it will evaporate through the stomata (pores) in its leaves. Water fills the vessels through which it moves, and this maintains the rigidity of soft tissues and the shape of leaves. If insufficient water is available in the soil or the rate of evaporation from stomata exceeds the rate at which the plant's vascular system can replace water lost from the leaves, the tree will experience *water stress*. Its cells will lose their rigidity, and the first obvious sign of water stress is the wilting of leaves on broad-leaved trees. Conifers show no immediate sign, but unless the water supply is restored quickly, both broad leaves and needles will die, turn brown, and be shed.

Water stress also causes leaf stomata to close. Photosynthesis is impossible if carbon dioxide cannot be absorbed and oxygen released. Closing the stomata reduces the loss of water, but it also prevents the exchange of gases, slowing and eventually halting photosynthesis. Wilting therefore halts the growth of the tree. Water stress also reduces the rate at which proteins are synthesized, so the entire metabolism of the tree slows. Continued drought may then affect twigs and later branches. As these are deprived of water, they start to die. Eventually the entire tree may die.

Forest Damage

Forests that grow on shallow soil or on soil such as wet clays where trees produce shallow roots experience drought damage before those on deeper soils, and some sites are more prone to summer drought than others. Slopes facing the noonday sun will receive more sunshine than shaded slopes, so they will be warmer and water will evaporate from them faster, rendering them more susceptible to drought.

Regardless of species, the first trees to be affected are the young seedlings. They are not old enough to have put down deep roots. After them, the taller saplings and immature trees suffer. Then it is the old trees. They are fully grown and already approaching the end of their lives. When the old trees die and fall, they create gaps in the forest canopy, but gaps that may take a long time to fill because of the shortage of young recruits. This can alter the composition of the forest by eliminating a substantial part of an entire generation of certain species.

Severe drought in a broad-leaved deciduous forest produces an effect resembling the fall. Trees shed their leaves and stand bare and dormant, as they do in winter. Many recover in subsequent years but by no means all of them, and those that do recover are likely to grow more slowly than they did prior to the drought. Of those that appear to recover, some will have suffered damage so severe that they die back slowly. The trees will continue to produce leaves year after year, but fungi invade them and little by little they die. Eventually there will come a year when they remain leafless. Then they will slowly disintegrate, losing a branch or two now and then and eventually falling.

During a summer drought, all forests become increasingly vulnerable to fire. This adds to the damage, but fire is rarely as harmful as it looks.

Wind, snow, ice, drought, and fire are natural phenomena. If a particular forest experiences any one of them repeatedly, the stress will act as an agent of natural selection. Those trees most severely affected by it will gradually disappear from the community, leaving the most tolerant species to predominate.

■ CLIMATIC REGIONS

Most school atlases include a map of the world showing the distribution of the various types of vegetation. It shows tropical rain forest in the Tropics and desert in the subtropics and indicates those parts of the world where grasslands and temperate forests might be expected to occur. Comparing this map with the one showing climatic regions, often on the next page, reveals that they are almost identical, although they may use different colors.

This will come as no surprise. It is mainly the climate that determines the type of vegetation that will grow in a particular place, so the two maps are bound to be very similar. What is more, the first attempts to classify climates did so on the basis of the vegetation characteristic of them. In other words, the climate determines the vegetation type, and the vegetation type defines the climate. Since the reasoning is circular, it is no wonder that vegetation and climate maps agree.

Why Is It Called the "Temperate Zone?"

Climate was not always linked to vegetation. Several Greek philosophers attempted to classify climates, and the most successful was Aristotle (384–322 B.C.E.). The Greek scholars divided the Earth into zones based on the amount of sunlight each receives. They called these zones *klimata,* from which we derive our word *climate.* There were three *klimata.* In the torrid zone, the Sun was almost overhead for most of the time. The frigid zone was the one receiving least sunlight. Between these two there lay the temperate zone. Aristotle's contribution was to define the boundaries of these zones. He worked out the location of the Tropics of Cancer and Capricorn, both of which he named, and of the Arctic and Antarctic Circles. In common with the other Greek scholars, Aristotle believed that the torrid zone was much too hot to be inhabited and the frigid zone much too cold.

With periodic variations, this division of the world into zones remained in use almost until modern times. Claudius Ptolemy (ca. 100–170 C.E.) divided the world into seven zones based on the duration of the longest day, and Edmund Halley (1656–1742) based his scheme on heat rather than sunlight, but the idea of zones survived.

People no longer talk of the torrid or frigid zones, but the term *temperate zone* is still used.

Ways of Classifying Climates

Throughout the 18th and 19th centuries, botanists were traveling to what were then the most inaccessible corners of the world in search of previously unknown plants. Thousands of specimens were shipped to Europe and America, and many were eventually bred to produce commercial crop plants and popular ornamentals. All botanists knew that certain types of plants grow in certain parts of the world, and some began to speculate about why this should be so.

By the middle of the 19th century, finding answers to this question involved taking account of new ideas about plant classification and evolutionary mechanisms. Improved classification allowed relationships among plants to be identified, and new ideas about the way evolution works improved understanding about how those relationships arose. Together, these new ideas helped to explain the geo-

graphical location of particular plants in a very general way, although the disjunct distribution of some plants remained a mystery (see "Trees That Grow on Either Side of a Vast Ocean" on page 11). Exploration, combined with studies of classification, or *taxonomy,* and evolutionary theories contributed to the emerging scientific discipline of what is now called *biogeography.* The scientists asking the questions were plant geographers, working in a field now known as *phytogeography* or geobotany. The early biogeographers linked climates and vegetation directly. They used such terms as *tundra climate, tropical rain forest climate, boreal forest climate,* and many more. Some, such as *savanna climate* and *desert climate,* still sound familiar. Others, such as *penguin climate,* were once used but have since been abandoned.

Two of the leading plant geographers of the late 18th and 19th centuries were the Swiss botanists Augustin-Pyramus de Candolle (1777–1841) and his son, Alphonse-Louis-Pierre-Pyramus de Candolle (1806–93). Both were professors at the University of Geneva, Augustin of natural history and, later, Alphonse of botany, and after Augustin died, Alphonse continued his work. In 1885 Alphonse published *Géographie Botanique Raisonnée en Exposition des Faits Principaux et des Lois Concernant la Distribution des Plantes d'Epoque Actuelle* (Botanical geography classified by exposing the principal matters and laws concerning the distribution of plants of the present epoch). This showed how plants are adapted to the environments in which they occur naturally. This work was taken further by the German botanist Andreas Franz Wilhelm Schimper (1856–1901), a professor at the University of Basel, Switzerland. His *Pflanzengeographie auf Physiologischer Grundlage* was first published in 1898 (and appeared in the English language as *Plant Geography on a Physiological Basis* in 1903). Schimper proposed that climate is the key to plant distribution, and it was from this idea that modern climate classification schemes developed.

The Köppen Scheme

The process culminated in the system for classifying climates devised by Wladimir Peter Köppen (see the sidebar on page 83), who was born in St. Petersburg, Russia, of German parents and worked mainly in Germany. He became a meteorologist, but he started out as a botanist and was very influenced by the ideas of the botanists of his day. Today there are several systems in use for classifying climates, but that of Köppen is the one most popular with geographers and biologists. It is the one that is most commonly used in school atlases.

Köppen moved to Graz, Austria, to work on the handbook. He died at Graz on June 22, 1940.

Everyone recognized that the type of vegetation depends on the temperature and rainfall. Temperate forests grow in temperate climates, with warm or at least mild summers,

Wladimir Peter Köppen (1846–1940)

Wladimir Köppen was born in St. Petersburg, Russia, on September 25, 1846, the son of German parents living in Russia. He attended school in the Crimea, Russia (now in Ukraine), and studied at the universities of Heidelberg and Leipzig, Germany. While he was still at school, Köppen became keenly interested in natural history.

His student dissertation, completed in 1870, dealt with the relationship between plant growth and temperature. After graduating, Köppen returned to Russia and joined the Russian meteorological service (1872–73). In 1875 he moved to Germany to head a new division in the Deutsche Seewart, based in Hamburg, where his team established a weather forecasting service for northwestern Germany and the adjacent sea areas. From 1879, with that service functioning, Köppen devoted himself to research.

He published his first map of climatic zones in 1884, and his system of climate classification first appeared in full in 1918. After several revisions, the final version was published in 1936. Köppen collaborated with his son-in-law, Alfred Lothar Wegener, in writing *Die Klimate der Geologischen Vorzeit* (The climates of the geological past), a book published in 1924. He also wrote *Grundriss der Klimakunde* (Outline of climate science), 1931, and in collaboration with Rudolf Geiger he began a five-volume work, *Handbuch der Klimatologie* (Handbook of climatology).

cool or cold winters that do not last too long, and moisture available for most of the year. This much is obvious. What Köppen did was to relate vegetation types more precisely to mean monthly temperatures and precipitation and to mean annual temperature.

What matters to plants is not the amount of precipitation, but the availability of moisture in the soil, and this is related to temperature. The higher the temperature, the greater the rate of evaporation, and, therefore, the higher the proportion of the precipitation reaching the ground that vanishes before plants can derive any benefit from it. This is called the *effective precipitation*, measured as precipitation minus evaporation.

Köppen began by dividing the world into six sets of climatic regions, all but one of them—dry climates—based on temperature. He had noted that trees will not grow where

the average summer temperature is below 50°F (10°C). Certain tropical plants will not grow where temperatures fall below 64.4°F (18°C). In middle latitudes, a mean winter temperature higher than 26.6°F (-3°C) indicates only a brief period of snow cover and a mesothermal climate. If the mean winter temperature is lower than this, snow cover lasts a considerable time and the climate is microthermal.

Köppen then refined this broad division by taking account of effective precipitation, generating a set of subdivisions. These allowed him to set boundaries to the climatic regions. There are many of these subdivisions, some defined by precipitation and some, in warmer regions, by additional features of temperature. In this classification, the boundary between steppe grassland and forest is defined by effective precipitation.

Geographers can use the Köppen climate classification in two ways. If they have details of the climate of a region, they can predict the type of vegetation that should grow there, and if they know the type of vegetation, they can deduce what the climate is like. Going further, they can summarize the system by drawing a map showing the distribution of vegetation over an imaginary continent. To do this, however, the map has to be based on certain assumptions. The imaginary continent must be low-lying because temperature decreases with increasing altitude. It must also be flat because when weather systems cross hill or mountain ranges, they lose moisture as air rises, making the climate on the windward side wet and producing a dry rain shadow on the lee side, often extending for a considerable distance.

Köppen made the further assumption that climate is the sole determining factor for vegetation. Certainly it is the most important one, but differences in soils also influence the distribution of plants. Nor are the boundaries of vegetation types, defined by minimum temperatures, entirely accurate or entirely constant. Those Köppen used were based on botanical surveys undertaken in the 19th century. These were only approximate then, and most will have shifted since.

Figures for the mean temperature and precipitation at a particular place are calculated over as long a period as the records permit, and temperature means are calculated separately for daytime and nighttime and published as average maximum and average minimum temperatures respectively. The records that are used commonly cover about 30 years, but the period may be much longer or shorter. Those for Portland, Oregon, for example, extend over 72 years, but those for Brussels, Belgium, cover only 10. In any case, averages include variations that can be quite wide. In Portland, the average daytime temperature in July, calculated over 72 years, is 77°F (25°C), but the highest recorded over this period was 107°F (42°C), and this is not exceptional. Most places record temperature and precipitation extremes that deviate from the mean by more than 30 percent and often by more than 40 percent. A sequence of years in which weather conditions deviate from the mean may be enough to shift the boundaries of the vegetation. The use of averages is unavoidable because of the wide variation possible in individual years, but averages can conceal gradual changes. In other words, the Köppen boundaries are too rigidly defined. Despite these criticisms, the classification remains useful as a general description of the relationship between vegetation and climate.

The Thornthwaite Scheme

The most widely used alternative to the Köppen classification is the one devised by the American climatologist Charles Warren Thornthwaite (1889–1963). One of the most distinguished climatologists of his generation, Thornthwaite was professor of climatology at Drexel Institute of Technology, Philadelphia, from 1941 to 1944 and president of the Section of Meteorology of the American Geophysical Union. In 1951 he was elected president of the Commission for Climatology of the World Meteorological Organization.

The Thornthwaite classification, which he first proposed in 1931, was based on the relationship between precipitation and evaporation. He abandoned entirely the use of vegetation boundaries to define climatic boundaries. Instead, he measured precipitation, temperature, and the rate at which water evaporated from a pan. From this, he was able to calculate what he called precipitation efficiency. Adding together the values for each of the 12 months produced a precipitation-efficiency (P-E) index from which he defined five moisture provinces. To this, Thornthwaite added a thermal-efficiency (T-E) index, based on the number of degrees Fahrenheit by which the mean temperature in each month is above freezing, a value of 0 indicating a frost climate and more than 127 a tropical climate.

Thornthwaite revised his system radically in 1948 by introducing an entirely new concept, potential evapotranspiration (PE, which should not be confused with P-E). He started with recorded mean monthly temperatures, corrected them for day length, and calculated how much water would be returned to the atmosphere by evaporation and transpiration if the amount of water were unlimited. This is what he called potential evapotranspiration (PE), and tables were compiled to make the calculations easier. He then allowed for the storage of moisture in the soil as surplus and deficit to produce a moisture index that he modified later to take account of different types of soil and vegetation. He was also able to derive thermal efficiency from the potential evapotranspiration because potential evapotranspiration is determined by temperature.

Putting all of this together, the Thornthwaite classification identifies nine moisture provinces and nine temperature provinces. As with the Köppen scheme, this broad classification is then refined by adding further details about variations in the availability of moisture through the year to allow for climates that have a seasonal distribution of precipitation. The Thornthwaite system corresponds reasonably well with the vegetation types found in temperate climates, although it is less reliable for tropical and semiarid areas.

The Thornthwaite and Köppen schemes are the two classifications most widely used, but there are many others. Between 1909 and 1970, nearly 30 were proposed, some more completely worked out than others. In 1969, for example, A. N. Strahler (born 1918) produced one based on air-mass source regions (see "Air Masses and Fronts" on pages 50–57), fronts, and general climatic conditions that identified 14 types of climate. Another scheme, devised by Hermann Flohn (1912–97) is based on air masses and the major global wind belts.

Climate Science and Weather Forecasting

As they have developed, the systems for classifying climates have increasingly tended to move away from the botanical origin of all such systems and have become essentially climatological. This change is entirely natural. When the first attempts were made, much more was known about the global distribution of vegetation than about climates and how they work. Early systems had little alternative but to be primarily botanical. The reason for this is very simple and has little to do with the introduction of more sensitive instruments for measuring temperature, humidity, and pressure and nothing at all to do with advances in mathematics. It is due to improvements in communication.

When botanists were exploring the world, identifying and collecting plants and mapping their distribution, they could take as much time as they needed and could afford (this was usually limited because most of them financed their expeditions from the sale of the specimens they shipped back to their home countries). They traveled by sea and moved overland by whatever means were available, often on horseback or foot or by boat. This was fine for collecting specimens and recording the general weather conditions they encountered, but it was useless for developing a broad understanding of how climates are produced. To do that, it is necessary to map weather systems on a scale large enough to cover entire continents and oceans, which is impossible unless conditions can be recorded by many observing stations at the same time and transmitted to a central point.

Until the middle of the last century, reports could travel no faster than a messenger could carry them on horseback, but in 1844 the first telegraph line opened, between Baltimore and Washington. Within a few years, the telegraph network had expanded widely, and the American physicist Joseph Henry (1797–1878) used the resources of the Smithsonian Institution, of which he was elected secretary in 1846, to gather weather reports from all over the United States. The system he devised formed the basis of the one developed later by the U.S. Weather Bureau, and in 1851 the first weather map was shown in public, at the Great Exhibition in London.

Once the telegraph network had expanded to cover North America, Europe, and parts of the other continents, it became possible to start issuing weather reports and then forecasts. Regular daily bulletins were first issued in 1869 from the Cincinnati Observatory, and in Britain the Meteorological Office published its first regular weekly report in 1878. Weather reports and forecasts were useful, and funding was provided for the work needed to compile them and for the scientific research needed to improve them. The scientific study of climates (climatology) developed alongside the study of weather (meteorology), and as understanding of atmospheric processes advanced, it became possible to define climatic regions and to classify them. This project, initiated by botanists, passed to climatologists.

All systems for classifying climates aim to provide a very general description. It is important to bear this in mind. The weakness of the Köppen scheme arises from its imposition of boundaries that are too rigid and too sharply defined. There are many places where the weather changes markedly over a distance as short as 20 miles (36 km), and if such an alteration is usual, it will amount to a climatological difference that will be reflected in the natural vegetation. The difference may be slight, but it is real, and it could alter the boundary between two major types of climate and vegetation. Generalizations about climates can never be more than approximate.

Approximations are sufficient for the general classification of climates, but nowadays attempts are being made to increase their precision. The primary aim now is to improve predictions of the regional consequences of possible climate change (see "Forests and the Greenhouse Effect" on pages 268–272) by taking account of the small scale on which such important events as cloud formation occur. As the techniques for achieving this become increasingly powerful, it will lead to ways of classifying climates in more detail, which means on a more local scale.

■ CONTINENTAL AND MARITIME CLIMATES

Ask people to name the place in the Northern Hemisphere that has the coldest winters, and (unless they already know the answer) the chances are they will suggest the North Pole or perhaps northern Greenland. At all events, it will be somewhere uninhabited. Their guess is entirely reasonable, but it is wrong. The coldest place in the Northern Hemisphere is Verkhoyansk in eastern Siberia, and although it lies in a sparsely populated region, it is far from uninhabited. It is a small town with a population of about 2,000, and in February, the coldest month, its citizens shiver in temperatures that have been known to fall to -90.4°F (-68°C). For comparison, and just to convince everyone that Verkhoyansk really is colder than Greenland, the lowest winter temperature recorded at Qaanaaq (formerly Thule) in northern Greenland, also in February, is -41°F (-40.5°C), which is cold enough for anyone, but still 49.4°F (27°C) warmer than Verkhoyansk. In case anyone suggests the atlas has got it wrong and that, despite appearances, Verkhoyansk is actually farther north than Qaanaaq, Verkhoyansk is in latitude 67.57°N and Qaanaaq is at 76.55°N, a difference that amounts to a distance of about 620 miles (998 km). In fact Qaanaaq is the most northerly town in the world, with a population of about 600 (and a large U.S. air base). Nor is the difference due to altitude. Lying 328 feet (100 m) above sea level, Verkhoyansk is somewhat higher then Thule at 121 feet (37 m), but the 207 feet (63 m) separating them should make Verkhoyansk the colder by only 0.7°F (0.4°C).

In summer, however, the people of Verkhoyansk win hands down. Their warmest month is July, when the highest temperature recorded was 98°F (37°C). In Qaanaaq, the highest temperature recorded (admittedly from data covering only three years) is 59°F (22°C).

Mean Temperatures and Temperature Ranges

These are extremes, of course, but Verkhoyansk is also colder on average. The mean annual temperature there is 1.1°F (-17.1°C) and at Qaanaaq it is 12°F (-11°C), and Verkhoyansk has a much wider range of mean temperature. There is a difference of 129°F (72°C) between the maximum and minimum mean temperatures, compared with a difference of 67°F (37°C) at Qaanaaq.

It is the range of temperature that is significant. Despite being farther south, Verkhoyansk has much hotter summers and colder winters than Qaanaaq. This is because Qaanaaq lies on the coast, albeit of a sea that is frozen, and Verkhoyansk lies in the middle of a vast continent. Qaanaaq enjoys a maritime climate, and Verkhoyansk has a continental climate. Obviously, all climates become cooler with increasing distance from the equator, but maritime climates are less extreme than continental climates.

Nothing much grows around Qaanaaq, but the most extreme continental climate in the world does not produce a desert. Indeed, Verkhoyansk lies in the heart of the boreal forest, or taiga, that extends for 6,000 miles (10,000 km) from the Baltic to the Sea of Okhotsk.

Cold Poles and the Thermal Equator

Verkhoyansk experiences the lowest temperatures in the Northern Hemisphere. It is therefore known as the *cold pole* to acknowledge this distinction while distinguishing it from the geographic pole. It is not the Northern Hemisphere's only cold pole because there are two continents in the Northern Hemisphere. The North American cold pole is at Snag, Yukon, in northwestern Canada in latitude 62.37°N. The average temperature at Snag in January is -18.5°F (-28.1°C), but in February 1947, it fell to -81°F (-63°C), which is the lowest temperature ever recorded in North America. In summer the average temperature is 57.0°F (13.9°C).

There is just one cold pole in the Southern Hemisphere. It is at Vostok Station—a Russian Antarctic research station—at latitude 78.46°S. The average winter temperature at Vostok in August, which is the coldest month, is –89.6°F (-67.6°C), but on July 21, 1983, the temperature was -128.6°F (-89.2°C). That is the lowest temperature ever recorded anywhere on the Earth's surface. In summer the temperature averages -25.7°F (-32.1°C). Vostok is colder than the South Pole. Vostok is 11,401 feet (3,475 m) above sea level. Its high elevation partly explains its low temperatures, but the main explanation is its latitude and the fact that it lies deep inside a vast continent.

The North and South Poles are not the coldest places on Earth, and the equator is not the hottest. The highest temperatures are found at the *thermal equator*. The tilt in the Earth's axis of rotation means that the belt exposed to the most intense sunshine moves in the course of the year (see the sidebar "Seasons" on page 36). The thermal equator moves between latitude 23°N and 10°–15°S. Its mean location is at 5°N.

Verkhoyansk and Vostok are colder even than Qaanaaq and Snag because they lie deep inside continents, whereas Qaanaaq and Snag are on the coast. They are extreme examples, but the difference is no less marked in lower latitudes. Compare, for example, San Francisco at latitude 37.78°N and 52 feet (16 m) above sea level with Kansas City at 39.12°N and 741 feet (226 m) above sea level. The difference in elevation should make Kansas City about 2.4°F (1.3°C) cooler than San Francisco, and, indeed, the mean annual temperature at Kansas City is 55°F (13°C), and at San Francisco it is 57°F (14°C). The range of mean temperature, however, distinguishes the continental from the maritime climate. At Kansas City the difference between the highest and lowest mean temperatures is 67°F (37°C), and at San Francisco it is only 24°F (13°C). Old World comparisons yield similar results. Ekaterinburg (formerly called Sverdlovsk) in Siberia and Oban on the west coast of Scotland both lie slightly

north of 56°N. In Oban, with a maritime climate, the mean temperature is 48°F (9°C) with a range of 28°F (16°C). Ekaterinburg has a continental climate, with a mean temperature of 32°F (0°C) and a range of 75°F (42°C).

Ocean Currents and Coastal Climates

Maritime climates are produced by air that has traveled a long distance over an ocean. This moderates its temperature but by an amount that depends on the temperature at the sea surface, and that is affected by ocean currents. Ocean currents follow circular paths, called *gyres* (see the sidebar on page 87), with boundary currents flowing away from the equator on one side of the ocean and toward the equator on the opposite side.

Air reaching western Europe crosses the warm Gulf Stream or its northwesterly extension, the North Atlantic Drift (also known as the North Atlantic Current). In winter this can raise the temperature of the air by as much as 17°F (9°C) between Iceland and northern Scotland, and it ensures that the Norwegian coast is free from ice throughout the winter.

California receives air that has been cooled by crossing the California Current, flowing south from the Arctic. As a consequence, the mean temperature in Lisbon, at 38.72°N, is 62°F (17°C). Lisbon has a climate several degrees warmer than that of San Francisco, which is in about the same latitude but a wider temperature range, of 36°F (20°C). Similarly, Atka in the Aleutian Islands at 52°N has a cooler climate than Oban, although it is a few degrees farther south. At Atka the mean temperature is 40°F (5°C), but the range of temperatures is exactly the same as that in Oban. The table sets out the mean temperatures and ranges for the places named.

Temperature				
	MEAN		RANGE	
	°F	°C	°F	°C
Continental				
Verkhoyansk	1.1	-17.1	129	72
Vostok	-67.1	-55.1	63.9	35.5
Ekaterinburg	32	0	75	42
Kansas City	55	13	67	37
Maritime				
Qaanaaq	12	-11	67	37
Atka	40	5	28	16
San Francisco	57	14	24	13
Oban	48	9	28	16
Lisbon	62	17	36	20

Gyres and Boundary Currents

On the northern side of the equator, the prevailing winds blow from the northeast, and on the southern side, they blow from the southeast. These are the trade winds, and they drive the surface water of the oceans from east to west on both sides of the equator.

These two surface currents, called the North Equatorial Current and South Equatorial Current, carry warm water across the ocean, but as they approach land—North and South America in the Atlantic and Asia in the Pacific—they are deflected, turning away from the equator. As the currents move into higher latitudes, the Coriolis effect (CorF) deflects them further. The CorF has no effect very close to the equator, but its magnitude increases with distance from the equator.

The farther the currents travel from the equator, the stronger the CorF becomes, turning the currents to the right in the Northern Hemisphere and to the left in the Southern Hemisphere. The effect continues until in middle latitudes the currents are flowing from west to east—the opposite direction to the equatorial currents. This brings them close to continents on the other side of the ocean, but they are still turning. The currents turn toward the equator, finally merging with the North and South Equatorial Currents.

The currents follow an approximately circular path in all the oceans. These paths are called gyres; the Greek *guros* means "ring."

Where the currents flow parallel to the coasts of continents, they are known as *boundary currents*. Boundary currents on the western side of each ocean are moving away from the equator and carry warm water into higher latitudes. These western boundary currents are narrow, fast-flowing, and deep. Eastern boundary currents carry cool water from high latitudes toward the equator. These currents are wide, slow-moving, and shallow. The map shows the ocean gyres and the names of the principal boundary currents.

The ocean gyres and ocean boundary currents. Boundary currents carry warm water along the eastern coasts of continents and cool water along the western coasts.

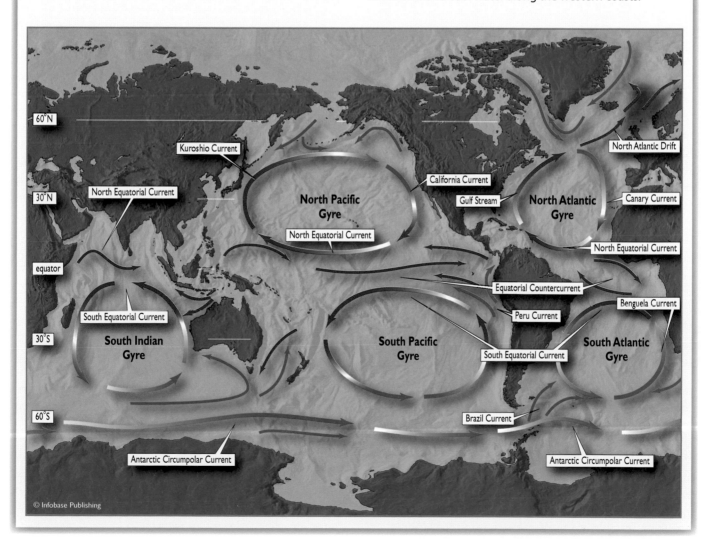

© Infobase Publishing

Average Annual Rainfall

	INCHES	MM	DISTRIBUTION
Continental			
Verkhoyansk	5.3	135	wet summer
Vostok	0.2	4.5	dry summer
Ekaterinburg	16.7	426	wet summer
Kansas City	37.3	947	dry winter
Maritime			
Qaanaaq	2.5	67	wet summer
Atka	69.3	1,761	even
San Francisco	22.1	563	dry summer
Oban	57.3	1,451	even
Lisbon	27.7	708	dry summer

Qaanaaq has a wider temperature range than other maritime regions because the sea around it is frozen. This reduces the moderating influence of the ocean. Elsewhere, it is the much higher heat capacity of the oceans compared with dry land that results in maritime climates experiencing a smaller temperature range than continental climates (see the sidebar "Specific Heat Capacity" on page 51).

Air masses that acquire their essential characteristics over the interior of a continent are drier than those that form over oceans. Obviously, this makes continental climates drier than maritime climates, but in temperate and high latitudes the difference is sometimes small or even nonexistent. In the examples mentioned here and summarized in the table, Kansas City has a wetter climate than San Francisco and Qaanaaq is drier than Verkhoyansk.

Mountain ranges running approximately north to south affect the characteristics of air masses crossing them. This effect is large in North America, where the western coastal belt has a maritime climate, but air is dried by its passage over the Rockies. To the east of the mountains, the continental climate begins abruptly and still not very far inland. In Europe, there is no comparable mountain range, and the Mediterranean extends far into the continent. Consequently, the oceanic influence extends far inland, and the change from a maritime to a continental climate is fairly gradual.

Measuring Climates— Continentality and Oceanicity

A gradual change from one type of climate to another implies degrees of *continentality* and of its opposite, *oceanicity*. There are several ways to measure continentality, usually by relating the annual temperature range to the latitude.

One widely used method is based on the equation: $K = 1.7 (A/\sin \Phi -20.4$, where K is the index of continentality, A is the annual temperature range in °C, and Φ is the angle of latitude. $K = 100$ indicates extreme continental conditions, and $K = 0$ indicates a fully maritime climate. Applying this equation makes it possible to rank places by their degree of continentality as: Verkhoyansk 112; Kansas City 79; Ekaterinburg 65; Qaanaaq 44; Vostok 41.2; Lisbon 34; San Francisco 16; Atka 14; and Oban 12.

Calculating the degree of continentality shows clearly the effect of the Rockies. Los Angeles has an index value of 40 and Phoenix of 91. In Europe, Bordeaux has a value of 37, Munich and Warsaw, both far from the nearest coast, each have a value of 44, and it is not until Moscow that the index rises to 59.

Measuring oceanicity is more complicated. The equations climate scientists use take account of the average annual precipitation and the number of days each year when the temperature is between freezing and 50°F (10°C). Consequently, the calculation calls for rather more detailed information about the climate. A value of 0 for the index of oceanicity indicates an extreme continental climate, and 100 an extreme maritime climate. The Faeroe Islands have an oceanicity index of 31, Fiji 67, and Omaha, Nebraska, 6.

Forests for Each Climate

Continental climates, because of their extremes of temperature and generally low precipitation, tend to favor coniferous forests. Ekaterinburg, on the eastern slopes of the Ural Mountains, and Verkhoyansk, much farther to the east and north, both lie in the taiga, where the predominant trees are pines (*Pinus*), spruces (*Picea*), firs (*Abies*), and larches (*Larix*), each species being dominant in certain areas.

Maritime climates, with less variation in temperature and generally higher precipitation, usually favor broad-leaved deciduous forest. Latitude affects the distribution because broad-leaved species are less tolerant than conifers of long, cold winters, which even maritime climates experience in high latitudes. Around Oban, for example, the forests are mainly of broad-leaved species, but inland, where the increasing altitude of the Scottish Highlands reduces temperatures, these give way to forests that include the only native British conifer, Scotch pine (*Pinus sylvestris*).

Both continental and maritime climates can support temperate forest, provided there is sufficient moisture. Where precipitation is very low, forests give way to steppe grassland in lower latitudes and to tundra in higher latitudes. Across Eurasia, it is the availability of water that determines these boundaries, but the transition is gradual. To the north and south of the taiga, there is a fairly broad belt in which the trees are more widely spaced, with other vegetation between them, forming forest–steppe and forest–tundra (called "taiga" by some western scientists).

■ TREES AS CLIMATIC INDICATORS

Visitors to almost any European forest will see many trees that are coated with ivy (*Hedera helix*). Ivy may also cover parts of the forest floor, and it also grows on walls. It is an Old World plant, unrelated to the American poison ivy, but it has been introduced in North America, where it is now grown widely. It is a woody climber and a broad-leaved evergreen, cloaking trees with dark foliage that is clearly visible in winter when the trees themselves are bare, and when the ivy dies, its stems often hang from the trees like the lianas that are so common in tropical forests.

Its resemblance to a tropical liana is no coincidence because that is really what ivy is. It belongs to a family (Araliaceae) comprising several hundred species of tropical plants. They are especially common in southeast Asia, but a very few species of ivy occur naturally in parts of Europe and now, as introductions, in North America. Ivy has never quite lost all its tropical preferences, however, and although in some places it is vigorous enough to be considered a weed, it remains very sensitive to temperature. It will not produce fruit and seed where the mean temperature in the coldest month of winter falls below 35°F (1.5°C), and a prolonged hard winter will kill it, nor will it thrive where summer mean temperatures fall below about 55°F (13°C).

This explains why it is so clearly visible in winter. It rarely occurs in evergreen coniferous forests because they grow naturally in regions with long, very cold winters. It thrives in broad-leaved deciduous forests. As befits so close a relative of tropical plants, ivy is also intolerant of drought. In the northern United States and southern Canada, it is not the hard winters that kill it but the hot, dry summers.

The Holly and the Ivy

Holly (*Ilex aquifolium*) is another plant that grows far from the tropical habitats favored by its many relatives in the family Aquifoliaceae. Like ivy, it is a broad-leaved evergreen, and it rarely produces seed where mean winter temperatures fall below about 31°F (-0.5°C) and summer temperatures below about 55°F (13°C). Today, European holly grows naturally from North Africa and western Turkey to the north of Scotland and along the coast of Sweden and Norway. Northern Europe marks the limit of its range. Between 1939 and 1942 there were several hard winters that killed most holly trees as far south as northern Germany. Where it thrives, it can form quite an extensive stand of trees, a "hollywood."

European holly has several North American close relatives, the one most like it in appearance being *Ilex opaca* which grows in southern Texas and around the Atlantic coast as far north as Massachusetts and inland along the valleys of the Mississippi and Missouri Rivers to Indiana.

Tropical American hollies are found in Central and South America.

Like ivy, holly is a tree associated with Christmas. The name *holly* is from the Anglo-Saxon *holegn,* the term for a holy tree that is used to decorate houses in winter. Probably the Romans used it in their Saturnalia rituals, and Christians continued the tradition. By retaining their leaves through the winter when the deciduous trees around them are bare, ivy and holly symbolize renewal and eternal life, reminding people that spring will return.

Holly cannot tolerate a continental climate. It is widespread in western Europe and in North America along coasts and major river valleys but absent from the interior of continents, where the climate is too dry and temperatures are too extreme.

Plants such as ivy and holly have fairly precise climatic requirements. This means that they can be used as indicators of climate. The presence of ivy and holly implies that the region has a mild, maritime climate. Holly can grow as far north as 62° along the Scandinavian coast but only because that coast is washed by the North Atlantic Drift extension of the warm Gulf Stream (see the sidebar "Gyres and Boundary Currents" on page 87). Places where mistletoe grows have warm summers.

Yew (*Taxus baccata*) also demands a maritime climate and cannot tolerate hard winters. Like holly, it grows as far north as 63° along the Norwegian coast, where the ocean current produces a moderate climate, but it is more widespread farther south and occurs only in the western, maritime, region of Europe.

Mistletoe

Mistletoe, a *parasite* of forest trees that is also associated with Christmas, is another plant limited by climate. Unlike holly and ivy, however, it can tolerate winter temperatures as low as 19°F (-7°C), but it must have warm summers, with mean temperatures no lower than about 63°F (17°C). The traditional mistletoe (*Viscum album*) occurs throughout Europe from Britain to northern Asia with about 100 other *Viscum* species that are found in southern Asia. Mistletoe is unusual in growing on both coniferous and broad-leaved trees and shrubs, although traditional mistletoe is most commonly found growing on apple, poplar, willow, lime (linden), and hawthorn trees.

North American mistletoes belong to the genera *Phoradendron* with 190 species and *Arceuthobium* with 31 species (and one European species most often found growing on juniper).

Lime Trees

Lime trees (*Tilia* species) need warmth, and small-leafed lime (*Tilia cordata*) has been planted along the streets of many North American cities as far north as Montreal to provide shade in summer. American lime or basswood (*T. americana*) grows as far north as southern Canada.

Between about 7,500 and 5,000 years ago, during a time that scientists call the Atlantic period, western Europe enjoyed a warmer, moister climate than that of today. Most of lowland Britain was blanketed in forest, and lime, especially small-leaved lime, was one of the most widespread forest trees. Lime trees still grow in Britain, but it is warm enough for them produce seed only in the south. Of all the trees that are native to Britain, the lime is the most warmth-loving.

As trees spread throughout Britain, however, the lime tree never reached Ireland or Scotland. It failed to reach Ireland because the lime arrived later than most other tree species. By the time it might have entered Ireland, the rising sea level had formed the Irish Sea, separating Britain and Ireland. The lime failed to reach Scotland because, despite the warm Atlantic climate, Scotland was too cold for it. As the climate grew cooler during the Sub-Boreal period that followed the Atlantic period, lime trees became less common, although their decline was also due to the activities of early farmers.

Dwarf Birch and Dwarf Juniper

Other trees are typical of cold climates. Dwarf birch (*Betula nana*), juniper (*Juniperus communis*), and dwarf willow (*Salix herbacea*), grow where the climate is too cold for other trees. They form part of the tundra vegetation of the Far North where other trees cannot survive. Tundra has not always been confined to the far north, however. During the ice ages (see "Ice Ages" on pages 91–92) tundra occurred to the south of the Great Lakes and across southern Britain.

Dwarf willow is rarely more than about two inches (5 cm) tall, dwarf birch forms a bush up to about three feet (1 m) tall, and although juniper often grows into a tall shrub, it can also creep over the ground. Keeping close to the ground shelters these plants from the drying effect of the wind.

Should the climate become warmer so that forests colonize the land, bigger trees would shade the dwarf birch and juniper, which disappear. Typical tundra plants, such as dwarf birch, juniper, and willow, grow where the weather is cold, and their presence indicates cold conditions.

Reading the Pollen Record

This seems to contribute little to our understanding of either botany or climatology. After all, scientists hardly need the presence of particular plants to tell them whether a climate is maritime or continental. They can measure it directly. Plants produce pollen, however, and pollen grains can remain in the soil long after the plants that produced them have vanished (see "Using Pollen to Study the Past" on pages 228–231). The

presence of pollen grains from ivy or holly in a region that now has a strongly continental climate shows that at some time in the past its climate was maritime. Knowing that, scientists can start looking for the pollen of other plants associated with maritime climates, and they can make an educated guess about the type of forest that once occupied the site.

Alder (*Alnus* species) and willow (*Salix* species) grow on wet ground and along riverbanks. Over the centuries since people began cultivating what had once been forest, much of what was wetland has been drained and now grows crops. The pollen survives, however, and alder produces unusually large amounts of it. Where alder and willow pollen are found, the land was once waterlogged and, therefore, the climate was moist.

Alder and willow can also indicate a change in climate. If the rainfall increased after a forest became established, water would drain toward the lower ground. In time this would produce areas where the ground was waterlogged. The dominant trees would be killed by the high water table, and their place would be taken by alder and willow. The forest would come to include glades of alder and willow growing on the wet ground, and the presence of their pollen in an area dominated by the pollen of other tree species would be evidence of a climate change of this kind. To this day there are mixed forests with glades of alder and willow in parts of Europe and the eastern United States.

Forests are their trees, of course, but herbs growing on the forest floor also produce pollen, and the herbs are characteristic of the types of forest in which they occur. Their pollen can also be used to reconstruct past climates and the vegetation associated with them.

Many flowering herbs that are now cultivated grew originally on the forest floor. They flower in spring, during the brief period of warming before the trees have had time to open their leaves and shade them. This is an event that takes place only in deciduous forests, so the presence of these herbs where no one has planted them implies that the area was once broad-leaved, deciduous forest. Where a road is bordered on both sides by cultivated fields, the profusion of spring flowers along its verges tells you that a forest once stood in that place.

The Dryas

Pollen gathered from soils that can be dated are used to reconstruct the changing climates of the past, and perhaps the best-known pollen comes from a plant that indicated a sharp, rapid, and long-lasting deterioration in climate.

As the last ice age was ending and the glaciers were in retreat, trees typical of cold climates, such as birch (*Betula* species), began to appear. A layer of soil from that time contains their pollen. Above that layer, however, there is a later layer containing pollen from mountain avens (*Dryas octopetala*), a flowering herb of mountainsides and the far north. The trees had died and the ground was carpeted with mountain avens, indicating that the climate had suddenly grown much colder. *Dryas* pollen is found all over the Northern Hemisphere, showing that the change was widespread.

Because it was the discovery of its pollen that led to the realization of what had happened, the period is called the Dryas. If fact, it happened twice, so there is an Older Dryas, which lasted from about 12,200 to 11,800 years ago, and a Younger Dryas, from about 11,000 to 10,000 years ago. During each Dryas, the glacial retreat halted and in places was reversed, and temperatures fell almost to their ice-age levels.

Tree Rings and Past Climates

Each year, trees lay down a layer of new cells around the trunk and branches. These annual layers are clearly visible in the wood as annual growth rings, or tree rings (see "How Wood Forms" on pages 113–116).

Counting the tree rings reveals the age of the tree, and this technique is used to determine the dates of past events. The science of measuring dates in this way is called *dendrochronology* (the Greek word *dendron* means "tree"). Although all trees produce growth rings, not all species have rings that are suitable for dendrochronological use. Oak (*Quercus* species) and pine (*Pinus* species) are the best trees for this purpose.

Tree rings also provide clues to past climates. *Dendroclimatology* is the science concerned with interpreting tree rings in this way. In years when the weather favors tree growth, the trees grow more vigorously and produce thicker growth rings than they do in poor years. The relative thickness of the rings tells dendroclimatologists whether growing conditions were good or bad during particular periods.

Scientists are very careful about interpreting tree ring data. They always compare the rings from several trees and date them by checking against a standard tree-ring index. They need to decide whether poor weather caused a period of reduced tree growth or it was due to some other cause, such as competition from other trees for sunlight and nutrients. Tree rings can also provide information about the ecological conditions—this science is called *dendroecology*.

■ HOLOCENE RECOLONIZATION

People who lived 18,000 years ago knew a world that was very different from the one we see today. Ice sheets covered almost half of Europe, northern Asia, and North America as far south as Seattle in the west and New York in the east, with a broad extension as far south as Cincinnati and Kansas City. The thickness of the ice varied from place to place, but on average

Annual growth rings are clearly visible when a tree is felled, as has happened here, but they can also be sampled by drilling a core through the bark to the center of the trunk. The thickness of the rings indicates growing conditions in that year, and the number of rings denotes the age of the tree (when allowance is made for missing years). *(Missouri Natural Resources Conservation Service)*

it lay 5,000 to 10,000 feet (1.5–3 km) deep on all the northern continents, completely burying entire ranges of hills.

Over North America, the Cordilleran ice sheet covered the northwest, and most of the northern part of the continent lay beneath the Laurentide ice sheet. Farther east, the Greenland ice sheet covered Greenland. The Fennoscandian ice sheet lay over Europe and Asia, with a smaller Siberian ice sheet to its east. Over most of Siberia, the climate was very dry, and the amount of snowfall was too small for large ice sheets to form, although there were several small ones. The map on page 92 shows the approximate area of land and sea covered by ice when the most recent ice age was at its maximum. The northern North Atlantic and Arctic Oceans and the North Sea were covered in thick ice all year round, of course, so it was possible to walk between Asia and North America and between Britain and the European mainland. In the Southern Hemisphere there was an ice sheet on each side of the Andes.

Ice Ages

During the last 2 million years—during the Pliocene, Pleistocene, and Holocene epochs—the ice sheets have advanced three or four times and then retreated again for interludes, called *interglacials,* lasting between 10,000 and 20,000 years (see the sidebar on page 93). At present we are living in an interglacial, called the Holocene (in Britain the Holocene is often called the Flandrian).

Ice ages, or *glacials,* have different names in different places. The most recent glacial is known in North America

Ice sheets:
1. Cordilleran
2. Laurentide
3. Greenland
4. Fennoscandian

© Infobase Publishing

The Northern Hemisphere ice sheets at their greatest extent, about 20,000 years ago

as the Wisconsinian, in Britain as the Devensian, in northern mainland Europe as the Weichselian, and in the Alps as the Wü. These names all refer to the same glacial, although there are local variations in the dates of its onset and end. It was at its maximum about 18,000 years ago when average temperatures over land were about 36°F (20°C) lower than they are

today. Because the ice sheets contained so much water, sea levels were as much as 525 feet (160 m) lower than they are now. In latitudes close enough to the equator to be free from ice, there were large areas of dry land that were inundated when the ice sheets melted and sea levels rose again.

Glaciers and Ice Sheets

Ice sheets flow. As more and more snow falls over the center, its weight compacts the snow beneath until it forms ice and then squeezes it outward so that the base of the sheet moves,

Pliocene, Pleistocene, and Holocene Glacials and Interglacials

APPROXIMATE DATE ('000 years BP)	N. AMERICA	GREAT BRITAIN	N.W. EUROPE
Holocene			
10–present	Holocene	Holocene (Flandrian)	Holocene (Flandrian)
Pleistocene			
75–10	Wisconsinian	Devensian Weichselian	
120–75	Sangamonian	Ipswichian	Eeemian
170–120	Illinoian	Wolstonian	Saalian
230–170	Yarmouthian	Hoxnian	Holsteinian
480–230	Kansan	Anglian	Elsterian
600–480	Aftonian Cromerian	Cromerian complex	
800–600	Nebraskan	Beestonian	Bavel complex
740–800		Pastonian	
900–800		Pre-Pastonian Menapian	
1,000–900		Bramertonian	Waalian
1,800–1,000		Baventian	Eburonian
Pliocene			
1,800		Antian	Tiglian
1,900		Thurnian	
2,000		Ludhamian	
2,300		Pre-Ludhamian	Pretiglian

BP means "before present" (present is taken to be 1950). Names in italic refer to interglacials. Other names refer to glacials (ice ages). Dates become increasingly uncertain for the older glacials and interglacials, and prior to about 2 million years ago evidence for these episodes has not been found in North America; in the case of the Thurnian glacial and Ludhamian interglacial, the only evidence is from a borehole at Ludham, in eastern England.

very slowly, downhill. There is no real difference between an ice sheet and a *glacier,* and scientists use the words interchangeably. The "river of ice" many people think of as a glacier is known technically as a valley glacier because it flows along a defined valley, confined by valley walls. A glacier flows into the sea or into a region where temperatures are above freezing and the ice melts, releasing meltwater to form rivers.

When a glaciation begins, glaciers begin to spread into lower latitudes. As they do so they scour away the soil beneath them. They may mix the soil with fragments of shattered rock or push it ahead of them. Glaciers carried soils from Canada south into the midwestern United States, where today farmers grow wheat and corn in them.

Obviously, no plants can survive beneath the ice sheet, and for some distance beyond its edges the ground is permanently frozen below the surface. This is permafrost. Its surface thaws briefly in summer—the layer that thaws is known as the active layer—and supports tundra plants, but large trees cannot grow in it. Tundra vegetation covered most of

Europe south of the ice sheets. South of the Laurentide ice sheet, there was a belt of tundra 60 to 125 miles (100–200 km) wide, but pine and spruce forests grew over much of the eastern United States as far south as Florida. Scientists believe that Japan remained forested throughout the glaciation. Temperate forests do not disappear. The expansion of the arctic region pushes the vegetational belts in lower latitudes closer to the equator. The temperate forests still exist during glaciations but to the south of their present location.

For many years scientists wondered what causes ice ages to begin and end. Some suspected that the cause is astronomical and is linked to variations in the amount of sunlight that falls on the Earth. The English astronomer Sir John Herschel (1792–1871) was the first to suggest a link of this kind, and in 1864 the Scottish climatologist James Croll (1821–90) proposed that reduced solar radiation triggers ice ages. These ideas were vague, however, and the Serbian mathematician and climatologist Milutin Milankovitch (1879–1958) decided to test it by studying records of astronomical

Milankovitch Cycles

In 1920 Milutin Milankovitch (1879–1958), a Serbian astrophysicist, proposed an astronomical explanation for why ice ages begin and end when they do. Milankovitch had examined regular changes that occur in the Earth's orbit about the Sun and in Earth's rotation on its own axis. He found there were three changes, or cycles, that periodically alter the amount of sunshine falling on the surface. When the three cycles coincide, an ice age either begins or ends. Most climate scientists now accept the Milankovitch theory. The three cycles affect orbital stretch, axial tilt, and axial wobble.

Orbital Stretch

Earth follows an elliptical path around the Sun, with the Sun occupying one focus of the ellipse. Over a period of 100,000 years, the orbital eccentricity changes. This means that the ellipse grows longer and then shorter again. When the orbit is at its longest, Earth is approximately 30 percent farther from the Sun at both its closest approach (perihelion) and its farthest point (aphelion) than it is when the ellipse is short—in fact, almost circular. At these times, the sunlight falling on the Earth is less intense.

Axial Tilt

Instead of being at right angles to the Sun's rays, Earth's axis is tilted, present by about 23.45°. Over a period of about 42,000 years, the angle of tilt varies from 22.1° to 24.5° and back again. The greater the angle of tilt, called the *obliquity,* the more intense the sunshine falling on high latitudes in summer, and the less these latitudes receive in winter.

Axial Wobble

The Earth's axis wobbles like a toy spinning top so that without changing the angle of tilt, the axis describes a circle, taking 25,800 years to complete one turn. This alters the dates of midsummer day, midwinter day, and the equinoxes when there are precisely 12 hours of day and 12 hours of night everywhere. At present Earth is at perihelion (closest to the Sun) on about July 4 and at aphelion on about January 3. In about 10,000 years from now, these dates will be reversed due to the axial wobble. The Northern Hemisphere will then receive more solar radiation in summer and less in winter, and the Southern Hemisphere will receive less radiation in summer and more in winter.

When the orbit is at its most eccentric, the axial tilt is at a minimum, and Earth is at aphelion in June or December, summer temperatures in one or other hemisphere may be low enough to trigger the onset of an ice age.

---

changes and climate cycles over hundreds of thousands of years. In 1920 he published his conclusion that three cycles affecting the Earth's orbit and rotation were responsible (see the sidebar). Today most climate scientists accept that the Milankovitch theory is correct, at least in general terms.

The Ice Retreats and then Returns

Starting about 14,000 years ago, the climate gradually grew warmer, and as the temperature rose above freezing around the edges of the glaciers, the ice commenced its long retreat. The warming was not a steady process. There were times when it halted or reversed, and there were the two major Dryas reversals (see "The Dryas" on pages 90–91), but by about 8,000 to 10,000 years ago, the edge of the permanent ice lay close to its present position.

Meltwater moistened the ground beyond the edges of the retreating glaciers, and the first plants appeared. At first, these were sedges and grasses. They helped soil to form (see "Soil Formation and Development" on pages 15–22), and when there was a sufficient depth of soil, the first trees arrived.

All the plants colonizing these bare landscapes arrived as seeds carried from the south. In North America and eastern Asia, there was no major geographical barrier between the newly exposed lands and the established grasslands and forests. In Europe, however, the migration of plants was restricted by the Alps and, to their south, by the Sahara. Redwoods (*Sequoia* species) and tulip trees (*Liriodendron* species) grew in Europe before the glaciation and then became extinct as the expansion of European glaciers trapped them, but they survived in the south of Asia and in North America. Following the glacial retreat, these and many other species were able to recolonize North America and Asia but not Europe. That is why to this day Asian and American temperate forests contain far more tree species than do European forests.

Glacier Bay

The sequence in which plants colonized the lands exposed by the retreating ice is known mainly from pollen, but there is one place where it has been observed directly. In southeastern Alaska, the edge of the glacier flowing into Glacier Bay has retreated by about 330 feet (100 m) during the last two centuries. John Muir (1838–1914), the American conservationist and glaciologist, visited Glacier Bay several times, first in 1879, and realized that the bay had not existed a century earlier when the charts he used were made.

Recolonization in Glacier Bay can be dated either by counting the rings on older trees or from recorded observations. The first arrivals were mosses and herbs, including mountain avens (*Dryas octopetala*), a plant that enriches the soil with nitrogen by means of bacteria around its roots and so encourages other colonizers. Then came willows (*Salix* species), followed by American green alder (*Alnus crispa*) and cottonwood (*Populus* species). Sitka spruce (*Picea sitchensis*) then entered, forming dense forest. Finally, the spruce forest developed into forest dominated by western hemlock (*Tsuga heterophylla*) and mountain hemlock (*T. mertensiana*).

This sequence seems to have been broadly typical of the way much of North America was recolonized. At first, scattered mosses and herbs established themselves on the bare ground. As sedges, grasses, and a few shrubs arrived, a tundra vegetation developed, interspersed with patches of grassland. By about 15,000 years ago, spruces arrived. These were scattered at first, but the gaps soon filled to produce forest dominated by spruce and tamarack (*Larix laricina*). This is a larch, and the forest must have looked much like the present-day taiga of Siberia. In the warmer regions to the south, oak (*Quercus* species) and balsam fir (*Abies balsamea*), the traditional American Christmas tree, replaced the spruce–tamarack forest. Then pines (*Pinus* species) appeared, including jack pine (*P. banksiana*). Pines require warmer conditions than most conifers, and maples (*Acer* species) and hickory (*Carya* species) joined them. The forest was changing from a boreal to a mixed conifer and broad-leaved deciduous type, and by about 8,000 years ago it was dominated by hemlocks (*Tsuga* species) and beech (*Fagus* species).

Return of the European Forests

In Europe, recolonizing plants began moving northward about 13,000 years ago. After the sedges and grasses, juniper (*Juniperus communis*) arrived, together with arctic willow (*Salix herbacea*) and dwarf birch (*Betula nana*), plants that grow close to the edge of the tundra (see "Trees as Climatic Indicators" on pages 81–89). By about 12,000 years ago these had been joined by downy birch (*B. pubescens*), silver birch (*B. pendula*), and aspen (*Populus tremula*), all of which need warmer conditions. At this stage the trees grew in the north as copses separated by tundra vegetation, and the landscape was quite open, like parkland. In the south, trees grew closer together as forest dominated by downy birch.

The northward migration was then interrupted by the Younger Dryas (or Loch Lomond) stadial. A *stadial* is a cold

period that is shorter and milder than a full glaciation (and a mild interlude shorter and less warm than an interglacial is an *interstadial*). During the Dryas, the West Highlands of Scotland were buried once more by an ice sheet hundreds of feet thick, and the birch copses survived only in isolated, sheltered places.

About 10,000 years ago, as temperatures started rising again, colonization resumed, and this time it was fairly rapid. Hazel (*Corylus avellana*) was abundant by 9,000 years ago, and over most of northern Europe it was soon followed by birch and Scots pine (*Pinus sylvestris*), producing birch–pine–hazel forest. This forest remained in northern England and Scotland, but farther south it gave way to forests of elm (*Ulmus* species) and oaks.

Then the climate entered an even warmer phase, and by 7,000 years ago, lime, or linden, trees (*Tilia* species) were established and spreading. Limes require a warm climate, and eventually they became widespread, with alder (*Alnus glutinosa*) growing on wetter ground.

It was during this period of warmth that the forests of Europe and North America developed the composition they had when humans first began to live in and exploit them.

■ RELICTS AND REFUGIA

Major climate changes never happen smoothly, and the Younger Dryas stadial was only the most dramatic of the fluctuations associated with the end of the last ice age. Throughout the Northern Hemisphere, during what is known as the Sub-Arctic period, tundra vegetation colonized land exposed by the retreat of glaciers. This gave way to cool, moist conditions in the Pre-Boreal period, when spruce (*Picea* species) and fir (*Abies* species) advanced in eastern North America and pines (*Pinus* species) and birch (*Betula* species) appeared in Europe. The Pre-Boreal was followed about 9,000 years ago by the Boreal period, when climates became warmer, drier, and more continental (see "Continental and Maritime Climates" on pages 85–89), favoring pines and oak (*Quercus* species) in North America and pines and hazel (*Corylus avellana*) in Europe. The Boreal period, marking the beginning of a long, warm interlude called the Hypsithermal, was followed about 7,000 years ago by moister, more maritime conditions in the Atlantic period, with oak and beech (*Fagus* species). The Sub-Atlantic period, starting about 3,000 years ago and marking the end of the Hypsithermal, was still moist, but cooler.

Rates of Colonization

Such changes affect vegetation patterns but at different rates, and there is always a delay before plants and animals respond to change. When the climate warmed vigorously at the start of the Boreal period, birch (*Betula* species) and aspen (*Populus* species) advanced into new ground at a rate of about half a mile (1 km) a year. Oak (*Quercus* species) and elm (*Ulmus* species) followed in the Atlantic period but rather more slowly, and beech (*Fagus* species) and spruce (*Picea* species) moved northward at less than 1,650 feet (500 m) a year. Cooling has a much more immediate effect because established trees fail to survive and may disappear from an area within a very few years.

Often, though, the changes are less than total, leaving places where plant communities and the animals associated with them survive from earlier times. Climatic warming allows new species to invade, but in mountainous regions, for example, the older species may migrate up the mountainsides and survive at a higher altitude. There are "sky islands" of this type high in the mountains of Arizona and New Mexico and at several places in the southern United States and in Nova Scotia, Canada.

In the southeastern United States, there are populations of eastern four-toed salamanders that are isolated from the main range of this population, which lies to the north. The map shows the approximate location of these groups. Probably the salamanders lived in the boreal forest that covered the south during the Wisconsinian glaciation and remained when warmer conditions caused the forest itself to migrate farther north.

More surprisingly, perhaps, there are also places where plants have survived a deterioration in climate. Even an advance of the ice sheets leaves some areas ice-free, and scattered plant and animal communities may survive in sheltered spots, close to the sea perhaps, where the proximity of water moderates the climate. As the surrounding land dries

Forest and animal refugia in North America

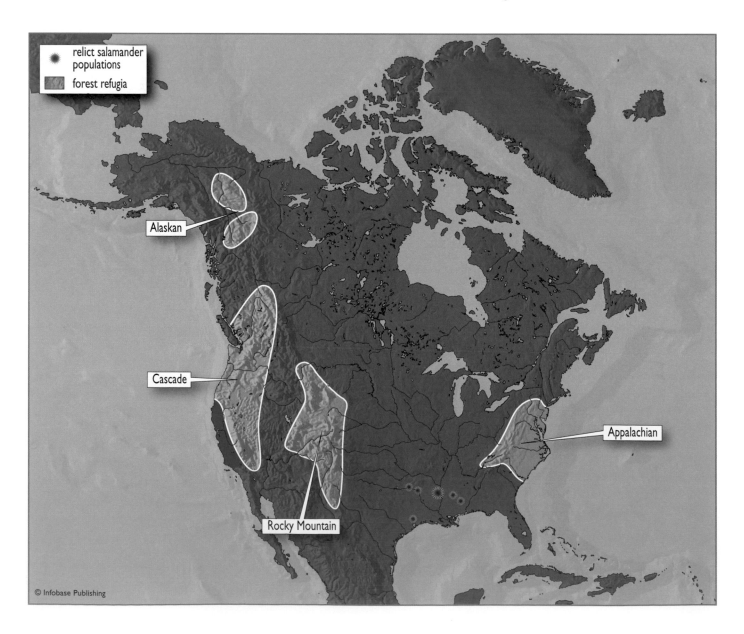

© Infobase Publishing

into desert, for example, isolated areas may remain where the ground is moist, and these may retain their vegetation pattern. A local area where the type of vegetation is typical of an earlier climatic period and is markedly different from the pattern surrounding it is called a *refugium* (plural *refugia*), and a plant species that survives from an earlier time is called a *relict,* or relic.

Living Fossils

It can even happen that one species survives the extinction of all other members of its genus. The maidenhair tree (*Gingko biloba*) survived in the gardens of Chinese monasteries and is the only remaining representative of a very ancient group of trees. The dawn redwood (*Metasequoia glyptostroboides*) is another relict species, the only member of its genus and, in the view of some botanists, sufficiently different from all other redwoods (for one thing it is deciduous) to warrant classification in a family of its own. It was first identified as a fossil in 1941 and believed to have been extinct for the last 26 million years. Then, in 1945, it was found growing in southern China. It is an attractive tree, now grown widely in parks and gardens, but it is a relict even though it is no longer rare.

The maidenhair tree and dawn redwood are *living fossils.* These are species that are almost identical to plants or animals that lived in the distant past and are known only from fossils. They are not completely identical to their fossil relatives; they have evolved extremely slowly, but their evolution has not halted.

Lusitanian Flora

The most southerly point in mainland Britain is the Lizard Peninsula, at the far tip of Cornwall. The Lizard is not forested, but its vegetation is unusual, containing many plants, such as Cornish heath (*Erica vagans*), that are rare elsewhere, even in other parts of Cornwall. This type of vegetation, called Lusitanian, also occurs in southwestern Ireland, where it includes the strawberry tree (*Arbutus unedo*), a plant otherwise found in the much warmer coastal regions of Portugal. Lusitania was the province of the Roman Empire that encompassed what are now Portugal and part of western Spain.

The Lizard and southwestern Ireland are refugia for a plant community that is believed to have survived the most recent glaciation in another refugium somewhere in southwestern Europe. That refugium has since vanished. The plants then migrated northward in the very earliest stages of the glacial retreat along European coasts that were later submerged as the sea level rose. The Lusitanian vegetation may once have been fairly widespread, but it survived subsequent cool episodes only in places that have an especially mild climate.

The most famous British refugium occupies about 80 square miles (207 km²) in Upper Teesdale, a wild, remote part of the north of England. Far from being forested, Upper Teesdale is remarkable because although some of the advancing trees probably reached it, they disappeared again before they could displace the vegetation that was already established. As a result, this upper part of the valley of the river Tees, at an elevation rising to about 1,500 feet (460 m), bears a cover of plants that scientists believe to be typical of those that grew over a much larger area just as the last of the glaciers were retreating. There are many grasses and flowering herbs, some of them rare, and dwarf birch (*Betula nana*) grows in Upper Teesdale. This tree also occurs on Lüneberg Heath, Germany, where it is also a relict species.

American Forest Refugia

There are four refugia in North America: in Alaska, the Appalachians, the Cascades, and the Rockies (see the map on page 96). These are places where coniferous forest survived throughout the Wisconsinian glaciation, some of them covering a much larger area than they do now.

Surrounded by tundra and steppe grassland, the Alaskan refugium occupied two separate ice-free areas and possibly included some offshore islands. The Pacific refugium, on the western slopes of the Cascade Range in Washington and Oregon, then extended south into California. The Rocky Mountain refugium covered a large area in Montana and Idaho, and the Appalachian refugium, centered on the middle of the Appalachian range, extended westward, south of the Great Lakes. During the glaciation, coniferous forest grew at a much lower altitude on mountainsides than it does now.

The Significance of Refugia

Upper Teesdale, the four North American areas of boreal forest, and to a lesser degree the Lusitanian flora of the Lizard and southwestern Ireland are major refugia and are easily recognized, but changing climates also leave more numerous, if subtler, clues. Plant seeds are dispersed in various ways, and if they fall on ground that suits them, they will germinate and grow. As a glaciation draws to an end and the climate warms, conditions at a high elevation will remain glacial for much longer.

Temperature decreases with height, but the decrease is counted from its sea-level value, so if the temperature rises at sea level, it will also rise high in the mountains. During a post glacial warming, this may change a mountain climate from one that is too cold to support plants at all to one capable of supporting plants of the type which grew at a lower level during the ice age. Even if the overall climate becomes temperate, it may remain glacial in exposed places at a higher level and be occupied by vegetation that was once more widespread. These are refugia. Small, scattered, and not easily recognized, such late-glacial survivals are found in many upland and mountain areas of Europe and North America.

Later, as the forests rapidly migrated northward, habitats unsuitable for the colonizing trees, perhaps because their soils were too acid, alkaline, or wet, would be surrounded and isolated. These, too, would become refugia.

Rising sea level and the inundation of low-lying ground forms offshore islands. If colonizing species are unable to cross the stretch of open water to reach an island, colonization of the island may be halted, and it, too, may become a refugium. Ireland is not a refugium, but if the sea level were to fall by 300 feet (90 m) it would be joined by land bridges to Wales and Scotland. Since the end of the last ice age, the sea has risen by much more than this, separating mainland Britain from Ireland, and Ireland has only about 70 percent of the plant species that are found in Britain. Britain is much larger than Ireland and has a more varied and somewhat less strongly maritime climate, which accounts for some of the disparity, but it is also partly due to the separation of the two islands before postglacial colonization was complete. Yet Britain supports far fewer species than are found in similar latitudes in a comparable area of Europe.

Relict species and refugia are of considerable scientific interest and, therefore, they are allotted a high priority when sites are selected for conservation. Much can be learned from relicts about the evolutionary history of plants, and refugia provide information about the individual components of types of vegetation that were once widespread. This may be especially important if, as many climatologists suppose, the climate is now changing (see "Forests and the Greenhouse Effect" on pages 268–272).

4

Biology of Forest Trees

Except when the wind rustles their leaves or a passing animal pushes them aside, the plants of the forest appear still. In fact they are moving—but too slowly for the human eye to detect. Their stillness may give the impression that plants are simple or at least that they are simpler than animals. The impression is misleading. Plants are highly complex, and this chapter begins by explaining what they are and how they have evolved. It describes how the first trees appeared and what the first forests were like.

The remainder of the chapter describes the way plants in general and trees in particular work. It explains the differences between coniferous and flowering plants. It tells how plants obtain nourishment, how they reproduce, and how they disperse their seeds. Green plants manufacture sugars from carbon dioxide and obtain water by the process of photosynthesis. The chapter contains a simple description of photosynthesis. Nitrogen is a key nutrient, which plants obtain with the help of bacteria. Trees produce wood, and the chapter describes how wood forms.

■ EVOLUTION OF TREES

Trees look very different from all other plants. Although it may be difficult to name the many species of trees, shrubs, herbs, and grasses seen on a visit to the countryside, it is easy to recognize a tree or, for that matter, a shrub, herb, or grass. These are easily distinguishable from one another, and a tree is the most recognizable of them all. A tree towers above the plants around it, and unlike herbs and grasses, its stem and branches are hard to the touch. It is tempting to suppose that trees are profoundly different from all other types of plant and that they represent some sort of pinnacle of plant development, an ultimate form to which the entire history of plants has led.

The truth is less dramatic. It is only its size that differentiates a tree from other plants, and the trees we see around us today are not the only plants to have grown tall. A tree is defined as a plant that is capable of growing to a height of 33

feet (10 m) or more, often with the lowest branch four feet (1.2 m) above the ground. It also typically has only one main stem, although there are exceptions to this, and it does not die at the end of each growing season, although it may shed its leaves. This definition is based on the appearance of the plant, especially on its size, and it says nothing directly about the structure of the plant or how it works. Any plant can be called a tree if it conforms to this description, regardless of its relationship to other plants that are also called trees. In fact, though, broad-leaved trees are more closely related to the flowering herbs growing around their bases than they are to coniferous trees.

The First Plants

Plants are distinguished from animals and fungi by the way they obtain food. Plants (and some bacteria) manufacture sugars from carbon dioxide and obtain water by the process of photosynthesis (see "Photosynthesis" on pages 116–122). Animals obtain their food by eating plants or other animals. Fungi absorb organic compounds (compounds containing carbon).

Life first emerged in water, and the earliest plantlike organisms were aquatic, single-celled, green algae. These are still widespread: They are what make stagnant water and some seawater green. In some algal species, cells join together into filaments. These inhabit shallow water, attached to a solid surface. Other algae grew much bigger into seaweeds. Then, about 425 million years ago, plants, descended from green algae, began to grow on land. Probably these were filamentous algae growing around coasts and lakeshores where tides or floods frequently immersed them.

Some descendants of these algae then acquired a cuticle—an outer coating like a skin—that covered them entirely and prevented them from drying out. They also developed protective coats for their reproductive cells. These adaptations helped plants to survive out of water. Plants of this type have flourished ever since. They are the mosses, liverworts, and hornworts, of which there are about 17,000 species living today.

Vascular Plants

In other plants certain cells joined together to form *vascular tissue*—bundles of tubes through which water and nutrients could pass. Some of this vascular tissue is visible as the veins of leaves. Plants could then grow much bigger and more complex because the plant itself could transport nutrients to tissues needing them, which meant each individual cell was no longer required to provide for itself. Not all the cells of a plant had to engage in photosynthesis, so some cells could live in permanent darkness, belowground or inside the plant.

Specialization then became necessary. Carbon dioxide and mineral nutrients are dispersed throughout water, so every part of an aquatic plant is bathed in them while also being exposed to sunlight. On land, however, water and mineral nutrients are located belowground. One part of the plant has to be responsible for obtaining them and also for anchoring the plant, and a different part of the plant has to perform photosynthesis. So roots came to be differentiated from the parts of the plant that grow aboveground.

Wood For Strength

Gravity then had to be overcome. Anyone who lives by the sea will be familiar with the untidy heaps of flattened seaweed that lie on sheltered beaches. As the tide rises, the seaweed floats up to stand erect, like a submarine forest. This is possible because the plant is buoyant, but buoyancy is no help out of water. Land plants had to stand erect by some other means in order to maximize their exposure to light.

For small plants, the rigidity of their water-filled cells, called *turgor* pressure, was sufficient, but it would not support the weight of a large plant. Greater rigidity was attained when some plants began to produce lignin inside the walls of their cells. *Lignin* is a hard material that remains after the death of the cell that produces it. Dead cells collapse, but the lignin remains. The process is called *lignification,* and the resulting material is wood.

With wood to provide rigidity and vascular tissue to transport water and nutrients driven by transpiration (see "Transpiration" on pages 48–50), plants could grow tall, and some began to do so at once. Growing taller raised them higher than the plants around them, so they were no longer shaded. This increased their exposure to light, and they developed leaves that presented large surfaces, thus maximizing their photosynthetic activity.

The First Forests

Since a tree is defined principally by its size, it was then that the first trees appeared and, with them, the first forests. These grew during the Carboniferous period, which began about 360 million years ago. Once established, the forests covered vast areas, and for more than 200 million years they remained the dominant type of vegetation.

These were not at all like modern forests, however, because the early trees were very different from the trees that grow today. They included *Calamites,* which grew with its roots in water and reached a height of about 50 feet (15 m). Its only modern relatives are about 15 species of horsetails (*Equisetum*), small plants of wet places. There was *Lepidodendron,* an even more impressive tree, with a very straight trunk up to 6.5 feet (2 m) in diameter, which grew more than 130 feet (40 m) tall. Its surviving relatives are the clubmosses, such as *Lycopodium* and *Selaginella.* Despite their name, these are not true mosses. The forests also contained tree ferns. Their descendants still exist and can often be seen growing in botanic gardens and nowadays in many private gardens. The commonest species is *Dicksonia antarctica.*

The forests comprised many species, of which these are only examples, and about 300 million years ago they grew close to the equator on flat, low-lying, swampy ground. When they and the plants around them died, their remains decomposed only partially in the airless swamp mud. Later, compression and heating turned them into the vast coal measures that supply a great deal of our fuel.

Most of these swamp species of the Carboniferous period became extinct about 280 million years ago when the climate became drier and the swamps turned into dry land. Among the forest species, though, there were less impressive plants that could tolerate drier conditions, and with the disappearance of the swamp giants their moment had arrived.

Gymnosperms

These plants were different because they produced seeds. During its life cycle, every plant passes through two distinct forms, each of which gives rise to the other. The process is called the *alternation of generations* (see "Alternation of Generations" on page 106), and the two forms are called *gametophytes* and *sporophytes.* In early plants, such as mosses, the gametophyte generation was the bigger of the two, but in later plants, including all present-day clubmosses, ferns, trees, and herbs, the visible plant is the sporophyte generation.

The new plants had gametophytes that were even smaller than those of other plants. Instead of being jettisoned to develop in the soil independently, the sporophyte retained the gametophytes within its own, moist tissues, where the embryo of what would grow into a new plant was surrounded with a store of food and enclosed in a covering. This was a *seed,* but at first it was not held in a chamber specialized for the purpose. The seed was naked. The Greek word for "naked" is *gymnos* and for "seed" *sperma,* so these plants are known as gymnosperms.

As the old forest trees disappeared, the gymnosperms took their place. Gymnosperms were better adapted for life on land because their reproduction was safer and more efficient (see "Tree Reproduction" on pages 105–111). As they continued to evolve, the gymnosperms diverged to form four groups. All of these survive to the present day (see "Why Seeds Are Better" on page 102), although only one has really prospered. The group that has flourished and is seen everywhere is that of the conifers. *Konos* is the Greek word for "cone," *phero* means "carry," so the conifers are "cone-bearers."

Flowering Plants

Despite the success of the gymnosperms, most of the plants we see around us today do not belong to this group. Palms, broad-leaved trees and shrubs, flowering herbs, and grasses are all angiosperms. The name refers to a container (Greek *angion*). The container is the ovary in which the seeds (*sperma*) are borne. The ovary develops from a flower, and so angiosperms are also known as the flowering plants.

Flowering plants first appeared 135 million–130 million years ago, and they quickly expanded to occupy all but the driest and coldest regions. Today approximately 235,000 species of angiosperms are known—the number of species varies slightly in different classifications and new species are discovered from time to time—compared with 721 species of gymnosperms. The deciduous forests, from the Tropics to middle latitudes, are composed mainly or entirely of angiosperms.

■ SEED PLANTS, CONIFERS, AND FLOWERING PLANTS

Ferns are very common, and there are many different kinds. In all, there are more than 12,000 species of ferns, including the tree ferns. About 300 million years ago, other species of much bigger tree ferns formed an important component of vast forests. Ferns grow today in many temperate forests, some on the ground and others on the branches of broad-leaved trees. In fact, they are more common now than they used to be. In the 19th century, ferns were fashionable as houseplants, and all of those that ornamented homes, hotels, and restaurants were collected from the wild, seriously depleting the natural populations of them. Ferns are still popular, though less so than they were then, but now they can be cultivated, so ferns sold in pots to take home have been grown in nurseries. Cultivation, combined with laws forbidding people from collecting wild plants, have allowed their numbers to recover.

A fern leaf, often called a frond, has veins. These are vessels carrying water and nutrients, and their presence means that ferns are vascular plants like the trees and herbs that grow in modern forests. The veins are branched, an arrangement that probably evolved through the development of small veins linking closely spaced larger ones. The botanical name for a leaf of this kind, found in ferns and in all seed-producing plants, is *megaphyll*. Many ferns have *compound leaves* in which each main frond is divided into many small leaflets. Each leaflet is called a *pinna*, and each division of a pinna is a *pinnule*. The main leaf stem is the *rachis*.

Producing Spores

On the underside of some of the pinna, there are what look like dark spots. Leaflets bearing these spots are known as *sporophylls,* and the spots themselves are sori (singular, *sorus*). In many species there are two parallel rows of sori.

Sporophylls are reproductive leaflets, specialized to form part of the reproductive mechanism of the plant. Each sorus is a cluster of tiny structures called sporangia (singular, *sporangium*), and each sporangium is a container filled with *spores*. All ferns produce sporangia, but not all of them bear their sporangia in clusters, so not all of them have sori. When conditions are right, a sporangium springs open and throws its spores into the air.

Spores form inside the sporangia from spore mother cells. Each spore mother cell divides meiotically (reduction division; see "How Cells Divide—Mitosis and Meiosis" on pages 106–107) to produce four *haploid* cells (each with one set of chromosomes, rather than the two sets in most *diploid* cells). These haploid cells, with toughened outer walls, are the spores. They are extremely small, and all of them are identical. They do not contain the embryo of a new plant, so they are quite different from seeds. Being so small and light, the wind can carry them long distances. Those that land in a favorable spot grow into the structure, forming the next stage in the life cycle of the plant.

Ferns, then, are vascular plants that do not produce seeds. They are classified in the plant division *Pterophyta,* by far the largest of the four divisions of seedless vascular plants surviving to the present day. Whiskferns, which are not really ferns, comprise the division *Psilophyta*. Clubmosses, or groundpines, which are neither mosses nor pines, form the division *Lycophyta,* and horsetails form the division *Sphenophyta*.

Plants That Rely on Spores

Spores work well enough. All the simpler plants, such as mosses and liverworts, reproduce by means of them, as do fungi and some bacteria. Spores evolved early in the history of life on our planet, and since they are still used, clearly they have stood the test of time.

The organisms that produce them pay a price for their simplicity, however. The new plant that develops directly from a spore must start at once to obtain its own food. All the spore provides is a set of genetic instructions. It contains no store of food to see the young plant through its first few days of life. Spores themselves can remain viable for long periods, and they develop only when they are moist, but if the new plant had even a small amount of protection, it would have a much better chance of surviving a brief spell of cold or dry weather while it was trying to establish itself. Most of the young plants die, and to compensate for this a fern produces hundreds of millions of spores every season. Each spore is so small that the plant needs to invest very little material or energy to produce it, but it loses much of this advantage by having to produce spores in such large quantities.

Why Seeds Are Better

By about 360 million years ago, one group of plants, the gymnosperms, had evolved a means of improving the survival chances of their young. Instead of releasing spores, they retained them until the new plant had grown the structures from which roots and leaves could develop quickly. This rudimentary plant is called an *embryo*. The embryo was surrounded by a store of food sufficient to keep it nourished until its leaves were big enough for it to produce its own food by photosynthesis (see "Photosynthesis" on pages 116–122), and both the embryo and its food store were enclosed in a coat tough enough to protect the embryo from harsh conditions. Together, the embryo, its food store, and the outer coating comprise a seed. Gymnosperms were the first seed plants.

There are four groups, or divisions, of gymnosperms: Cycadophyta; Ginkgophyta; Gnetophyta; and Coniferophyta. Cycads form the division Cycadophyta. They are sometimes called sago palms and are rather like palms in appearance, but botanically they are very different. The ginkgos, forming the division Ginkgophyta, are reduced to only one species, the maidenhair tree (*Ginkgo biloba*), which is grown quite widely as an ornamental. The gnetae, division Gnetophyta, comprises a number of species, including trees, which are quite different from each other. The division does include one of the most remarkable of all plants, *Welwitschia mira-bilis* (or *W. bainesii*), that grows only in the deserts of south-western Africa. It produces the biggest of all plant leaves, huge, straplike structures that grow more than five inches (12.7 cm) a year and wear away at their tips. *Welwitschia* obtains all its moisture from dew and fog and is said to live for more than 1,000 years. The fourth group is the division Coniferophyta, the conifers.

Conifers and Their Cones

Conifers are by far the most abundant and familiar of all gymnosperms. The pines, spruces, firs, larches, hemlocks, and redwoods of the coniferous forests all belong to this division. As their name suggests, all of these trees produce cones. In fact, conifers produce two types of cone. Pollen cones consist of very small sporophylls (reproductive leaves) with large numbers of sporangia. Haploid cells produced in the sporangia develop into pollen grains that are dispersed by the wind. Ovulate cones are composed of scales, each containing two ovules, and each ovule contains a sporangium, called the *nucellus,* that is enclosed by walls called *integuments* that have a single opening, the *micropyle.* A pollen grain falling onto an ovulate cone enters the ovule through the micropyle and fertilizes eggs that are formed in the ovule. This leads to the development of the seed (see "Conifers and Their Male and Female Cones" on pages 107–108).

Flowers, Fruits, and Seeds

No one knows just how angiosperms appeared. The earliest angiosperm fossils of two plants named *Archaefructus liaoningensis* and *A. sinensis* that lived between 145 million years ago and 125 million years ago were found in the 1990s about 250 miles (400 km) northeast of Beijing, China, but the ancestry of angiosperms remains uncertain. What is certain is that by about 65 million years ago angiosperms had become far more numerous than any other type of plant, and they remain so still. There are fewer than 1,000 species of gymnosperms living today, but there are at least 260,000 species of angiosperms extant.

Botanists classify all angiosperms as the infraphylum Angiospermae, within the phylum Tracheophyta—the vascular plants. There are two types of angiosperm: monocots and dicots. The seeds of monocots contain a single *cotyledon* or seed leaf, and those of dicots contain two. Monocots and dicots were formerly classed as separate divisions, the Monocotyledonae and Dicotyledonae, but these names are no longer used. Grasses, cereals, sugarcane, lilies, bamboos, and palm trees are monocots. The broad-leaved trees of temperate forests are dicots, as are most of the herbs growing in association with them. What all angiosperms share in common is the production of flowers. A flower is a reproductive structure in which ovules are contained inside an ovary. The ovary protects the ovules and the seeds that develop from them.

Many flowers have brightly colored petals, and many mature ovaries become fruits with the seeds hidden inside them. Bright colors attract insects and other animals to a store of nectar, which is a sugary liquid. Fossils have been found of insects that lived about 125 million years ago and that appear to have fed on nectar. As they move about inside the flower consuming the nectar, the insects also transfer pollen, thus allowing fertilization to take place.

Edible fruits also attract visitors, including humans, who help distribute the seeds they contain (see "Fruit Eaters" on pages 112–113). There are other differences between angiosperms and gymnosperms (see "Xylem" and "Phloem" on pages 104–105) and not all angiosperms produce colored flowers or edible fruits. Those, such as grasses, that are pollinated by the wind, have drab, inconspicuous flowers, and many trees produce inedible seeds. The important distinguishing feature of angiosperms is their protection of ovules inside an ovary.

All three groups of vascular plants continue to live side by side. Ferns that produce and release spores are common and widespread. Gymnosperms, pollinated by wind and producing naked ovules, dominate the forests of high latitudes. It is the angiosperms that have proved most successful, however, and it is they that provide us with the great majority of our ornamental plants and our food, as well as with the trees of our most beautiful forests.

HOW A TREE WORKS

In spring, a deciduous forest seems an enchanted place. In that brief interval between the thawing of the ground and the bursting of the leaf buds on the trees, sunlight dapples the forest floor. There is enough light for the small herbs to flower and produce their seeds, and in the larger clearings, the grasses are in the bright green flush of their first growth of the year. Gazing at the grasses, the flowers, the small shrubs and tree saplings, and the trees towering high overhead, a visitor may well imagine that all these plants function in many, quite different ways. Yet it is only its size that distinguishes a tree from any other plant. The oaks, beeches, hornbeams, redwoods, and all the rest work in exactly the same way as the smallest of the herbs growing around their bases.

Every land plant grows and repairs its tissues by absorbing nutrients from the ground. Sunlight, harnessed by photosynthesis (see "Photosynthesis" on pages 116–122) in its green leaves and released by respiration, supplies the energy it needs to grow, repair itself, and reproduce. It uses water to transport nutrients in a constant stream from the roots to the leaves where it evaporates (see "Transpiration" on pages 48–50) and uses another system of vessels to transport the products of photosynthesis to all of its tissues.

When a seed germinates, the emerging plant forms two structures. One will become the stem and eventually the main trunk of a tree; the other, called a *radicle,* will develop into the root system. Almost from the start, the plant divides itself into two distinct parts, the shoot system aboveground and the root system below. These two main parts of the plant, shown in the illustration, depend on one another. The roots

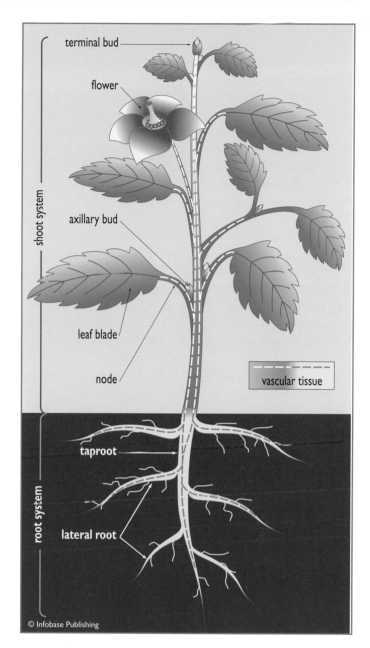

The aboveground and belowground parts of a flowering plant

being belowground and, therefore, in perpetual darkness cannot perform photosynthesis to synthesize sugars. These must be supplied from the leaves that are aboveground. The shoot system cannot obtain mineral nutrients, which are found only in the soil, so these must be supplied by the roots.

Roots

A tree root is woody, like the trunk and branches, but roots and shoots differ in their structure and the way they grow. Many of the smaller dicots (see "Flowers, Fruits, and Seeds" on pages 102–103) produce a large taproot, often penetrating to a considerable depth. In some plants this also doubles as a food-storage organ. Carrots, sweet potatoes, and parsnips are taproots of this kind. Trees, as well as all monocots and gymnosperms, produce roots that grow to the sides and are shallower.

The lateral roots branch into smaller and smaller side roots and eventually into a mass of fine fibers no more than 0.04 inch (0.1 cm) thick. The volume of soil occupied by the root system may be greater than the volume of air occupied by the parts of the tree that are above the ground.

As the growing root pushes its way through the soil, its tip is protected from damage by a tough cap that is constantly renewed as it wears away, and the root extends itself by the division of cells behind the cap. Some plants have roots with very fine hairs through which they absorb water and nutrients, but most plants, including almost all the trees growing in temperate forests, rely on soil fungi that grow on and below the surfaces of the roots. The resulting structure of roots and fungi together is called a *mycorrhiza,* from the Greek words *myco* meaning "fungus" and *rhiza* meaning "root."

The main part of a fungus is the mycelium, which is a very extensive network of fine fibers through which the fungus collects nutrients. Mycorrhizal fungi supply the plant with nutrients, some in the form of large organic molecules. The plant supplies the fungus with carbon that it produces by photosynthesis. This kind of close relationship between two organisms of different species that benefits both is called *mutualism.*

Fungi reproduce by spores that they release from a fruiting body. The fruiting body is the visible part of the fungus, and the mushrooms and toadstools that appear on the forest floor, especially in the fall, belong to mycorrhizal species living in close partnership with the trees. These fungi are known as basidiomycetes, and they feed on decomposing plant and animal matter.

Nitrogen and Root Nodules

Dig up a pea or bean plant complete with its roots, carefully brush away some of the soil, and you will see masses of white or gray nodules clinging to the roots. Each is about as big as the head of a map pin. The nodules are colonies of bacteria that are able to convert gaseous nitrogen, present in the air between soil particles, into nitrate (NO_3), some of which passes into the plant root. In return, the bacteria obtain carbohydrates from the plant. Peas, beans, lupins, and a large number of related plants are called *legumes,* a name originally derived from the Latin verb *legere,* "to pick," because they could be picked by hand. The relationship between legumes and nitrogen-fixing bacteria has been known for a long time (see "Nitrogen Fixation and Denitrification" on pages 122–125).

Nitrogen is a key ingredient of amino acids, from which proteins are made. This makes it an essential nutrient element for all living organisms. Air is about 78 percent nitrogen, but plants are unable to absorb nitrogen in its gaseous form; it must be combined with another element. In the soil, nitrogen is present as an ingredient of the amino acids in dead organic matter, and when these dissolve, it occurs as dissolved organic nitrogen (DON). As the amino acids break down, the nitrogen occurs as ammonium (NH_4), which is then converted to nitrate (NO_3). Most plants can use nitrogen as DON, NH_4, or NO_3, but some have clear preferences. White spruce (*Picea glauca*), for example, absorbs NH_4 but has difficulty utilizing NO_3. This tree is grown extensively for pulp production to make paper, but when it is planted on ground that has been disturbed by clear-felling a previous tree crop, it often grows weakly and may even be displaced by other species. Its weakness is probably due to disturbing the soil and thereby stimulating chemical reactions that convert NH_4 to NO_3.

How Soil Organisms Help Plants

Relationships between plants and other organisms are not confined to legumes. Alder trees (*Alnus* species), which are not legumes, have similar nodules on their roots, and most plants depend on soil microorganisms to process nutrients into forms their roots can absorb.

About 95 percent of all vascular plants, including almost all trees, form associations between their roots and fungi called mycorrhizae, meaning "fungus roots," in which root cells link with fungal hyphae. This arrangement unites the plant roots with the fungal mycelium and greatly increases the surface area through which nutrients can be absorbed. The fungus supplies the plant with a range of nutrients that it obtains as compounds that it absorbs from decomposing organic material. In return, the plant passes a substantial proportion of the carbon it fixes by photosynthesis directly to the fungus.

Soil organisms also require nutrients, of course, so plant roots must compete to obtain those they need, and plant roots exude chemical compounds as well as absorbing them. The *rhizosphere,* which is the soil around the roots of a plant, is like a chemical factory. It contains a vast array of fungi, bacteria, and other microorganisms, as well as the roots themselves, all engaged in processes that are so complex that scientists have not yet been able to describe fully how it all works.

Xylem

Once inside the root, nutrients, dissolved in water, enter the xylem, which transports them to every other part of the plant. In gymnosperms, xylem consists of specialized cells called *tracheids.* These are long, approximately cylindrical cells with tapered ends lying end to end, so that they overlap, as shown in the illustration on page 105. Tracheid walls are strengthened with rings or spirals of lignin, the tough

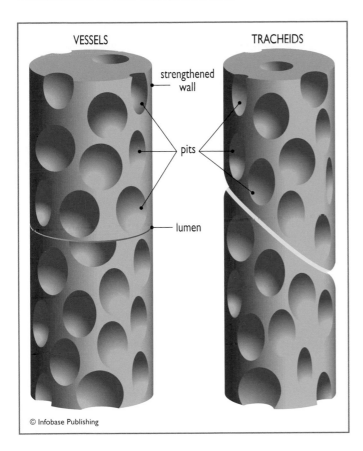

VESSELS   TRACHEIDS

strengthened wall

pits

lumen

© Infobase Publishing

Vessels (angiosperms) and tracheids (gymnosperms)

material that gives wood its strength. Lignin does not cover the entire exterior of the cells, however. Gaps remain as pits where the wall is thin enough for the nutrient solution to cross, and the solution moves along from one tracheid to the next through the pits in the overlapping sections.

Angiosperms also have tracheids, but in addition they possess a more efficient type of cell, called a *vessel element.* These are generally rather wider and shorter than tracheids. They have pits, but their end walls are perforated, forming a *lumen,* or opening, so that the cells provide an open tube called a *vessel* through which the solution flows freely.

Xylem tracheids and vessels run from the tips of the roots to every part of the plant and form the channels through which water and mineral nutrients are distributed. Carbohydrates, produced in the leaves by photosynthesis, are transported by another set of tubes, called the *phloem.* Xylem cells are dead, but phloem cells are alive, although they lack nuclei and other components found in most cells.

Phloem

Each phloem cell, called a *sieve element,* is cylindrical in shape, and its ends, called *sieve plates,* are perforated. The sieve elements are joined at their sieve plates to form sieve tubes.

Although they lack nuclei, in angiosperms each sieve element is closely linked to one or more companion cells to either side of it. These possess both nuclei and *ribosomes* and are the sites of intense metabolic activity. Ribosomes—tiny granules of RNA (ribonucleic acid) and protein—are where proteins are synthesized. Botanists believe they may provide the sieve elements with the energy and nutrients they need.

In gymnosperms, the sieve elements are not associated with companion cells, but they are surrounded by *parenchyma* cells, which may serve the same purpose. Parenchyma cells are unspecialized plant cells, which manufacture and store a range of products. In leaves, photosynthesis takes place in parenchyma cells, and elsewhere in the plant they store starches and sugars. The edible parts of root vegetables and fruit consist mostly of parenchyma cells.

Sieve tubes, tracheids, and, in angiosperms, vessels together form the vascular system of the plant, transporting water and materials from where they are produced or obtained to where they are needed. A cut plant stem exudes sap. This is material being transported by the vascular system and it is of two kinds. Phloem sap is rich in sugars and often tastes sweet. Xylem sap is mostly water, in which mineral nutrients and some amino acids are dissolved. The two systems, of xylem and phloem, lie side by side in vascular bundles, and there are many vascular bundles in the trunk and main branches of a tree. As the vascular system extends into the smaller twigs and then through leaf stalks and finally becomes the veins of the leaves, the number of vascular bundles decreases.

■ TREE REPRODUCTION

All trees are able to reproduce sexually, but for many species sexual reproduction is not the only method available. When you see a clump of trees all of the same species and growing close together surrounded by trees of a different species, there is a good chance that the clump has arisen asexually. Genetically, all the trees are identical and form a *clone.*

Some trees reproduce in this way more readily than do others. Elms (*Ulmus* species) and limes (also known as lindens or basswoods, *Tilia* species) are especially prone to this type of vegetative reproduction. Near the climatic boundary of their range, where the summer is not always warm and sunny enough for them to produce viable seed, it is the strategy that allows them to survive in what would otherwise be an inhospitable region.

Plants are able to reproduce in this way because they grow by the repeated division of unspecialized cells in tissues called *meristem.* Meristem cells can continue dividing indefinitely, provided they have the nutrients and warmth they need, and the tissues they produce may specialize later. That is why cutting a branch from a tree or cutting the whole tree almost at ground level often produces a mass of new twigs, a fact that has been exploited for centuries as a way to produce thin poles (see "Coppicing" on page 291). Similarly, a healthy, cut twig will produce a mass of undifferentiated cells, called a *callus,* at the cut end from which new roots will grow, producing a new plant. Apart from meristem tissue, much of the bulk of a plant consists of parenchyma cells. These are also undifferentiated and can be induced to develop into whole new plants.

Many trees can be "layered." In this technique a young, pliable branch is bent over until it touches the ground, and is then held there by a peg. In a little while it will produce roots and shoots from the point where it touches the ground. The new tree can then be cut free from its parent.

Valuable tree species, mainly ornamentals or orchard trees, are usually propagated vegetatively, nowadays often by culturing a batch of cells until there is a young plant big enough to be planted. Because all such progeny are identical, the technique ensures that they possess all the desirable characteristics of the parent plant, and clones can be produced cheaply in large numbers.

Tree growers often propagate basswoods and limes by layering, but this is not the method used naturally by elm trees. They produce new shoots at intervals along their lateral roots, so a clump of elms is not simply a clone: Often the trees are literally a single plant, all sharing the same root system. Several dicot species spread in this way, and because they can continue to do so for much longer than the lifetime of any individual tree, this process allows them to achieve something approaching immortality. In the Mojave Desert, California, there is a clone of creosote bushes that is believed to be 12,000 years old.

Alternation of Generations

Most trees reproduce sexually, and to do so their life cycle takes them through two distinct forms, a progress known as the alternation of generations. One generation is called the gametophyte, the other the sporophyte, and each gives rise to the other.

The two generations differ in structure and appearance, and one generation is much bigger and more prominent than is the other. In mosses, liverworts, and hornworts, the visible plant is the gametophyte generation. In all larger plants, including trees, the visible plant is the sporophyte, and the gametophyte is tiny and retained within specialized organs of the sporophyte.

The dominance of the sporophyte generation, involving the protection of the gametophyte within the body of the sporophyte, is one of the ways in which plants have evolved for life out of water. On land, organisms are exposed to more intense solar radiation, including radiation at ultraviolet (UV) wavelengths, than they are in water, which absorbs UV radiation. UV radiation can damage chromosomes, so possessing two copies of each gene may mean that losing one copy is not disastrous. It makes sense, therefore, to expose only diploid cells—cells with two copies of each chromosome—and to protect haploid cells, which possess only one set of chromosomes. In earlier plants, fertilization involved sperm swimming through water toward eggs. Mosses can manage this if there is just a film of water covering the plant, but it would present serious difficulties for larger plants and, of course, it would make the colonization of dry habitats impossible. Gymnosperms and angiosperms have evolved an alternative: They use pollen to deliver their sperm, making water unnecessary, and the fertilized egg then develops into a seed (see "Fruit and Seeds" on pages 110–111), rather than the spore produced by simpler plants. Seeds are protected against drought, cold, and other harsh conditions, and they can be dispersed over land more or less efficiently.

How Cells Divide—Mitosis and Meiosis

The essential difference between the generations lies in their *chromosomes,* the threadlike structures carrying the genes that are found in the nucleus of every cell. Most cells carry two copies of each chromosome and are called diploid. Sporophytes are diploid. Cells carrying a single copy of each chromosome are said to be haploid. In animals, including humans, spermatozoa and ova (egg cells) are haploid, and all other body cells are diploid. In plants, the gametophyte is haploid. Gametophytes cannot develop into sporophytes directly but produce sperm and egg cells. These reproductive cells are known as *gametes* (in animals as well as in plants), and the diploid sporophyte develops from the union of gametes to form a fertilized female gamete called a *zygote.*

All living cells increase their number by dividing. Diploid cells do so by the process of *mitosis,* in which each chromosome replicates itself, producing a pair of sister chromatids. As the nucleus divides in two, the chromatids separate, one going to each emerging nucleus, so the process results in two daughter cells, each with a full diploid complement of chromosomes.

Haploid cells are produced from diploid cells by *meiosis,* a different process. The chromosomes replicate once, but the cell divides twice, producing four daughter cells, each with a single set of chromosomes. When haploid cells unite at fertilization, their chromosomes join, and the resulting cell is diploid.

Pollen grains, produced by the male part of the plant and containing sperm, are male gametophytes. Eggs, produced by the female part of the plant, are female gametophytes. In some species, male and female gametophytes are produced on the same plant, in others they are produced on separate male and female plants. Some angiosperms produce male and female gametophytes in the same flower, others in separate male and female flowers.

A species is said to be *monoecious* if it bears male and female organs in separate structures but both on the same individual plant. This arrangement reduces the risk of the plant pollinating itself and is often found in plants, such as grasses, that are pollinated by wind. If male and female organs are borne on separate plants, the species is said to be *dioecious*. The terms are derived from the Greek *oikos*, meaning "house," so monoecious means "single-house" and dioecious "double-house."

Conifers and Their Male and Female Cones

The alternation of generations can be traced fairly easily through the life cycle of gymnosperms, illustrated in the diagram. Gymnosperms produce both male and female cones.

In some species, such as European larch (*Larix decidua*) and Norway spruce (*Picea abies*), these occur on different branches of the same tree, but in others, such as Scotch pine (*Pinus sylvestris*), they form on the same branch. Male cones are small and insignificant, although in some species they are a different color from the leaves and look a little like flowers (although they are not true flowers). The big, handsome cones that people collect to decorate their homes or predict the weather, and that foresters collect when they need seeds, are female cones. (Cones collected for their seeds should be gathered while they are still green.) Female cones form at the tips of branches. It takes several years for them to mature,

The life cycle of a gymnosperm

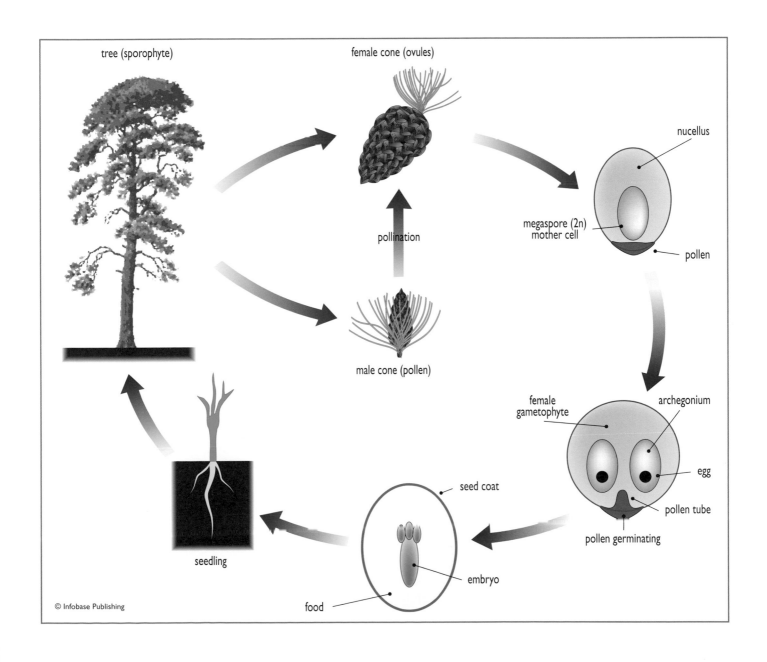

tree (sporophyte)

female cone (ovules)

nucellus

megaspore (2n) mother cell

pollen

pollination

male cone (pollen)

female gametophyte

archegonium

egg

pollen tube

pollen germinating

seed coat

seedling

embryo

food

Female pine cones consist of woody scales, each of which contains two ovules. This is a cone of the loblolly pine (*Pinus taeda*).   *(Vanderbilt University)*

however, during which time the branch continues to grow, so mature cones are often some distance from the tip.

The mature tree is the sporophyte generation. Its male cones comprise very small sporophylls, or reproductive leaves, each of which contains a large number of sporangia. A sporangium is a protective capsule containing diploid cells. These cells divide by meiosis, each cell producing four haploid daughters. The daughter cells develop into pollen grains, and the pollen grains are male gametophytes, although they are immature until they start to grow.

Female cones are made of tough, woody scales, and each scale contains two *ovules* surrounded by protective tissues, called integuments. An ovule comprises a sporangium, called the nucellus, and a (diploid) megaspore mother cell. There is a single opening to the nucellus, called the micropyle.

How Gymnosperms Reproduce

When the pollen is ready, the male sporangia split open. At this stage, each pollen grain consists of a single cell attached to two air-filled sacs, which make the pollen grain very light. Pollen leaves the sporangia, and some of it falls onto female cones where a single pollen grain is drawn into each ovule through the micropyle. The pollen grain germinates and starts to grow a pollen tube that advances into the nucellus. Pollination is then complete.

While this is taking place, the megaspore mother cell divides by meiosis to produce four haploid cells, only one of which survives. The surviving cell is called a *megaspore*, and it divides repeatedly, forming the immature female gametophyte. Inside the female gametophyte, two or three *archegonia* develop. These are very simple structures, each containing an egg.

Pollination usually occurs in spring. Some trees produce seeds in the fall of the same year, but in many species more than a year elapses between pollination and fertilization. During this time the archegonia develop and the pollen grain matures. The pollen tube continues to grow into the nucellus until it reaches the megaspore. While it is growing, two sperm cells are produced inside the pollen grain, making it a mature male gametophyte. When the pollen tube reaches the megaspore, the sperm travel along it and unite with the eggs. This is *fertilization*. The female gametophyte contains two or three eggs. All of them may be fertilized, but only one is likely to develop further.

Fertilization unites the haploid sperm and egg cells to form a diploid zygote, a cell that divides repeatedly by mitosis, eventually becoming an *embryo*. The embryo is destined to become the new sporophyte generation. It has a whorl of two, three, or many rudimentary leaves, called *cotyledons* or seed leaves, the number varying according to species, and a *radicle,* which will become the root. The embryo is surrounded by a store of nutrients derived from tissues of the female gametophyte, and both the embryo and its food store are enclosed in a seed coat produced by the parent plant, the seed coat and its contents comprising the seed. Although the seed coat protects the embryo, the seed as a whole is still located on the scale of the cone and is not enclosed by any other structure.

In due course, the seed will be released and dispersed (see "How Trees Disperse" on pages 111–113), usually by the wind. Those seeds that reach a favorable spot will germinate into seedlings, which grow into mature trees, thus completing the cycle.

The Variety of Flowers

Angiosperms produce flowers. Like all plants, their complete life cycle involves an alternation of generations, but in flowering plants the gametophyte generation is contained within the flower. The plant you see is the sporophyte.

Flowers come in an almost infinite variety of sizes, shapes, and colors. *Rafflesia arnoldii,* a parasite of the roots of tropical trees, has flowers that are up to 32 inches (80 cm) in diameter. These are the biggest of all flowers (and they smell of rotting meat, to attract the flies that pollinate them). At the other extreme, there are many plants with flowers no more than about one-tenth of an inch (0.3 cm) across, though such tiny flowers (*florets*) usually occur in large clusters (*inflorescences*) so they look bigger. A sunflower, for example, is actually a composite of hundreds of small florets.

Despite their huge variety, all flowers can be described in relation to a generalized type, shown in the illustration. Working inward from the outside of the plant, a flower comprises sepals, petals, stamens, and carpel. *Bracts* are small leaves, often modified, that grow where the flower is attached to the plant stem. *Stamens,* each consisting of a long *filament* topped by a clublike *anther,* are the male part of the flower. The *stigma, style,* and ovary, together forming the *carpel,* comprise the female part of the flower. The enlarged end of the *peduncle* (flower stalk), to which the carpel and stamens are attached, is the *receptacle.* If the ovary is attached to the receptacle above the point of attachment of the filaments (as in the flower shown here), it is said to be superior. An inferior ovary is attached below the filaments.

The flower shown in the drawing possesses both stamens and carpel. Such a flower is said to be perfect. Not all flowers are perfect. Some lack stamens. These are female flowers and are described as carpellate because they have only a carpel. Others lack carpels. Having only stamens and, therefore, being male, they are called staminate. Carpellate and staminate flowers are imperfect. The flower in the illustration is also complete, in that it possesses sepals and petals as well as stamens and carpel. Some flowers are incomplete. Grasses, for example, produce flowers lacking petals.

Sepals, which are usually green, are the leaflike structures that formerly enclosed the flower bud; collectively the sepals comprise the *calyx.* Petals evolved with a variety of

The parts of a flower. Not all flowers possess all of these parts; the flower shown here is complete and perfect.

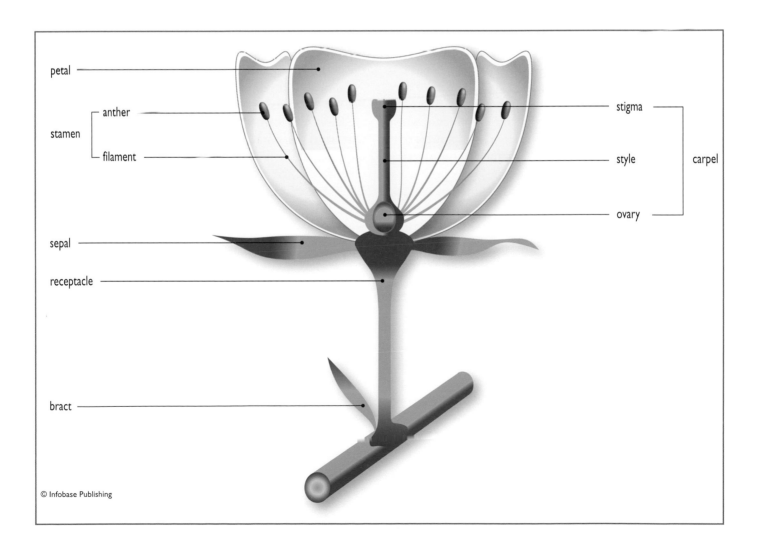

shapes and bright colors to attract pollinating insects or, less commonly, birds. Pollinators recognize the shape and color of their preferred flowers, inside which they can feed on a sugary liquid, called *nectar,* which is secreted by the nectary, an organ at the base of the flower (not shown in the diagram). Flower scents evolved for the same reason. Many insects are acutely sensitive to odors and can locate flowers by following a gradient of increasing odor intensity.

How Flowering Plants Reproduce

The angiosperm reproductive cycle follows the same general course as that of gymnosperms but with a few important differences. Inside each of the anthers are several sporangia, called pollen sacs, each sporangium containing many diploid cells called microsporocytes. These cells divide by meiosis; each microsporocyte produces four haploid microspores that divide by mitosis to produce pollen grains enclosed in very tough, almost indestructible, coats. These are the immature male gametophytes.

In the carpel, the ovary contains one or more ovules, the number varying according to species, and each ovule contains a diploid cell called a *megasporocyte.* This cell divides by meiosis, producing four haploid cells, only one of which survives. The survivor, known as the megaspore, is the female gametophyte.

Pollination occurs when pollen grains are deposited on the stigma, which is sticky so as to hold them once they arrive. Plants that are pollinated by animals are arranged in such a way that pollen grains adhere to the pollinator when it enters the flower and stick to the stigmas of flowers that it visits later. This is an economical method of transferring pollen, and the plant can achieve pollination by producing only modest amounts of pollen. Wind pollination is much less reliable because it is a matter of chance whether or not pollen falls onto stigmas. Consequently, wind-pollinated plants must produce truly vast quantities of pollen to have any real hope of success. People who are unfortunate enough to suffer from hay fever know only too well that when the grasses are flowering; the air is filled with microscopically small pollen grains.

A perfect flower produces both male and female gametophytes, which means that there is a risk that the flower might pollinate itself. Over many generations this would be genetically disadvantageous. The advantages of sexual reproduction arise from the mixing of genes from separate parents, and without such mixing, harmful recessive genes will be expressed more frequently in the offspring.

Some plants allow self-pollination, but most do not, and they have three ways of avoiding it. Within an individual flower, the male and female gametophytes may be produced at different times, making self-pollination absolutely impossible. The structure of the flower may make it almost impossi-

ble for a pollinating animal to collect pollen from the anthers and transfer it to the stigma of the same flower. Finally, the plant may produce a chemical block that inactivates any of its own pollen grains that happen to reach its stigma.

Once it has stuck to the stigma, a pollen grain develops two sperm, the male gametes, thus becoming the mature male gametophyte, and starts to extend a pollen tube down between the cells of the style to and then into the ovary. When it reaches an ovule, the pollen grain discharges sperm.

In all gymnosperms except for *Ephedra,* a shrub that grows in the American desert, only one sperm cell is involved in fertilization, but an angiosperm uses two, for double fertilization. One sperm fertilizes the egg, which develops into the diploid zygote. The other fertilizes two nuclei inside the ovule, forming a cell with three sets of chromosomes. This cell divides to form *endosperm,* the nutritive substance that will sustain the embryo.

Fruit and Seeds

Most monocots produce enough endosperm to nourish the young seedling after the seed has germinated. In many dicots, all the endosperm is absorbed into the cotyledons (seed leaves). The zygote, developing into a embryo, together with the endosperm and the seed coat surrounding them comprise the seed. In angiosperms the seed remains inside the ovary, which develops into the fruit.

Not all fruits are edible, and although they may be edible, not all "fruits" are really fruits at all. Apples, for example, are false fruits formed from the swollen receptacle (with the true fruit) that contains the seeds at its center, which forms the core. As the fruit develops at the base of the flower, the other flower parts die and are shed.

Think of fruit, and it is probably an orange, pear, or plum that springs to mind, but fruits come in many kinds. Nuts are fruits, usually with a single seed and a woody outer coat, with the exception, among nuts that are widely eaten, of Brazil nuts. These are seeds. The fruits, each containing between 12 and 24 seeds (the edible nuts), are woody and inedible. Cereal grains, from which we obtain flour and other types of meal, are also fruits, in this case of a type that botanists call a caryopsis. A caryopsis is a dry fruit that contains a single seed and does not break open of its own accord. Peas and beans are seeds, and the pods that contain them are fruits.

Seeds are dry, but with most plants water can pass through their protective coats. When it does so, the absorbed water makes the seed swell, bursting the coat, and the water also activates enzymes that start digesting the endosperm or contents of the cotyledons. Nutrients move to the growing part of the embryo, and before long the radicle emerges and starts to grow downward, followed by the shoot, which grows upward, breaks through the ground surface, and spreads its cotyledons to commence photosynthesis. A new plant has appeared, and the reproductive cycle is complete.

HOW TREES DISPERSE

Once seeds are ripe, they are released, but this presents a difficulty. Trees could simply detach their seeds, allowing them to fall to the ground, but if they did that, the vast majority of seeds would land around the base of their parent. Some might germinate, but the emerging seedlings would find themselves competing for light, water, and nutrients with their overwhelmingly bigger parent. One or two young trees might manage to become established, but even those would be unable to grow very tall until the parent tree died and fell.

As they fall, however, seeds are affected by the wind. This will carry at least some of them away from their parent. In addition, tree seeds are produced on branches, not on the main trunk, so they are released a little way from their parent—although most will still fall on ground, exploited by the parent's root system. Where seeds simply fall to the ground, a few land very close to their parent; most, and a short distance away, and a few are carried rather farther, the seed density decreasing rapidly with distance.

A Few Big Seeds or Many Small Seeds?

The seeds of most trees need to germinate fairly quickly. Some years ago scientists counted the tree seeds in the soil in part of a forest that had been standing in the White Mountains, New Hampshire, for 95 years. They found 1,150 seeds of yellow birch (*Betula lutea*), 530 of paper birch (*P. papyrifera*), and more than 10 seeds of other tree species in every square foot of the forest floor ($961/m^2$, $445/m^2$, and $78/m^2$ respectively). All of these seeds were waiting in the ground for an opportunity to germinate and grow, but most were waiting in vain. They will have been seeds that were released within the previous year, and by the following year few of them will still have been capable of germinating. Seeds of most broad-leaved trees remain dormant for six to 18 months, but few can survive for much longer than this. While the seeds remain viable, they make up a valuable store, called a seed bank, with seeds ready in position to provide a new tree to fill the smallest space that appears in the forest. Their relatively short period of viability is not a disadvantage, of course, because each season a fresh crop of seeds replenishes the seed bank.

There are exceptions, and some trees produce seeds that remain viable for several years. Often they do not release them until the parent tree falls or is destroyed by fire (see "Adaptation to Fire" on pages 153–154). Many pine trees (*Pinus* species) produce seeds of this kind, dispersing them when the surrounding ground is clear and they have the best

possible chance of germinating. Until then the seeds remain stored, still in their cones, on branches high up in the tree crown. This storage of seeds in a forest canopy is called *serotiny* (from the Latin *serotinus,* meaning "late").

Obviously, the smaller and lighter the seed is, the farther the wind will carry it. This might seem to favor small, light seeds, but it is not quite so simple. Seeds carry a store of food with them, and the larger the store, the better the start in life it gives the young plant. If trees discard some of this baggage in order to produce seeds that travel light, the seeds will go farther but only to arrive relatively ill-equipped, and many of their seedlings will die. The tree can resolve this dilemma either by producing very large numbers of small seeds or by producing a smaller number of well-endowed seeds, carrying heavy loads of food. Such seeds may not travel far, but they are more likely to survive after germination.

Winged Seeds

In either case, the parent tree has to supply the materials to make its seeds, and this involves an expenditure of energy. Everything in nature has to be paid for, and in the case of reproduction, the bill is paid by the mother in terms of the material from which the seed and structures surrounding it are made. Since there is an unavoidable cost, many trees have evolved a different way of allocating resources. They produce medium-sized seeds in fairly large quantities and equip them with "wings." Winged seeds are very common.

Most gymnosperm trees, although not all of them, produce winged seeds. All of them have a single wing. They vary considerably in size, but that of Sitka spruce (*Picea sitchensis*), shown in the illustration on page 112, is typical. Winged seeds evolved first in gymnosperms, but many angiosperms have retained them. Again they vary greatly in size, but in angiosperms the seeds usually have two wings, like those of silver birch (*Betula pendula*) and field maple (*Acer campestris*) shown in the illustration on page 112.

Wings, made from material supplied by the parent tree, slow the rate at which the seed falls to the ground—it remains airborne for longer than it would if it lacked wings. In still air, the seed will fall approximately vertically, but if there is even the slightest wind, slowing its descent increases the length of time during which the seed is available to be carried by it, and so the wing acts like a sail. A winged seed travels farther than a seed without wings. Even so, most winged seeds land within a short distance of the tree that released them.

Air transport is a form of passive dispersal that involves only the tree itself, and because all tree seeds are fairly large and, therefore, heavy, its efficacy is rather limited. Flowing water might offer an alternative means of passive dispersal that would carry seeds much farther, but very few plants of any kind make use of it. Even most aquatic plants produce their seeds out of the water. Obviously, some seeds from

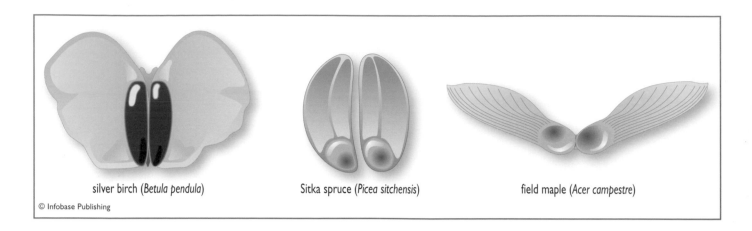

silver birch (*Betula pendula*)   Sitka spruce (*Picea sitchensis*)   field maple (*Acer campestre*)

© Infobase Publishing

Winged seeds of two broad-leaved trees (silver birch and field maple) and one coniferous tree (Sitka spruce)

riverside trees fall into the water, but since they have no sure way of returning to dry land, they are almost invariably doomed to fail.

Animals That Hoard Food

There is an alternative. Trees can and do exploit certain mammals and birds as dispersal agents by offering them a reward. Tree squirrels are famous for burying acorns and other tree seeds. In North America, the eastern gray squirrel (*Sciurus carolinensis*) prefers to feed on the seeds of hickory, beech, and oak but will take other food if these are scarce. In Britain, where this animal is known simply as the gray squirrel, its preferred foods are seeds of oak beech, sweet chestnut, and hazel. The Eurasian red squirrel (*Sciurus vulgaris*), which is the native British squirrel (the gray was introduced in the late 19th and early 20th centuries), prefers conifer seeds and cones.

Squirrels take tree seeds in order to eat them. Clearly, this does not benefit the tree, but in years when the weather is favorable, trees are able to produce vast quantities of seeds, presenting squirrels with far more food than they can eat. They continue collecting food but hoard the surplus in their nests, in holes and cracks in trees, and in the ground where they bury several seeds at a time, an inch or two (2–5 cm) below the surface. The eastern gray is much keener on storing food than the Eurasian red is in Britain, but in eastern Europe, where the winters are harder, the red is also a great hoarder.

Their food hoards keep the squirrels supplied through the winter (they do not hibernate), but they do not remember where they have hidden food and have to locate it by scent, which is possible only when the ground is moist. Inevitably, many of the seeds they store remain undis-turbed and are able to germinate far from the tree that produced them.

Some birds also hoard seeds. One of the most remarkable is the jay (*Garrulus glandarius*), a member of the crow family that occurs in most of temperate Eurasia but not in North America. It also buries acorns, carrying up to five at a time, the biggest one in its bill and the others in its throat and esophagus. The more acorns the bird carries, the farther it flies with them, taking some more than a mile (1.6 km) from the tree where it collected them. Then it buries them individually, unlike squirrels, which bury seeds in groups. In the course of a season a jay may bury much more than 4,000 acorns. What is remarkable is that when the bird needs to eat them, it does not search by scent (most birds have at best a very poor sense of smell) but remembers where it buried each acorn. It rarely needs all of its stored acorns, so some of them are allowed to germinate.

In these cases, both the animals and the trees benefit. The animals find food, and in doing so they disperse the tree seeds. Edible seeds and the habit of hoarding food have evolved together for the mutual benefit of the participants. This is an example of coevolution. Hoarding behavior does not always benefit the trees, however, and squirrels and jays may be exceptional. A North American bird, the acorn woodpecker (*Melanerpes formicovorus*), also hoards acorns, but it hides them in holes it drills in dead wood, such as fence posts, where they have no chance to germinate.

Fruit Eaters

Exploiting the hoarding behavior of certain animals is an unreliable way to disperse seeds, not least because the seeds are themselves the bait and many will be eaten and destroyed. This is especially serious in years when the weather is poor and seeds are few. Some angiosperms have evolved a much better method. They produce fruit that is edible, often rich in sugar to provide food energy for any animal that eats it, with the seeds concealed inside. Generally, the seeds have tough coats that make them not only unpalatable but also indigestible. Just to help matters along, fruits are often brightly colored, making them easy to find, and their color changes as they (and the seeds inside them) ripen, so fruit-eaters need not waste time with unripe fruit and trees need not risk losing seeds before they are ready to germinate.

Anyone who watches the way birds feed in late summer and fall or who talks to a fruit grower will see just how effective this strategy is. If the seed is large, like the stone in a cherry or plum, the bird or mammal eating the flesh of the fruit will simply discard the seed. If the seeds are small, like those of an apple or rowan (*Sorbus aucuparia*), they will be swallowed, travel unaltered through the animal's gut, and be deposited with the feces. The feces will also provide nutrient for the young plant. This explains why fruit trees with small seeds such as apples and pears can spring up in the most unlikely places.

Producing edible and nutritious fruit allows trees to disperse their seeds very widely. Seeds that are swallowed may be carried several miles before being deposited on the ground. Even those seeds that are discarded as the fruit is eaten may be carried well clear of the parent tree because animals often take fruit away to eat it out of range of competitors who might try to steal it.

HOW WOOD FORMS

Trees and shrubs have stems and branches that are woody. Other plants do not produce wood, and so plants are often described as being "woody" or "nonwoody." It is wood that gives woody plants their rigidity and mechanical strength, and, of course, wood is the most useful material that people obtain from trees.

Coniferous trees and all gymnosperms produce *softwood*. Broad-leaved trees produce *hardwood*. These names refer to the working properties of the wood and give an idea of such matters as how easy the wood is to cut and how quickly it blunts saws, planes, and chisels. The information is generally reliable, but there are exceptions. Balsa, the softest of all wood, comes from *Ochroma lagopus*. This is a broad-leaved tree, so balsa is classified as a hardwood. Wood from most larches (*Larix* species) is fairly hard, on the other hand, but since larches are gymnosperms, their wood is automatically listed among the softwoods. In gymnosperms, the vascular system consists only of tracheids, and in angiosperms it also includes vessel elements (see "Xylem" and "Phloem" on pages 104 and 105), but this is the only essential difference in the cellular structure of softwoods and hardwoods.

Primary and Secondary Growth

After a tree has been felled, the face of the stump left in the ground shows that the trunk, or stem, is made entirely of wood. It also reveals that the wood is not quite the same color throughout. In some parts, it is darker than in others and is marked with many concentric dark and light rings. One pair of light and dark rings forms each year, so the number of rings indicates the age of the tree. Around the outside, where there are no rings, the woody trunk is enclosed in a layer of bark. This is darker than the wood and has a very rough, uneven, exterior surface.

At the tips of its twigs, a tree produces a terminal bud, a dome-shaped mass of cells that divide to form new tissue. This process lengthens the twigs and, therefore, makes the tree bigger. It is known as primary growth, and if that were all that happened, the tree would quickly turn into a rather curious plant: Its branches would grow longer, and the tree as a whole would grow taller, but neither its trunk nor its branches would grow any thicker. It would be so spindly that it could not support its own weight, and it would straggle across the ground more like a bramble or a wild rose than the spruce or oak it was supposed to be. In fact, of course, the trunks and branches of trees also grow thicker as well as longer, and their thickening is known as secondary growth.

The drawing on page 114 shows a cross section through the trunk of a tree. It is clear from this that there are two types of wood: heartwood and sapwood. Between them these account for all but the outermost layer of material, the bark. As the drawing shows, the structure is not quite so simple, mainly because the bark is a much more complex substance than it may seem.

Vascular Cambium

The key to secondary growth and the production of wood is the layer of vascular *cambium*. Cambium is a layer of tissue in a woody plant comprising cells that give rise to a different type of tissue on one or both sides. In most trees this layer is no more than one cell thick, but the cells forming it are parenchyma cells (unspecialized, all-purpose plant cells) that have acquired the capacity to divide in such a way as to produce specialized cells. In botanical terms they have become meristem tissue.

The cambium cells occur inside the vascular bundles (see "Phloem" on page 105) and in the pith rays. Pith rays are layers that run radially from near the center of the stem to the bark. Each pith ray comprises a few vertically stacked parenchyma cells, usually acting as starch stores. In the xylem, pith rays also provide channels for the lateral transport of water and nutrients.

A meristem cell is sometimes called an initial, and any cell derived from an initial is known as a derivative. In the case of the vascular cambium, each initial first divides to produce two cells like itself (two initials). Then one of the two initials becomes a different kind of cell. The diagram on page 115 shows how this works. Starting on the left of the

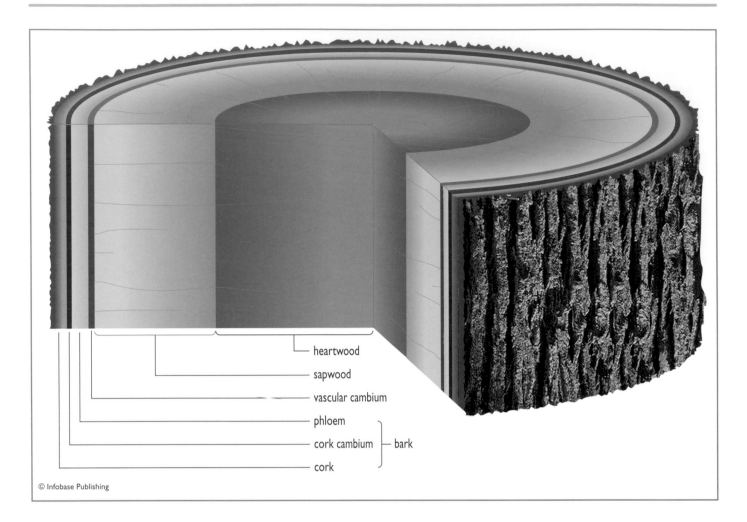

A cross section through a tree trunk

diagram, one cambium (initial) cell divides. There are then two cambium cells. The one on the inside (nearest the center of the tree) turns into a xylem cell, and the other cambium cell divides to produce two more cells like itself. The outer one (nearest the outside of the tree) turns into a phloem cell and the other divides. The inside cambium cell turns into another xylem cell, the other divides, the outer turns into a phloem cell, and so it continues. A central layer of cambium, never more than two cells thick, generates phloem on the outside and xylem on the inside. These are the two types of vascular tissue, which is why this is known as the vascular cambium. Cambium cells in the pith rays divide in a similar fashion to extend the rays.

The diagram of the tree trunk shows the layer of phloem. Xylem, comprising the inner layer to the right of the vascular cambium in the drawing, is what is called the sapwood. Phloem and xylem that result from cell division in the vascular cambium form part of the secondary growth which thick-ens the trunk and branches of a tree, and they are known as secondary phloem and xylem. Phloem comprises the tissues through which nutrients are transported from the leaves to other parts of the plant, and, therefore, it is essential to the survival of the tree. If it is cut through all the way around the trunk so its nutrient transport ceases, the tree will die. The phloem forms the innermost layer of the bark, and damaging a tree in this way is called girdling or ring-barking.

Cork Cambium

Outside the phloem there is a second region of cambium called the cork cambium. It gives rise not to vascular tissue but to epidermis, the outermost layer of the plant, which is made from cork cells. A meristem initial in the cambium changes into a derivative. The derivative lays down a layer of a waxy material called suberin in its walls, and then the cell dies. It has become a cork cell, dead and with a waxy coating. The layer of cork cells protects the interior of the trunk or branch from losing water and from damage and invasion by insects.

In most species the layer of cork cells is quite thin, but there is one famous exception, the cork oak (*Quercus subur*) of southern Europe and North Africa. It produces a very

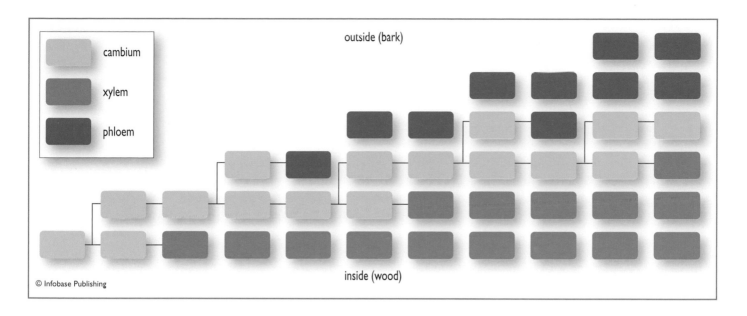

outside (bark)

cambium

xylem

phloem

inside (wood)

© Infobase Publishing

How cambium cells divide to produce secondary growth

thick cork layer, and the cork can be stripped every eight or 10 years for a century or so without harming the tree. *Q. subur* is the source of all commercial cork.

Unlike the vascular cambium, the cork cambium is not a permanent feature. It does not form a layer surrounding the trunk or branch but occurs as a cylinder of cells. After a few weeks these cells lose their meristemic ability, turn into cork cells, and die, and the cork cambium disappears. As the trunk or branch grows thicker, however, more epidermis is needed to contain it. The existing skin splits, producing the characteristically rugged appearance of the outside of a tree, and new cork cambium forms to produce cells to fill the gaps between the cracks. Eventually no meristemic cells are left. Parenchyma cells in the secondary phloem then become meristemic and turn into a cylinder of cork cambium and the process repeats.

Sapwood and Heartwood

Inside the vascular cambium, the secondary xylem transports water from the roots to the rest of the tree, and each year a new layer of xylem is added. As the tree grows, more xylem vessels are needed to keep it supplied with water, but since each new layer forms as a cylinder laid outside the cylinders of previous years, the xylem circumference grows continually and the number of vessels in the xylem increases each year. The growth of new secondary xylem keeps pace with the increasing demand for water.

As new layers of xylem are added around the outside, layers on the inside die. While they lived, the tracheids and (in

angiosperms) vessel elements that comprise the xylem tissue produced lignins in their cell walls. Lignins strengthen the living cells, and when the cells die, the lignins remain, forming a central core of strongly lignified dead cells. Often, these cells fill with various waste products, altering their color and making this wood easily distinguishable from the sapwood.

The central core of dead xylem, no longer transporting water, comprises the heartwood, a column of tough material that adds greatly to the overall strength of the tree.

How Growth Rings Form

In temperate regions, plant growth ceases for a time in winter. This is obvious in respect of primary growth, which is clearly visible, but it also applies to secondary growth. It is this annual cessation and resumption of secondary growth that produces the annual growth rings that can be seen in any cross section through a tree trunk or branch.

When growth commences in spring, the vascular cambium starts to produce tracheids and vessel elements. They grow rapidly and consequently have large diameters and thin walls. These cells appear pale, and because of their size, they form a fairly wide band around the secondary xylem. By later summer, growth has slowed, and the cambium is producing cells of smaller diameter with thicker, darker walls. These cells pack more densely into a narrower band. When growth ceases for the winter, the year's accumulation comprises a wide, pale band on the inside and a narrow, dark band on the outside. The two together represent the growth for that year, and the following spring a new band of big, pale cells will be deposited outside the dark band from the previous year.

Counting annual growth rings reveals the age of the tree, but the rings can reveal much more information than that. If the weather was good during the growing season, the tree will have grown well, and the growth ring will be wide. If the weather was poor, there will have been less growth, and the ring will be narrow. Conditions may have been so poor in some years that the tree put on no secondary growth at all. When that happens, there is no pair of rings corresponding to that year, so a straightforward count of the number of rings will give an incorrect age for the tree. Tree rings must be interpreted carefully. The scientists who perform this task, called dendrochronologists, check the rings from their specimens against standard tree-ring chronologies that allow them to detect missing years. The study of tree rings is highly skilled and specialized, but it is valuable. Knowledge of the precise climatic preferences for a particular tree species, combined with a careful study of its annual growth rings, can be made to reveal climatic changes over the years.

PHOTOSYNTHESIS

Spring is the green time. In a broad-leaved deciduous forest it is the season when buds burst open and the trees hide their branches beneath young, bright green leaves. Even in the evergreen coniferous forest, there is some intensification of the green. Broad-leaved shrubs and herbs, the undergrowth that borders paths and fills the open spaces, produce its fresh leaves and the conifers, too, and resume their growth, producing new needles, brighter than their older neighbors.

In fall, the colors change. Before they are shed, many of the useful substances that leaves contain are absorbed into the tree branch, where they are stored until the following spring when they will be supplied to new leaves. As summer gives way to fall, leaves cease making chlorophyll. Its green color fades, and the colors of other compounds become visible. Leaves turn yellow, red, or brown before they wither with the loss of their water and nutrient supply, then they are detached from the tree, and discarded.

Green is the color of the most important type of chlorophyll, and chlorophyll is the compound that allows plants to manufacture sugars from carbon dioxide and water. The process is called photosynthesis, literally the assembly, or synthesis, of a complex substance using the energy of light—for which the Greek is *phōs phōtos*. Photosynthesis is the process on which most life depends.

Light, Shade, and Warmth

Obviously, photosynthesis requires light, but not too much light. If the light is too intense, photosynthesis slows down. The effect is called *solarization* and is probably due to reactions that destroy chlorophyll or stop certain essential reactions. Some plants are more sensitive to this than others. Those adapted to life in the shade, where they can photosynthesize at relatively low light levels, suffer badly if they find themselves exposed to full sunlight. Such shade-loving plants grow on the floor of a temperate forest. If trees fall, exposing areas of ground to bright sunlight, the shade-lovers disappear, and opportunistic light-lovers take their place until young trees have grown up to shade the ground once more and allow the shade-lovers to return. Plants that grow in open habitats, on the other hand, tolerate quite high light levels.

Photosynthesis works best and produces the largest amount of sugars if periods of light and darkness alternate. It is as though during the light periods some substance is produced faster than it can be utilized, so dark periods are needed for that substance to be absorbed by other reactions.

Photosynthesis also requires warmth. It ceases altogether when the temperature falls below 21°F (-6°C) and proceeds very slowly at temperatures close to freezing, but between 32°F (0°C) and 95°F (35°C) its rate doubles with every 18°F (10°C) temperature rise. Above 95°F (35°C) the rate decreases rapidly, and when the temperature exceeds about 113°F (45°C), photosynthesis ceases, and within a fairly short time, most plants adapted to temperate climates die.

Chlorophyll and Other Pigments

Chlorophyll is the chemical compound at the heart of the process of photosynthesis. It absorbs light in a way that makes the energy possessed by units of light, called photons, available to power the chain of photosynthetic chemical reactions. Chlorophyll absorbs light energy at quite precise wavelengths, and slight variations in chlorophyll molecules allow them to absorb at different wavelengths. Other compounds also absorb light energy and can pass it on to chlorophyll molecules. Substances that absorb light are called pigments, and the color of a pigment is the color of the light it reflects, not of the light it absorbs.

One type of chlorophyll, known as chlorophyll *a*, is also called P700 because it absorbs most strongly at a wavelength of 700 nanometers (nm) in the far red part of the spectrum. (A nanometer is one-billionth of a meter and is equal to approximately 0.00000004 of an inch.) Another type of chlorophyll *a*, known as P680, also absorbs red light but most strongly at 680 nm. Chlorophyll *a* is blue-green because this is the color of the light it reflects, and both P700 and P680 versions play important parts in photosynthesis in what are called photosystem I (PSI) and photosystem II (PSII) respectively.

Chlorophyll *b*, which is chemically very slightly different from chlorophyll *a*, is yellow-green in color. The most important of the other photosynthetic pigments are the xanthophylls, which are yellow, and the carotenoids, which are various shades of red through orange. Each absorbing at a different wavelength, these pigments expand the waveband of light plants can use.

Both PSI and PSII contain these nonchlorophyll pigments. PSI also contains P700 and PSII contains P680.

Chloroplasts, Where Photosynthesis Happens

Chlorophyll is found inside structures called *chloroplasts* within plant cells. Chloroplasts can reproduce by dividing, independently of the cell containing them, and they have their own DNA. Their possession of DNA that is distinct from the DNA in the cell nucleus, the manner in which they divide, and the structure and chemistry of their inner membranes give chloroplasts the features of simple organisms that live independently. Most biologists believe that the chloroplasts in plant and algal cells are the descendants of prokaryotic organisms that once lived independently but that entered and became incorporated into bigger cells (see the sidebar below). This merger probably happened more than 2 billion years ago after chloroplasts had been powering themselves by photosynthesis for some hundreds of millions of years. The story began, obviously, with the synthesis

Prokaryotes and Eukaryotes

All living organisms are made from units called cells. Plants and animals contain billions of cells, but there are also organisms that consist of just a single cell.

Scientists used to believe that there are two basic types of cell, known as prokaryotes or prokaryotic cells and eukaryotes or eukaryotic cells. Fungi, algae, plants, animals, and protozoa consist of eukaryotic cells. These are eukaryotic organisms. Bacteria are single-celled, prokaryotic organisms.

The prokaryotic cell is the simplest type. It has no nucleus and its DNA exists as a loop inside the cell. Some species also contain plasmids, which are small circles of DNA that can move out of one cell and into another. The DNA is not associated with RNA or with proteins. The cell contains globules of fat (lipids) as a food reserve and small ribosomes, which are the sites of protein synthesis. Enclosing the cell, there is an inner membrane involved with respiration and in some species with nitrogen fixation and photosynthesis. This membrane is surrounded by a cell wall and a protective outer layer called a capsule. Prokaryotic cells divide by dividing into two. Some also engage in conjugation, a process in which two or more cells link together through cellular extensions called pili. The cells pass DNA through the pili, altering the genetic constitution of the recipient cells, then separate and divide into two, each daughter cell carrying the new genetic constitution.

A eukaryotic cell is much bigger than a prokaryotic cell. Most of its DNA is linear rather than a loop, is associated with RNA and proteins, and it is contained inside a nucleus enclosed by a double-membrane envelope. Its ribosomes are bigger than those of prokaryotes and some of them are bound to structures called endoplasmic reticulums, which transport proteins and stores, carbohydrates, lipids, and other nonprotein products of cell metabolism. As well as the cell nucleus, the cell contains mitochondria and some cells contain chloroplasts. These structures are known as organelles and each type performs a particular function. Mitochondria and chloroplasts possess DNA of their own, suggesting that these organelles have descended from cells that were free-living. The cells of algae and plants are contained within a wall containing cellulose. Fungal cell walls contain *chitin*. Animal cells have no wall. Eukaryotic cells reproduce by mitosis or meiosis.

In about 1980 a third type of single-celled organism was discovered. This group included cells living in extreme environments (and called extremophiles), methanogens (cells that release methane as a metabolic by-product), and cells that respire by reducing sulfate. These organisms are apparently prokaryotic and at first were thought to be a category of bacteria. They are now known to be related to eukaryotes more closely than to bacteria, however, and this has given rise to a revision in the classification of living organisms. All organisms now belong to one of three domains. The domain Archaea includes the extremophiles, methanogens, and sulfate-reducing organisms. The domain Bacteria includes all bacteria and cyanobacteria. The domain Eukarya includes all eukaryotic organisms.

Using the names *prokaryote* and *eukaryote* suggested that eukaryotes evolved from prokaryotes. The profound differences between the three domains of life clearly show that this is not so. Consequently, the term *prokaryote* is misleading and in years to come it will probably fall into disuse.

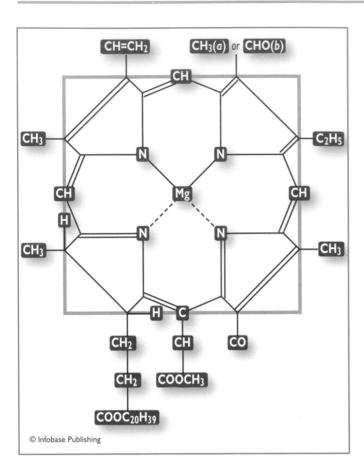

The chlorophyll molecule

of chlorophyll, and as a side effect of photosynthesis, the atmosphere gradually accumulated its by-product, oxygen.

A chlorophyll molecule is very large and complex. The diagram shows its general structure. The pigment makes up the "head" of the molecule, shown enclosed inside a gray border. Chemically, it is a porphyrin ring centered on an atom of magnesium (Mg). Chlorophylls *a* and *b* differ in the group shown in the upper right corner; chlorophyll *a* has a CH_3 group at this position, and chlorophyll *b* has CHO. At its base, abbreviated in the figure as $COOC_{20}H_{39}$, there is a long "tail" of carbon and hydrogen. This tail is water repellent and attaches the chlorophyll molecule to the membrane in the chloroplast.

A chloroplast is lens shaped, about 2–4 μm by 4–7 μm, and comprises a double outer membrane containing a dense liquid called the stroma, and stacks, called grana, of membranes, called thylakoid membranes, that form disk-shaped capsules each enclosing a thylakoid space and keeping it separate from the stroma. The chlorophyll molecules of types *a* and *b* as well as xanthophyll and carotenoid molecules are located in the thylakoid membranes. The stroma also contains starch grains and fat molecules. Chloroplasts

are concentrated in the cells, called mesophyll cells, which form the interior of leaves. Each mesophyll cell contains between 30 and 40 chloroplasts.

The Light-Dependent Stage

When a chlorophyll molecule, held in the thylakoid membranes of chloroplasts, is exposed to light, it is excited. The molecule absorbs a photon of light that possesses an amount of energy precisely equal to the amount needed to raise one electron from its usual orbital, called its ground state, to a higher-energy orbital, called its excited state. The excited electron escapes from the chlorophyll molecule, but it is immediately captured by a neighboring molecule, known as a primary electron acceptor. This acceptor molecule passes (the technical term is *donates*) an electron (not necessarily the same one) to another acceptor molecule and so on along an electron-transport chain.

The overall result, shown in the diagram, is that the energy supplied by the photon is used to split a water molecule into hydrogen and oxygen. The oxygen is released into the air, and the hydrogen passes on to the next stage in the process. The splitting of water molecules to provide hydrogen can take place only when chlorophyll is exposed to light. For that reason, this sequence of reactions is known as the light-dependent stage, or light reactions, of photosynthesis. It is the light-independent stage, or dark reactions, in which the hydrogen is combined with carbon dioxide to produce sugars. These reactions also take place while the plant is exposed to light, but they are able to proceed in darkness. They do not depend on exposure to light. The entire sequence can be summarized by a very simple equation:

$$6CO_2 + 6H_2O + \text{light energy} \xrightarrow{\text{chlorophyll}} C_6H_{12}O_6 + 6O_2\uparrow$$

In the presence of light energy and chlorophyll, carbon dioxide reacts with water to yield sugar and gaseous oxygen. $C_6H_{12}O_6$ is glucose, also called dextrose; it is a simple sugar. The arrow pointing upward indicates that the oxygen is released into the atmosphere.

In reality, the process is rather more complicated. The diagram on page 119 shows that the light-dependent stage involves the splitting of water to provide hydrogen. This is true, but it is not all that happens. Water molecules sometimes split of their own accord into hydrogen (H^+) and hydroxyl (OH^-) ions, so there are always a few of these ions inside a cell. When the absorption of photons causes chlorophyll molecules to release electrons (carrying negative charge), some of these electrons combine with hydrogen ions to form hydrogen atoms. Free hydrogen atoms can

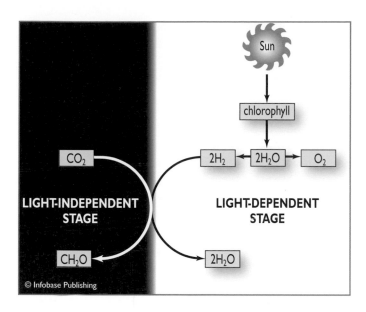

A simplified summary of the two stages in photosynthesis

ATP) and releases it again as that bond is relinquished (ATP becomes ADP). This is the principal mechanism by which energy is transported around every living organism as molecules of ATP and released where it is needed.

So the light energy striking the chlorophyll molecule provides chemical energy as well as hydrogen. Eventually, the last of the electron carriers passes the electron back to the chlorophyll molecule, completing the cycle. Because it uses light energy to establish high-energy phosphate bonds, this type of phosphorylation is called photophosphorylation, and because the process repeats itself for as long as light absorption by chlorophyll continues to liberate electrons, it is known as cyclic photophosphorylation.

The electron from the hydroxyl (from water) also travels along an electron transport chain, in this case one involving both photosystems (I and II). Photosystem I (PSI) is at a higher energy level than photosystem II (PSII). Light striking PSI releases an electron that is passed to NADP, and it is replaced in PSI by an electron passed from PSII to plastoquinone (one of a group of complex organic molecules) and then to PSI. As the electron passes along the electron transport chain, its energy is increased, and this allows the formation of more ATP. This is also photophosphorylation but known in this case as noncyclic photophosphorylation to distinguish it from the other, cyclic, form of photophosphorylation.

The Light-Independent Stage and the Calvin Cycle

The photosynthetic action then moves from the thylakoid membranes to the stroma, which is where the light-independent reactions occur. The "photo" stage gives way to the "synthesis" stage. The light-dependent reactions release energy, but the light-independent reactions absorb it, and the energy is supplied as ATP. Carbon is reduced, the hydrogen for the purpose being provided from the splitting of water during the light-dependent stage and carried by NADPH. Carbon and hydrogen are then used to construct sugars in a series of steps discovered by a team at the University of California at Berkeley led by the American biochemist Melvin Calvin (1911–97; see the sidebar on page 120), for which he received the 1961 Nobel Prize in chemistry. The process is known as the Calvin cycle. It is a cycle because one of its key materials, ribulose biphosphate (RuBP), is broken down at one point and then reconstituted at another, so the process can begin again.

The cycle commences when a molecule of carbon dioxide is attached to RuBP with the assistance of an enzyme, RuBP carboxylase (usually called rubisco). Because photosynthesis is so widespread and rubisco is so abundant inside the stroma of chloroplasts, this enzyme may be the most

attach themselves to a hydrogen acceptor, nicotinamide adenine dinucleotide phosphate (NADP), so NADP becomes NADPH (sometimes called $NADP.H_2$). The NADP has been reduced, and in this form, it enters the light-independent stage, in the course of which it is oxidized by losing its hydrogen and can be used again in the light stage.

The chlorophyll molecule has lost an electron in this reaction, leaving it with a positive charge. The hydroxyl ion (OH^-) has an extra electron, which it donates to the chlorophyll. This returns both the chlorophyll and hydroxyl to their neutral state, and hydroxyls can then combine to produce water with the release of oxygen: $4OH \rightarrow 2H_2O + O_2$. That is how the oxygen comes to be released.

Cyclic and Noncyclic Photophosphorylation

At the same time that the light-dependent stage is proceeding, other electrons released by the chlorophyll are accepted by ferredoxin (one of a group of proteins containing iron) and then pass along a series of carriers, all of them at slightly different energy levels. Energy that is removed from the electrons, causing them to fall to a lower energy state, is transferred to molecules of adenosine diphosphate (ADP), allowing the ADP to take up an additional phosphate group and become adenosine triphosphate (ATP). The process of acquiring a phosphate group is called phosphorylation.

The reversible change between ADP and ATP absorbs energy as the third phosphate bond is formed (ADP becomes

Melvin Calvin (1911–1997)

Melvin Calvin, the chemist who discovered the sequence of reactions by which green plants convert carbon dioxide and water into carbohydrates and oxygen, was born in St. Paul, Minnesota, on April 8, 1911. His parents were immigrants from Russia. Calvin was educated at the Michigan College of Mining and Technology, from which he graduated in 1931, and he obtained his Ph.D. in 1935 from the University of Minnesota. He then spent two years working at the University of Manchester, England, before moving to the University of California, where he spent the rest of his career. He was appointed University Professor of Chemistry in 1971.

Calvin began to study photosynthesis in 1949, while he was director of the bioorganic chemistry group at the university's Lawrence Radiation Laboratory. He used radioactive carbon-14 to trace the steps by which carbon dioxide is converted to starch. Using the single-celled green alga *Chlorella,* Calvin showed that the light-independent reactions comprise a cycle, now known as the Calvin cycle. It took him almost 15 years to work out all the reactions in the cycle. It was for this work that Calvin was awarded the 1961 Nobel Prize in chemistry. He won many prizes in addition to the Nobel.

From 1960 until 1980 Calvin was director of the Laboratory of Chemical Biodynamics at Berkeley. During this time he turned his attention to the origins of life on Earth and to the possibility of life elsewhere in the universe. He also worked on the development of alternatives to fossil fuels.

Melvin Calvin died at Berkeley, California, on January 8, 1997.

plentiful protein on Earth. RuBP has five carbon atoms, so the addition of carbon dioxide produces a 6-carbon compound. This is very unstable and immediately divides into two molecules of a 3-carbon compound, phosphoglyceric acid, also called 3-phosphoglycerate (PGA).

An enzyme then transfers an additional phosphate group from ATP to each molecule of PGA, producing 1,3-diphosphoglycerate. NADPH donates two electrons. These reduce the carboxyl group (COOH) on 1,3-diphosphoglycerate to a carbonyl group (CO), an arrangement that stores more energy, and the product is six molecules of glyceraldehyde phosphate, a 3-carbon sugar. In a series of steps, one of these molecules is built into a 6-carbon

sugar that can be converted to starch for storage. The remaining five molecules are used to reconstruct ribulose biphosphate (RuBP), allowing the cycle to be repeated, and energy for the reactions is supplied by the conversion of ATP to ADP.

The cycle is rather complicated, but it is easier to understand if you follow the way carbon atoms are assembled. The illustration names the intermediate compounds, with the number of molecules, and beside each name it indicates the number of carbon atoms present at that stage. The input to the cycle at the top of the diagram on page 121, consists of carbon dioxide, with one carbon atom, and the output, at the bottom of the diagram, is a 3-carbon sugar that is then processed further to make starch, fats, and proteins. The cycle must be repeated three times in order to incorporate the three carbon atoms from three carbon dioxide molecules. The energy driving the cycle is supplied by the ATP-to-ADP mechanism, using nine ATP units. Six NADPH molecules supply the high-energy electrons needed to bond together the atoms comprising the sugar.

Three molecules of carbon dioxide join three molecules of the 5-carbon ribulose biphosphate, forming unstable 6-carbon molecules that divide. There are now 18 carbons (three 6-carbon molecules which become six 3-carbon molecules). Three of these leave the cycle as its output, in the form of one molecule of glyceraldehyde phosphate, leaving 15 carbons to be reconstituted as the three molecules of ribulose biphosphate which complete the cycle.

C3, C4, and CAM Plants

The first product in the Calvin cycle is 3-phosphoglycerate, a compound with three carbon atoms, and for this reason plants which use this version of the light-independent stage of photosynthesis are known as C3 plants. There are also C4 plants in which the addition of carbon dioxide (a process called carboxylation) to an acceptor molecule, phosphoenolpyruvic acid (PEP), yields a 4-carbon molecule, oxaloacetate (OAA). OAA is then converted to other 4-carbon compounds. This leads to a much more efficient version of photosynthesis, and C4 plants evolved more recently than C3 plants. The C4 pathway requires more ATP and more water than the C3 pathway, but under suitable conditions it yields more sugar for a given leaf area, so C4 plants grow faster than C3 plants. They can also tolerate higher light intensities and lower atmospheric carbon dioxide concentrations. There are many C4 plants, but most are grasses, including sugarcane and corn (maize), or desert plants. All trees, including those of temperate forests, are C3 plants. C4 plants have not replaced C3 plants because they perform better than C3 plants only when the light is intense and the temperature high. In temperate latitudes, C3 plants retain the advantage.

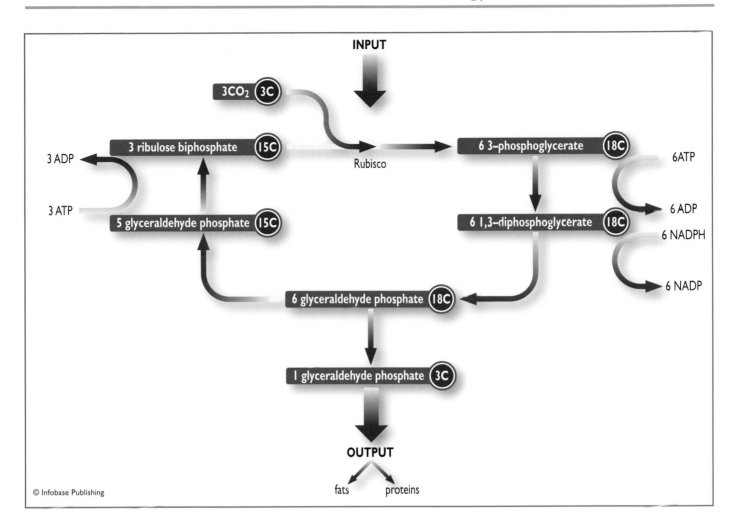

INPUT

3CO₂ 3C

3 ribulose biphosphate 15C

Rubisco

6 3-phosphoglycerate 18C

6ATP

3 ADP

6 ADP

3 ATP

5 glyceraldehyde phosphate 15C

6 1,3-diphosphoglycerate 18C

6 NADPH

6 glyceraldehyde phosphate 18C

6 NADP

1 glyceraldehyde phosphate 3C

OUTPUT

fats   proteins

© Infobase Publishing

The Calvin cycle, which makes up the light-independent stage in photosynthesis

Another group of plants of dry climates, including cacti and the pineapple, close their stomata during the day which conserves water and open them at night. This is the opposite of the pattern followed by most plants, and it requires a different mechanism, called the crassulacean acid metabolism (CAM) after the plant family in which it was first recognized, the Crassulaceae (the family that includes the stonecrops and houseleeks). CAM plants absorb carbon dioxide at night, store it in the form of organic acids until daylight, then close their stomata, use sunlight as a source of energy for producing ATP and NADPH, and release the stored carbon dioxide into the Calvin cycle.

Photorespiration

Photosynthesis is a complex process, but it is not very efficient. Chlorophyll captures only a very small proportion of the sunlight falling on plants, and once carbon dioxide has been fixed, a considerable amount is lost again by a process called *photorespiration*. C4 and CAM plants avoid photorespiration, but in some C3 plants it can rob the plant of as much as half of the carbon entering the Calvin cycle.

The trouble arises because rubisco, the enzyme that facilitates the union of ribulose biphosphate and carbon dioxide, is able to accept oxygen instead of carbon dioxide and has no particular preference for one rather than the other. During daylight, the light-dependent stage of photosynthesis releases oxygen as a by-product, so leaf cells contain oxygen that escapes into the outside air when the stomata are open. They also contain carbon dioxide, which enters through the stomata and is incorporated in the Calvin cycle, or light-independent stage. How much of each gas the leaf cells contain depends largely on the light intensity. The brighter the light, the more photons chlorophyll molecules capture. The flow of electrons increases, more water molecules are split, and more oxygen is produced. Its dependence on light intensity is why this form of respiration is called *photo*respiration, and it is most extreme on bright, sunny days in summer.

If the cells contain more oxygen than carbon dioxide, rubisco will gather the oxygen and combine it with RuBP. Instead of the two three-carbon molecules of the Calvin cycle, this produces one three-carbon molecule which continues in the cycle and one molecule of glycolate, a two-carbon compound. This leaves the Calvin cycle and is broken down elsewhere in the cell, in a series of stages, releasing carbon dioxide. Respiration is the oxidation of carbon to carbon dioxide, but in this case it confers no known benefit on the plant. Photorespiration produces no ATP, so it consumes energy rather than releasing it, and it robs the Calvin cycle of carbon, thereby reducing the amount of sugar produced by photosynthesis.

C4 and CAM plants do not suffer from photorespiration losses because instead of ribulose biphosphate they use phosphoenolpyruvate (PEP) to fix carbon dioxide in a reaction catalyzed by the enzyme PEP carboxylase. PEP carboxylase cannot react with oxygen, so oxygen and carbon dioxide cannot compete for it.

Photosynthesis and the Food We Eat

Inefficient and wasteful it may be, but photosynthesis is the process on which almost all life depends (there are some ecosystems around volcanic vents in the deep oceans and in sulfur springs and boiling muds where light never penetrates and no organisms practice photosynthesis), and its overall output is impressive. Biologists have calculated that each year green plants synthesize about 180 billion tons (160 billion tonnes) of carbohydrates.

When the young leaves burst forth in spring and even the conifers resume their growth and produce pale, new needles, it is a sign that the frozen ground has thawed, the temperature has risen, and photosynthesis is beginning once again. Photosynthesis makes plant growth possible. It does so partly by assembling carbohydrates to be used structurally but mainly by providing energy. Sugars, converted to starch, store energy that is released by the process of *respiration* (not photorespiration). That energy is used to convert ADP into ATP. ATP is transported to every part of the organism and released to cells that need energy by converting ATP back into ADP. This is the mechanism that drives the plant metabolism by which mineral nutrients from the soil are combined with carbon, oxygen, and hydrogen obtained by photosynthesis to form the complex organic (carbon-based) compounds from which cells and tissues are constructed. Photosynthesis is the suite of processes by which light energy is captured and held in order for the necessary materials to be assembled for growth and reproduction.

Herbivorous animals live by eating plants, and carnivores by eating herbivores. Dependence on plant-eaters means that even the most committed carnivores such as cats and weasels could not survive without plants. Fungi, which are neither plants nor animals, obtain nourishment by absorbing organic compounds. Green plants synthesized those compounds, so fungi are no less dependent on green plants than animals are. It is the process of photosynthesis that feeds the planet.

■ NITROGEN FIXATION AND DENITRIFICATION

Life on this planet is based on proteins. Sugars and fats supply animal bodies with fuel, but proteins are the building blocks from which all plant, animal, and fungal tissues are constructed. Enzymes and hormones, which control chemical reactions, are also proteins.

Proteins are arrangements of amino acids. About 80 amino acids occur naturally, and about 20 of these commonly occur in proteins. All amino acids comprise an amino group (NH_2) and a carboxyl group (COOH), both of which are attached to the same carbon atom, to which a side chain is also attached. The side chain is different for each amino acid.

Amino acids are synthesized by plants, in which they are assembled to form plant proteins. Herbivorous animals obtain their amino acids by eating plants and carnivores by eating herbivores. Animals can alter many of the amino acids they obtain from plants to construct proteins that suit their own requirements, but ultimately it is plants that supply the raw materials. Those raw materials include carbon, oxygen, and hydrogen, and also nitrogen. Carbon, oxygen, and hydrogen, the ingredients of carbohydrates, are obtained in the first instance by photosynthesis (see "Photosynthesis" on pages 116–122) from atmospheric carbon dioxide and water. Nitrogen is also obtained from the air but by a different route. Essential though it is, plants cannot obtain their own nitrogen. They need help.

Why Plants Cannot Use Nitrogen Gas

It is not that nitrogen is scarce. Gaseous nitrogen comprises about 78 percent of the Earth's atmosphere. Air-breathing animals inhale it and then exhale it again unaltered. Nitrogen gas (N_2) is very reluctant to engage in chemical reactions, which means that plants and animals cannot assimilate it directly. Plants can absorb nitrogen only in the form of ammonia (NH_3), ammonium (NH_4), or nitrate (NO_3). Once nitrogen has been induced to combine with oxygen or hydrogen, its character changes. Nitrogen compounds will readily participate in reactions, and these can be used in the synthesis of amino acids. The trick is to persuade nitrogen to join another element in the first place. Fixation is the technical term for the capture of an element such as nitrogen or carbon and its incorporation in a compound that living cells can use.

Enough energy will do it. Nitrogen fertilizer is made in factories by heating air to 750–930°F (400–500°C) under about 200 times atmospheric pressure in the presence of a catalyst. Lightning supplies enough natural energy to oxidize nitrogen in a series of steps that produce nitric acid (HNO_3) that is washed to the ground in rain (contributing to the natural acidity of all rain) where it forms nitrates (NO_3) that can be used directly by plants. Lightning fixes a considerable amount of atmospheric nitrogen, but soil microorganisms fix very much more, and it is their activities that make available most of the nitrogen entering plant roots.

Colonies of *Rhizobium* bacteria in nodules attached to the roots of leguminous and some nonleguminous plants (see "Nitrogen and Root Nodules" on page 104) fix large amounts of gaseous nitrogen, but they are not alone. There are also free-living bacteria that fix nitrogen, those of the genera *Azotobacter* and *Clostridium* being the most closely studied. Certain sulfur bacteria, including members of the genera *Chromatium*, *Rhodospirillum*, and *Chlorobium*, also fix nitrogen in the soil, and in aquatic systems there are cyanobacteria, including *Nostoc* and *Anabaena* species, that also do so.

Soil Bacteria That Make Ammonia

Some of nitrogen-fixing organisms require oxygen (are aerobic); others cannot tolerate oxygen (are anaerobic); some perform photosynthesis, and others do not. The nitrogen-fixing talent is distributed among a disparate group of microorganisms. What they are all believed to have in common is nitrogenase, an enzyme that makes nitrogen react with hydrogen at ordinary environmental temperatures and pressures.

Nitrogenase catalyzes a sequence of reactions that produces ammonia (NH_3). The sequence can be summarized as:

$$N_2 + 8e^- + 8H^+ + 16ATP \rightarrow 2NH_3 + H_2 + 16ADP + 16P$$

The bacteria use the nitrogen they have fixed to synthesize the amino acids they need, but the process is expensive in its use of energy. Producing each molecule of ammonia requires the expenditure of 8 molecules of ATP in the ATP → ADP reaction. The summarized reaction indicates this by showing that the conversion involves attaching a nitrogen molecule to 8 electrons (e^-) and 8 hydrogen nuclei (H^+). This requires 16ATP units of energy. The reaction yields 2 molecules of ammonia, one molecule of hydrogen, 16 molecules of ADP, and the 16 phosphate groups (P) that were detached from ATP.

Bacteria That Seize Ammonia

Plants are able to use ammonia, but aerobic bacteria seize most of it. As a group, they are known as nitrifying bacteria because of the way they process nitrogen. In the soil, free ammonia, dissolved in water, combines with carbon dioxide

to form ammonium (NH_4) carbonate ($(NH_4)_2CO_3$). Certain bacteria then oxidize the ammonium carbonate, a chemical reaction that releases energy the bacteria can use:

$$(NH_4)_2CO_3 + 3O_2 \rightarrow 2NHO_2 + CO_2 + 3H_2O + energy$$

This is how *Nitrosomonas* and *Nitrococcus* bacteria live. The somewhat unstable nitrous acid that they produce quickly combines with magnesium or calcium to form a nitrite (Mg or $Ca(NO_2)_2$) and this is the wherewithal for another group, *Nitrobacter*, which obtains its energy by oxidizing the nitrite to nitrate (Mg may take the place of Ca):

$$2Ca(NO_2)_2 + 2O_2 \rightarrow 2Ca(NO_3)_2 + energy$$

These are not the only bacterial species involved. There are several others that oxidize ammonia to nitrate. All of these processes are known as nitrification.

Nitrogen Recycling

Nitrogen fixation is the original source of all soil nitrogen, but once fixed, the nitrogen is recycled repeatedly and returns to ammonia in the course of its cycle (see the sidebar "The Nitrogen Cycle" on page 124). Bacteria and other soil organisms die. So do plants and animals that live on the surface. All living organisms produce waste. Wastes and the remains of once-living organisms are deposited on or in the soil. All of this dead organic matter contains nitrogen in the form of proteins or as nitrogen compounds that are the end product of protein metabolism. Some of this nitrogen returns to the soil as ammonia. Fungi and bacteria that obtain their nutrients from dead organic matter convert the remaining nitrogen compounds into ammonia (NH_3) or ammonium (NH_4). From there the nitrogen is taken up by nitrifying bacteria, converted to nitrate (NO_3), and taken up again by plants in that form.

Plants have preferences. Most coniferous trees take up ammonium more readily than nitrate, for example, although, like most plants and microorganisms, they are able to utilize ammonium, nitrate, or amino acids dissolved from dead organic matter. At Hubbard Brook Experimental Forest, a broad-leaved deciduous forest in New Hampshire, scientists have measured the fixation of nitrogen by soil microorganisms as 12.5 pounds of nitrogen per acre per year (14 kg/ha/yr). In addition, they discovered that 5.8 pounds per acre per year (6.5 kg/ha/yr) was washed to the ground in rain. Of the total, about 85 percent of the nitrogen was in the form of nitrate and the remainder as ammonium, so most of the nitrogen being taken up by trees was in the form of nitrate.

Once nitrogen enters the cycle, living organisms hold onto it tenaciously. When nitrogen is returned to the soil, as dead organic matter, it does not remain there long before

The Nitrogen Cycle

Like many other chemical elements, nitrogen moves between air, soil, water, and living organisms in an end-less cycle. These are known as biogeochemical cycles, and those cycles involving essential nutrient elements, such as nitrogen, are called nutrient cycles. The illustra-tion shows the principal stages in the cycle.

Nitrogen is by far the most abundant gas in the atmosphere. It enters the cycle and becomes available to plants and animals by the process of fixation. This occurs through bacterial action, by oxidation in the air using the energy of lightning, and also industrially, through the manufacture of fertilizers. Nowadays the pressures and temperatures achieved in the cylinders of vehicle engines also fix a significant amount of nitrogen.

Bacteria fix nitrogen by attaching it to hydrogen to form ammonia (NH_3). Ammonia is also the compound produced by the Haber process for making fertilizer ($N_2 + 3H_2 \rightarrow 2NH_3$). Ammonia is soluble in water. One group of soil-nitrifying bacteria converts ammonia to ammo-nium (NH_4). Ammonium reacts with carbon dioxide (CO_2) to form ammonium carbonate (($NH_4)CO_3$). Nitrify-ing bacteria convert ammonium carbonate into nitrites (NO_2) with the release of energy, and a second group of bacteria converts nitrites to nitrates (NO_3).

Nitrogen is fixed by lightning and in car engines by oxidation. Nitrogen oxides dissolve in rain and reach the ground as dissolved nitrates, such as nitric acid (HNO_3).

Nitrates dissolve in water, and plants take up nitrates from the soil solution, using them in the synthesis of amino acids. Plants, animals, and fungi construct pro-teins from amino acids.

Plant, animal, and fungal waste products and dead matter decay in the soil. The bacteria and fungi respon-

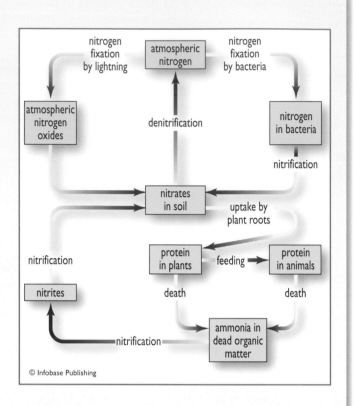

The nitrogen cycle

sible for decomposition release nitrogen as ammonia. Most of the ammonia returns to the soil cycle, but some vaporizes and returns to the atmosphere.

A proportion of the nitrates in the soil is utilized by denitrifying bacteria. Certain species of these organ-isms release nitrogen gas (N_2), completing the nitrogen cycle.

plant roots absorb it again, and in a forest the trees them-selves hold a large amount of nitrogen in their tissues. At Hubbard Brook, part of the study involved measuring the effect of felling all the trees in a particular area. This greatly reduced transpiration, of course (see "Transpiration" on pages 48–50), consequently much more of the rain drained directly into the groundwater, which increased the rate at which nutrients were leached from the soil. The effect on the loss of nitrogen was the most dramatic, however. The rate at which nitrogen left the area, as nitrate dissolved in water, increased 60-fold. This indicates the amount that is con-stantly being cycled in a temperate forest and should serve

as a warning of the water pollution that may result from widespread forest clearance. Nitrate stimulates plant growth and in still or slow-moving water it can encourage blooms of algae; respiration by the bacteria decomposing dead algae uses up the oxygen dissolved in the water on which aquatic animals depend.

Evergreen plants use nutrients, including nitrogen, more efficiently than deciduous plants. This is due partly to their more efficient mechanisms for withdrawing and stor-ing nutrients from leaves before they are shed and partly to the fact that evergreen leaves live much longer than do deciduous leaves.

Denitrification—Returning Nitrogen to the Air

A visitor from another galaxy, viewing the solar system from afar but able to observe it in fine detail, might conclude that this star system was dead, except for the third planet. That planet possesses some very curious features, one of which is the large amount of nitrogen in the atmosphere.

On Mars and Venus, the two neighboring planets, the atmosphere contains almost no nitrogen—2.7 percent on Mars and 3.5 percent on Venus—and there is a very straightforward reason why that is how it should be on Earth. Lightning oxidizes nitrogen into compounds that are both soluble in water and stable. Rain washes these compounds to the surface, and there they should remain. Even if the atmosphere started with a large amount of nitrogen, it would take no more than a few million years for all of it to disappear into the oceans.

It was this type of reasoning that led the British chemist James Lovelock to conclude that the chemical composition of its atmosphere can reveal whether or not a planet supports life. Put simply, by removing substances from their environment and returning their metabolic waste products to the environment, living organisms alter the chemical composition of their surroundings. On a lifeless planet, the laws of physics and chemistry will drive all the constituents of the atmosphere and oceans to a state of equilibrium, where they will remain. Living organisms intervene in ways that prevent the attainment of this equilibrium, so on a planet that supports life the atmosphere and oceans will be chemically unstable. This is the idea Lovelock developed as the Gaia hypothesis.

Clearly, therefore, not only is atmospheric nitrogen being brought down to the surface, but sufficient nitrogen is also being returned to the atmosphere to maintain a constant concentration. In other words, as much nitrogen returns to the air as is taken from it. The organisms responsible for this part of the cycle are known as denitrifying bacteria, and the process is called denitrification.

The bacteria concerned, and there are several species, have no interest in the welfare of the planet, of course, or in the neat balancing of the nitrogen budget. They reduce nitrogen compounds in order to obtain their oxygen, which they use to oxidize sugars or, in the case of *Thiobacillus denitrificans,* sulfur compounds. This reaction releases energy that the bacteria use to synthesize the organic compounds they need for their own maintenance and reproduction.

Respiration is the process by which cells release energy through the oxidation of carbon, usually in the form of glucose ($C_6H_{12}O_6$). Where oxygen is abundant, respiration is aerobic (requiring oxygen). The process is the opposite of photosynthesis:

$$C_6H_{12}O_6 + 6O_2 \rightarrow 6CO_2 + 6H_2O + energy$$

Denitrification takes place in soils where levels of oxygen are too low for aerobic respiration. Aerobic organisms cannot survive without oxygen, but certain bacteria can switch from being aerobic to being anaerobic (not requiring oxygen); these are known as facultative anaerobes. Anaerobic respiration is less efficient than aerobic respiration, but it allows these organisms to inhabit places aerobic organisms find intolerable. Anaerobic respiration uses nitrate or sulfate as a source of oxygen, and where nitrate is used, the effect is to reduce nitrate in a series of reactions, eventually to dinitrogen oxide (N_2O) and finally to nitrogen:

$$NO_3 \rightarrow NO_2 \rightarrow NO \rightarrow N_2O \rightarrow N_2$$

Thiobacillus denitrificans releases gaseous nitrogen (N_2). Other bacteria, including species of *Pseudomonas, Micrococcus,* and *Clostridium,* produce ammonia. Ammonia is very soluble in water, but it boils at -29°F (-34°C) so it vaporizes at once if it comes out of solution. This returns a little to the air where some of it may be oxidized, but most is captured by nitrifying bacteria and remains in the soil until *T. denitrificans* succeeds in converting it to unreactive pure nitrogen. In waterlogged soils, denitrifying bacteria can remove as much as half of all the nitrogen present.

The amount of nitrogen fixed and returned to the atmosphere each year is very small when compared with the amount held firmly by living organisms and cycled from one to another within ecosystems. Nevertheless, nitrogen fixation, nitrification, and denitrification all depend on bacteria and, to a lesser extent, on fungi that use these processes to obtain energy for themselves. Without them, no plants could grow because they would have no access to nitrogen in a form that they could absorb, and nitrogen is an essential ingredient of proteins. Without plants, there could be no animals. Forests and all the plants and animals that inhabit them are able to exist only because of the activities of bacteria. These prokaryotes (see the sidebar "Prokaryotes and Eukaryotes" on page 117) are so small they that can be seen only with the help of a powerful microscope and so ancient they have been the only form of life throughout most of the history of our planet.

5

Ecology of Temperate Forests

A forest is a community of living organisms. As well as the trees, there are smaller plants, animals, fungi, and microorganisms. Each group affects all the others in many ways. A community of living organisms is called an ecosystem, and the study of ecosystems is called ecology. This chapter outlines the ecology of temperate forests.

The chapter begins with an outline of the history of ecological ideas and brief biographical notes about some of the scientists most closely involved in their development. These scientists had to find ways to study ecosystems. To do this they developed methods for measuring what goes on inside such a community and tracing all of the relationships that drive ecological processes. The chapter explains the most important ecological concepts. It then describes the structure of a temperate forest and the different small communities found within it. Among the organisms living in a forest, there are some that harm trees, causing diseases or physical damage that can be fatal. The chapter describes some of these organisms and their effects. Finally the chapter describes the difference between natural forests that have developed without any outside intervention and commercial tree plantations.

■ HISTORY OF ECOLOGICAL IDEAS

A forest is an extensive group of trees, but it is also much more than that. In the first place, even a plantation forest usually contains more than one species of tree, and a natural forest will contain several. Pisgah Forest in New Hampshire is fairly typical of a mixed, old-growth forest. It comprises five principal tree species and several less common ones. In Europe, the Białowieża Forest in Poland contains oak, hornbeam, lime, and spruce in some areas and alder, ash, willow, elm, and pine in others. It is a large forest, occupying 482.5 square miles (1,250 km²) of a plain, most of it inside the boundaries of a national park, and its composition varies from place to place.

It is the trees that define the forest, but other plants are also present. There are shrubs and herbs growing in the shade of the trees, climbers such as ivy using the trees for support, ferns, mosses, and liverworts in moister places, and lichens growing on exposed rocks and on the branches of some of the trees. All the plants depend on soil organisms, including fungi, and at the right time of year, the visitor may find the large fruiting bodies produced by some of those. Then there are the insects, birds, mammals, aMPHibians, and reptiles that inhabit the forest. The birds advertise their presence by their calls, but many of the mammals and reptiles are extremely secretive. Some leave traces by which they can be identified, but anyone wishing to catch a glimpse of them must know where to look and be prepared for a long wait.

The forest, then, is a living community of plants, animals, fungi, and microorganisms, all of them interacting with one another, and all of them ultimately reliant on the energy and nutrient elements captured by the photosynthesizing green plants. Since all of these organisms, living side by side within an area that has a clearly marked boundary separating it from adjacent areas with a distinctly different character, react with one another, it is possible to study them as a community, rather than as isolated individuals.

The Economy of Nature

The possibility of studying entire communities of organisms grew out of ideas about evolution that were being discussed in the 18th and 19th centuries. These ideas branched in several directions, but one line pursued the concept of the "economy of nature." According to this, organisms all live together in harmony, endowed by God with the means to satisfy their biological needs, and are thus able to supply human needs. The economy of nature was closely linked to the idea of natural theology, a theory with a long history. Natural theology held that plants and animals had been designed by God to coexist harmoniously, each performing its ordained function, and that through the study of nature the divine purpose might be revealed.

Natural theology led to the romantic, and very misleading, idea of a balance of nature. This is the notion that internal relationships between organisms maintain natural communities in a stable condition that must not be disturbed lest the community structure collapses (see "Primeval Forest, Ancient Woodland, Old-Growth Forest, Plantations" on pages 144–147). Natural theology also led to more rigorous studies of communities. These were first proposed by the eminent German zoologist Ernst Heinrich Haeckel (1834–1919), who sought to work out all the implications of the Darwinian theory of the evolution of species by natural selection (see the sidebar). In 1866, Haeckel published a two-volume work called *Generelle Morphologie der Organismen* (General morphology of organisms) in which he coined the word *Ökologie* to describe the study of a kind of large-scale economy of nature, in which all species are engaged. In effect, the new science (for that is what it amounted to) would concern itself with the details of how organisms reacted among themselves to manage the "household" in which they lived. The modern spelling, *ecology*, was adopted at the International Botanical Congress in 1893.

Ecology and Ecosystems

One of the earliest British ecologists, and to this day one of the most eminent, was Professor Sir Arthur George Tansley (1871–1955; see the sidebar on page 128). Tansley began the introduction to his book *Practical Plant Ecology* (published in 1923) with a definition.

> The word ECOLOGY, as is well known, is derived, like the common word *economy,* from the Greek oikos, *house, abode, dwelling.* In its widest meaning ecology is the study of plants and animals *as they exist in their natural homes;* or better, perhaps, the study of their *household affairs,* which is actually a secondary meaning of the Greek word. (The italics are his.)

In 1935, in an article in the *Journal of Ecology,* Tansley coined another word that has become no less familiar, and he introduced it to a wider readership in *Introduction to Plant Ecology,* a revised second edition of his *Practical Plant Ecology* that was published in 1946. He grouped all the living organisms together with the physical and chemical aspects of the climate and soil that affect them and described them as component parts of a single system which he called an *ecosystem.*

Plant Sociology

Nowadays students are taught ecology as a subject in its own right so that they can start their professional lives as qualified ecologists. This is a very recent development, however, and most ecologists started out as botanists. In mainland Europe, one of the leading early figures was a Swiss botanist, Josias

Ernst Heinrich Philipp August Haeckel (1834–1919)

Ernst Heinrich Philipp August Haeckel was a German zoologist and anatomist who was an enthusiastic supporter of Charles Darwin's theory of evolution by natural selection. Haeckel constructed evolutionary trees for many organisms. He also coined the word *ecology.*

Haeckel was born on February 16, 1834, in Potsdam, Prussia (now part of Germany), the son of a lawyer. He attended school in Merseberg and studied medicine at the University of Würzburg, graduating from the University of Berlin in 1857. It was while studying medicine that he became interested in zoology. After graduating, Haeckel spent some time traveling and then practiced medicine for one year. In 1861 he was appointed a lecturer at the University of Jena, and in 1862 he was made Extraordinary Professor of Comparative Anatomy at the Jena Zoological Institute. He became a full professor in 1865, remaining in the post until he retired in 1909.

In 1866 Haeckel met Darwin. After their meeting, Haeckel developed his law of recapitulation. This held that in the course of its development, an animal embryo passes through the evolutionary stages of its ancestral species. This is not true, although the idea became very popular during the 19th century. Biologists now know that the embryos of different species resemble one another during the early stages of their development but diverge rapidly in the later stages.

It was also in 1866 that Haeckel developed a method for representing the evolutionary history of animals in treelike diagrams. The positions of different species on a tree reflect the closeness of the evolutionary relationships between them. His method is still used. Haeckel also believed that life arose from the chemical ingredients and physical conditions of the environment and that the simplest forms of life developed by crystallization.

Haeckel's third major achievement of 1866 was the publication of his two-volume work *Generelle Morphologie der Organismen* (General morphology of organisms). It was in this book that he introduced the word *Ökologie,* which he derived from the Greek words *oikos* meaning "house" and *logos* meaning "discourse."

Ernst Haeckel died at Jena on August 8, 1919.

Arthur George Tansley (1871–1955)

Arthur Tansley was an English botanist who pioneered the science of plant ecology and through his teaching and writing brought ecological ideas and concepts to a wide audience. He edited and contributed to *Types of British Vegetation,* published in 1911, which was the first major book on the plants of the British Isles. Tansley later rewrote and enlarged this work. The new version, *The British Islands and their Vegetation,* published in 1939, was Tansley's greatest intellectual achievement. A shorter, more popular version was published in 1949, called *Britain's Green Mantle.*

Tansley was born in London on August 15, 1871. He attended Highgate School from 1886 to 1889 and then studied science at University College, London. It was there that he began to learn botany. In 1890 Tansley went to Trinity College, Cambridge. During his final year he combined his studies with teaching at Trinity College. After graduating in 1894, he remained at Trinity until 1906 as a demonstrator and assistant to his former teacher. He became a lecturer in botany in 1907. In 1900 and 1901 Tansley took time out to study the plants of Sri Lanka, the Malay Peninsula, and Egypt. On his return

he founded a journal called *The New Phytologist* in order to publish his findings (phytology is the study of plants) and remained its editor for 30 years. In 1913 the scientists who produced *Types of British Vegetation* founded the British Ecological Society, and Tansley was its first president. From 1916 to 1938 Tansley edited the Society's *Journal of Ecology.*

After World War I, Tansley abandoned plant science for a time and studied psychology in Austria under Sigmund Freud. In 1927, however, he was appointed Sherardian Professor of Botany at the University of Oxford and a Fellow of Magdalen College. He remained in this post until he retired in 1939.

After his retirement, Tansley continued to be active in conservation. He was chairman of the Nature Conservancy (a British government agency) from 1947 to 1953 and also president of the Council for the Promotion of Field Studies (later renamed the Field Studies Council).

Tansley was elected a Fellow of the Royal Society in 1915 and received the Gold Medal of the Linnean Society in 1941. He was knighted in 1950. Sir Arthur Tansley died at Grantchester, near Cambridge, on November 25, 1955.

Braun-Blanquet (1884–1980). Braun-Blanquet worked first at Zürich and then became the first director of the Station Internationale de Géobotanique Mediterranéenne et Alpine, at Montpellier, France.

From about 1913, Braun-Blanquet and his colleagues became known as the Zürich–Montpellier (or ZM) School. They set about classifying plant communities based on the smallest area a particular association of plants could occupy. For oak woodland, for example, this minimal area was about 240 square yards (200 m²), and a minimal area was calculated for every type of vegetation. Then they would mark out an area inside a stand that occupied not less than the minimal area for its type and record all the plants inside the marked area known as a relevé or Aufnahme, together with the area each species covered and the way it grew. This defined what the school called the plants' "sociability." Relevés were then grouped into classes, called phytocoena, which could be compared. The ZM approach led to the foundation of an entire scientific discipline called phytosociology, or the sociology of plants.

At about the time the ZM School was developing its approach, ecologists at Uppsala, Sweden, were devising a somewhat similar scheme, though (confusingly for students) they introduced their own terminology. The Uppsala

School was led by J. Rutger Sernander (1866–1944) and Gustaf Einar Du Rietz (1895–1967).

Succession and Climax

In the United States, ecologists were moving in a rather different direction. For a time they were strongly influenced by Eugenius Warming (1841–1924), a Danish botanist who maintained that plants have particular physical qualities that allow them to grow in some places but not in others. Warming also recognized that the development of a community of plants is strongly affected by other organisms, such as parasites.

Then a group of botanists broke away from the Warming school, not so much because they thought it mistaken as because they could not apply it in their own work. Led by Frederic E. Clements (1874–1945; see the sidebar on page 129), they were studying the prairie, a task made urgent by the expansion of farming into natural grassland. The classification of types of vegetation works well enough when an experienced botanist looks at the trees in a forest or even the shrubs comprising heathland or tundra and can quickly identify the most important components of the ecosystem. Prairie grassland is composed of plants all of fairly similar size forming

Frederic Edward Clements (1874–1945)

The founder of the climax theory of vegetation, Frederic E. Clements was born in Lincoln, Nebraska, on September 16, 1874, the eldest of the three children of Ephraim George Clements and Mary Angeline Scroggin. Ephraim was a photographer with a studio in Lincoln. Frederic studied botany at the University of Nebraska, graduating in 1894 and obtaining a master's degree in 1896 and a Ph.D. in 1898, also from Nebraska University. On May 30, 1899, he married Edith Gertrude Schwartz, whom he had met while they were both students. Mrs. Clements gained a Ph.D. in botany in 1904.

Clements began to teach botany at the University of Nebraska in 1897 and became a full professor in 1905. In 1907 he left to become head of the botany department at the University of Minnesota in Minneapolis. He gave up teaching in 1917 when he joined the Carnegie Institution of Washington as a research associate based in Tucson, Arizona, until 1925, when he moved to the Institution's coastal laboratory at Santa Barbara, California. Clements founded the Institution's Alpine Laboratory in Angel Canyon on Pikes Peak, Colorado, which is where he spent his summers doing fieldwork.

Clements developed his ecological theory from observations he made of the vegetation of his native Nebraska. He noted that plant communities change over time and suggested that these changes could best be understood as a sequence of stages leading to a mature climax. If the vegetation is disturbed, under ideal conditions, it will grow back to that climax state. His theory was strongly criticized by Henry Allan Gleason (1882–1973), Sir Arthur Tansley (1871–1955), and others. Nevertheless, Tansley maintained that Clements was by far the greatest individual creator of the modern science of vegetation, and despite all the criticisms, the climax theory was highly influential during the early decades of the 20th century.

Frederic E. Clements died at Santa Barbara on July 26, 1945.

of each species growing inside the quadrat. This technique is still widely used by ecologists.

Clements did more. He cleared all the vegetation from some quadrats and then recorded the order in which plants recolonized the bare ground. This led him to propose that plants colonize an area in a predictable sequence, which he called a succession, and also known as a sere, composed of seral stages. Eventually, he maintained, the succession reaches a *climax*. This is the final stage in its development and, provided it is not severely perturbed from outside for example by a substantial change in climate, it will undergo no further modification. This image of groups of plants of different types succeeding one another until a stable condition is established proved attractive. As originally proposed it was rather too simple, however, and later modifications allowed for different types of climax arising from a similar starting point, but the theory of climaxes became firmly established in Europe as well as in North America. In temperate regions, for example, it came to be widely believed that forest is the natural climax and that when humans abandon an area of land, eventually forest will appear and remain until some outside factor alters or destroys it.

Not all ecologists accepted this theory. Sir Arthur Tansley was very critical of it and soon after Clements proposed his theory Henry Allan Gleason (1882–1973), another American ecologist, cast doubt on the whole idea. Gleason pointed out that plants scatter seed all over the place and will grow wherever they can. If similar plant communities appear in two places, it is because conditions in those places are similar. He utterly rejected the idea that successions proceed in anything like an orderly fashion or that the outcome can be predicted.

Ecology is a young science. It is little more than a century since its name was coined, and all the research on which its accumulated knowledge is based has been conducted in this century. Ecologists still have much to learn, and many of their ideas are somewhat fluid, constantly under review and liable to be altered in the light of new findings. In some ways this is unfortunate because ecologists are now called on to advise on a wide variety of environmental matters, and there are many questions to which the answers are still unknown.

■ FOOD CHAINS AND FOOD WEBS

A forest, any forest, can be described as an ecosystem. This means that a line can be drawn around the forest. On one side of the line there is forest, and on the other side of the line there is some other kind of vegetation, such as farmland. This other type of vegetation may also constitute an ecosystem.

In the real world, it is not usually quite so simple, except on seacoasts, the banks of large rivers and lakes, or where humans have defined the boundaries by erecting a fence

an exceedingly complex mixture. It cannot be characterized simply by looking at it.

Instead, Clements devised a method in which measured areas, called quadrats, were marked off and then cataloged by listing every plant species and the number of individuals around the forest and then building or plowing right up the edge of the fence. Left to itself, the forest does not so much end as fade away. The trees become more widely spaced. More shrubs and undergrowth grow in partly shaded clearings. In open spaces there are grasses and herbs. Eventually, the forest gives way to parkland, which is mainly grassy, but with isolated trees or groups of trees (called copses).

Natural ecosystems grade into one another, with an area of overlap, which in some cases is quite wide. This overlap is called an *ecotone,* and it often supports a larger variety of plant and animal species than the main ecosystems to either side of it. Ecotone areas are often rich in species, but their richness may be based on commonplace plants thriving at the expense of species that cannot tolerate the conditions prevailing in the ecotone (see "Forest Communities" on pages 161–167).

What Is a System?

Ecosystem, the word coined by Sir Arthur Tansley, sounds as though it is short for "ecological system," although Sir Arthur did not say so straight out. The meaning and derivation of *ecological* had already been established (see "The Economy of Nature" on pages 126–127). *System* is a word that means different things to different people. A geologist knows what a system is, but it is not at all the same thing as the system a chemist understands, and a chemical system is not in the least like a biological system. It is not uncommon for a word to have quite different meanings in different scientific disciplines. In this case, since ecology is one of the biological, or life, sciences, it would be natural to assume that ecologists use the term in its biological sense, but this is mistaken. The concept of an ecosystem is most closely related to engineering. To an engineer, a system is an assemblage of components that interact in such a way as to form a coherent whole, and that regulates its own performance. In other words, a system is a kind of engine.

An engine has many parts, all of them essential, assembled in such a way that the engine as a whole forms a discrete unit. It is clearly visible under the hood of a car or on a factory floor. Many engines that operate machinery in factories are designed to run at a constant speed. If the engine runs faster, a device called a governor partly closes a valve, restricting the fuel supply and slowing the engine. If it runs too slowly, the governor makes the valve open wider. This is what engineers call a feedback mechanism. In this case the feedback is negative because when the engine speed changes, the change is corrected. Positive feedback can also occur. Then, when the engine runs faster, the valve opens, making it run faster still, and when it runs slowly, the valve closes, slowing it further.

This is the sense in which an ecosystem, such as a forest, is a system. It has inbuilt mechanisms for self-regulation and, therefore, it is fairly stable. Those mechanisms arise from the relationships among the species comprising the forest. The dominant species are trees, of course, but they are not alone. Animals feed on the trees and are hunted by other animals, and dead organic matter feeds soil animals, fungi, and bacteria, which return nutrients to the trees (see "Decomposers" on pages 135–137). Finally, there are parasites that feed on all the organisms in the ecosystem, including the bacteria and one another. The relationships on which feedback mechanisms are based are feeding relationships.

Food Chains

At first glance the way an ecosystem works appears quite straightforward. Trees produce leaves, and there are caterpillars that feed on those leaves. If there are so many caterpillars that they eat all the leaves, many of the caterpillars will starve. Few of them will grow into butterflies and moths, so fewer eggs will be laid and the next year there will be fewer caterpillars. The trees will produce new leaves, and fewer of those leaves will be attacked, allowing more of the caterpillars to mature and lay more eggs that hatch into a larger number of caterpillars the following year.

Caterpillar numbers rarely increase to this extent, however, because there are small birds that feed on caterpillars. The more caterpillars there are, the more food there will be for hatchling birds, and when the young leave their nests, there will be more birds to eat up the caterpillars. Sadly for them, small birds also have enemies. There are hawks watching them from above and waiting for a chance to swoop and grab them. Once again, the more small birds there are, the more food there will be for baby hawks, but if the hawks eat too many small birds, they will reduce their own food supply and some of them may starve.

All of the feeding relationships are based on feedback, and the leaves, caterpillars, small birds, and hawks form a chain called a food chain. Each link in the chain affects the links to either side. If there are more leaves, there can be more food for the leaf-eating animals, so their numbers will increase, and in the end there can be more hawks. In this example the chain has four links: leaves → caterpillars → small birds → hawks. The arrows indicate the direction in which food travels along the chain. Green plants such as trees are called primary producers because they convert carbon dioxide, water, and mineral nutrients into leaves. The herbivorous caterpillars are primary consumers because they feed directly on the products made by the plants. The insectivorous small birds are secondary consumers, and the carnivorous hawks are tertiary consumers.

All of the links in a simple chain of this type are easy to find. In a forest in summer there are caterpillars feeding on leaves and small birds feeding on the caterpillars. Occasionally a hawk may swoop down to seize an unwary small bird and carry it away. This is a much rarer sight, however, because there are far more small birds than there are hawks (see "Pyramid of Biomass" on page 124 to find out why).

Limitations of the Food-Chain Concept

A little more thought reveals that life in the forest cannot possibly be quite so simple as the food chain suggests. Caterpillars are around for only a short time, so how do the small birds manage for the rest of the year? The answer is that some of them eat buds or seeds while others rummage among the dead leaves on the forest floor for the invertebrate animals that live there. The birds are versatile in their food preferences, as they must be if they are to survive. Not all the insectivorous birds feed on insects all the time. Nor do the hawks feed exclusively on insectivorous birds. They also eat seed-eating birds, and when times are hard, they will eat the insects themselves.

Perhaps a student of forest life might notice some berries that look delicious and eat a few of them while waiting for birds to catch caterpillars or hawks to capture small birds. That would make the student a temporary part of the forest ecosystem, but where do people fit in? Humans eat berries, leaves, roots, and nuts, but they also eat meat. Anyone trying to fit people into a food chain will find them turning up in several different places.

This food chain is also incomplete in another way. The chain runs in only one direction, from leaves to hawks. That seems logical, but there is also another chain that begins with dead leaves and other organic waste and ends with simple inorganic compounds that can be absorbed by tree roots. Add this part of the chain, and what began as leaves → caterpillars → small birds → hawks is transformed into the rather more complicated pattern shown in the diagram, where the members of the aboveground chain become leaves once more. Even this is an oversimplification, however. It shows that all the members of the upper part of the chain contribute dead organic matter for recycling, but the organisms engaged in decomposition (called *saprobes* or *saprotrophs*) also produce dead organic matter, of course. Consequently, what is shown as a neat straight chain is really a confusing tangle of loops.

The concept of the food chain is useful. It shows that a forest, or any other ecosystem, is regulated by feeding relationships, and it illustrates how nutrients are recycled. It can also be used to demonstrate how certain poisons, such as chemically stable insecticides, can be concentrated as they are passed along it. This is called biological magnification. Using the food-chain diagram as an example, if each caterpillar absorbs a dose of insecticide that is too small to kill it, the bird feeding on caterpillars absorbs that small dose with every caterpillar it eats—and a small bird eats many caterpillars. Being chemically stable, the insecticide may accumulate

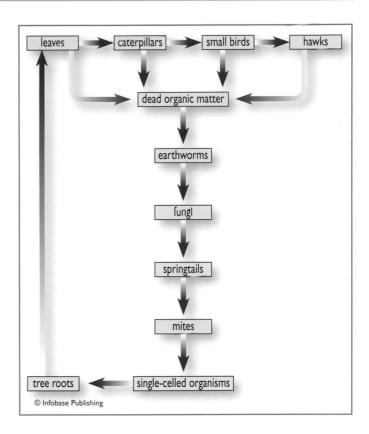

A food chain that includes the organisms involved in the breakdown of dead organic matter and the recycling of its constituents

in the body of the bird. Hawks then absorb the accumulated dose from each of the small birds they eat. At each stage in the food chain, the insecticide concentration increases, possibly reaching levels high enough to harm animals at the top of the chain, in this case, hawks.

Food Webs

It is impossible to take the food-chain idea any farther than this, and a careful look at what goes on in any ecosystem shows why. Examine the leaves or pine needles in a forest, and it is soon obvious that they provide food for more than just caterpillars. Beetles, aphids, and other insects also eat them, several species of birds feed on the insects, and in addition to the insects and birds, the forest contains mammals and, probably, reptiles and amphibians. The leaves may also be diseased. Black spots often seen on tree leaves are made by parasitic fungi, and leaves that are curled up or abnormally colored are probably suffering from a viral infection.

Try to represent all of this, and the diagram ceases to be of a food chain and becomes one of a food web. The diagram on page 132 shows a food web for a broad-leaved deciduous forest in eastern North America. It is quite difficult to trace all the relationships in it, yet despite looking so complicated, the

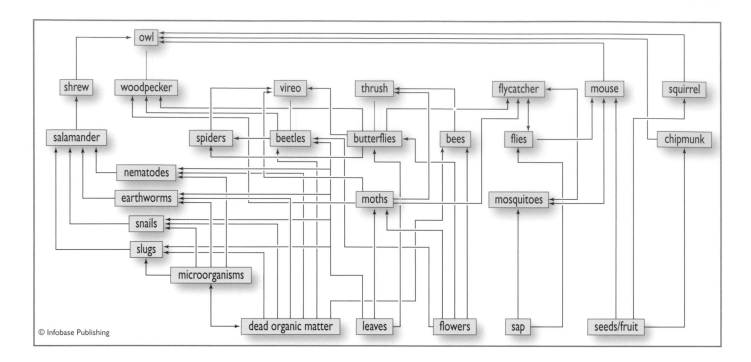

A food web. Despite the complexity of the links between organisms, this diagram greatly simplifies the relationships in a real ecosystem.

diagram in fact is greatly oversimplified. Thrushes as well as salamanders eat snails, for example, and a hungry mouse will not turn up its nose at the chance of snack comprising a beetle, nor is the owl likely to be the only predatory bird. There will probably be hawks, and almost certainly there will be snakes, weasels, badgers, martens, and cats. Despite its limitations, however, the food web gives a much better impression of the relationships within an ecosystem than does a food chain.

■ ECOLOGICAL PYRAMIDS

It would be an impossible task to figure out what every organism living in a forest eats, and even if someone were able to accomplish such an heroic feat, the result would be hopelessly confusing. There are hundreds of species, their numbers fluctuate with the seasons and from year to year, and everything is busily eating everything else. An hour or two spent trying to count and measure all of it might leave an investigator with the firm conviction that a forest is an incomprehensible mess.

Fortunately, there is another way to approach the task. First the investigator must identify the organisms at each level. That is to say, she or he should identify the plants, which are the primary producers, and then identify the animals that feed on the plants (primary consumers), the carnivores that feed on the herbivores (secondary consumers),

and the carnivores that feed on other carnivores (tertiary consumers). The investigator has then divided all the members of the ecosystem into groups according to the *way* they feed, rather than precisely *what* they eat. These groups are called trophic levels, from the Greek *trophe,* meaning "nourishment." If there are omnivores, such as humans, including them requires measuring or estimating the proportion of their food they obtain from plants and how much from herbivorous and carnivorous animals, and allocating some of the omnivores to each trophic level accordingly.

Difficulties in Counting Plants

The next step is to count the number of individual organisms at each trophic level. Obviously, this is very tedious, but at least it is simple and there is a shortcut. Working through the entire forest counting absolutely everything would take a very long time. It would also be unreliable because it would take so long that by the time the investigator finished, the numbers near the start would have changed. Instead, the investigator takes samples, using a sampling technique that makes the sample statistically reliable. This means that the samples must be taken randomly.

It sounds much more difficult than it is. One way is to use a large-scale map. The investigator superimposes a grid over the map; then he or she uses a set of random number tables to select the coordinates of points within the grid, which he or she marks on the map, marking as many points as he or she will need samples. Map and possibly compass in hand, he or she then marches confidently into the forest and finds each of the sample sites marked on her map. At each site he or she marks out an area, called a quadrat, using string and pegs. The quadrat can be any size, but in sampling a forest, it usually has an area of 12 square yards (10 m²) or 24 square yards (20 m²). It can be rectangular or circular. Then the investigator counts and records every plant inside each quadrat. It is still a slow job, but it is better than counting the entire forest!

Alternatively, the investigator can take a random walk, which saves using a map and grid. She or he walks into the forest and picks a place from which to start. This can be anywhere. Then she or he uses random numbers, from a random number table, to find out how many steps to walk in a straight line in any direction (provided she or he does not walk out of the forest, of course). Where that walk ends, she or he marks the spot, usually with a striped stick that is easy to find. Then she or he turns 90°, tossing a coin to decide whether to turn left or right and always using the same rule, such as heads for left, tails for right. She or he takes the next random number from the table, walks that number of steps in a straight line, marks the spot to which that takes her or him, turns left or right, and keeps repeating the process until she or he has as many sample sites as needed. Then she or he marks out a quadrat on each site and counts the plants inside it.

How to Count Animals

Quadrats are all very well for counting plants, but animals are mobile and other methods are needed. Mammals can be trapped, marked, and released and total numbers calculated by comparing the number of marked and unmarked individuals trapped subsequently, using a standard algebraic equation. This does not harm the animals, but the technique must be used with care. Some individuals hate the traps, become "trap shy," and having been caught once will always avoid being trapped again. Others enjoy the food and the nest provided in the trap and keep returning for a free meal and quiet rest. Such individuals can distort the findings because the traps are no longer taking a random sample of the population.

Insects can be brushed from low vegetation, and pupae can be collected and counted. Birds can be observed from the ground and identified by their songs. Sampling soil organisms requires special equipment, but the techniques are standardized and not too difficult.

Pyramid of Numbers

At the end of the project, the investigator has collected data from which to compile a table of species, grouped according to the way they feed, and the number of individuals in each group. She may decide to represent her results in the form of a graph, and there is a type of graph devised especially for the purpose. It is rather like a histogram laid on its side, with bars, all of similar width drawn horizontally, and set one above another.

The lowest bar represents the primary producers, its length proportional to the number of individual plants. Centered on top of it is the second bar, its length being proportional to the number of primary consumers (herbivores). A bar for secondary consumers (carnivores) is laid above that one, and a bar representing the tertiary consumers (top carnivores) above that. The result is often something like the diagram below (1 in the illustration), in the shape of a stepped pyramid—although it is very important to note that the illustration is not drawn to scale. Because of its shape, the graph is usually called an ecological pyramid. It was invented by a very distinguished

Two ecological pyramids. The upper pyramid (1) suggests that a large number of primary producers feed diminishing numbers of primary and secondary consumers. The lower pyramid (2) might represent the situation in a forest if this is a pyramid of numbers. The base of primary producers is small because trees are big, so they are few in number.

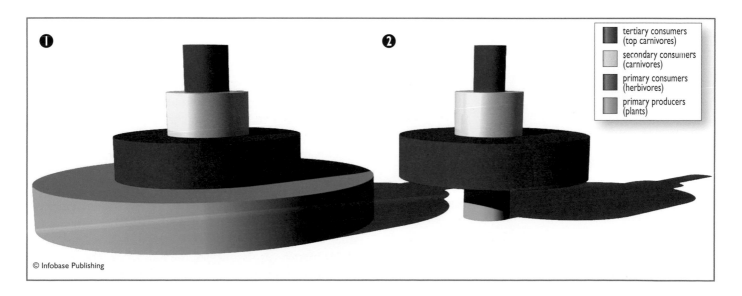

tertiary consumers (top carnivores)

secondary consumers (carnivores)

primary consumers (herbivores)

primary producers (plants)

British zoologist and ecologist, Charles Sutherland Elton (1900–91; see the sidebar), so it is sometimes known as an Eltonian pyramid.

Charles Sutherland Elton (1900–1991)

One of the most influential ecologists of the 20th century, Charles Elton was born on March 29, 1900, at Withington, a district of Manchester, England. He attended Liverpool College from 1913 until 1918, and in 1919 he began to study zoology at New College, University of Oxford, graduating in 1922. In 1923 he was appointed a demonstrator in the zoology department at Oxford University. He was made university reader in animal ecology and a senior research fellow of Corpus Christi College, Oxford, in 1936.

Elton took part in three Oxford University expeditions to Spitsbergen in 1921, 1923, and 1924. In 1925 Elton was made a biological consultant to the Hudson Bay Company. He was a founding member of the Oxford University Exploration Club, formed in 1927. Later Elton spent three more seasons in Spitsbergen, funded in 1929 by the Royal Geographic Society. In 1930 he took part in an Oxford University expedition to Lapland.

His first important paper, "Periodic Fluctuations in the Number of Animals: Their Causes and Effects," appeared in 1924 in the *British Journal of Experimental Biology*. Elton published his most influential book, *Animal Ecology,* in 1927. In 1932 he founded the *Journal of Animal Ecology*. Also in 1932 Elton founded the Bureau of Animal Population at the University of Oxford and was its director from 1932 until 1967.

Elton published several other important books, including *Voles, Mice and Lemmings: Problems in Population Dynamics* (1942), *The Ecology of Invasions by Animals and Plants* (1958), and *The Pattern of Animal Communities* (1966). He was made an honorary member of the New York Zoological Society in 1931 and was elected a fellow of the Royal Society in 1953. He was made an honorary member of the British Ecological Society in 1960 and a life member and eminent ecologist of the Ecological Society of America in 1961. In 1967 he received the Gold Medal of the Linnean Society and in 1970 the Darwin Medal of the Royal Society.

Charles Elton died at Oxford on May 1, 1991.

In fact the shape is fairly obvious. For example, if there are sheep eating grass, there must be more grass plants than there are sheep, or some of the sheep will go hungry. Similarly, if hawks hunt small birds, there must be more small birds than there are hawks. In the case of a forest, however, a pyramid that counts the number of individual organisms at each level may not have this apparently obvious form. It is more likely to resemble the diagram (2) in the illustration on page 133. This is because trees are big, so there are not many of them. One tree can provide food for many small herbivores, so the pyramid may be much narrower at the base than it is at higher levels. On the other hand, if it were a subtropical forest and the primary consumers were elephants, the pyramid would change shape again because there would be only a few elephants.

Pyramid of Biomass

Using the number of individuals is not very satisfactory because there is no way to compensate for the difference between ecosystems containing large numbers of small organisms and those with small numbers of big ones. For this reason, the pyramid of numbers, as this pyramid is known, is little used.

One alternative is the pyramid of biomass. This solves the problem of size. Instead of counting the number of individuals in each quadrat or sampling area, all the individuals are collected and sorted according to their feeding type, dried in an oven, and then weighed. It is necessary to dry them because organisms vary greatly in the amount of water they contain, and this can also vary over time. Drying removes the water, so only solids are weighed. This sounds very cruel, but the investigator does not really have to kill and oven-dry all the animals in the sample. It is enough to identify the species and count the numbers. There are standard tables for the biomass of most animal species, so once they have been recorded the animals are released.

Biomass is the weight, or mass, of living matter in a given area. That area can be any size; it is possible to talk of the biomass of the planet Earth. It can also refer to all the living organisms present or to just some of them. To construct a pyramid of biomass for an ecosystem, it is necessary to calculate the biomass at each trophic level.

A pyramid of biomass makes no distinction between mice and deer. Size does not matter, only total mass. The resulting pyramid is shaped like the first one in the illustration on page 133, and it conveys much more information than a pyramid of numbers because it compares like with like. It begins, at the base, with the primary production, a figure representing all the leaves, stems, branches, roots, fruits, seeds, and so forth produced by green plants and measured as dry matter. Above this bar, the second illustration shows the total amount of dry matter in all the primary consumers.

Higher levels also represent total dry matter. This makes it possible to measure how much of the dry matter at each level becomes incorporated into the organisms at the level above. As the pyramid shows, rather little of the matter that is produced by plants forms the bodies of herbivores, and rather little "herbivore stuff" forms the bodies of carnivores. The amount varies, but on average no more than about 10 percent of biomass is transferred from each level to the one above.

That explains why an ecological pyramid seldom has more than four trophic levels. A fifth level would contain very few individuals indeed, and a sixth layer would probably contain too few to comprise a viable breeding population because the chance of encounters between adult males and females would be too small. That is why top predators such as hawks (see "Food Chains" on pages 130–131) are always fairly uncommon, and if predators are individually large, like tigers, a huge area of land is needed to supply them with food. Big predators have always been rare and always will be. (Food is more plentiful in some places than in others, but a tiger needs a territory of 25–250 square miles [65–50 km²] to provide it with enough to eat.)

The small amount of matter that each level incorporates from the level below also explains why it is impractical to draw a pyramid of biomass to scale. If the bar representing the primary producers were the width of a page, say 7 inches (17.8 cm), the bar representing the primary consumers would be 0.7 inches (1.78 cm) wide. The bar for the secondary consumers would be 0.07 inches (0.178 cm) wide, and the bar for the tertiary consumers 0.007 inches (0.0178 cm) wide. If there were a fifth level, its bar would be so narrow it would be very difficult to print and to see.

Pyramid of Energy

Despite being a great improvement on the pyramid of numbers, the pyramid of biomass does not illustrate the most fundamental basis of the relationship between trophic levels. Primary productivity begins with photosynthesis (see "Photosynthesis" on pages 116–122). Photosynthesis is powered by sunlight, so it can be understood as plants capturing energy. A proportion of that energy passes to animals that eat plants, a proportion of their energy goes to the carnivores that eat them, and so on.

The energy being absorbed at each level can be measured (see "Energy Flow" on pages 167–169), and when it is represented graphically, the result is a pyramid of energy, which looks very much like the pyramid of biomass and like the upper pyramid in the illustration. Ecologists consider the pyramid of energy to be the most fundamental of the ecological pyramids and the most informative.

The amount of energy present at each trophic level is proportional to the biomass at that level. It is this fact that accounts for the similar shapes of the pyramids of biomass

and energy. Of the energy present at each level, only about 10 percent passes to the next level above. The remaining 90 percent is released by the process of respiration and used by organisms for their own growth, maintenance, and reproduction. Because the amount of energy decreases so sharply from one level to the next, the impracticality of drawing a pyramid to scale applies to the pyramid of energy just as it does to the pyramid of biomass. These pyramids illustrate the principle of trophic relationships, but they are symbolic, not literal, representations.

■ DECOMPOSERS

Pyramids of numbers, biomass, and energy portray trophic relationships that begin with living matter. Plants produce food for herbivores, which are eaten by carnivores. It all happens above ground because that is where plants conduct photosynthesis.

Belowground there is another community, and it, too, consists of groups of organisms that pass food and energy from one trophic level to another. The species comprising the soil community are very different from those living aboveground, but the underground community is no less complex. It includes herbivores and carnivores, and it can be described by food webs and ecological pyramids.

Since the aboveground pyramids are what appears to be the right way up, with the widest band at the bottom, it is tempting to draw the pyramid for the underground communities the other way, with the broadest band at the top. This would indicate that one pyramid describes relationships aboveground and the other those belowground. If it is a pyramid of numbers, this is indeed the way it may form because as nutrients pass from one trophic level to the next, the organisms feeding on them become smaller and, therefore, more numerous. Contriving to turn the underground pyramids upside down is misleading, however. It suggests that the trophic levels are arranged spatially on top of one another, with the implication that the top predators live up in the sky somewhere in the aboveground pyramids and beneath the other levels in the underground pyramids. This is clearly absurd and a misunderstanding of the meaning of the pyramids. The underground pyramids should be shown like any other pyramid, with the first band at the bottom, whether or not it is the widest. The difference between the aboveground and underground pyramids arises only from the material comprising the lowest band. Instead of fresh, living, primary produce, belowground it is dead organic matter.

Litter on the Forest Floor

A forest floor is covered with dead leaves and conifer needles, twigs, branches and, here and there, whole trees that have fallen. These are the most obvious and by far the most abundant forms of dead organic matter, but they are not the only ones. Insects molt, discarding their old exoskeletons, and snakes shed their skins. Mammals drop hairs, birds discard worn-out feathers and eggshells, deer shed their antlers, and, of course, all animals produce urine and feces. All of this is dead organic matter, and it forms the primary resource for the soil community. It is called litter.

Fungi and bacteria are often the first to arrive. This is because they are everywhere, as spores floating in the air and resting on the dead material itself even before it falls to the ground. Fungi feed on sugars and starches. Removing these weakens the structure of the material so it tends to break apart. What remains is more resistant to attack by fungi and bacteria, but the fungi themselves are food to some animals, including certain species of ants, termites, beetles, and nematodes.

The Soil Community

A search among and beneath the dead leaves on the ground quickly exposes wood lice, snails, slugs, earthworms, millipedes, and various insect larvae. All of these animals feed on the litter. Between them they shred leaves and similar plant and animal material into tiny pieces. They drag the fragments below the ground surface, moving the material from the place where it fell and spreading it throughout the uppermost soil layer, where smaller animals can reach it.

The smaller soil animals include worms that are much smaller than the earthworms, as well as mites and springtails. None is more than 0.08 inch (0.2 cm) long, so a good magnifying glass is needed in order to see them clearly.

They are far from being the smallest animals. When the worms, mites, springtails, and similar-sized animals have eaten their fill, what remains passes to nematode worms, rotifers, and single-celled protozoans, none of which is more than 0.004 inch (100 μm = 0.0001 cm) long. These members of the soil community are visible only with the help of a microscope. Then the material returns to the bacteria, prokaryotes (see the sidebar "Prokaryotes and Eukaryotes" on page 117), 0.000004–0.0004 inch (0.1–10μm = 0.00001–0.001 cm) across.

Large soil animals, called the megafauna, are fairly mobile. Rummaging among the dead leaves or digging a few inches into the soil will usually reveal a few of them, but others sense the disturbance and quickly burrow out of reach. Smaller animals of the mesofauna are barely visible to the naked eye, and the smallest of all, the microfauna, cannot be seen at all. Their small size makes the soil and even the top layer of litter seem sparsely inhabited. Nothing could be further from the truth. In a temperate forest, beneath every square foot of the soil surface, there may be close to 95 species of animals (1,000 species/m^2),

never mind the fungi, bacteria, and others, and those animals are present in vast numbers. Each square foot of forest floor may conceal more than 890,000 nematode worms, 8,900 springtails, and 8,900 mites, as well as about 4,500 animals of assorted other species (10 million, 100,000, and 50,000/m^2 respectively). In broad-leaved deciduous forest, if the soil is not too acid, there may be several hundred earthworms beneath each square yard of surface, but their numbers can vary greatly over quite short distances. These larger animals are sometimes called detritivores because they feed on waste material, or detritus.

Plant Material Is Tough

Plant material is very tough. Plant cell walls contain cellulose, and woody material contains lignin, which is the principal component of dead heartwood. There are also some very complex proteins, as well as suberin, from which cork is made. These substances protect or strengthen the cells, making it very difficult for consumers to gain access to the nutritious contents of the cells. They are broken down chemically in reactions assisted by enzymes. Certain species of bacteria and protozoa produce cellulases, the enzymes that catalyze the breakdown of cellulose. Fungi secrete most of the enzymes catalyzing the breakdown of other tough compounds. In order to work, enzymes must come into direct contact with the compounds that will react. Different fungi produce different enzymes, so many species contribute to the breaking down of all the resistant compounds, and it is the soil animals that break material into smaller, increasingly ragged pieces, so increasing the surface area on which the chemical reactions take place.

The megafauna and mesofauna play a more important part in the decomposition of material in a forest than they do in other ecosystems. Grasses and herbs are softer than the leaves and wood of trees, so they are broken down more easily by the microfauna. On a forest floor, however, the smaller organisms are more reliant on the megafauna and mesofauna to chew plant fragments into ever smaller pieces, making the material more easily accessible to them. As they eat, the larger animals also drop feces, and their fecal pellets keep the material on which they fall moist, which facilitates fungal and bacterial attack and also makes it easier for material to be washed below the surface by rain or dragged there by animals. Worm casts, which are often found lying on the surface early in the morning, are the most familiar example of feces produced by the megafauna, and they have usually disappeared back belowground within a few hours. Charles Darwin calculated that every year, the earthworms in grassland near his home brought about 20 tons of their feces to the surface of each acre (50 tonnes per hectare). Their casts are what remains of material the worms have dragged below the surface and passed through their bodies.

■ ANIMALS THAT RECRUIT ASSISTANTS

A few animals, including some species of earthworms, snails, and insect larvae, can produce their own cellulases. Those unable to do so recruit assistants, leading to mutualism. Mutualism is a partnership between members of two different species that benefits both. Because such partnerships are mutually beneficial, organisms tend to form them quite readily and to accept new partner species in addition to those they have already. Consequently, there are countless examples. Termites carry communities of bacteria and protozoa in their guts. The bacteria and protozoa produce the enzymes needed to break down cellulose and lignin, allowing the termites to feed on dead wood, which is plentiful. Woodlice also harbor gut microorganisms that produce cellulases. Other animals, including certain species of springtails, ants, and termites, wait for microorganisms to break down tough material; they then feed on the microorganisms. There are even ants and termites that cultivate fungal gardens. They gather leaf fragments, chew them, assemble them in special places, sow them with fungal spores, and then feed on the fungi.

Hunters in the Soil

While these animals are grazing, others are hunting them. The soil community includes many carnivores. They include pseudoscorpions, small spiders, harvestmen, and centipedes, all of them formidable predators of animals smaller than they are themselves.

Pseudoscorpions (order Pseudoscorpiones) resemble scorpions but lack the long abdomen and sting of a scorpion. They are tiny, rarely more than 0.3 inch (0.8 cm) long, and many of the 2,000 known species live among the litter on the forest floor. They feed on mites and springtails, which they kill or paralyze with poison injected from glands in their claws.

Harvestmen (order Opiliones, also known as harvest spiders and daddy longlegs), have bodies 0.2–0.4 inch (0.5–1.0 cm) across, but with very long legs. They live partly as scavengers, feeding on dead material, but they also hunt small spiders, and there are some species that feed on snails.

Dysderid spiders (family Dysderidae) are one of the few predators that hunt woodlice. Woodlice are crustaceans, related to crabs and lobsters, which have taken to life on land, although they must remain in moist surroundings. Their tough carapace protects them from most hunters, and those that penetrate the carapace are likely to be deterred by the foul taste of a woodlouse. Dysderids are the exception. They specialize in hunting woodlice—a food few others will touch. Up to about 0.6 inch (1.5 cm) long, a dysderid spider has huge fangs, which it uses to impale woodlice. The spider is not repelled by the taste of most woodlice, but some species are too revolting even for this predator. *Dysdera crocata*, a species found in many parts of the world, is striking in appearance. The foreparts of its body are brick red, and the abdomen is pale gray and tubular in shape.

Centipedes (class Chilopoda) are related to scorpions, spiders, and insects. There are approximately 3,000 known species, found all over the world. They require moist conditions, and many species inhabit temperate forests where they hunt among the litter and in the soil belowground. They vary in size, but most of the species found in North America and Europe are 1.2–2.4 inches (3–6 cm) long. Most temperate species are red-brown in color. Centipedes have a head with antennae, mouthparts with teeth, and large claws that inject poison into their prey. The body is long and consists of identical segments, each with one pair of legs. The hindmost legs are longer than the others, and the animal can defend itself with them by pinching an attacker. Centipedes can tell the difference between light and dark, but that is the extent of their vision—they are virtually blind, using their antennae to locate prey. Fast runners, depending on their size, centipedes feed on nematodes, insects, earthworms, or snails.

Eating One Food at a Time

Some of the chemical compounds in dead organic matter being more difficult to digest than others, there is an order in which they are broken down and disappear. Sugars are the first to go. These are easy to digest, and some are soluble, so they are washed away by the rain.

Starches are next to go, and then hemicelluloses. The hemicelluloses comprise a group of sugars that occur in the cell walls of plants in association with cellulose. They are easily hydrolyzed; hydrolysis is a reaction between a substance and the hydrogen and hydroxyl ions in water, which splits molecules of the substance in two. Pectins, carbohydrates that also occur in plant cell walls, disappear at about the same time as hemicelluloses, and so do proteins.

After that, the really tough compounds start to decompose. Cellulose is first, followed by lignins and then suberins, the fatty compounds that make cork water-resistant. Finally cutin is digested. This is another fatty substance. It occurs in cell walls and helps to waterproof them.

Rate of Decomposition

All plants are made from these ingredients, and this is the order in which the ingredients are consumed, but the proportions vary. After all, not all plants are the same. Humans eat plants, and people have no trouble distinguishing the taste and texture of one vegetable from those of another. Not surprisingly, therefore, some plant remains vanish more quickly than others.

Fruits do not last long. Mammals and birds enjoy them, but even while they are nibbling and pecking, the fungi and bacteria are busily at work stripping out the sugars and starches. In most cases, when the sugars and starches have gone, not much of a fruit remains. Where oak and beech trees grow together, fallen oak leaves disappear faster than those of beech. Conifer needles are eaten only slowly.

Decomposition continues through the year, but it proceeds more vigorously in summer when the ground is warm. This alters the contents of the larder on the forest floor as the year progresses. The change is much more marked in a deciduous forest where leaves fall all together than it is in an evergreen one where leaves are shed at a constant rate that does not vary with the seasons.

Mostly small and hidden below ground, the decomposers are easily overlooked. Like all organisms, they feed, reproduce, and die, but in doing so they make life possible for others. At the end of the processes of decomposition, dead organic matter has been completely taken apart and transformed into simple chemical compounds. Dissolved in water, these compounds can pass through the walls of root cells and so sustain living plants. It is its trees that define the forest, but the trees appear by arrangement with the soil community.

HABITATS AND NICHES

Many people live in the city. At night, streetlights illuminate the street outside their homes, making it safer for them if they are out walking after dark. There are shops not too far away where they can buy groceries, clothes, and everything else they need to sustain them. When their homes or anything in them needs repair, skilled builders and engineers will respond to their calls. Should someone fall sick, there are doctors, dentists, and hospitals to help, supported by an ambulance service in case of emergencies. A fire engine will arrive in minutes should a home catch fire, and a police force and courts uphold the rule of law. The city provides employment. It has schools, museums, libraries, theaters, art galleries, and concert halls, as well as swimming pools, parks, gymnasiums, and sports fields.

Life in a real city is not quite so stress-free as this makes it sound, of course. Things go wrong, services break down, some people are unemployed, there is bad housing and poverty. All the same, for the whole of human history people have believed that cities are good places to live. Our word *city* is from the Latin *civis*, citizen, and so are the words *civil*, *civilian*, *civility*, and *civilization*. Literally, a civilized person is a city-dweller. That is what the word means.

Cities, then, are hospitable places within which people can find anything they might need, at least in principle. An ecologist might say that the city is a good habitat. Temperate forest is also a good *habitat*, as is the soil beneath the forest floor. The word *habitat* has the same meaning in all three cases, but clearly there are differences in these habitats. Humans have always found forests extremely difficult places to live (see "Forests in Folklore and Literature" on pages 236–241). Domestic cats fare well enough in cities, but the lynx (*Felis lynx* in Europe and Asia, *F. canadensis* in North America) would find city streets, shopping malls, and even the parks very inhospitable. So would a family of beavers (*Castor canadensis* in North America and *C. fiber* in Europe and Asia).

Whose Habitat Is It?

Habitats must be defined in relation to the species they accommodate. It is meaningless to describe an area simply as "good habitat" without naming some organism that might find it so. Golden eagles, woodchucks, giant redwoods, seaweeds, whales, and walruses all have quite different requirements, and what is good habitat for one would be intolerable for another.

Animals and plants do not live in isolation, of course. What is good habitat for one species will be just as hospitable for several more, but each will experience it differently. Cities provide good habitat for humans. When people think of a city, they usually imagine its buildings, streets, transportation systems, services, and amenities—the ability of the city to satisfy human needs—but cities also support a surprising amount of wildlife, comprising many other species. The city must appear very different to the feral pigeons that roost on its high buildings and feed on its squares, to the rats that inhabit the dark, secret places, to the house mice, the sparrows, or the cockroaches. They experience it differently, but it is their habitat just as much as it is human habitat.

Similarly, a forest provides habitat for a wide variety of species, but it does so by satisfying their quite distinct needs. The insect larvae feeding on leaves in the canopy provide food for insectivorous birds, which are hunted by birds of prey, as are the small rodents rummaging in the litter on the forest floor. To each of these and to countless others aboveground and belowground, the forest represents something different. The only thing on which all these species would agree is that within this specified area, they are able to find whatever it is they need.

Stable, Unstable, and Ephemeral Habitats

The realization that every species has its own particular requirements and that a suitable habitat is a place where those requirements are to be found may seem to leave *habitat* as a word that is impossible to define. Like beauty, it is in the eye of the beholder. Ecologists are not left entirely helpless, however, because no species lives in isolation. A forest—or riverbank, backyard, or any other area that supports life—is a community of organisms of many different kinds. So it is possible to classify habitats by type—as temperate forest habitat, for example—and by quality. The quality of a habitat depends on the variety of organisms it supports. A rich habitat supports many species, or to put it another way, it is a habitat in which many species find the food, space, shelter, and reproductive opportunities that they need.

It is also possible to describe a habitat as stable, unstable, or ephemeral. If it is stable, conditions for those species at home in it will remain hospitable all the time. If it is unstable, conditions will change, so the habitat is more hospitable at some times than it is at others. If the habitat is ephemeral, hospitable conditions will occur only occasionally and then disappear quickly. From this point of view, a city appears to be a stable habitat because the shelter, protection, food, services, and amenities which it provides are available at all times. A temperate forest, on the other hand, is unstable. Conditions in it vary widely between summer and winter.

Closer inspection reveals that matters are not quite so simple. The city may be stable so far as humans are concerned, but its nonhuman inhabitants are affected by the seasons, just like forest-dwellers, and for some organisms, the part of the city or forest they occupy appears and vanishes rapidly and unpredictably. The half-eaten burger thrown in a trash can and the dead mouse on the forest floor are valuable resources to a host of flies and bacteria, but before long the trash can is emptied, the corpse is gone, and the hospitable habitat has disappeared. A puddle of stagnant water left after a heavy shower will attract a variety of insects and other small organisms, but in a matter of hours or a day or two, warm sunshine will evaporate the water, and the habitat will be gone.

Humans are adapted to life in a stable habitat, and the cities people have been building for thousands of years reflect that adaptation. People need that stability, and this fact may lead them to suppose that less stable habitats are necessarily inferior. So they are for people, but not for species that are adapted to them. Adaptation to unstable or ephemeral circumstances requires an organism to grab as much as it can of the briefly abundant resources before they disappear, and the way to do that is by adopting a particular reproductive strategy.

A Few Offspring or Many?

Where the habitat is stable, it is best for a species to maximize its competitive advantages within its habitat. It can do this by producing few offspring and caring for them over a fairly long period, so by the time they must fend for themselves, they are nearly fully grown and well equipped to do so. This is known as *K-selection*. Humans reproduce in this way; ours is a *K*-species. In the logistic equation K represents the largest number of individuals of a given species that an environment will support—often called the carrying capacity. The logistic equation is a mathematical description of the rate of growth of a population in a confined space with limited resources.

K-selection would not work in unstable conditions. Essential resources would disappear before the young were strong and experienced enough to move away and find sustenance elsewhere. If the habitat is unstable, a better strategy is to maximize reproductive efficiency. The way to do this is to produce a large number of offspring quickly, the moment conditions become favorable. The young enter the world small and vulnerable, but they are able to feed rapidly and they grow fast, so they seize the resources while they are available. This is called *r-selection*. In the logistic equation, r stands for the maximum rate at which a species can produce offspring. This strategy ensures that by the time the resources vanish, enough offspring will have matured to guarantee that there will be another generation as soon as the good times return. Rats, house mice, and locusts reproduce in this way. Many *r*-species are pests, exploiting food that people provide in vast amounts but only briefly, such as farm crops and full granaries.

Reproductive strategy is one way in which a species adapts to its habitat, but what matters is the stability or instability of its own portion of the habitat, not of the habitat as a whole. Evidence of this can be seen whenever a piece of meat is left for a few days in warm weather. The rotting meat becomes infested with fly larvae, which mature rapidly into a swarm of flies. The flies then depart to find rotting meat elsewhere where they can lay their eggs. The meat is an unstable portion of the larger habitat of house and city, which is stable. Forests last for many years. They are stable, but within a forest there are many parts of the overall habitat that are unstable or ephemeral. Fruits, seeds, mature pinecones, and the fruiting bodies of forest fungi are available only in certain months, for example, and a puddle of water may evaporate in a matter of hours.

This type of adaptation can also determine how often members of a species breed, a choice of conditions known as semelparity and iteroparity. Which strategy works best depends on several factors, including the size of individuals and whether or not large size confers an advantage in a particular habitat. Annual plants and many insects reproduce only once in their lifetime. Species using this "one-shot" reproductive strategy are said to be semelparous. Other plants, including forest trees and almost all animals, are iteroparous, meaning that they reproduce more than once in their lifetime.

Niches

The statement that Caroline and Harry live in the city or that the gray squirrel (*Sciurus carolinensis*) lives in the forest conveys no information at all about how Caroline, Harry, and the squirrel make their livings. The statement reports where they live, but not how they live. Where they live or their habitat is their address. Admittedly, the mail carrier might need more detail than "G. Squirrel, The Forest," but at least he or she would not go looking for the squirrel in the middle of a lake or on the seashore.

If its habitat is the address of a species, the way it makes a living at that address is its ecological *niche.* Harry drives a bus. That is his job, so an ecologist could say it is his niche in his city habitat. His job pays him wages that enable him to buy food and clothes and to pay the rent. The gray squirrel feeds mainly on seeds, leaves of trees, and fungi. In spring and summer its diet is more varied, and it spends most of its time high in the forest canopy. In fall, when the leaves are less nutritious but seeds are appearing, it spends more time on the ground, storing food in caches, and in winter it is on the ground a good deal, feeding on its stores. Bus driving is how Harry earns his living, and foraging in the forest is how "G. Squirrel" earns his.

An ecological niche can be described as the presence of a range of factors. These factors might include, for example, the daily and annual range of temperature, annual precipitation and its distribution through the year, the pattern of light and shade, the availability of shelter and nesting sites, and the existing population of plants, fungi, and animals. Taken together these amount to an opportunity. If a particular species is able to travel to the place where the opportunity exists and then to take advantage of it, it will have occupied its ecological niche.

Which species is first on the scene is a matter of chance, but the one that arrives will define the niche. There are many ways resources can be exploited, and no two species will exploit them in precisely the same way. In other words, a niche does not exist until it is occupied and the species that moves into it also creates it. Each ecological niche is unique.

◼ TREE ADAPTATIONS TO CLIMATE

It is easy to take familiar surroundings for granted, to assume that the commonplace plants and animals, those seen by the roadside and in woods and fields close to home, are just as common everywhere. Tropical species do not occur naturally in northern Ontario, of course, so obviously there are limits, but within a general climatic region where the type of vegetation is much the same over large areas, it seems reasonable to suppose that the species comprising that vegetation are also the same everywhere. The assumption may be reasonable, but it is profoundly mistaken.

There are certain places where hickory trees (*Carya* species) are fairly commonplace. *Carya illinoensis* produces the pecan nut, which is believed to have played a very important part in American history. Someone living in a place where there are many hickory trees might be tempted to think they are common all over the world, at least in temperate regions. As the map on page 141 shows, however, this is far from the truth. About 20 species of *Carya* are native to North America, two more are native to China, and they comprise all the hickory species there are in the entire world. Hickory trees may be common in some places, but in the world at large they are rare. Of course, people have planted useful or attractive trees in many places where they do not grow naturally, but some trees, like hickories, occur naturally in only a few places.

What is true for hickories is also true for most other tree species, though to a lesser degree. Elms (*Ulmus* species), for example, occur widely in temperate regions, and in Asia they extend almost to the equator, but there are large gaps in their distribution (which may be widening because of disease; see "Tree Predators and Parasites" on pages 173–180). Birch (*Betula* species) is much more widespread. It grows naturally as a component of broad-leaved deciduous and coniferous forests, and dwarf birch (*B. nana*), which reaches a height of only 20–40 inches (50–100 cm), grows in the tundra, inside the Arctic Circle. Among the gymnosperms (see "Gymnosperms" on pages 100–101), the spruces (*Picea* species) are probably the most widespread.

Pecans are grown commercially on a large scale, so hickory trees can be found growing outside the area shown in the map. Birches are rarely grown outside their natural range, but this is only because the area is already so large. Many birches are attractive trees, grown for ornament. Numerically, Sitka spruce is now the most common tree in Britain, but spruces do not occur naturally in western Europe. They are grown for their timber.

It is possible to cultivate trees outside the area in which they occur naturally, and throughout history people have been taking their favorite trees with them whenever they moved, planting them around their new homes. Ecologists distinguish between species that are native to a region and those introduced by humans, some of which have become naturalized (see "Primeval Forest, Ancient Woodland, Old-Growth Forest, Plantations" on pages 144–147).

Climatic Limits

Climate is the most important factor allowing some species to thrive but not others, and this limitation is very evident in Britain. Lying between latitudes 49.97°N (Lizard Point) and 58.62°N (Dunnet Head), the mainland of Great Britain (the Orkney and Shetland Islands lie farther to the north) is at the climatic limit for many tree species. The map of Great Britain on page 142 uses three widely grown native species to illus-

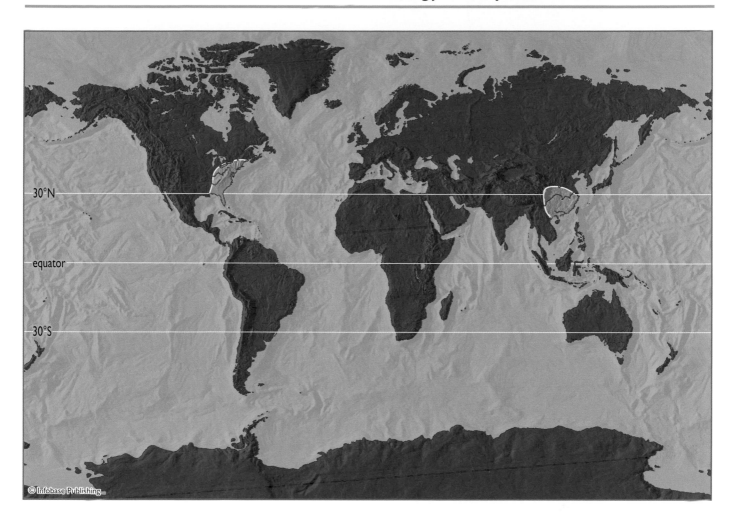

The global distribution of hickory

trate the point. Field (or common) maple (*Acer campestre*) grows naturally only in the southern half of England and in east Wales, hornbeam (*Carpinus betulus*) in the southeast quarter of England, and small-leaved lime (*Tilia cordata*) in central and eastern England and east Wales. These are the climatic limits, although all three trees are cultivated extensively much farther north.

Trees, like all green plants, photosynthesize carbohydrates to obtain the energy they need and photosynthesis is a process limited by temperature (see "Photosynthesis" on pages 116–122). Photosynthesis ceases altogether when the temperature falls below 21°F (-6°C), and it works most efficiently at temperatures higher than 41°F (5°C). Throughout the temperate region, winter temperatures regularly fall below 41°F (5°C) and often to below 21°F (-6°C), so there is a period during which photosynthesis is so restricted that trees cannot grow. Increasing distance from the equator produces even lower winter temperatures, but these are not important in themselves. When it is already too cold for

photosynthesis, a still lower temperature makes no difference. What matters is the length of the cold period, and the higher the latitude, the longer that period will be.

Flowering plants produce flowers and, from fertilized female flowers, they develop seeds (see "Flowers, Fruits, and Seeds" on pages 102–103). This is an effective method of reproduction, but it has one disadvantage: It takes time. A tree cannot produce flowers until its own metabolism becomes active, although it does not have to wait for its leaves to commence photosynthesis. The hornbeam produces flowers in April before its leaves open. The field maple and small-leaved lime flower after their leaves open, the field maple in May or June, the small-leaved lime in June or July. In all three, seeds ripen by September or October. The birch flowers in April, the male flowers (catkins) having developed the previous fall and hung waiting through the winter, and the seeds are ripe by July or August. For all broad-leaved trees the entire reproductive process takes place between spring and late summer or early fall. An unusually long winter, late frosts that damage early flowering species such as hornbeam or a cold, dull summer that slows growth can severely reduce seed production. Trees native to northern Europe, including Britain, are not able

The climatic limits of field maple, small-leaved lime, and hornbeam in the United Kingdom

to produce seed every year, and when they do, the seed is often sterile. Lime trees (*Tilia* species) require warmth during the growing season, a factor that limits their northward spread. Some species, including limes and elms, reproduce mainly by vegetative means (see "Tree Reproduction" on pages 105–111). Even when abundant seed is produced, it must wait in the ground at least until the following spring and often for more than a year before it can germinate. During that time it represents a food source for many animals, and much of the seed is eaten.

Birch has such a wide range, extending so far to the north, because its season is very short, and its seeds, produced in copious amounts, are usually viable and germinate quickly. Birch is often the first tree to appear on ground in high latitudes that has been cleared of other vegetation.

Flowering plants reproduce more efficiently than conifers, provided they have the length of growing sea-

son they need. They are adapted to short winters and warm summers.

Conifers—Thriving Where the Growing Season Is Short

Where the growing season is short, conifers have the advantage. They take much longer to produce seeds, but unlike flowers their cones are tough and not damaged by cold. Coniferous trees are adapted to long winters and can tolerate cool summers. Except for the larches (*Larix* species), they bear leaves (as needles or scales) throughout the year. Even during winter, conifer leaves photosynthesize on sunny days, and the first hint of spring finds them ready and waiting. The trees do not have to produce fresh foliage before they can begin to grow. On average, a deciduous tree—one that sheds all its leaves and is bare for part of the year—can maintain its annual growth only if there are 120 days when the temperature rises above 50°F (10°C). An evergreen coniferous tree needs only 30 warm days.

Cool winters present all trees with a further problem. Water freezes and precipitation falls as snow and then remains lying on the ground surface. Roots can absorb water only in liquid form, so prolonged freezing temperatures are equivalent to drought. Drought also occurs in lower latitudes, of course, as a dry season that is the equivalent of the high-latitude winter.

Trees have adapted to seasonal water shortage in two ways. The first involves severely restricting water loss. Leaves

The leaves of most coniferous trees are reduced in size, either to needles or to small scales. They are tough and (with a few exceptions) the tree does not shed all of its leaves at the same time. Thus the tree does not need to grow new leaves before it can commence photosynthesis in spring, allowing it to take full advantage of a short growing season. These are needles of red spruce (*Picea rubens*).   (*Vanderbilt University*)

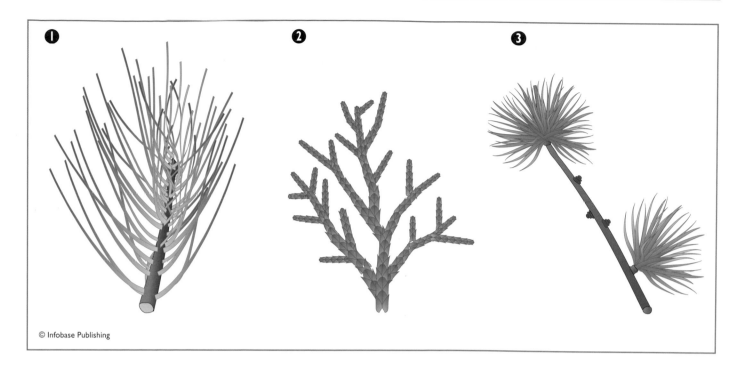

© Infobase Publishing

Leaves of three common conifers: pine, western red cedar, and larch

are covered with a thick, tough cuticle (outer skin), and the stomata (pores through which gases are exchanged) are reduced in number and sheltered in pits or crevices in the leaf surface. In conifers, the leaves are also reduced in size, which also helps them retain water. They have the form of long, narrow needles, as in pines (*Pinus* species) and larches (*Larix* species) or are like tiny, overlapping scales, as in cypresses (*Cupressus* species) and the western red cedar (*Thuja plicata*). These principal forms are illustrated in the drawing.

Broad-Leaved Evergreens— Surviving Drought

It is not only gymnosperms that have adapted to drought by evolving leaves protected by a thick, often waxy cuticle through which water loss is much reduced. So have the broad-leaved evergreen trees of lower latitudes. Many of these leaves also have sharp prickles around their edges to discourage browsing animals. Holly (*Ilex aquifolium*) is a familiar example that grows well in higher latitudes. Many ornamental varieties are cultivated for their attractive foliage, but holly is the dominant tree in some woods (hence the name *hollywood*).

Hollies are trees of the sclerophyllous forests. Most of these forests have been cleared, but this was the type of vegetation natural to the Mediterranean region, central and southern California, parts of Chile and South Africa, and southern and western Australia, where the summers are long, hot, and dry.

Trees That Shed Their Leaves Each Year

The alternative to producing tough, hard-wearing leaves is to grow leaves rapidly early in spring, expose them to sunshine all summer, and shed them in the fall. Parts of a plant or animal that are shed all together in this way are said to be deciduous. Ants have deciduous wings that they shed after mating, most species of deer have deciduous antlers, and human children have deciduous teeth (the milk teeth). Many trees growing in temperate climates have deciduous leaves.

Shedding all of the leaves seems wasteful, but it works well. Deciduous forests occur where temperatures range from about -22°F (-30°C) in winter to 86°F (30°C) in summer and where precipitation is spread fairly evenly through the year. The growing season lasts five or six months. That is as long as the leaves need to last, so they can be thinner and flimsier than the leaves of evergreen trees. Fewer resources are used in producing them.

As summer draws to a close, the days grow shorter and the temperature starts to fall. Deciduous leaves respond to these changes by progressively reducing their production of auxins, which are chemical substances that stimulate plant growth, and increasing their production of ethylene, which inhibits cell growth. No more chlorophyll is produced, and many of the nutrient elements in the leaves are transported back into the branch or trunk of the tree and stored there until they are needed the following spring. There is a layer of parenchyma cells near the base of the leaf stalk (the petiole).

This layer gradually weakens until a gust of wind is enough to detach the leaf. The process by which a deciduous tree loses its leaves is called *abscission*.

Abscission incurs some cost, in energy and materials, but this is minimized. It helps that nutrients are withdrawn before leaves are discarded, and once the leaves are on the forest floor, their decomposition is fairly swift. Deciduous leaves disappear much more quickly than conifer needles or the leaves of broad-leaved evergreen trees, and the nutrients they contain are recycled the following spring.

Growing deciduous leaves is a highly sophisticated adaptation to climate. It works extremely well over a large part of the temperate regions.

PRIMEVAL FOREST, ANCIENT WOODLAND, OLD-GROWTH FOREST, AND PLANTATIONS

Forests are dynamic communities of plants and animals, changing constantly in countless small ways in response to changing circumstances. Over long periods, these small changes can accumulate, making a forest today substantially different from the forest that occupied the same site in earlier times.

People used to believe in the "balance of nature" and some environmentalists still do. It is an essentially religious concept, holding that God created natural systems in such a way that their internal checks and balances maintain them permanently in a stable condition. The difficulty with the concept is its implication of stasis. It suggests that a naturally balanced system remains always the same. Not only is that mistaken but it is impossible. Unless systems are able to adjust to changes in the conditions around them, those changes may destroy them. Any climatic change, for example, will cause difficulties for certain species and their populations will decrease in number. This will allow other species, better suited to the changed conditions, to flourish and, perhaps, eventually to replace the disadvantaged species. If the system is a forest, it will remain forest, but by becoming a different kind of forest. Were there to be a static balance of nature so that the disadvantaged species could not be replaced, the forest would disappear, presumably leaving bare ground. There is a balance in the sense that all complex systems regulate themselves, but it is not one that precludes change. An unchanging system could not survive for long.

What Does "Natural" Mean?

Change is natural, but recognizing the fact presents a further problem. Looking at an area of forest, how can anyone tell whether its development has been affected by human activities and, if so, to what extent? Did its trees grow with no help from humans, grow from seed produced by neighboring trees, or were they planted? Perhaps the whole forest was planted long ago and there is no record of the planting. The forest comprises certain tree species. Did these arrive of their own accord, advancing northward as the climate grew warmer after the ice sheets had retreated, or did people bring them? Sweet chestnut (*Castanea sativa*) grows in many English forests, but it was the Romans who brought the species to Britain. The "English" walnut (*Juglans regia*) grows naturally nowhere closer to England than southeastern Europe. It was brought to Britain, probably in the 15th century, and *Acer pseudoplatanus,* the tree known in Britain as the sycamore, was introduced at about the same time, or a little later. It has invaded many forests and is now so common that some people consider it to be a weed.

What does it take, then, for a forest to be described as natural? In practice, forests (and other ecosystems) are classified according to their naturalness, but the classification is difficult and rather vague, although its underlying concept is simple enough. This is that while deer, beavers, and many other animals live in and affect the development of forests, only humans deliberately manage forests, thereby altering them in planned ways, so the degree of naturalness is inversely proportional to the amount of human intervention.

It can be argued, of course, that humans are as much a part of nature as are any other animals and it is entirely natural for them to live the way they do. After all, ants, termites, and prairie dogs build elaborate shelters for themselves, prairie dogs deliberately kill woody plants around their colonies, and grazing animals, such as deer and cattle, kill tree saplings, thereby encouraging the growth of grasses and herbs. Why should people consider themselves superior (or inferior)? Taking that point of view renders the word *natural* at best relative. Nevertheless, most people would likely agree that a backyard, growing flowers and vegetables, is in some sense more natural than the street and the farmed countryside more natural than the city. Naturalness requires the presence of growing plants and living animals.

Value of Undisturbed Habitat

Naturalists, however, attach importance to the idea of naturalness, mainly because they believe that a habitat left undisturbed by humans and allowed to develop in its own way is likely to support a wider range of species than a managed habitat. This is often the case. Ecologists also regard a natural habitat as being of greater scientific importance than one modified by human activity because it affords them greater opportunity for learning how natural systems function. The understanding they acquire is of practical use in evaluating the likely consequences of different types of land management in adjacent areas or areas similar to the forested area.

Inevitably, efforts to protect areas designated as natural often lead to conflict. These most commonly arise over competing uses for land, but there are also disputes over the desirability of managing established forests. In 1996, environmentalists managed to halt an operation in the Headwaters Forest, California, in which the Pacific Lumber Company planned to remove dead, diseased, and dying trees from groves of redwoods and Douglas firs in an area of old-growth forest. The protest was triggered by a worker who accidentally knocked down a healthy hemlock tree. Here, the conflict was between those who sought to manage the forest and those who wished it to be left undisturbed. No one had proposed clearing the forest. There were already plans to take the forest into public ownership. On September 29, an agreement was reached to this effect, and the matter was finally resolved on March 1, 1999, when the U.S. Bureau of Land Management and the State of California purchased the Headwaters Forest and the surrounding land from the Pacific Lumber Company. The forest reserve covers 7,400 acres (2,995 ha), and it is of great conservation value. Some of its redwoods are 300 feet (90 m) tall and more than 2,000 years old. There is a great diversity of herbs, ferns, and lichens. Rivers flowing through the forest contain coho or silver salmon (*Oncorhynchus kisutch*), and in spring and summer the forest is home to the marbled murrelet (*BrachyraMPHus marmoratus*), a bird of the auk family (Alcidae). Auks are seabirds—the Northern Hemisphere equivalents of the penguins of the Southern Hemisphere. All 22 species nest in colonies on coastal cliffs except for the marbled murrelet. It nests up to six miles (10 km) inland and up to 100 feet (30 m) aboveground on the branches of coniferous trees, and while nesting it is very sensitive to human intrusion.

Defining Primeval Forest and Wilderness

The controversy over the Headwaters Forest centered on its management and not on whether the forest is natural. Conflicts arising from a desire to protect natural forest cannot be resolved unless there is an agreed definition of natural forest. At one level, this is not too difficult. Until the first humans reached it, the development of a forest must have been guided entirely by nonhuman events. Clearly, that primeval forest was natural. In 1981, in the first edition of his book *Woodland Conservation and Management,* George F. Peterken proposed that this forest be called "original-natural."

Had the forest, or some part of it, never been invaded by humans, even though they lived not far away, it would have remained undisturbed. By now it would be substantially different from the original-natural forest because over the thousands of years since humans arrived, the climate has changed several times and soils have also developed. Nevertheless, the changes have not been wrought directly by humans. Peterken suggested calling this kind of forest "present-natural."

In areas remote from human habitation, such present-natural forests still survive as wilderness. By definition, a wilderness area is one in which humans have never lived permanently or only as isolated individuals or families, and that has never been managed or exploited economically. It is not an area that was formerly occupied and has since been abandoned. Defined in this way, there are areas of true wilderness in North America and Europe, but whether they comprise truly primeval forest is more doubtful. Prehistoric peoples almost certainly exerted some influence, and in the United States ecologists prefer to classify forests that have remained unchanged since the arrival of Europeans as "pre-settlement." Europeans are more generous with their use of the word *primeval.* The Germans, for example, consider some of their forests to be primeval (*Urwald*).

Despite the changes over the centuries, a present-natural forest may still retain similarities to the original-natural forest. It may be possible to tell, for example, by studying pollen found belowground (see "Using Pollen to Study the Past" on pages 228–231), that the forest has descended directly from the primeval forest. Peterken calls this kind of forest "past-natural."

What Happens When Land Is Abandoned?

Abandoned land can never become true wilderness, but plants and animals will colonize it in an entirely natural way, and no one would deny that a forest established under these circumstances is more natural than one in which all the trees were planted by foresters. Take any area of land, suppose that all the people left, that their buildings and roads were demolished and the debris removed, and that the area transformed itself instantly into a mature forest. This forest would be what Peterken calls "potential-natural." The idea sounds fanciful, but it is extremely important. Obviously (and fortunately), no one has the power to remove all the people in this way and to clear away every last shred of evidence of their presence. Even if that were to happen, plants would return a few at a time, and it might take centuries for a mature forest to develop. Nevertheless, it is valuable to be able to predict what that forest would be like, assuming the climate and soils remained unaltered. This knowledge reveals what it is that those homes, roads, and so forth are replacing. When a change in land use is proposed anywhere, a calculation of the potential-natural state of the area helps in evaluating the ecological effects of that change.

Plants are quick to start colonizing bare ground. Their bright flowers are cheerfully but tragically conspicuous in cities devastated by war. Left undisturbed, the first arrivals are soon replaced and slowly the plant community comes to be dominated by woody plants and in many places eventually by trees (see "Succession and Climax" on pages 147–151). If the abandoned area is large enough, it may become forested. This forest, which Peterken calls "future-natural," will not contain those trees of the original-natural forest that have become extinct, and it will contain introduced species that have become naturalized and invasive. It will be forest but one that is quite unlike the original-natural forest.

"Ancient," "Recent," and "Old-Growth" Forests

Most forests fail to qualify in any of these categories because humans have deliberately managed them, but since all forests support some wildlife, all of them are somewhat natural. This has led to alternative ways for classifying forests.

Forestry, which is the deliberate planting of entire forests for commercial exploitation, began in the 17th century, and that is also when the size and location of woods and forests were first shown on British maps. In Britain, therefore, a forest or woodland that is known from records to have been in existence continuously since 1600 or earlier is called "ancient." Woodland younger than this is described as "recent." If an ancient forest or woodland is known to have been in existence prior to the clearance of the surrounding primeval forest, it is classed as "primary forest." If it came into existence later than that, it is "secondary forest."

North American ecologists are reluctant to describe their forests as "virgin." Instead, they use the classification "old-growth." Unfortunately, "old-growth" is no more easily defined than any of the other terms. Designation of an old-growth forest is based on assessments of the sizes of the trees and fallen wood and the number of tree species present. While forests conforming to these criteria could develop on abandoned farmland or from forests that were once managed, in fact most are presettlement forests and probably are

Old-growth forest is difficult to define, but in most cases it is likely to be primeval. This old-growth forest is dominated by hemlock (*Tsuga* species).   *(Vanderbilt University)*

primeval. Indeed, many old-growth forests are likely to be older than the German *Urwälder*.

Naturalness is a difficult concept, and no one would attempt to use it in classifying forests were it not for the usefulness of the resulting classification. It allows ecologists to identify those forests likely to be of greatest scientific value and it helps in assessing the environmental consequences of proposed developments.

■ SUCCESSION AND CLIMAX

When an old building is demolished, sometimes the rubble is cleared away and then the site is left undisturbed for a few weeks or months. Except in the depths of winter, a few plants will start to appear within days. At first they will be small and hard to see, but before long they will cover the whole area and then they will flower. One of the most poignant sights in a city devastated by bombing is the gaudy mass of flowers that quickly arrive, seemingly from nowhere, to brighten the ruins of what once were homes. In Britain, November 11 is called Poppy Day when people wear artificial common (or field) poppies (*Papaver rhoes*), sold to raise funds to care for veterans, and thus decorate wreaths with them to commemorate the lives lost in the two world wars. This red poppy is used because during World War I, it carpeted the battlefields of Flanders.

Bare ground seldom remains bare for long. The plants do not really come from nowhere, of course. Some arise from tiny, light seeds that ride the wind in their billions and occasionally fall on fertile ground. Others, including the common poppy, have seeds that remain viable for decades and germinate when some disturbance removes competing plants, allowing them access to light and moisture. They are opportunists, those first plants with their cheerful flowers, and they do not last. Other plants arrive to displace them. Leave the demolition site longer, and grasses will appear and then brambles. It is as though one group of plants prepares the way for the next in a fairly orderly sequence.

The decomposition of dead organic matter proceeds in a rather similar fashion. Larger animals, fungi, and bacteria feed on the material, one group of decomposers following another. Indeed, the sequence is so reliable that it has given rise to forensic entomology, a scientific discipline whose practitioners determine the age of a corpse from the insect species feeding upon it.

Changing vegetational sequences of this kind are not confined to city demolition sites. Leave a garden untended, and weeds will soon overpower many of the cultivated plants. Fields left untilled and pasture left ungrazed will be invaded in the same way. Within a few years farms and gardens will turn into impenetrable scrub. The poet and naturalist Henry David Thoreau (1817–62) observed this kind of change happening around his home village of Concord, Massachusetts, where natural forest was gradually being converted into a mixture of managed woodland and pasture.

Thoreau was among those 19th-century naturalists who recognized that there was order in the sequence by which one group of plants followed another on disturbed ground—that the sequence was in fact a succession. At first this idea was fairly vague and applied only to particular areas in which it had been recorded, but in the early years of the 20th century the concept was made more general and more precise.

The Plants around Chicago

One of the first contributions to this process came from the American botanist Henry Chandler Cowles (1869–1939). In 1899 and 1901 Cowles published two papers in the *Botanical Gazette* (which he edited from 1925 until 1934) describing the vegetation around Lake Michigan and Chicago. His major statement on long-term succession appeared in 1901 as "The plant societies of Chicago and vicinity," published by the University of Chicago Press in the *Bulletin of The Geographic Society of Chicago*. In this paper he described the plant communities that had succeeded one another since the final retreat of the ice sheets in Illinois, as water drained from the huge, glacial Lake Chicago, leaving the present much smaller Lake Michigan and the Chicago plain.

Cowles reported in his *Botanical Gazette* papers that stable sand dunes around Lake Michigan were colonized by basswood (*Tilia americana*). As the climate became increasingly moderate, the community dominated by basswood gave way to others that were better adapted to these conditions, until finally there was a mesophytic community (one comprising plants that thrive where there are no extremes of temperature and moisture is available through the year). Maple (*Acer* species) and beech (*Fagus grandifolia*) dominated the mesophytic community. As Cowles saw it, the vegetation pattern changed constantly, but with the mesophytic community it reached a kind of optimum, or climax, state.

Attaining this state was not the end of the process, because the pattern could change further away from the climax and then return to it. His study of the draining of Lake Chicago and the emergence of the Chicago plain described the way a landscape had developed during many thousands of years. Vegetation patterns were related to landscape changes, which in turn were related to climatic changes, and so the ecological succession of plants was unending. It could never reach a final state in which it would remain forever because as it approached a kind of mesophytic ideal, the climate or landscape would change. If a succession can be described as a path toward a target, the target keeps moving, so it can never be reached.

Clements and the "Superorganism"

Frederic Edward Clements (1874–1945; see the sidebar "Frederic Edward Clements" on page 129), a botanist and ecologist initially at the University of Nebraska and from 1907 at the University of Minnesota, took the theory of succession a step further. His monograph "Plant succession: analysis of the development of vegetation," which appeared in 1916 in the *Publications of the Carnegie Institution, Washington,* was so well received that Clements was appointed to a post at the Carnegie Institution. He moved from Minnesota to Washington, D.C., where he remained for the rest of his life. In the years that followed, Clements wrote or coauthored close on one thousand papers describing successions in various sites all over the country. It was a remarkable achievement for anyone and especially for a man who was unwell for much of the time during his later years. Assisted by his wife Edith, Frederic Clements worked so hard and wrote so much that his ideas became very influential, and his textbook, *Plant Ecology,* first published in 1929, brought his views to several generations of students.

Anyone who sets out to describe a plant succession has to devise some means of classification. Without this, it is impossible to show any change at all. At one time there must be a community of one kind and later a community of a different kind, but those kinds must be defined and named and then allotted a position within a classificatory system. This need arises only because it is almost impossible to discuss things unless those things have names. Once suitable names have been attached, the discussion may deal accurately enough with the overall process, but there is a problem. Because they have names, the stages of the process sound fixed, as though one morning an ecologist might arrive to discover that the "stage 1" noted a week ago had gone and the site was now occupied by a "stage 2" community. In reality, however, the change is seamless. It is possible to observe it at all only by visiting the site at quite widely spaced intervals to take snapshots that freeze the process at particular points and allow the community present at one time to be compared with the community present at a later time. It is helpful to think of these communities, separated by intervals of time, as distinct stages, but there is a danger that they may come to resemble organisms in their own right, or "superorganisms."

Up to a point, the idea of a superorganism is quite persuasive. In the northeastern United States, for example, most of the broad-leaved deciduous forests are composed of various combinations of a limited number of species of beech, birch, maple, and oak, with their associated shrubs and climbers. Over much of Britain, oak (*Quercus robur*) and hazel (*Corylus avellana*) dominated the original, primeval forest. Large tracts of boreal forest contain just one or two species of trees. Certain species do seem often to occur together and it is convenient and not wholly misleading to think of them as though they were a superorganism. The most extreme example of a superorganism occurs in some interpretations of James Lovelock's Gaia hypothesis. The hypothesis proposes that on any planet supporting life, the living organisms modify their environment, and the unintended consequence of their modification is to maintain conditions favorable to themselves. The more extreme interpretation holds that this means that the entire global community of living organisms—the whole biosphere—is a single living organism, or superorganism.

Climax Theory

Clements worked initially on the ecology of prairie grassland. His first major publication, written in collaboration with his Nebraska colleague Roscoe Pound (1870–1964), was *The Phytogeography of Nebraska,* published in 1898, which became a standard botany textbook (phytogeography is the scientific study of the spatial distribution of plants). Clements believed that the grassland community really was a superorganism, an assembly of plants and animals locked together in a network of inevitable relationships, among themselves and with the nonliving factors in their environment, in such a way that the assembly behaved as though it were a single organism. It reached its final state, the climax, by passing through a succession of previous states called a sere, each seral stage of which was also like a single organism but one that matured, died, and was replaced. If the sere begins on bare ground, it is known as a primary sere, or prisere. Prior to the climax, no seral stage could endure, because its component species competed with one another while at the same time altering the physical and chemical conditions around them in ways that were eventually detrimental to them.

Nevertheless, the succession could advance in only one way, toward a single climax. As the succession progressed, so its composition became increasingly stable. The stability arose because as the succession proceeded, certain plants came increasingly to dominate the other species and also the conditions in which all the plants grew. If these conditions were sometimes extreme at the start when the site was bare ground, as the succession proceeded, they became more moderate.

The climax was reached when the dominant mesophytic plants prevented the establishment of invading species that might later become dominant. This was a stable condition. The climax composition was very predictable because Clements believed that the succession was a process of development, and as he explained in *Plant Succession and Plant Indicators,* published in 1928, this must inevitably lead to the adult form in a community of plants as much as in an individual plant. According to Clements, only one climax is possible, so this is sometimes known as the mono-climax theory.

Clements's account of a sere is sometimes called the interactive theory of climax because it advances through the interactions of all the species in the succession. Clements seems to have pictured the succession as something like a single organism, which is born, grows, and matures when it reaches its climax condition. Clear away the vegetation from an area and in time it will reform with exactly the same species in exactly the same proportions. North American grasslands, he believed, had reached their climax state once the ice sheets had disappeared and the climate warmed to its present level. Where the natural grassland had been removed, it could be restored at any time simply by leaving the land undisturbed.

The reestablishment of the prairie was inevitable provided there was natural prairie close enough to supply seeds, and what applied to the prairie also applied to any other type of climax. Clear a forest, leave the site undisturbed, and in time the forest will return. Eventually the area that was cleared will be indistinguishable from the rest of the forest. This can be seen in many parts of Massachusetts, New Hampshire, and Maine, where much but not all of the presettlement forest was cleared to provide farmland between about 1700 and 1850. After 1850, farming declined and land was abandoned. The trees returned, now as secondary forest very similar to the original type, with remains of the old field walls running here and there through it.

Clements and his colleagues wanted to help farmers by showing them the effect their agricultural practices had on natural assemblages of vegetation. Understanding the natural succession and climax would allow ecologists to advise farmers on ways to improve their methods. Farming would then become less ecologically disruptive, and crop yields would be more dependable. At the same time their ecological understanding would convince farmers that the natural vegetation is very resilient. It can be cleared away, but given the opportunity it will return of its own accord. Clements had good news for both farmers and conservationists.

Can Superorganisms Die?

If it is true that a sere resembles the growth of a single organism, there is something distinctly odd about it. This organism appears or is born, it grows and matures, but that is as far as it goes. Real organisms do not stop there. They grow old and feeble, and eventually they die. The monoclimax theory makes no allowance for this. Cowles did not believe that a final equilibrium could ever be reached, but Clements disagreed, although the monoclimax that he proposed could not evolve. Once attained, it was changeless, and his optimistic message was based on that idea of changelessness.

In Britain, the North York Moors are renowned for their scenic beauty, and they form a national park covering 553 square miles (1,432 km²). As its name suggests, much of the area is now moorland, dominated by heather and shrubs, but it was not always so. Beneath the pale, acid, relatively infertile, surface soil there is a forest soil: The moorland was once forest. Gradually, during the centuries, the trees were cleared to provide pasture; the grassland was managed inefficiently; and the area changed from forest, which was one climax, to moor, which is another. Many ecologists consider moorland inferior to forest and not the true climax. The British ecologist Sir Arthur Tansley (1871–1955) rather dismissively called it a subclimax.

This might seem to undermine and perhaps demolish the monoclimax theory, but it does not. According to that theory, two things have happened to the North York Moors. In the first place, people cleared the forest, a little at a time. They did so to encourage grasses, but by removing the trees, which were the dominant plants, they created a vacancy, an opportunity for a new kind of dominant. At the same time, the farmers altered the soil, and their livestock trampled and killed tree seedlings. So the clearance of the forest was followed by a different succession to a different but no less stable climax. Had those farmers abandoned the area soon after they cleared the forest, the forest would have returned. Instead, they and their livestock prevented the natural succession and unwittingly imposed a different one. If the forest monoclimax was a superorganism, the farmers destroyed it by clearing the forest and preventing its return. In this case, therefore, the superorganism died, and in time a different superorganism replaced it. But this happened only because of human intervention; it was not a natural process.

Natural changes in climate can also modify a succession or its climax, however. A climate which is hotter or drier than that of the region as a whole, or that becomes hotter or drier than it used to be, produces a "preclimax." A cooler or wetter climate produces a "postclimax," and other disturbances may interrupt the succession or modify the climax. So it is possible for the superorganism die naturally, at least in the sense that one monoclimax changes into another that is markedly different.

An observer walking through most forests will find the composition of the forest varies from one place to another. This is not surprising. Some places are relatively low-lying and wet; others are high and exposed. Soils wash down hillsides, so the soils at the bottom of a slope are often a little different from those at the top. Humans have interfered with most forests but have been more active in some places than in others. All of these very local variations are reflected in differences in the succession and climax. The area as a whole may be occupied by a forest climax, but within that overall climax there will be many local climaxes, all a little different and all equally stable. Sir Arthur

Tansley called this a polyclimax, meaning a climax composed of many smaller climaxes, and it is an alternative to the monoclimax theory.

Polyclimaxes are more difficult to categorize than the monoclimax. Are they components of a larger superorganism? Or are they small superorganisms, making the forest into a community of superorganisms? Perhaps ecologists needed to look again at the whole idea of the monoclimax.

The Individualistic Hypothesis

Frederic Clements was very influential, but his monoclimax theory was quickly challenged, most strongly by Henry Allan Gleason (1882–1975; see the sidebar), who developed what is known as the individualistic hypothesis. The individualistic hypothesis says that a climax plant community comprises species that have come together largely by chance; it is an assembly of individuals. Those individual plants just happened to be the ones that arrived at a site where the soil, illumination, temperature, and moisture suited them, and because they found the habitat hospitable, they thrived, coming to dominate the site to the exclusion of would-be invaders. Clear all the plants away and, of course, the site will be recolonized, but there is no guarantee that the climax community that develops there will be identical to or even closely resemble the original community.

If the interactive and individualistic hypotheses are meant to be universal, applying to any site at any time, clearly they cannot both be correct (although they could both be wrong). Deciding between them was not easy, because neither Clements nor Gleason was a pure theorist. Both were keen observers who spent much time in the field, and they based their theories on their observations. It was not easy to decide between them, but neither was it impossible.

The individualistic hypothesis predicts that species will be distributed according to gradients in the resources they need, such as light, warmth, moisture, and nutrients. The more abundant the resources, the more members there will be of the species needing them, and the assemblages of species will comprise those which just happen to be in that area. If Clements was right, the community of plants found in a particular area will comprise species that are dependent on one another and are almost always found together. Studies of actual plant communities tend to support the individualistic hypothesis. Plants grow where conditions suit them, and where there is an environmental gradient, the plants found at each point along the gradient are those best suited to the conditions at that point.

Henry Allan Gleason (1882–1975)

Henry Gleason was born in Dalton City, Illinois, on January 2, 1882. When he was 10, the family moved to the town of Decatur. Gleason was 13 when he began to study botany, and he published his first scientific paper in *The American Naturalist* while still at high school. He enrolled at the University of Illinois when he was 15. At the time this was one of the very few educational establishments in the United States where ecological ideas were being discussed and taught. Gleason graduated in 1901 with a thesis on "The Flora of the Prairie" and received his master's degree in 1904 with a thesis on "The Vegetation of the Ozark Region in Southern Illinois." In his early years he was strongly influenced by Henry Chandler Cowles (1869–1939).

After leaving the University of Illinois, Gleason spent one year as a research fellow at the University of Ohio and a summer as an invertebrate zoologist working on a survey sponsored by the University of Michigan of Isle Royale, Lake Superior. He then moved to Columbia University to study taxonomy, graduating with a Ph.D. in 1906. He conducted his research mainly at New York Botanical Gardens.

Gleason taught for a time at the Universities of Illinois and Michigan and then spent a year studying tropical vegetation in the Philippines, Java, and Sri Lanka. He published an account of his travels in 1916. In 1917, now back at Michigan, Gleason developed his individualistic concept of the plant community.

In 1918 Gleason delivered a lecture on his studies of tropical plants to the Torrey Botanical Club. After the talk, his supervisor from Columbia University, Nathaniel Lord Britton, offered him a permanent position at the New York Botanical Garden. Gleason moved to New York and remained at the Botanical Garden for 30 years until he retired in December 1950. During this time he served as curator, head curator, and assistant director, and for 19 months he was acting director.

On January 25, 1971, the Illinois Nature Preserves Commission dedicated the Henry Allan Gleason Memorial Nature Preserve in an area known locally as Devil's Neck. That is where Gleason conducted much of his early research. Henry Gleason died on April 12, 1975.

Are There Such Things as Successions and Climaxes?

Ecologists still talk of successions and climaxes, and the terms are used widely in textbooks and even more in the writings of environmental campaigners, but in recent years they have become unfashionable among scientists. In damaging the interactive hypothesis, the individualistic hypothesis also weakened itself, and now the whole idea of climaxes and the seres leading to them appears to generate more difficulties than it solves.

Gleason showed that a vegetation pattern changes smoothly along a gradient. This means that it cannot be regarded as a superorganism because its boundaries cannot be clearly defined. They are much too variable, both spatially and over time. Nor is it always true that each stage in a succession prepares the way for the next. Sometimes that is what happens, but at other times it does not, and there seems to be no way of predicting whether it will be so in a particular instance.

Even the central idea, of the stability of the climax, turns out to be illusory. Stability can be measured only in a forest that has remained completely undisturbed for centuries. It is doubtful whether such a forest exists anywhere in the world. Disturbance is entirely natural and in some cases essential for the continuation of the forest (see "Fire Climax" below), so the forest climax cannot be stable in the sense intended by Clements, Tansley, and Gleason. There is some evidence to suggest that in certain ecosystems (but not forests), the greater the number of plant species, the more stable the climax will be, but the link may not be very strong.

Relationships within communities of plants are much more complex than the early ecologists imagined, and their development is affected by many more factors than climate, soil, and the plants themselves. Nevertheless it remains true that starting from bare or disturbed ground plants will appear and disappear until a community develops that is capable of enduring much longer than any of those preceding it. Ecologists may prefer not to call this a climax, but so far they have found no other name, and the idea of the sere leading to it also remains useful, provided it is understood as a very general description, rather than a detailed account of what really happens.

■ FIRE CLIMAX

Smokey the Bear used to warn people of the danger of forest fires, and *Bambi*, the Disney movie enjoyed by millions of people, underlined the danger. The movie depicted forest animals fleeing in terror from a fire that was advancing almost as fast as they could run. Dramatic and highly believable, this was the popular view of forest fires, and there was a time when many foresters and wildlife conservationists shared it. A forest fire is a terrifying sight, after all, and is extremely dangerous for people living in its path. It makes obvious sense to take great care not to start a fire accidentally, and if one should start, to extinguish it as quickly as possible before it has a chance to injure anyone, damage any property, or kill any wild animals.

For many years the management policy for natural forests was to prevent fires and, if that failed, to extinguish the fires rapidly. Then, in 1988, a fire started in Yellowstone National Park, and despite the heroic efforts made to check it, the fire raged out of control until the first winter snows extinguished it (see the sidebar on page 152).

The Yellowstone fire seemed like a major disaster, but by spring of the following year, new plants were emerging, and it was clear that a new succession had begun. After a few years, not only had the forest recovered, but it had also been invigorated. Meanwhile, the management policy was being altered. Since 1992, this calls for fires to be suppressed in some parts of the park, but if surface litter accumulates to more than a certain extent, fires will be started deliberately in order to remove it. Elsewhere in the park, naturally occurring fires are controlled but not prevented. It is now recognized that, in certain types of forest, fires are a natural phenomenon to which the plant and animal communities have adapted. Fire serves a useful purpose, in fact, and sometimes it is used as a management tool.

Lightning and Litter

In the dry climate of Yellowstone, decomposition is a slow process. Litter accumulates on the forest floor, and from time to time a fire removes it. While it is important not to start forest fires, it is lightning, not people, that causes most forest fires. Dry electrical storms with lightning but no rain are common in the western United States. Whether a particular lightning strike will ignite tinder-dry material depends on the length of time the electric current is sustained (see "Hot Lightning and Cold Lightning" on page 67).

Not all forests suffer equally from fires. Fires are uncommon in broad-leaved deciduous forests. This is partly because such forests occur in fairly wet climates where litter on the forest floor is usually too damp to burn and decomposes quickly. Nor are the trees themselves very combustible. Their leaves contain enough water to make them slow-burning and difficult to ignite, and their wood also burns slowly.

Coniferous forests are much more vulnerable. Their wood contains resins, which are highly flammable; their needles contain much less water than do broad-leaves; and the forests tend to grow in climates that are drier than those of broad-leaved forests. Studies of forests in northern Sweden

The Yellowstone Fire

Dead plant material on the forest floor dries out during the hot, dry summer in Yellowstone National Park. From time to time, lightning ignites the surface litter, and every summer there are several fires. There were 235 fires in the park between 1972 and 1987, all of which died down naturally. Since 1992 it has been the policy to keep fires under control but not to extinguish them.

During the 1980s there was a series of unusually wet summers, and fires were less frequent. Plant material accumulated on the forest floor and continued to do so until 1988. The spring of 1988 was very wet. There were a few fires but not sufficient to clear most of the litter. Then, in June, a severe drought began. The litter dried, and by the third week in July, fires were breaking out in many parts of the forest. The park authorities decided to bring the fires under control, but this proved impossible: There was simply too much fuel. August 20 was the worst day for fires. On that day more than 150,000 acres (60,700 ha) of the forest were ablaze.

The fires continued to burn vigorously until September 11, when the first substantial fall of snow began to suppress them. They continued to smolder until November, when the last of them was extinguished. By then the fires had burned almost 1 million acres (400,000 ha) of the forest, amounting to about 45 percent of the area of the national park.

Young trees could be seen emerging in the spring of 1989, and within a few years Yellowstone had recovered. In fact, the fires had invigorated the forest by leaving a layer of ash across the surface. Rich in potassium, ash is a valuable fertilizer, and the first rains washed it into the soil to feed the new seedlings.

As this photograph shows, the 1988 fire in Yellowstone National Park was severe. Nevertheless, the forest quickly recovered.   *(Yellowstone National Park)*

found that prior to 1900, when it became official policy to prevent and fight fires, about 1 percent of the forest area burned each year and that the same areas would catch fire at intervals of about 80 years. Not all the fires started naturally; for centuries local people had been setting fires periodically to remove trees and provide pasture for their livestock.

Once a fire is blazing, of course, how it started makes no difference to its progress. In the United States, a huge forest fire at Yacholt, Washington, destroyed nearly 250,000 acres (about 100,000 ha) of forest in 1902, and in 1933 a fire in Tillamook, Oregon, burned a similar area. Some forests in Minnesota have burned repeatedly, and prior to European settlement about 80 percent of the total area burned every century.

Many parts of North America have a climate dry enough for fires to be fairly common in coniferous forests, and the hot, dry Australian summers make wildfires fairly common. In March 1965, one huge storm produced 75 separate lightning strikes, starting fires that burned 741,300 acres (about 300,000 ha) of Australian forest, but the worst wildfires Australia has ever experienced happened in 1983. They were called the Ash Wednesday fires because they began on February 16, which that year was Ash Wednesday in the Christian calendar. About 180 fires broke out on that day in Victoria and South Australia. The fires were quickly brought under control, but there were so many of them that they burned a total of 1,384 square miles (3,585 km²). Hundreds of people were injured and 75 lost their lives.

Types of Forest Fire

Not all forest fires turn into raging infernos. Most burn only a small area of the forest before dying of their own accord. Those that spread more widely do so across or beneath the surface or through the canopy. Surface fires burn through the litter, spreading rapidly and destroying herbs and some shrubs. They scorch the trunks of large trees, usually without causing them serious harm, although they can kill seedlings and small trees. Ground fires produce no flames, burning below ground and spreading quite slowly. These fires are hot and can continue smoldering for a long time. They consume the roots that penetrate the burning layer, killing most of the plants. Wet ground and even bogs can burn in this way because a ground fire advances slowly enough to dry out the area ahead of it.

The spectacular forest fires that feature in movies are crown fires. These are driven by the wind and can spread very rapidly through the forest canopy, leaping from tree to tree. Blazing cinders are carried upward by convection, blown ahead of the fire by the wind, and then fall back into the canopy, sparking new fires. The temperature in the fiercest fires may be so high that material some distance from the flames ignites by radiation, the way, without touching it, a sheet of paper will char and then catch light when it is held close to a fire. The heat may be so intense as to cause firestorms. These happen when the rapid convective rise of hot air causes air to converge strongly near ground level. As it converges, the air starts to turn cyclonically (counterclockwise in the Northern Hemisphere) and is accelerated, producing a twisting wind that spirals upward, like a tornado. The converging air picks up loose material and sweeps it into the vortex, adding fuel to the flames. After a crown fire has passed, the forest is reduced to charred sticks rising from a thick layer of ash. The 1988 Yellowstone fires were of this type.

Adaptation to Fire

Big forest fires make the TV news, and anyone watching a crown fire leaping and roaring through a forest might wonder how any living thing could survive a conflagration of this kind. Yet clearly, many do. North American forests have burned repeatedly throughout their history, but after each fire they have established themselves once more.

Trees that grow naturally in fire-prone areas have adapted to fire in various ways. Opportunist species, which are quick to colonize open or disturbed ground, are killed by fire, but their seeds are not. These species produce seeds in prodigious amounts (see "A Few Offspring or Many?" on page 139). The seeds disperse widely and germinate rapidly. It is a strategy that allows species such as quaking aspen (*Populus tremuloides*), balsam poplar (*P. balsamifera*), and paper birch (*Betula papyrifera*) to occupy sites cleared by fire in the northern and western United States and to mature and produce seed before the conifers grow taller and shade them. The early colonizers are intolerant of shade, so they disappear as the conifers grow. They depend on fire for their chance to grow and reproduce.

The coast redwood (*Sequoia sempervirens*) and Sierra redwood (*Sequoiadendron giganteum*) have spongy bark that is very difficult to ignite and thick enough to insulate the sensitive tissue beneath the outer layer. Provided that its temperature remains below about 149°F (65°C), the vascular cambium (see "Vascular Cambium" on pages 113–114) will not be destroyed. The bark of these trees is so spongy that it is possible to punch the trunk quite hard without injuring the hand, and the tree can remain unscathed by most fires. A redwood that is burned so badly that it falls often sprouts vigorous new growth from the stump.

Some trees regenerate rapidly from snags (standing dead trees), provided that the roots survive the fire. Pitch pine (*Pinus rigida*) is one. Other pines not only survive fires, but they also positively depend on them. These are the species such as longleaf pine (*P. palustris*, also called pitch pine) and jack pine (*P. banksiana*) that produce serotinous cones. Serotiny is the retention of seeds, often for several years, until some disaster triggers their release. Serotinous cones remain on the tree and are tightly closed until the heat of the fire causes them to open and release their seeds, which fall into the ash on the forest floor. There they are able to germinate in soil enriched with ash on a forest floor bathed in direct sunlight, the conditions that give the seedlings the best possible start in life. Fire-prone forests also harbor seeds

Crown fires, like this one in Montana, are driven by the wind and spread rapidly through the canopy. *(Ned Hetinger)*

of certain shrubs and herbaceous plants that are similarly adapted, germinating only when a fire has heated them.

Most animals, but not all of them, can also survive fire. In the 1988 Yellowstone fire, about 250 elk or wapiti (*Cervus elaphus,* known in Britain as the red deer) died from smoke inhalation, but they were the only casualties out of a population of about 31,000 elk present in the park at the time. Birds and many mammals simply move out of the way of the fire. Others shelter in burrows or beneath logs. Those animals that do leave face other dangers because predators patrol just ahead of the fire front on the lookout for fleeing prey. In

general, *Bambi* painted a very misleading picture. In the real world, forest fires appear to cause few animal casualties.

Fire and the Soil

During a fire in the surface litter, temperatures at ground level are commonly between 194°F and 248°F (90°C to 120°C), but they fall sharply below the surface, and at a depth of a few centimeters, the temperature is tolerable for most organisms. After the fire has died down, the ground is covered with ash. This is rich in certain nutrient elements, especially potassium, phosphorus, calcium, magnesium, and sodium, which are washed down into the soil by the first rain, temporarily increasing soil fertility.

Below the surface, microbial activity in the soil often increases after a fire. Some plants produce chemical compounds that remain in the soil and that inhibit bacteria. These include phenols, found in such substances as tannins, and the fire will burn off a proportion of them. The fire also produces charcoal, and scientists now believe that charcoal absorbs the phenolic compounds. This removes the inhibiting factor, allowing bacterial activity to increase, and the activity includes nitrification (see "Nitrogen Fixation and Denitrification" on pages 122–125). Together, the mineral nutrients washed down from the layer of ash and the additional nitrate produced by the invigorated soil bacteria strongly boost soil fertility.

At the same time, fire clears away plants that formerly shaded the soil, and in places it blackens the surface, which increases the ground's capacity for absorbing heat. Since more sunlight can penetrate to the surface where more of it

Quaking aspens (*Populus tremuloides*) are quick to occupy sites cleared by fire and can produce seed before recovering conifers grow tall enough to shade them. *(Fogstock.com)*

is absorbed, the soil is warmer after a fire than it was before. This, too, stimulates microbial activity, the germination of seeds, and plant growth.

Fire Climax

Where forest fires occur repeatedly at fairly regular intervals, they come to determine the type of climax vegetation, which is known as a fire climax, or pyroclimax (from the Greek *pur* meaning "fire"). A fire climax comprises species that are tolerant of fire or even dependent on it, as well as the opportunist species that are capable of waiting for a chance to occupy temporarily vacant sites. Within the forest, repeated burning tends to produce stands of trees all of similar age, although in a large forest this pattern is quite complex because different areas within the forest may burn at different intervals.

In regions that are not subject to regular burning, fires are much more harmful. Where species are not adapted to fire, they are less likely to survive or regenerate. Nevertheless, in regions where they are common, forest fires cause good as well as harm, and the forests soon recover.

■ ARRESTED CLIMAX

Farmers and gardeners spend a great deal of time fighting weeds (see the sidebar). Once people start to grow crops, rival plants appear, and if these are not removed, the best the grower can hope for is a drastically reduced yield. The grower who really does nothing at all about the weeds is quite likely to lose the entire crop. Nowadays there are herbicides to help in the fight and genetically modified crop varieties that allow growers to apply herbicides in smaller amounts but at the stage in plant growth when they are most effective. For most of history, though, weeding has been a slow, tedious, and often backbreaking task. No wonder that a break from weeding with a hoe was an occasion for a party, or hoedown. In parts of Africa the tradition of hospitality calls for a visiting stranger to be given a meal and a bed and on the following morning to be handed a hoe.

A weed is simply a plant that grows in the wrong place and competes for nutrients, water, and sunlight with the plants sown by people. Ecologically there is not the slightest mystery about what weeds are or why they make so much hard work for folk, but to understand what is happening, it is necessary to imagine the land as it might have been before humans started to cultivate it.

At that time what is now the farm or garden would have been covered with plants, forming a community of temperate forest, perhaps, that changed little over the years. In order to grow crops, the trees and undergrowth had to be cleared. Forest clearance left bare ground but bare ground that held countless seeds, just waiting for an opportunity to

What Are Pests and Weeds?

Farmers grow crops and raise livestock in order to provide food, fiber, and other products. Any animal that eats or damages produce intended for human use is competing with the farmer and, in a broader sense, with society at large. If the amount that the animal takes or damages causes the farmer significant economic loss, then the animal is a pest.

A weed is a plant that is in the wrong place. It may be an exotic plant that has been introduced from another part of the world and that flourishes at the expense of the native plants it displaces.

It may also be a cultivated plant growing among a crop of a different species. Seeds that are spilled in the course of harvesting may germinate along with those sown for the following crop. Tubers that are missed during potato harvesting often produce new plants the next year when the land is growing a different variety or different crop. Cultivated plants that become weeds in this way are called volunteers, and they can be very troublesome because they are often similar to the crop plant—barley growing among wheat, for example. They are difficult to remove, and at harvest, their seed may contaminate the seed from the crop.

Weeds on farmland are more often plants that occur there naturally as opportunist species producing seeds that germinate when the soil is disturbed. The plants themselves then compete with the crop plants for nutrients, water, and light.

The concept of a pest or weed is entirely economic. They are simply wild animals or plants until their numbers increase to such an extent that they begin to cause harm that is serious because it is costly. Their numbers are prone to increase, however, because farmers provide ideal conditions for them. A crop growing in a field or a forest plantation represents an almost limitless food supply to certain species of animals. If those animals are opportunists they will reproduce rapidly to take advantage of the resource before it disappears. That is when an animal becomes a pest.

Similarly, opportunist plants rapidly colonize disturbed ground. If they manage to establish themselves before the crop plants have time to grow tall enough to shade them, they can seriously reduce the rates of crop germination and growth. That is when wild plants become weeds.

germinate. Bare ground will soon be colonized by pioneer plants, growing from that stored seed or from seed carried there on the wind, and a new succession will have begun (see "Succession and Climax" on pages 147–151).

Pioneer plants are usually annuals. They complete their life cycles, from germination to seed production, within a single season, which means that they must germinate and grow fast. That gives them a clear advantage on a bare site. Perennial species grow more slowly, so they appear later, but once they have appeared they put down permanent roots from which they produce new growth every year. This allows them to crowd out most of the annuals, so the perennials dominate the site until they are crowded out in their turn by larger perennials that shade them.

Ecology of Farmland

A prisere, which is a succession starting on bare ground, begins with annual species, and during a succession the total amount (the biomass) of plant material steadily increases; once a climax is reached, the total biomass remains more or less constant. Most crop plants are also annuals. There are exceptions, such as fruit trees and bushes, but cereals and vegetables are grown from seed (seed potatoes are actually tubers grown specially for planting) sown into bare ground for each crop. As annuals, they grow fast, and on bare ground they have no competition for light, water, and nutrients. That is how agriculture and horticulture work.

Ecologically, therefore, annual crop plants are equivalent to the pioneer species that begin the prisere. Not surprisingly, other annuals soon arrive to join the pioneers. These are not too difficult to control provided they can be killed or removed before they produce seed, but among them, and growing more slowly, there are the perennials. These are much more difficult to control because most of them can propagate vegetatively (see "Tree Reproduction" on pages 103–105) and will grow a new plant from just a fragment of root. The perennials represent the next seral stage in the progress toward a new climax. Their appearance is unavoidable, and unless they are controlled, they will spread to replace most of the annuals, and the crop will vanish. Weeds are a fact of life with which farmers and gardeners must live.

Season after season, the ground is tilled, the prisere commences, and by suppressing weeds the succession is prevented from advancing. The sere, or succession, is held in check. Arable fields and garden plots are ecologically unstable because without human intervention, their floristic character would change radically. This does not mean that farming and gardening are unsustainable, of course, but only that annual crop plants must be tended and protected from the weeds that are capable of outcompeting them.

A field crop—or forest plantation growing just a few species of commercially valuable trees—also attracts animal pests. These, too, are opportunists seeking to exploit a virtually limitless food supply. As inevitable as weeds, pests must also be controlled, for a severe infestation can strip a crop or plantation of so much of its foliage as to kill the plants.

Ecology of Pastureland

It is not only annual crops that farmers grow. They also tend livestock, and, traditionally, sheep and cattle have always been fed on grass (although today many are fed other diets). Grass is also a crop, and fields of pasture occupy a seral stage in which perennial herbs are the dominant plants. The pasture remains in this state indefinitely because grass tolerates constant grazing by livestock, but woody plants are destroyed by trampling or by being bitten off close to ground level. Sheep and cattle, as well as rabbits in parts of Britain and throughout most of Australia, hold the ecological succession in check.

Provided grazing is sustained, grasslands remain unchanged. They are stable enough to constitute an ecological climax, but this is not the climax that would exist in the absence of grazing. This was demonstrated during the 1970s on the English Downs. These are chalk and limestone hills in southern England from which the forests were cleared thousands of years ago. The largest areas are the North Downs and South Downs in the southeast of the country. The map shows their location. Nowadays grazing by sheep and rabbits maintains the Downs as grassland. During the 1970s, however, sheep farming declined for economic reasons, and the rabbit population was drastically reduced by myxomatosis, a fatal disease of rabbits. Grazing ceased over large areas of downland, and within a few years, scrub invaded. Had events been left to take their course (and conservationists controlled the scrub to protect the rare downland herbs), the Down grasslands would have reverted to forest. After a few years sheep farming became profitable again, the rabbit population recovered, grazing was resumed, and the grassland returned.

Permanent grassland of this kind represents an arrested climax, which is a type of plagioclimax (from the Greek *pla-*

The English North and South Downs lie to either side of the Weald of Kent, in southeastern England.

gios, meaning "oblique"). A plagioclimax is a climax resulting from human intervention. It may be stable, and may comprise only native species, but it nevertheless results from a sere that has been deflected from the path it would otherwise have followed. If the succession has been arrested at one of its stages, removing the inhibiting factor will allow it to resume, but it will not necessarily proceed to the climax that would have existed had the sere not been interrupted.

Effects of Land Management

Plantation forest is also a plagioclimax (see "Natural Forest and Plantation Forest" on pages 180–182). Usually trees are grown commercially on land that was formerly covered by natural forest, although where agricultural land is scarce, forestry may be pushed into marginal lands that were formerly low-quality pasture because agricultural land commands a higher price than forested land. Forests that are planted on formerly forested land are substantially different from the original forests. Fast-growing conifers are often preferred to the broad-leaved deciduous trees which once grew naturally on the site, and were the plantation to disappear, a new succession might lead to some kind of broad-leaved climax.

Managing the land can change it irreversibly, leading to plagioclimaxes that are as stable as true climaxes. This is a major fear in some areas of tropical forest, and there are temperate examples illustrating what can happen. Most European heathlands have resulted from the repeated burning of the original forest. This accelerated the leaching of nutrients from the upper soil horizon in a process called podzolization; this produces an infertile, acid soil that is unsuitable for trees. Once the heathland becomes established, it may be impossible for the forest to return.

Farming also alters soils. If it is good farming, of course, the soil fertility improves, and whether it is good or poor, regular tillage will produce a soil that is substantially different from a soil that remains uncultivated. Should farming cease, the succession will resume, but the changed soil is one of the factors that may prevent the replication of the original climax.

Recolonization of Abandoned Farmland

In northern Vermont, most of the original forest was cleared in the second half of the 18th century either for arable farming or for use as pasture, and boreal forest, growing in the area but at a higher altitude, was logged for timber. Later, farming ceased and the forest returned. The forest that has developed on the abandoned farmland comprises the same tree species as the original forest but in very different proportions. Beech (*Fagus grandifolia*) is much rarer now because it grows mainly from old roots and is slow to regenerate, so it suffered badly from clearance. Maple (*Acer*

species), birch (*Betula* species), pine (*Pinus* species), and poplar (*Populus* species) are more abundant than they were. Even the boreal forest changed as a result of logging, with spruce (*Picea* species) becoming rarer and fir (*Abies* species) and hemlock (*Tsuga* species) more abundant. The forest has returned, but it is not the same forest.

In North Carolina, on high ground near the foothills of the mountains, cultivated fields were abandoned twice, some toward the end of the 19th century and others after 1930. All of the old fields have reverted to pine forest, but there is a marked difference between the older and newer forests. In the forests that have grown up on the first fields to be abandoned, shortleaf pine (*Pinus echinata*) is more common than loblolly pine (*P. taeda*); on those abandoned in the 20th century the positions are reversed. No one knows why this should be so. It may be that at the time of the first colonization, the nearest forest was some distance away, and seeds of shortleaf pine were better able to make the journey. Shortleaf pine seeds are small and disperse widely. Later, there were plantation forests nearby, and loblolly pine is extensively cultivated, so its seeds were available. Shortleaf pine grows on poor soil, but loblolly pine grows faster. Perhaps the poorest fields were abandoned first, favoring shortleaf pine, or the use of manures and fertilizers in the 20th century favored the loblolly pines on the fields abandoned later.

So many explanations are possible that it is pointless to speculate. What matters is the clear example that this story provides of the way that successions can diverge. They appear to be highly sensitive to small variations in the conditions affecting them. This means that a plagioclimax will always differ from the climax that would develop in the absence of human intervention and that, while the cessation of human intervention will allow a succession to resume, it is very unlikely to continue on the course it was following prior to its interruption. This does not mean that a plagioclimax or arrested climax is necessarily inferior to the climax that would have developed otherwise. It may support as many species and be no less resilient, but it will be different, and over large areas of the world, human activity throughout the centuries has undoubtedly had a profound effect.

■ FOREST STRUCTURE

Some plants are more tolerant of shade than others, and this tolerance extends to trees. Tolerant species, such as field maple (*Acer campestre*), sugar maple (*A. saccharum*), hornbeam (*Carpinus betulus*), and eastern hemlock (*Tsuga canadensis*), can survive in the shade. Some species of tolerant trees will continue to grow slowly but steadily whether or not they are exposed to direct sunlight. Other species will grow to the size of seedlings or small saplings and remain at that size until the light intensity increases when they grow rapidly. Eastern hemlock falls into the former category, sugar maple into the latter.

Tolerant species can grow beneath a closed forest canopy, even if they are present only as seedlings. Most species maximize their light absorption by having many layers of leaves. The trees grow slowly, but they are long-lived. Their strategy is to wait, if necessary for many years, until one or more of the surrounding trees fall, creating a hole in the canopy. Then they grow rapidly to their full size, filling the gap, and flower and produce seed.

Seeds of many tolerant species are not dispersed very widely, so the trees appear fairly late in a succession and they are typical of climax forest, but there are exceptions. Holly (*Ilex aquifolium*), for example, is a tolerant species, but it sometimes appears on open ground as a pioneer because birds find its berries attractive and distribute its seeds over a wide area.

Intolerant species, such as downy birch (*Betula pubescens*) and silver birch (*B. pendula*), are often pioneers, arriving early in a succession, growing rapidly, and dispersing seeds widely, but there are also slow-growing intolerant trees. These thrive by establishing themselves in situations that in some respect are too harsh for their competitors.

There are also tolerant species that produce widely dispersed seed and intolerant species that grow very slowly so that they can take advantage of gaps when these appear. Whether or not a species is tolerant depends on its sensitivity to solarization (see "Photosynthesis" on pages 116–122) and on its compensation point. The *compensation point* is the intensity of light at which the rate that a plant absorbs carbon dioxide for photosynthesis is exactly equal to the amount it releases as a by-product of respiration. It results from the fact that green plants absorb carbon dioxide and use it in photosynthesis and from the fact that they also respire, which involves the oxidation of carbohydrates and the excretion of carbon dioxide as a waste product.

The compensation point varies from one plant species to another. Plants are able to grow at light levels above the compensation point because they gain more carbon than they lose, but growth is impossible at levels below the compensation point because there is a net loss of carbon. Tolerant plants have a low compensation point, and they are often very sensitive to solarization. Intolerant plants can continue to photosynthesize in very bright light, and they have a high compensation point.

Spring Flowers

Spring is a time when flowers bloom with bright colors, marking an end to the drab tedium of winter. It is the season of primroses (*Primula* species), snowdrops (*Galanthus nivalis*), daffodils (*Narcissus* species), crocuses (*Crocus* species), violets (*Viola* species), wood anemones (*Anemone nemorosa*), and many more. Primroses are usually the first to appear. Their name is from the Latin *prima rosa,* first rose. Gardeners love these plants, and there are countless cultivated varieties of them, many of which have escaped and now grow in the wild.

All of these plants share certain features. They are small. The daffodil is the tallest, but the wild daffodil stands no more than about 12 inches (30 cm) high—many cultivated varieties are taller. They appear abruptly and flower soon after they emerge. Once they have flowered, the plants die down and quickly disappear.

These traits betray the plants' origins. They are woodland plants that grow naturally in broad-leaved deciduous forests.

During winter the forest trees are bare and sunlight reaches the forest floor, but the temperature is too low for plant growth. In summer the trees are in full leaf, and the forest floor is shaded. There is too little light at ground level for most flowering plants.

In early spring, however, there is a brief period when the weather is warm enough for plants to grow, but the trees have not yet opened their leaves. That is when the Sun warms the forest floor, and flowering plants are able to expand their leaves and then produce flowers and set seeds.

Time is short, and the plants must produce their flowers as quickly as possible. They have no time to grow to a large size and must invest their resources in the production of flowers and seeds. Their flowers must attract the pollinating insects that are starting to emerge from their winter dormancy. They cannot produce big, spectacular flowers, but instead they attract insects with bright colors. The plants are competing for pollinators, of course, and the brightest flower is the one most likely to catch the attention of a passing bee.

Then the show is all over. Leaf buds on the trees open in a great burst of leaves, and very soon the ground lies in deep shade. The spring flowers that brightened the forest floor have completed their life cycle. They will wait belowground as bulbs or seeds until the warmth of the next spring stimulates them to emerge once more.

Forest Shade and the Forest Canopy

In a mature forest, the foliage of each fully grown tree touches that of its neighbors, so together the tree crowns form a closed canopy, completely covering the sky and shading the ground. The cover is complete throughout the year in an evergreen forest, but in a deciduous forest it is absent in winter. This absence allows more light to reach the forest floor in winter when the temperature is too low for plant growth and also in early spring when temperatures are rising. Many plants exploit this fairly brief exposure to light, and so a deciduous forest usually supports a richer variety of smaller plants than does a coniferous forest (see the sidebar "Spring Flowers" on page 158).

There are gaps in all forests, however, where trees have fallen, and in those gaps, or clearings, light can penetrate to the surface. As the illustration shows, the amount of light reaching the forest floor is a matter of simple geometry. In the top drawing, there is a large gap between trees. As the Sun crosses the sky, the whole of the floor between the trees is illuminated, and although nowhere is lit for the whole time, the total period of illumination for some part of the clearing, indicated by the distance between the two suns, is fairly long. The gap on the right is much smaller, so the Sun takes a much shorter time to cross it. The two gaps each end with a single tree, but a wider area is also illuminated, less intensely because of partial shading, and the size of that area is also greater around the large gap than around the small one. An area that is illuminated more brightly or for longer than another will also be warmer. This, too, affects the rate of tree growth.

Rain falls in much the same way as does light because the wind usually causes it to descend diagonally. Trees shelter the ground as well as shade it. Where water is often scarce, a large gap will receive more of it than a small gap, and, again, tree growth will be accelerated.

Suppose the sapling of a tolerant species were growing at the center of each gap. It would grow much faster in the gap on the left, where it was fully illuminated for longer each day, than it would in the gap on the right. This means that the bigger the gap, the faster that trees will grow to fill it, making it vanish.

Forest Layers

As trees fall and are replaced, gaps come and go and conditions in the forest slowly change. Seeds germinate, plants grow and each time that the conditions around them temporarily improve, they grow a little bigger. By the time the largest trees have grown to their full size, the forest has acquired a distinct vertical structure, with plants growing to several different heights. These heights form recognizable strata, or stories.

Shade and gap size. The wide gap in the upper diagram allows a large area to be illuminated for the fairly long time it takes the Sun to cross the gap. The smaller gap in the lower diagram exposes a smaller area to sunlight and for a shorter time.

Botanists usually describe the commonest plant species in an area, or the species that gives the area its distinctive character, as the dominant. Foresters use the word differently, applying it to trees that are taller than their neighbors, regardless of species. The tallest trees in the forests, or dominants, are all more or less of the same height. Between them, not reaching to quite the same height, but still forming part of the canopy, there are smaller, *codominant* trees. These belong to the same species as the dominants, and should an adjacent dominant fall, a codominant will grow just a little larger and take its place. Smaller and lower than the codominants, there are *subdominant* trees, also of the

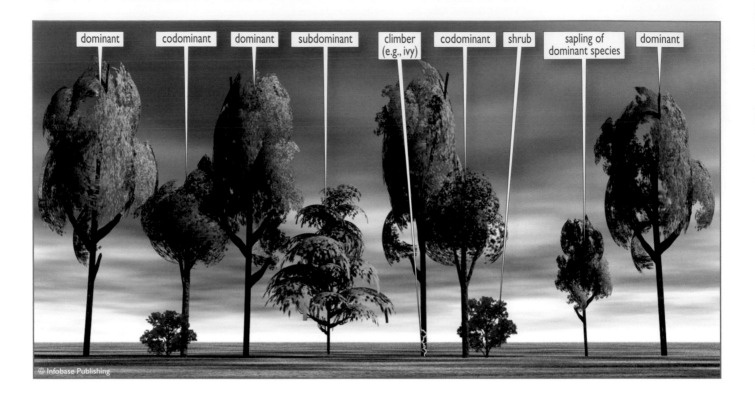

dominant | codominant | dominant | subdominant | climber (e.g., ivy) | codominant | shrub | sapling of dominant species | dominant

© Infobase Publishing

The structure of a temperate forest, showing the dominants, codominants, subdominants, saplings, shrubs, and climbers. The small plants growing on the forest floor comprise the herb layer.

same species. Together, the dominants, codominants, and subdominants comprise the main story of the forest, and their crowns touch, forming the forest canopy, as illustrated in the diagram.

Shaded by the branches of the trees comprising the main story, there are young trees of the same species as the dominants that cannot penetrate all the way into the canopy until the tree that is shading them falls (although it may bring down the smaller tree with it). These trees are said to be suppressed. Along with smaller trees that are more than 33 feet (10 m) tall but not tall enough to reach the canopy, they comprise the understory.

Technically, a tree is a woody plant that is more than 33 feet (10 m) tall when fully grown. Plants that are shorter than this (and usually bushier) are called shrubs. Below the understory of trees, therefore, is the shrub layer. Shrubs can be up to 33 feet (10 m) tall, but some are very much smaller, so the shrub layer comprises plants that are more than three feet (1 m) tall. Woody plants (shrubs) that are less than three feet (1 m) tall, such as bramble (*Rubus fructicosus*), form a dwarf shrub layer.

Beneath all the trees and shrubs there are the non-woody plants. The tallest of these, which may stand more than three feet (1 m) high, comprise the tall herb layer.

Some grasses grow taller than this, but they are not usually counted as members of this layer. To make the distinction, the tall herb layer is sometimes known as the *forb* layer, a forb being a broad-leaved (that is, not a grass), herbaceous plant. Herbs that are less than three feet (1 m) tall make up the herb layer, sometimes called the field layer, and below them there is the ground layer, consisting of very low-growing plants such as mosses. In addition to these there are the *lianas,* or climbers, such as ivy (*Hedera helix*), which cannot be placed in any layer, and the *epiphytes* (plants that grow on the surface of bigger plants, such as some ferns that grow on the branches of trees).

Not all forests possess all of these layers, and the number of layers often varies from one part to another of the same forest. There is rarely an understory in temperate forests, for example, unless one has been planted (see "Natural Forest and Plantation Forest" on pages 180–182), and the development of the lower layers is critically dependent on the amount of light penetrating the canopy. Although a mature forest has a closed canopy, some trees produce a denser canopy than others. Sometimes, especially in coniferous forest, the ground is too deeply shaded for most plants, even tolerant ones.

Distribution of Animals

Animals are also distributed throughout the vertical structure. Each layer of the forest has its own microclimate with small variations in temperature and humidity that attract some animals and repel others. A study of the birds living in a European forest dominated by oak (*Quercus* species) and hornbeam (*Carpinus betulus*) illustrates this effect. The study found that of all the bird species present, 15 percent nested on the ground, 25 percent in the herb and shrub layers, 31 percent on or in the trunks of canopy trees, and 29 percent in the canopy itself. That is where the birds nested, but 52 percent foraged for food on the ground and 23 percent foraged in the foliage of the trees.

Squirrels spend part of their time on the ground and part in the canopy, but most mammals are confined to the ground or to the surface and to burrows beneath the surface. They share this habitat with many invertebrates, although some invertebrates, such as snails and millipedes, will climb into the herbs or up the trunks of trees. Most insects are quite particular about the conditions they demand, and the species distribute themselves among the forest layers, although there are exceptions, such as ants and some beetles, which climb readily and are active at all levels.

Its complex vertical structure makes a mature forest one of the richest of temperate habitats.

■ FOREST COMMUNITIES

Forests look much the same from every point inside them. Indeed, historically their reputation has been partly based on the ease with which people can become severely disoriented in them. It is possible to walk among the trees for hours without seeing any recognizable change of scenery, and unless there are well made paths and signposts to provide guidance, there is a real risk of becoming lost.

This lack of obvious change or visible landmarks throughout a vast forest might suggest that when the forest changes as a whole, it does so evenly. The whole forest changes together, with every part being affected at more or less the same rate and in the same way. In fact, this is far from being the case, and the apparent uniformity of the forest is illusory.

Lumpiness

Research has shown that no ecosystem is that simple, including the temperate forests. In 1997 Professor C. S. Holling, a Canadian mathematical ecologist then at the University of Florida, discovered that the factors controlling change in ecosystems vary in scale and timing in big jumps, with nothing between the jumps, and this makes ecosystems lumpy rather than even. Since then evidence of the lumpiness of ecosystems has continued to emerge, and scientists are exploring the possible mathematical reasons why lumpiness develops.

In a spruce forest, for example, the population size of insects that eat needles varies seasonally, so there is a sea-

sonal variation in damage to the trees from this cause. Damage in the canopy varies on a scale of tens of years, and competition between spruce trees and broad-leaved species produces changes that are measured over about a century. This lumpiness should accumulate to produce cycles, some of them long and others short, in seed output, populations of birds and mammals, and the prevalence of fires.

Lumpiness is also evident in the body sizes of vertebrate land animals (amphibians, reptiles, birds, and mammals). They occur as groups of certain body weights, with very few individuals of intermediate size.

Animals Divide the Forest between Them

The existence of lumpiness suggests that a forest or any other ecosystem is made up of compartments, each with its own occupants. In effect, the animals partition the forest among themselves, and because of the vertical structure of the forest (see "Forest Layers" on pages 159–160), the partitioning is repeated in each story. This can often reduce competition between species because although two or more closely related species may feed on very similar food, they may not compete for it if they obtain it in different ways.

Birds living in North American deciduous forest that has developed in a moderately moist climate (known as a mesic environment) provide a convincing example. Black-capped or Carolina chickadees (*Parus atricapillus* or *P. carolinensis*) feed on insects in the shrub layers, while tufted titmice (*P. bicolor*), close relatives of the chickadees, feed on the same diet in the understory and main story. There are two species of vireos that feed on insects in the canopy, but the red-eyed vireo (*Vireo olivaceous*) searches for them on leaves and the yellow-throated vireo (*V. flavifrons*) finds them on twigs.

Seasonal migration causes substantial changes in the size and composition of the bird population. A study of a broad-leaved deciduous forest in Ohio found that there were about 2.5 to five permanent bird residents to each hectare (one or two per acre). This number changed little from season to season, although it was closer to the lower value in summer and to the higher value in fall, winter, and spring. Winter visitors started to arrive in September and began to depart in April. Their population reached a maximum of about three per acre (7.5 per hectare) in November and December. In April the summer visitors began to arrive. Their population reached a maximum of about seven per acre (17.5 per hectare) in May and June but then started to decline. In late April and early May and again from September to early November, migrants were breaking their journeys by stopping in the forest for a time to rest and feed. Their numbers peaked twice in spring and fall, each time at around 15 birds per acre (37.5 per hectare).

Birds are highly mobile, but they are not the only partitioners of the habitat. Among the mammals, white-footed deer mice (*Peromyscus leucopus*) and flying squirrels (*Glaucomys sabrinus*) nest and feed 33 feet (10 m) or more above the ground, although flying squirrels often descend to the floor to see what food they can find among the litter and around decaying logs. Flying squirrels are not capable of powered flight like bats, of course, but glide from a high location on one tree to a lower position on another tree or to the ground. Eastern chipmunks (*Tamias striatus*) also forage for food aboveground and on the forest floor, but since they cannot glide, they do not climb as high as flying squirrels. Deer mice (*Peromyscus maniculatus*) live and feed in the shrub layer, on the ground, and in burrows belowground. Red-backed mice (in fact they are voles, genus *Clethrionomys*) and shrews (*Sorex* species) dig more permanent tunnels, and hairy-tailed moles (*Parascalops breweri*) spend most of their time belowground.

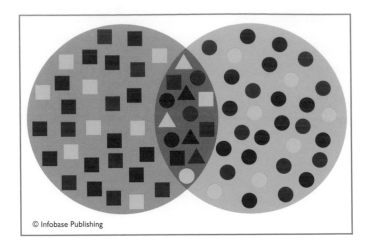

The edge effect between two ecosystems each of which contains three species (red, yellow, and blue squares and circles). The area of overlap contains all six of these species, plus three (triangles) that occur nowhere else.

Ecosystem Made from Smaller Ecosystems

Animals partition the resources available to them, and these resources are provided in the first instance by the vegetation, which varies from place to place within the forest, depending on such local factors as differences in the soil, moisture, and exposure to wind and light. Local habitat differences that affect plants and partitioning by the animals feeding on the plants and on each other combine to compartmentalize the forest in many ways. The resulting structure is often extremely complex.

In fact, the ecosystem of a forest can also be regarded as a mosaic of smaller ecosystems. These adjoin one another, as every ecosystem must adjoin another, and there is some overlap. A boundary where two adjacent ecosystems overlap slightly is called an ecotone, and it often supports a markedly wider variety of plant and animal species than can be found in the ecosystems on either side of it.

Edge Effect

This diversity of species found in an ecotone is the result of what is known as the edge effect, illustrated in the diagram. The illustration shows two overlapping ecosystems, drawn as circles. For simplicity, each is shown as supporting only three species, illustrated as red, yellow, and blue squares in one ecosystem and red, yellow, and blue circles in the other. One ecosystem contains only squares and the other contains only circles, but the area where they overlap contains both squares and circles. That area also provides conditions peculiar to itself. At the natural boundary between a forest and grassland, for example, the forest does not end sharply.

The trees become more widely spaced, more shrubs grow between them, and the forest thins gradually, the thin forest providing conditions quite different from those of either the full-canopy forest to one side or the open grassland to the other. The area of overlap thus supports species that find its conditions congenial. In the diagram three of these are shown as red, yellow, and blue triangles. The final tally is that the two main ecosystems each support three species, but where they overlap there are nine species.

Extending this two-dimensional image to the three-dimensional mosaic of habitats in a forest gives a sense of the ecological complexity of a temperate forest. It also reveals the fact that the richest habitat—the habitat supporting the greatest diversity of species—is often to be found at the forest edge in the ecotone between an area that is forested and an adjacent area that is not.

Harmful Edge Effects

Since an ecotone area supports a greater diversity of species than an area deep inside an ecosystem and since ecotone areas occur along the borders between adjacent ecosystems, it might seem that the way to increase species diversity is to increase the area of ecotone. This appears to imply that clearing part of a forest can increase the number of plant and animal species by increasing the length of forest edge, and for many years this was the prevailing view among conservationists.

The diagram illustrates how this might work. The square on the left represents undisturbed forest, with edges only at the boundary between the forest and the surrounding ecosystem. Then a number of roads are built through the forest.

Obviously, road construction causes considerable disturbance, but once the roads are completed and the workers and machinery have departed, the area quickly recovers. The roads permit full sunlight to reach the ground. In this respect they resemble forest clearings, but with the difference that forest clearings are temporary and that they disappear when young trees grow to their full size and close the canopy once more. A road is permanent, and there is a well-illuminated strip of land on either side of it where the climate is quite different from the climate of the forest interior. The roadside strip is forest edge and will support species that are not found in the deep shade inside the forest. Building roads and clearing patches of forest undoubtedly increases the length of forest edge. So it looks like a good idea. But is it?

Like many rules of thumb, it is only partly true, and if applied too enthusiastically it can be highly disruptive. Nowadays, when forest ecologists speak of edge effects, they do so disapprovingly.

What the concept ignores is that the habitat mosaic includes regions deep inside a forest that are inhabited by species that benefit from the protection afforded by the forest surrounding them. There are many species that cannot thrive close to the edge, so as the forest is carved into ever smaller blocks, their habitat is reduced. Eventually they may disappear altogether. In contrast, the species found along the forest edge tend to be commonplace. They are the plants that grow abundantly on open ground anywhere. The fact that the edge of the forest supports more species than the interior means little if the edge species are of no conservation value. The edge effect does nothing to increase or protect biodiversity (see "What is Biodiversity and Why Does It Matter?" on pages 183–184).

Around the edge of the forest, more light penetrates, and the climate is warmer and drier than the climate of the interior. Plants that are intolerant of shade find it congenial, so they flourish at the expense of shade-loving plants. The edge climate is also windier because inside the forest the trees shelter one another. Lacking such protection, trees at the forest edge are at greater risk of being blown down by the wind. When windblown trees fall, they may bring down others with them directly or indirectly (by falling against them and causing fatal damage). With each violent windstorm the edge area, with its attendant species, extends a little farther into the forest.

The vulnerability of deep-forest species is also partly due to the fact that grazing herbivores, predators, and parasites can enter the forest from the edge. White-tailed deer (*Odocoileus hemionus*), for example, browse on low-growing vegetation, but they are uncommon deep inside a large forest. Where the forest has been fragmented by local clearances or the building of roads, the deer have access to more of it, and their overbrowsing can kill some plant species and prevent others from regenerating. Canada yew (*Taxus*

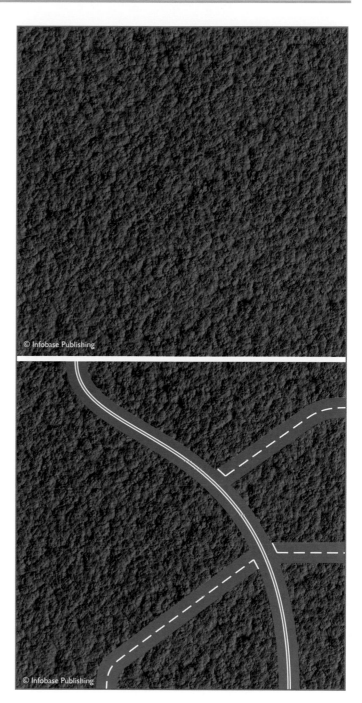

Increasing the length of forest edge. The square on the top represents undisturbed forest. Its edges occur only around the square itself. Several roads are then built through the forest. Forest edge now occurs around the outer boundary but also to either side of each road.

canadensis) has been greatly reduced by white-tailed deer in some forests in Canada and the northern United States. Other species of deer have caused similar harm elsewhere in North America and in Europe. Ground-nesting birds may be particularly at risk, and the danger to them increases still further if they find the forest-edge habitat attractive. This is because foxes and domestic cats and dogs also find the area attractive, not least for the opportunities it affords for dining on eggs, fledglings, or even adult birds.

A Large Forest That Is All Edge and No Middle

Most travelers see only as much of a forest as is visible from the road. People may stop the car and walk in among the trees a little way, but few visitors venture even as much as half a mile (1 km) into the forest. They see only the forest edge, and the plants and animals they admire are those that thrive along the edge. If the forest is big enough to possess an interior far away from the edge, there a visitor would find a different range of plants and animals, but if there are so many clearings and roads that no part of the forest is far from the edge and anyone walking half a mile into the trees will arrive at farmland or another road, the interior may have been lost. In that case, the entire forest may consist of edge, and the familiar roadside species will be the ones occurring throughout it. This may be so even though the forest looks big on the map and driving through it takes a long time. The habitat loss results not from wholesale destruction but from increasing the area of edge at the expense of interior and in this way reducing the richness of the mosaic.

An undisturbed forest has edges, and therefore it possesses an ecotone border supporting more species than the interior, but the rate at which clearance and road building can alter the relative proportions of forest interior and edge is dramatic. It is a matter of simple geometry. Suppose there is a circular forest with an area of 100 (the units are unimportant). Its circumference (the edge) will be 35.45. Two forests, each with an area of 50, will amount to the same total area (100), but a circle with an area of 50 has a circumference of 25, so two such circles have a circumference of 50. Halve the area again, to 25, and the edge length is 17.7, so for four such circles with a combined area of 100, it is 70.8. It is not difficult to see how injudicious clearance can produce a forest that is nothing but edge, despite occupying a large total area. When this happens, the benefits of increased diversity at the edge are lost because the interior habitat is lost.

Edges formed by clearance can also allow a secondary succession to commence. In temperate regions these pass through a stage where they are all but impenetrable scrub. This reduces the value of the habitat for many species, as well as destroying its amenity value for people. When the succession reaches its climax (see "Succession and Climax" on pages 147–151), the result will be an area of woodland that is different from the main forest and ecologically rather isolated from it. Interestingly, the impenetrable thicket resulting from secondary growth that follows forest clear-

ance in India and is often seen there along roadsides is known as "jungle" (and this is the correct meaning of the word). *Jungle* is from the Hindi *jangal,* and the Hindi word comes from the Sanskrit *jangala,* which can mean desert as well as forest.

Typical North American Forest Types

Broad-leaved deciduous forest in North America occurs naturally in the eastern United States and in southeastern Canada from southern Ontario and Québec to New Brunswick, Prince Edward Island, and Nova Scotia. In New England, most of the original forest was cleared in the past for farming, but when farming proved unprofitable, the forest was allowed to regenerate, although with a somewhat different composition.

The original Canadian forest is similar to the forest around the Great Lakes in the United States, extending eastward from Minnesota and known as the northern hardwood forest. It consists of white pine (*Pinus strobus*) together with sugar maple (*Acer saccharum*), beech (*Fagus grandifolia*), and yellow birch (*Betula alleghaniensis*) as well as varying amounts of hemlock (*Tsuga canadensis*). This forest forms an ecotone boundary between the fully broad-leaved forests to the south and the coniferous, boreal forest to the north. Moving northward through the ecotone, the maple, beech, and birch become more scattered, their places taken by white spruce (*Picea glauca*) and balsam fir (*Abies balsamea*).

To the south of the northern hardwood forest, in Wisconsin and Illinois, the natural forest consists of sugar maple and basswood (*Tilia americana*), sometimes with red oak (*Quercus rubra*). To the south of that, through Wisconsin and Indiana as far as Texas, the forest is dominated by black oak (*Quercus velutina*), white oak (*Q. alba*), pignut hickory (*Carya glabra*), and shagbark hickory (*C. ovata*), with flowering dogwood (*Cornus florida*) forming the understory in many places. Farther south again, into the Gulf states, there is oak–pine forest, dominated by white oak and loblolly and shortleaf pine (*Quercus alba, Pinus taeda Quercus alba,* and *P. echinata* respectively).

In the east, this type of forest extends northward in a broad band as far as the southern tip of Massachusetts. To the east of it, as far north as Massachusetts and from the Great Smoky Mountains on the border between Tennessee and North Carolina to the coastal plain, the original forest was oak–chestnut, with red oak, chestnut oak (*Quercus prinus*), and tulip tree (*Liriodendron tulipifera*). The chestnut has now all but disappeared, due to chestnut blight (see "Tree Predators and Parasites" on pages 173–180), and oak has replaced it.

West of the oak–chestnut forest there is mixed forest, varying in composition from place to place. To its north, around the southern borders of the Great Lakes but extending south into Illinois, Indiana, and Ohio, there is beech–

maple forest, dominated by beech and sugar maple, with basswood also occurring in the canopy.

Typical European Forest Types

Temperate forests are usually described by up to about three of their dominant tree species, the most important being named first. Over much of northern and part of central Europe, beech (*Fagus sylvatica*) is by far the most common forest tree. In England, however, except for the southeast, the pattern is different. Beech gives way to small-leaved lime (*Tilia cordata*), or hornbeam (*Carpinus betulus*) in a small area of East Anglia, and in northern England and Scotland to wych elm (*Ulmus glabra*), although this species has largely succumbed to Dutch elm disease (see "Tree Predators and Parasites" on pages 173–180) and disappeared.

In the mountainous regions of Europe, beech grows with silver fir (*Abies alba*), producing beech–silver fir forests. There are also birch–oak forests (*Betula pendula* with *Quercus robur* and *Q. petraea*) on the poorer, acid soils in the lowlands. On moister soils, the forests are more often hornbeam–oak.

On sunny sites in the warmer climate of southern Europe, the forests are of sessile, pubescent, and pedunculate oak (*Quercus petraea, Q. pubescens,* and *Q. robur* respectively) mixed with field maple (*Acer campestre*), small-leaved lime, and ash (*Fraxinus excelsior*). Broad-leaved evergreens, such as holm oak (*Quercus ilex*), form forests with several species of pines on lowland sites bordering the Mediterranean.

Farther north, the European boreal forests are composed of various combinations of Norway spruce (*Picea abies*), Scotch pine (*Pinus sylvestris*), birch, aspen (*Populus tremula*), and rowan (*Sorbus aucuparia*), often with juniper (*Juniperus communis*) in the understory.

Temperate Rain Forest

On the Pacific coast, from California to Alaska, the climate is mild and moist, and in some places fog is frequent. In Vancouver, British Columbia, for example, the average daytime temperature in January is 41°F (5°C), and in July and August, which are the warmest months, it is 74°F (23°C). Rain falls throughout the year but is rather heavier in winter than in summer, with an annual average of 57.4 inches (1,458 mm). In this climate, the natural vegetation is temperate rain forest.

Temperate rain forest is dominated by huge conifers such as Douglas fir (*Pseudotsuga taxifolia*), western hemlock (*Tsuga heterophylla*), western red cedar (*Thuja plicata*), and the redwoods. The forest canopy is between 197 feet and 230 feet (60–70 m) tall in many places, but some of the redwoods are taller. *Sequoiadendron giganteum,* the tree known as the Sierra redwood, big tree, giant sequoia, mammoth tree, and wellingtonia, grows to 330 feet (100 m) and the coast redwood (*Sequoia sempervirens*) to 394 feet (120 m). In Washington, sitka spruce (*Picea sitchensis*) also forms part of the forest.

Coast redwoods (*Sequoia sempervirens*) are among the world's tallest trees and thrive in a mild, wet climate. *(John Reynolds)*

There is also temperate rain forest in southern Chile. The Valdivian Rain Forest, to the west of the Andes, is dominated by broad-leaved evergreens. These include Chilean laurel (*Laurelia serrata*) and Winter's bark (*Drimys winteri*, named after a Captain Winter who sailed with Sir Francis Drake and used the bark from this tree to treat scurvy among the ship's crew). These species are mixed with the monkey-puzzle tree, or Chilean pine (*Araucaria araucana*), and Patagonian cypress (*Fitzroya cupressoides*).

Temperate rain forest also occurs in southern Japan and southern China, where it is dominated by evergreen oaks and magnolias with some pines. Around the Black Sea and as far south as the Caspian, in Transcaucasia, the Colchic Forest is a relict of a type of vegetation that existed more than 2 million years ago before the Pleistocene ice ages. It is also temperate rain forest and rich in tree species, although much of it has been cleared to provide land for growing tea.

Parts of eastern Australia and Tasmania as well as much of New Zealand, where the climate is mild and moist, also support temperate rain forest. The dominant trees are the southern beeches (various species of *Nothofagus*) and coachwood (*Ceratopetalum apetalum*). In Australia these grow alongside conifers such as bunya bunya (*Araucaria bidwilli*), Moreton Bay pine, also called hoop pine (*A. cunninghamii*), and various species of kauri pine (*Agathis*). In New Zealand the forests are dominated by *Nothofagus*, growing with red pines (*Dacrydium* species) and podocarps (*Podocarpus* species) as well as the kauri or cowdie pine (*Agathis australis*).

It is important to remember that these forest types are the ones that occur naturally. It does not follow that someone visiting a particular region will see the forest that is said to be native to that region or, indeed, any forest at all. People have been clearing and altering forests for many thousands of years.

Temperate rain forest in New Zealand   *(University of Wisconsin, Stevens Point Study Abroad Program)*

ENERGY FLOW

According to some engineers, the bicycle is the most efficient means of mechanical transport ever invented. It is efficient because it greatly augments the power that human leg muscles can deliver, so a cyclist travels farther and faster than a walker or runner on the same amount of fuel. The fuel, of course, is the food a person eats, and a cyclist needs to eat no more food to prepare for a bike ride lasting an hour than she or he would need for a walk lasting an hour.

People use food partly as fuel to provide their bodies with energy. The energy is released by the oxidation of carbon contained in the sugars, starches, and fats that are a major part of food. Carbon oxidation is a chemical reaction that releases energy. It is the same reaction as the one that warms people's homes and powers their cars when they burn coal, natural gas, fuel oil, or gasoline. In animal bodies the bringing together of oxygen and a source of carbon, the oxidation reaction itself, and the excretion of the reaction product, carbon dioxide, comprise respiration.

All living organisms need energy, and all of them obtain it by respiration, although anaerobic bacteria do not use oxygen (but carbon loses electrons and so the reaction is oxidation, nevertheless). Oxygen is easily obtained from the air, and since oxygen is slightly soluble in water, aquatic organisms use dissolved oxygen. Carbon is also present in the air as carbon dioxide, but photosynthesis (see "Photosynthesis" on pages 116–122) is the only biological process for reducing it (adding electrons and separating its carbon and oxygen) and storing the carbon in a form that allows it to be oxidized once more. Only green plants can store carbon in a useful form, so animals, fungi, and protozoans must obtain the carbon they need from plants or from one another.

Relationships within an ecosystem are based on feeding. Those relationships that link species can be represented as food chains or webs (see "Food Chains and Food Webs" on pages 129–132) and those between entire trophic levels as ecological pyramids (see "Ecological pyramids" on pages 132–135). This approach was never wholly satisfactory because it dealt with the structure of an ecosystem but tended to overlook what happened in decomposition and said very little about the ecosystem as a whole.

How Energy Moves

In 1942, the American ecologist Raymond L. Lindemann proposed a more comprehensive way of studying ecosystems. His paper, "The trophic–dynamic aspect of ecology," published in the journal *Ecology*, became a classic, and a new field of scientific study developed from his idea. The field is called ecological energetics, and what Lindemann proposed was the study of the flow of energy through an ecosystem.

Energy, after all, is needed to power any ecosystem, so all the relationships within the system are based on the flow of energy. When animals eat, most of their food is used to provide their bodies with energy. It is the energy that matters, and according to the first law of thermodynamics, energy can change its form (from potential to kinetic, for example) but it can be neither created nor destroyed. Within any ecosystem there must be an original energy source, a route by which it passes from organism to organism, and a sink into which ultimately it is poured. In a forest ecosystem the Sun is the original source of energy, and eventually all of the energy captured by photosynthesis is expended in respiration. Also, in conformation with the second law of thermodynamics, it is finally converted into heat, warming the air. Clearly, Lindemann was proposing something very sensible, but it was also helpful because the flow of energy is surprisingly easy to measure, at least in principle.

Solar energy is captured by photosynthesis and is used to manufacture carbohydrates and, by other reactions, proteins and the other compounds from which living tissues are composed. The tissues, therefore, are where the captured energy is stored. Weigh those and the result will be proportional to the amount of energy stored in them. Alternatively, their stored energy can be measured directly by burning the material in an oven equipped with a calorimeter, which is a device for measuring the amount of energy released. The material is dried before being burned because the water content of biological material is highly variable and would make it impossible to compare one sample with another accurately. On average, 1 ounce (28.4 grams) of oven-dried plant material releases 114–128 kcal (474–534 kJ), and 1 ounce of animal tissue releases 142–156 kcal (594–653 kJ). (See the sidebar on page 168 "Energy, Work, and Heat".)

All ecosystems can be studied in this way, and because the techniques and energy units are standard, ecosystems and parts of them can be compared directly. The total quantity of living organisms in an entire ecosystem or in some part of it is known as the biomass, or standing crop, and it can be reported in units either of mass or of energy because they both come to the same thing.

Entire systems do not have to be dried and burned, of course, but only samples from which the biomass can be calculated and biomass samples taken at different trophic levels allow the flow of energy to be traced through the system. Lindemann proposed in his paper that the efficiency of energy transfer can be calculated by comparing the amount of energy assimilated at one trophic level with the amount assimilated at the preceding level. This is known as Lindemann's efficiency, but measuring it in the field is difficult. That is because consumer organisms seldom occupy only one trophic level. Sheep, which are herbivores and, therefore, primary consumers, also eat significant amounts of insects and other invertebrate animals clinging to the

There are other techniques for comparing trophic levels, but they all encounter this difficulty. Nevertheless, comparisons can be made under laboratory conditions and, with care, from field samples.

How Efficiently Is Energy Used?

Measurements of energy flow show that most of the energy assimilated at each trophic level is used for respiration at that level and is not available to the level above. On average, between 10 percent and 20 percent of the energy assimilated at one level is passed to the level above. In other words, the ecological efficiency of ecosystems is 10–20 percent.

Respiration accounts for much more of the energy that is assimilated by *endotherms* (animals such as mammals and birds, which use some of their food to maintain a constant body temperature) than of *poikilotherms* (animals such as fish and invertebrates, which have no metabolic means for regulating body temperature). Herbivorous endotherms have an ecological efficiency of 0.3–1.5 percent and herbivorous poikilotherms of 9–25 percent. The efficiency of carnivorous endotherms is 0.6–1.8 percent and of carnivorous poikilotherms 12–25 percent. Most of the animal biomass in any ecosystem consists of poikilotherms, so they strongly affect the overall efficiency of the ecosystem.

Animals eat primarily to obtain fuel for their bodies. If only 10 percent of the energy at one trophic level is available to the level above, the amount being transferred soon dwindles to insignificance, as the illustration shows. The pyramid of energy is the third of the ecological pyramids devised by Sir Charles Elton (see "Ecological Pyramids" on pages 132–135), and the majority of ecologists consider it the most useful. The diagram demonstrates the fact that the amount of energy reaching the third trophic level, of carnivores (secondary consumers), is so small that it is

A pyramid of energy

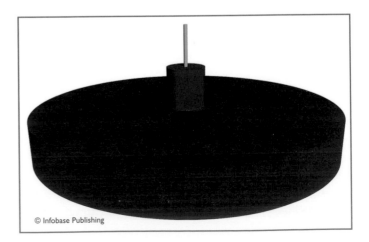

herbage. This makes them secondary consumers (consuming primary consumers) when they swallow herbivorous insects but places them at a much higher trophic level when they consume predators such as spiders and centipedes.

uncertain whether the system could support a fourth level, although some do support higher levels. The clear message the pyramid conveys is that big carnivores such as tigers must always be rare.

Productivity

Once biomass of any kind can be measured in terms of the energy it represents, it becomes possible to study ecosystems in very much greater detail. The amount of new biomass that is synthesized by the plants of an ecosystem in a given period (usually a year) is known as the primary productivity (or production) of that system. The total amount of new matter is called the gross primary productivity (GPP). This is adjusted to the net primary productivity (NPP) by deducting the amount of energy that is used by the producers in respiration. When the amount of material consumed at the higher trophic levels is also allowed for, the rate at which biomass accumulates is the net community productivity (NCP).

NCP is related to production (P) divided by respiration (R). If $P/R = 1$, then NCP = 0 and the ecosystem is in a steady state. If $P/R > 1$ (NCP > 0), the ecosystem is growing (accumulating biomass), and if $P/R < 1$ (NCP < 0), it is deteriorating (losing biomass). Measuring the NCP and P/R ratio makes it possible to determine the state of an ecosystem. During a sere, both will have positive values, but when a climax is reached, NCP will be at or close to 0 and $P/R = 1$.

NPP provides the best measurement for comparing ecosystems, and combined with the biomass (standing crop) per unit area, the NPP measure can be used to calculate the ratio of annual production to biomass (P/B). Regardless of the size or type of ecosystem, this ratio reveals the ecosystem's productivity in a way that permits direct comparisons because the greater the P/B ratio, the more productive the system is. In an upstate New York oak–pine forest, for example, the annual NPP of the trees was measured as 0.04 ounces per square foot (1,060 g/m²) and the tree biomass as 31.6 ounces per square foot (9,700 g/m²), giving a P/B of 0.11. This ratio varies according to the type of forest and climate. Forests in the Great Smoky Mountains have ratios of 0.02 to 0.03.

Everyone knows that tropical forests are the richest of all ecosystems on land, and because they are so lush and their plants grow so vigorously, people tend to assume that they are also the most productive. Indeed, in tropical rain forest, the mean annual NPP is 7 ounces per square foot (2,200 g/m²), which is the highest of any ecosystem on land. The biomass is also high, however, and $P/B = 0.05$, about half that of the oak–pine forest in New York. Temperate forests of all types (broad-leaved evergreen, broad-leaved deciduous, and boreal) average $P/B = 0.04$, which is not much lower than the value for tropical rain forest. Not surprisingly, per-

haps, the highest P/B for any land system is 0.65 for cultivated farmland.

These ratios do not alter the fact that the tropical forests contain considerably more biomass than temperate forests and that their NPP is higher in total. Over the world as a whole, the annual NPP for tropical rain forests and seasonal forests is 54.3 billion (10⁹) tons (49.4 billion tonnes), and for temperate forests it is 27 billion tons (24.5 billion tonnes). Both types occupy about the same total area, tropical forests covering 9.5 million square miles (24.5 million km²) and temperate forests 9.3 million square miles (24 million km²), so tropical forests are cycling carbon and mineral nutrients much faster than are temperate forests.

■ NUTRIENT CYCLING

There is a classic school science demonstration in which the teacher lights a candle, covers it with a bell jar, and the class watches to see what happens. After a little while, the candle flame falters and dies. The demonstration shows that when the candle burns, carbon in the wax from which it is made is oxidized, using oxygen from the air inside the jar: $C + O_2 \rightarrow CO_2$. The amount of oxygen in the jar decreases as the amount of carbon dioxide increases, and the process continues until the jar contains insufficient oxygen to sustain the oxidation. That is when the flame flickers and dies. Flames will burn only so long as there is enough oxygen to feed the oxidation, and the way to extinguish flames is to smother them with blankets, water, foam, or a nonflammable gas such as carbon dioxide or nitrogen and so cut off their oxygen supply.

The bell-jar demonstration is extreme, but outside in the natural world, the composition of the air really does change from time to time on a very local scale. People become aware of it when the air feels stuffy, stifling, or oppressive. Usually the air feels this way because its humidity and temperature both increase. The remedy is to open a window or to step outdoors into the fresh air.

Deep inside a forest, the trees sometimes have a similar problem. Photosynthesis absorbs carbon dioxide, transpiration releases water vapor, and up in the canopy this happens on a very large scale because that is where most of the leaves are. The leaves are there because that is where the sunlight is brightest, so the canopy is rather like a factory, manufacturing carbohydrates to sustain the trees. Not surprisingly, the tree factory affects the air. It removes carbon dioxide and adds water vapor, and the air in the canopy grows quite warm because it is held there, moving little, and is warmed by contact with the sunlit trees. Stuffiness is harmless to humans, but if the amount of carbon dioxide in the air decreases, it is possible that leaves will spend more time with their stomata open. Stomata are the pores through which carbon dioxide enters and oxygen and water vapor leave. If they are open for longer to obtain the carbon dioxide they need, it is inevitable that more water vapor will leave through them, and this can harm the tree. So carbon dioxide depletion can be quite serious for a tree.

The wind blows freely in the air above the canopy, but the air is fairly still where it is held in the canopy among the leaves of the tree crowns and between crowns. This results in a marked wind shear due to the difference in wind speed inside the canopy and in the air immediately above it. This shear causes gusts and eddies that bring down air from above into the canopy. Air from outside, which is relatively cool, dry, and rich in carbon dioxide, replaces the warm, moist, canopy air that is depleted in carbon dioxide.

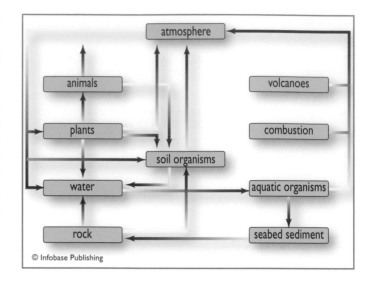

The carbon cycle

The Carbon Cycle

Photosynthesizing leaves seize carbon dioxide molecules. This removes carbon dioxide from the air, but the moving air brings more. There is a constant supply. Eventually, of course, photosynthesis would strip the air of all carbon dioxide were there no route by which it could return, but it does return, or at least most of it does. Its carbon is incorporated into every cell of the plant that captured it and every cell of every animal that eats it, but some is lost again rapidly. Carbon is used in respiration to supply energy, and respiration comprises a sequence of reactions that oxidize the carbon again to carbon dioxide that is returned to the air.

Carbon that is used to make structural materials, such as proteins and celluloses, is held for longer, but eventually cells and organisms die and decompose. Decomposition (see "Decomposers" on pages 135–137) is performed by organisms that consume those structural materials, break them down, and oxidize their carbon to provide energy. Again, carbon dioxide is formed and returns to the air. Photosynthesis captures carbon dioxide molecules, and respiration releases them once more.

There is some loss. Carbon dioxide dissolves in water to form carbonic acid (H_2CO_3). Aquatic plants also photosynthesize, using carbon dioxide obtained from the water, and aquatic animals graze the plants. There are also plants and animals that combine calcium with bicarbonates (HCO_3) to make shells and plates of calcium carbonate ($CaCO_3$). In water, too, the carbon is oxidized by respiration, but a small proportion of shells, plates, and dead organic matter accumulates in seabed sediments before it can be fully decomposed. These sediments are compressed into rocks. Crustal movements return some to the surface, millions of years later, allowing chemical weathering to return them to living cells, and some are subducted down into the mantle. This loss is made good by volcanoes, which release into the air carbon dioxide from rocks carried upward from below the Earth's crust.

Like all the chemical elements used to make and sustain living organisms, carbon is constantly recycled. Following the journey of an individual carbon atom reveals that it spends some of its time in the air, some as part of living cells and some dissolved in water that flows beneath the surface in groundwater, in rivers, and finally reaches the ocean. The diagram illustrates the steps in the cycle.

The amount of carbon present in the air varies. Burning plant material and fossil fuels, which releases heat energy by oxidizing carbon, just as in respiration, has been responsible for a general increase in the atmospheric concentration of carbon dioxide during the past century (see "Forests and the Greenhouse Effect" on pages 268–272). The concentration also varies seasonally. Respiration continues at a constant rate throughout the year, but photosynthesis takes place most strongly in spring and summer. Consequently, the atmospheric concentration of carbon dioxide increases through the late fall and winter as photosynthesis slows and then almost ceases in middle and high latitudes and decreases as photosynthesis increases again in spring to reach a minimum in late summer.

Nitrogen

Nitrogen is also an atmospheric gas. It is an essential ingredient of proteins, compounds that are used in the construction of living tissues and which perform a wide variety of other functions. Hormones and enzymes are proteins, for example. Atmospheric nitrogen is oxidized by lightning, and nitrogen oxides are converted into nitrate (NO_3), ammonia (NH_3), and ammonium (NH_4) compounds by nitrogen-fixing bacteria in the soil, some of them living in close association with plant roots. When dead organic matter decomposes, most of the nitrogen from large, complex molecules is converted into simple compounds that dissolve in the soil water and enter plant roots once more, but denitrifying bacteria convert some of it into gaseous nitrogen. Denitrification returns nitrogen to the air, completing the cycle (see "Nitrogen Fixation and Denitrification" on pages 122–125).

For part of the time a nitrogen atom is in the air, for a time it is a component of living tissue, and for a time it is dissolved in flowing water. Like carbon, it is engaged in an endless cycle, called a biogeochemical cycle or a nutrient cycle, because it is an essential nutrient element that is being cycled. The cycles are chemical because they involve chemical elements, compounds, and reactions. They are also geological because those chemical substances move through the nonliving components of the Earth. The elements pass through water and air. On a longer time scale, the cycles involve the rocks from which mineral elements are taken, the sediments in which they are deposited and from which they are returned to living organisms through weathering processes, and the removal of elements by the subduction of crustal plates and their return by volcanism. The cycles are also biological because for part of the time the elements are incorporated in living tissues. Rock, water, air, and living cells are like compartments into which biogeochemical cycles can be divided.

Every year, lightning and bacteria between them fix about 165 million tons (150 million tonnes) of nitrogen by combining it with oxygen or hydrogen. This is the rate at which nitrogen moves naturally from the air (geological) to the living (biological) compartment of the biogeochemical cycle and, therefore, the rate at which the element is made available for new plant growth.

It takes a great deal of energy to oxidize nitrogen. Lightning can do it and nowadays so can car engines and certain industrial processes. Cars and factories release nitrogen oxides as air pollutants that contribute to acid rain (see "Acid Rain" on pages 225–226) but that are also plant nutrients.

Farmers need nutrients in addition to those that are supplied naturally because agriculture alters the cycle. Crop plants, animals, and animal products such as milk and wool are removed, so that people can eat or wear them, rather than being allowed to die and decompose in the soil where they grew. Unless the nutrients they contain can be replaced, in time the stock of them in soil will be depleted, the soil will become infertile, and crop yields will fall. Therefore, farmers grow legumes such as beans and alfalfa, plants with their own nitrogen-fixing bacteria. When the legumes die or are removed as crops, their roots remain in the soil, enriching its nitrogen content for the crop that follows. More importantly, farmers add fertilizer, manufactured in factories where nitrogen is made to combine with hydrogen.

Human activities, especially the use of fertilizers containing nitrogen, are now fixing nitrogen at about the same rate as lightning and naturally occurring bacteria. In addition, by draining wetland and plowing grasslands to provide more land for cropping or space to build homes, people are accelerating the natural cycle still more. Each year lightning and bacteria fix approximately 154 million tons (140 million tonnes) of nitrogen, making it available to plants. Each year human activities fix approximately 231 million tons (210 million tonnes). In other words, industry and agriculture have increased the rate at which nitrogen is entering the biological compartment of the cycle by approximately 50 percent. No one knows what effect this will have in the long term, although scientists suspect that it may be making forests grow faster in Europe and North America. If so, those forests will be removing more carbon dioxide from the air than they would do otherwise, and that may partly offset global warming due to increased carbon dioxide concentrations.

Phosphorus

Within living organisms, energy is used to attach a phosphate group (the process is called phosphorylation) to adenosine diphosphate (ADP), making it into adenosine triphosphate (ATP), the form in which it is transported. The phosphate contains phosphorus (P), making this another essential nutrient element and one that also forms part of nucleic acids and animal bone. It is needed only in small amounts. Plants consist of no more than about 3 percent of phosphorus by weight. It is often the limiting element, however, because most soils contain even less in a form that can enter plant roots. When fertilizer runoff from farmland causes algal blooms in lakes and reservoirs, it is almost invariably phosphorus that causes the problem, not nitrogen.

Phosphorus enters the biogeochemical cycle from phosphate rock. As the name suggests, this is a rock composed of phosphate minerals such as apatite $(Ca_5PO_4)_3(F,Cl,OH)$ in which calcium (Ca), phosphorus (P), and oxygen (O) are combined with fluorine (F), chlorine (Cl), or hydroxyl (OH). Chemical weathering (see the sidebar "Weathering" on page 18) releases phosphorus from this insoluble rock as orthophosphate $(H_2PO_4, HPO_4,$ and, if conditions are very alkaline, PO_4). Orthophosphate is soluble, and this is the form in which phosphorus enters plants, but the soil water around plant roots contains only about 0.000003 percent orthophosphate, and orthophosphate that is not taken up by plants may be lost to them. If the pH of the soil is below 5.5 (see the sidebar "Acidity and the pH Scale" on page 259), it will form insoluble compounds with iron and aluminum, and if the pH is above 7.0, orthophosphate will form insoluble compounds with calcium and other elements.

Such phosphorus as does enter the biological compartment of the cycle tends to remain there for some time. It is so valuable to living organisms that they recycle it rapidly. In a fairly remote part of New Hampshire, there is an area of broad-leaved forest drained by the Hubbard Brook. The chemical composition of the brook water has been monitored for many years, so there is a reliable inventory of the rate at which substances leach out of the forest soil. Combined with analyses of the soil, this shows that each year weathering adds 1.34–1.60 pounds of phosphorus to each acre of soil (1.5–1.8 kg per hectare). The decomposition of organic material releases 8.9 pounds per acre (10 kg per hectare), but Hubbard Brook carries away only 0.006 pound per acre (0.007 kg per ha).

In southern Ontario, forests on soils derived from igneous rocks were found to be losing an average of 0.04 pound of phosphorus from each acre (0.05 kg per hectare) through streams draining the area. Forests growing on soils derived from sedimentary rocks were losing an average of 0.09 pound per acre (0.1 kg/ha). Where the area was partly forest and partly pasture, the comparable figures were 0.1 pound per acre (0.1 kg/ha) and 0.3 pound per acre (0.3 kg/ha) respectively. Igneous rocks are hard rocks, such as granite, that are formed by the solidification of cooling magma. Sedimentary rocks are softer and are formed by the compression and heating of sedimentary particles.

As with nitrogen, the amount of phosphorus available naturally is sufficient for the needs of intact ecosystems but not for farm crops, and farmers have been adding phosphate fertilizers since the 1840s. These are made by grinding phosphate rock to a powder and subjecting the powder to treatments that convert some of the phosphate to soluble forms. Treatment with sulfuric acid (H_2SO_4), for example, yields a product containing 18–22 percent of phosphoric acid P_2O_5, which is soluble. Further treatment can more than double the P_2O_5 content, and reacting P_2O_5 with ammonia (NH_3) produces a range of ammonium (NH_4) phosphates that are completely soluble in water and provide nitrogen as well as phosphorus.

Eventually, phosphorus reaches the ocean, but as the Hubbard Brook and Ontario studies show, ecosystems release it only very slowly, and forests release it at about half the rate at which it moves out of pastureland. Once it does reach the ocean, the phosphorus is virtually trapped because it readily forms insoluble compounds that collect in the seabed sediment. Marine animals, which incorporate phosphorus in their bones and excrete the surplus, return some of it to the land, most notably as the droppings of seabirds that collect in some places as guano. A small amount returns from the air as dry particles or in rain. Measured as orthophosphate (PO_4), sites in the southeastern United States receive annually an average of about 0.1 pound of phosphorus per acre (0.1 kg/ha) in rain and 0.2 pound per acre (0.3 kg/ha) as dry particles.

The main route by which phosphorus moves from the sea to the land is geological. Changes in sea level sometimes isolate arms of the sea, which then dry out, leaving the mineral constituents of the original seawater as evaporite deposits, often rich in phosphate. Alternatively, seabed sediments are compressed and heated to form sedimentary rock, and crustal movements raise those rocks above the surface, where chemical weathering can release their phosphates once more. Within a forest, the phosphorus recycles rapidly, but phosphorus reaching the ocean enters a cycle that takes millions of years to return to its starting point.

Potassium and Sulfur

Potassium, another nutrient that plants need in relatively large amounts, is widely distributed in rocks and is released by chemical weathering. In plants it is involved in the synthesis of proteins, in the maintenance of a water balance, and in the opening and closing of stomata. In animals it is also involved in the functioning of nerve cells. Potassium is recycled by plants, and very little moves out of the soil. Hubbard Brook carries away about 1.5 pound per acre (1.7 kg/ha) each year, but the net loss is much smaller because precipitation and the deposition of dry particles supply about 1.0 pound per acre (1.1 kg/ha).

Sulfur is an essential component of some amino acids and therefore of proteins. Volcanoes release it into the air as sulfur dioxide (SO_2). Sulfur also occurs in many rocks and, in a few places, as the element itself. Some plants can absorb it in gaseous form as SO_2 taken directly from the air, but most absorb it through their roots as sulfate (SO_4) dissolved in soil water.

In recent times the burning of fossil fuels (especially some coals) that contain relatively large amounts of sulfur compounds has increased the amount of SO_2 present in the air, where it is oxidized in several steps and dissolves in water to form sulfuric acid (H_2SO_4). This is carried to the surface in precipitation or is deposited directly onto surfaces from dry air. It contributes to acid rain, but it also supplies soils with a valuable plant nutrient. Now that steps are being taken to reduce the amount of SO_2 in smokestack gases, a few years from now farmers may find it necessary to use more sulfur-based fertilizers.

Sulfur leaches from soils fairly readily, accumulating, mainly as calcium sulfate ($CaSO_4$) only where the climate is very dry. Industrial pollution apart, for many years there was something of a mystery about how sulfur that has drained into rivers and been carried to the sea returns to the land. It was believed that large, organic molecules containing carbon found their way into muds in swamps, bogs, and most of all in estuaries. There, anaerobic bacteria broke down the molecules and released the sulfur as hydrogen sulfide (H_2S) gas. Hydrogen sulfide then bubbled into the air and was oxidized into sulfate. Calculations revealed, however, that this process could account for nowhere near all of the sulfur moving through the cycle. Furthermore, if most sulfur were to be converted into H_2S, people living nearby would certainly be aware of it because that gas has the smell of rotten eggs and is poisonous in fairly low concentrations.

What really happens turned out to be much more interesting. It was the British chemist James E. Lovelock, the author of the Gaia hypothesis, who first discovered the route. Certain species of marine algae (phytoplankton) absorb sulfur from seawater. The algae convert the sulfur to dimethylsulfonioproprionate, a compound that prevents the salt content of their cells from rising to levels that might be harmful. When the cells die, the dimethylsulfonioproprionate decomposes to dimethyl sulfide (DMS, $(CH_3)_2S$). Some of the DMS is oxidized in the water, but some escapes into the air. In very dilute form, it gives the sea its characteristic smell. In the air, the sulfur in DMS is oxidized in a series of steps to sulfate (SO_4). One oxidation product is sulfuric acid (H_2SO_4). Water vapor readily condenses onto minute droplets of H_2SO_4. Over the open ocean, thousands of miles from land where the air is almost completely free from dust, DMS is by far the main source of the condensation nuclei that lead to the formation of cloud. Some of the cloud, containing the sulfur that came originally from DMS released by algae, is carried over land, and the sulfur is returned in precipitation that is acid but naturally so.

Trees, like all living organisms, are made from a range of chemical elements. In addition to hydrogen, carbon, and oxygen, which are obtained from the air and water, their tissues contain relatively large amounts (measured in pounds per acre or kilograms per hectare) of nitrogen, potassium, calcium, phosphorus, magnesium, and sulfur. They also contain much smaller amounts (ounces per acre or grams per hectare) of iron, manganese, zinc, boron, copper, molybdenum, and cobalt, as well as even smaller amounts of other elements, including silicon and selenium.

When plants and animals die, their chemical constituents are recycled. Forests, with their huge root systems, recycle elements very efficiently, but inevitably there is some loss. Substances that escape from the forest then continue around the great biogeochemical cycles that take them from living organisms through freshwater, seawater, air, and sometimes rocks, each element following its own path.

Without these cycles, after thousands of years, even the smallest loss would leave soils too depleted to support plants at all. Life on land would not be possible. There would be no forests and no people to admire them. The biogeochemical cycles will endure. Nothing is lost from them, not even a single atom, for there is nowhere it can go to escape from the Earth.

■ TREE PREDATORS AND PARASITES

At one time, the chestnut was an important component of deciduous forests in eastern North America. *Castanea dentata,* the American chestnut, is a handsome tree, on favorable sites growing rapidly to a height of about 100 feet (30 m), and its timber had many uses. Then, in about 1904, some chestnut trees were imported from China, and they brought with them a fungus, *Cryphonectria* (formerly *Endothia*) *parasitica.*

In Asia, trees infected with *Cryphonectria parasitica* suffer only mild symptoms, hardly qualifying as those of a disease. In America, however, it was different. On Long Island, New York, the fungus infected American chestnuts and killed them. Infected trees produced cankers—local areas where the tree produces no bark. The sickness was fatal and spread fast, also killing European sweet chestnuts (*C. sativa*), which were being grown in orchards (this species provides the best chestnuts, or marrons). The disease was called chestnut blight. Not only was the infection invariably lethal, but its spread was uncontrollable. Desperate attempts were made to halt its advance, but by about 1940 the American chestnut had been almost eliminated, leaving thousands upon thousands of dead trees. Some individuals did prove resistant to chestnut blight, so the tree has not vanished completely, and European sweet chestnuts are still cultivated mainly in California and the Pacific Northwest, but chestnut trees ceased to form an important part of American forests.

There are stretches of double-stranded RNA, rather like viruses but lacking the protein coat that all viruses have that are known as hypoviruses. When *Cryphonectria parasitica* is infected with a hypovirus, it becomes much less virulent, producing far fewer spores asexually, and completely losing female sexual fertility, and the fungus produces smaller cankers. Once infected, the fungal parasite rarely kills its host tree. Hypoviruses had been used for several decades to control *Cryphonectria parasitica* infections in Europe. Experiments in the United States showed that hypovirus effectiveness varied considerably from one strain to another and that after a time the fungus became more tolerant of certain strains. Nevertheless, treatment with hypovirus offers hope of bringing this disease under control.

Chestnut blight was certainly among the worst of all forest disasters, but it was neither the most widespread nor the most devastating.

Sudden Oak Death

In 1995 foresters in the western United States noticed a new disease that was killing trees. They called it sudden oak death, although later it was found to attack other tree species. It is also called ramorum leaf blight and ramorum dieback.

Sudden oak death is caused by *Phytophthora ramorum*, a funguslike organism also known as a downy mildew or water mold. It can infest: big leaf maple (*Acer macrophyllum*); black oak (*Quercus kelloggii*); California bay laurel (*Umbellularia californica*); California buckeye (*Aesculus californica*); California coffeeberry (*Rhamnus californica*);

California honeysuckle (*Lonicera hispidula*); coast live oak (*Q. agrifolia*); coast redwood (*Sequoia sempervirens*); Douglas fir (*Pseudotsuga menziesii* var. *menziesii*); huckleberry (*Vaccinium ovatum*); madrone (*Arbutus menziesii*); manzanita (*Arctostaphylos manzanita*); rhododendron and azalea (*Rhododendron* species); Shreve's oak (*Q. parvula* var. *shrevei*); tanoak (*Lithocarpus densiflorus*); and toyon (*Heteromeles arbutifolia*).

Phytophthora ramorum also occurs in Germany and Denmark, where it attacks rhododendron and viburnum, and there have been cases of it in Britain. In the United States it has been found so far only in coastal areas of California to the north and south of San Francisco and in one location in southwestern Oregon. Foresters in the eastern United States have also been warned to look out for it, however, because two eastern trees, northern pin oak (*Q. palustris*) and northern red oak (*Q. rubra*), have been found experimentally to be highly susceptible.

Dutch Elm Disease

Europe as well as North America has suffered severely and repeatedly from Dutch elm disease, a fatal disease of elm trees. Like chestnut blight, this disease is caused by a fungus that may have originated in Asia.

Dutch elm disease was noticed in France in 1918, and the fungus responsible was identified in the Netherlands in 1919. This was unlikely to have been the first outbreak of the disease, however. John Claudius Loudon, the 19th-century authority on gardening, referred to it in 1838. The disease came to be called Dutch elm disease because it was Dutch scientists who made the first accurate identification of the fungus. Confusingly, this led some people to suppose it to be a disease of Dutch elms, of which there are several, all hybrids, or a disease that originated in the Netherlands. Neither is true.

An outbreak of the disease in the 1920s reached North America and caused a great deal of harm. By about 1940 it had died down in Europe, although not in the United States, where the native elms (*Ulmus americana*) are more susceptible and more species of insects were involved in transmitting the infection. In the 1960s the disease returned to Europe with increased virulence. There were then about 30 million elm trees in Britain. The 1960s epidemic killed some 25 million of them. Their loss profoundly altered the appearance of the countryside because as well as growing in forests, elms grew in many of the hedges bordering fields, so they formed a visually prominent component of the landscape. In fact, the disease spread more slowly in forests than it did along hedgerows and among elms planted in city parks. That is because the wych elm (*Ulmus glabra*), which is the species most often found growing in European forests, is rather more resistant to the disease than the hedgerow

and field elms (*U. procera* and *U. minor* respectively). North American forests also suffered less than isolated patches of elm woodland, in this case because forest elms tend to be spaced farther apart and their roots are not in direct contact, whereas clumps of elms are likely to be growing from the same root system.

Elms propagate mainly vegetatively from shoots that grow from the roots, and although Dutch elm disease killed the trees, in many cases their roots survived and, gradually, elms began to reappear. Then, in the 1990s, Europe began to suffer a new outbreak. Dutch elm disease now occurs almost everywhere that elm trees grow. It has spread across Europe and Asia as far as the Chinese border. In North America it began in the Great Lakes region and spread both east and west, all the way to both coasts. *O. himal-ulmi* is a related fungus that attacks elms in the Himalayan region.

The first culprit was a fungus, *Ophiostoma* (formerly *Ceratocystis*) *ulmi,* which blocks the xylem vessels. The second outbreak, in the 1960s, was caused by *Ophiostoma novo-ulmi,* a much more virulent species. Fungal spores are spread by small beetles that burrow into the bark of elms to make nuptial chambers in which adults mate. The larvae feed on the sapwood, emerging as adults, which then feed on elm sap. Several beetle species behave in this way. In Europe, the fungus is carried mainly by *Scolytus scolytus,* a beetle about 0.2 inch (0.5 cm) long and *S. multistriatus,* about 0.1 inch (0.3 cm) long. When the beetles attack a tree they do so in vast numbers, and their larvae are capable of killing a tree without any help from the fungus.

It was *S. multistriatus* that was carried, along with the fungus, from Europe to America in logs to be made into veneer for furniture. Being smaller than *S. scolytus,* it can carry fewer fungal spores and so is less efficient at transmitting the infection, but in North America another bark beetle, *Hylurgopinus rufipes,* is also a vector for the infection, so between them they sustained the epidemic.

Ophiostoma novo-ulmi is vulnerable to diseases of its own, caused by sections of mitochondrial RNA known as d-factors and similar to the hypovirus that attacks *Cryphonectria parasitica.* It may have been the spread of d-factors that checked the first outbreak. Scientists have isolated about 40 d-factors and hope to be able to use the more effective of them to combat the disease. At the same time they are hoping to modify elm trees genetically so that they produce substances that kill either the bark beetles or the fungus. Plant breeders are also trying to develop elms that are resistant to the fungal attack.

Ambrosia Beetles and Bark Beetles

There is more to the elm bark beetles than the appalling damage they cause. These members of the family Scolytidae, together with the families Platypodidae and Lymexylonidae, are known as ambrosia beetles. Their name refers to the fact that they cannot survive without ambrosia fungi. Ambrosia was the food of the gods in Greek mythology, so that is what the fungus is called and why the beetles are known as ambrosia beetles. There are several genera of ambrosia fungi, and each species of beetle ordinarily uses only one species of fungus. Depending on the medium on which they are growing, the fungi can change their form. Under certain conditions, they are fluffy like the mold that grows on stale bread, and under other conditions, they are much denser, like yeast.

Every ambrosia beetle has on its front legs a small pocket called a mycangium that always contains spores of its preferred fungus. *Scolytus* beetles feed on *Ophiostoma ulmi* and *novo-ulmi*, which is why it was inevitable that the *Scolytus* beetles that crossed from Europe to America would also carry Dutch elm disease. When it bores into a tree, the beetle spreads spores from its mycangia along the sides of its tunnel. The spores germinate rapidly, and soon the tunnel is lined with a velvety, yeastlike, fungal coating, the fungus feeding on nutrients it obtains from the wood. The beetle larvae then feed on the fungus. The beetles tend the ambrosia and keep their tunnels clean by removing all the fragments of wood from their tunneling as well as beetle droppings. The beetles deposit all this material outside the tunnel entrance, where it sometimes accumulates. Ambrosia beetles rarely eat wood (unlike termites). The fungus itself stains the wood a dark color, and a discoloration around a tiny hole is an indication of an attack by ambrosia beetles.

Clearings in natural European coniferous forests are often caused by more limited attacks from bark beetles. First the insects kill one or two trees and then move to a few more adjacent to them. This creates an opening that wind damage may widen. Bark beetles also extend gaps made when windstorms have blown down trees, by feeding on the fallen timber and moving from there into standing trees nearby.

Tramp Species

Chestnut blight and Dutch elm disease illustrate the devastation that can be caused when species are introduced to parts of the world where they find new opportunities. The North American epidemic of Dutch elm disease is milder than that in Europe because *Scolytus multistriatus* is a much less effective vector than *S. scolytus*. Should *S. scolytus* cross the Atlantic, however, the disease might become much more severe. Similarly, a North American disease that causes oak trees to wilt, caused by the fungus *Ceratocystis fagacearum*, is not serious, but should the fungus cross to Europe its character might change. In Europe there is an oak bark (ambrosia) beetle, *Scolytus intricatus,* which could disseminate the fungal spores very efficiently, causing a serious epidemic.

Harmful species are not introduced deliberately, of course. They are carried, unobserved, in cargoes transported by sea or air, in ballast carried by ships, and in some cases even on the soles of the shoes worn by passengers. Humans have unwittingly spread many species around the world in this way. The black rat (*Rattus rattus*) and brown rat (*Rattus norvegicus*) are famous examples, and the familiar house mouse (*Mus musculus*) has traveled all over the world in the same way. Its ancestors lived in the Middle East and may have accompanied people who migrated westward in pre-Roman times. Animals, plants, and fungi that hitch rides in this way are known as tramp species, and they are hitching rides still.

Great Spruce Bark Beetle

Great spruce bark beetles (*Dendroctonus micans*) apparently originated in Asia, and over the course of a century or so, they spread westward from Siberia to France. Then, in August 1982, they were found in a spruce plantation in Shropshire, England. The species is now well established throughout Europe, including the British Isles. Many insect pests prefer to feed on dead or dying trees, which limits the harm they do, but others, including the great spruce bark beetle, attack healthy trees. Streams of resin flowing from the trunk are usually the first sign of an infestation. All species of spruce (*Picea*) are susceptible, and the beetle is sometimes found on other conifers.

Adults bore into and through the bark, where the females lay 250 or more eggs in one or more brood chambers. The larvae feed beneath the bark, breaking and blocking xylem vessels as they do so. On a large enough scale, this damage reduces the vigor of the tree and may eventually girdle it, breaking all the vessels and thus killing the tree. When the larvae mature into adults, they disperse. They have functional wings but are unable to fly in temperatures below 73°F (23°C) because their muscles cannot produce sufficient power. The adult beetles do not allow cool weather to confine them to the place where they emerge, however, and usually move by crawling.

The beetle arrived in Britain without its principal enemy, a predatory beetle, *Rhizophagus grandis*, which controls the pest population very effectively in Russia, Belgium, and probably elsewhere, so British scientists imported a stock of this species from Belgium. *R. grandis* beetles locate the places where great spruce bark beetle larvae are feeding. They wound the larvae and then lay their own eggs nearby. When the *R. grandis* larvae hatch, they feed on the *D. micans* larvae, mature into adults, and then go off in search of more *D. micans*.

These predatory beetles are now reared industrially and released into great spruce-bark-beetle infestations. It is an example of biological pest control, but one that is often used in conjunction with other methods. Bark may have to be stripped from infested parts of trees and the exposed area sprayed with insecticide, and if the attack is extensive, a whole area of forest may have to be clear-felled and the timber treated and then removed.

Aphids and Adelgids

Like Dutch elm disease but less severe, beech bark disease is caused by a fungus carried by an insect vector. As the name suggests, it attacks beech trees, and both the European (*Fagus sylvatica*) and American (*F. grandifolia*) species are vulnerable. The insect is the felted beech coccus (*Cryptococcus fagisuga*). There is no mistaking its presence because it covers large areas of the trunk and branches with what looks like waxy wool. This can allow the fungus, *Nectria coccinea,* to establish itself. Small, black spots appear on the bark and start to exude a dark liquid; then the fungus produces fruiting bodies on the bark. These are tiny, bright red, roughly spherical, and clearly visible. The disease retards the growth of trees and can kill them.

The felted beech coccus is one of many species of woolly aphids. They feed on sap, and most are pests specializing on one or just a few plant species. *Elatobium abietinum,* for example, feeds on the needles of Sitka spruce (*Picea sitchensis*) and can defoliate a tree completely, although the tree usually recovers.

The Adelgidae is a family of insects with a way of life very similar to that of the aphids (Aphididae), but its members occur only on coniferous trees. They always reveal their presence by covering themselves with a white, woolly coating, and like aphids, some produce honeydew. Adelgids are serious pests of forest trees. If the canopy in a stand of larch (*Larix* species) is obviously discolored, the culprit may well be *Adelges laricis,* an adelgid that causes dieback in which the tips of branches die first and the disease spreads back into the main body of the plant. Some adelgids cause the formation of galls. *A. abietis,* for example, causes pineapple gall on Norway spruce (*Picea abies*), the species most often grown in Europe as a Christmas tree.

Adelges piceae is a European species. In Europe it attacks the stems of fir trees, especially the giant fir (*Abies grandis*). The giant fir is a North American species that is grown extensively in Europe for its timber. Ironically, giant fir was introduced to British plantations partly because it was no longer practicable to grow its European equivalent, the silver fir (*Abies alba*), commercially. Silver fir occurs naturally in the mountains of central and southern Europe, and the species was planted widely in Britain in the 19th century. Then *Adelges nordmannianae* (or *nuesslini*) was accidentally introduced from eastern Europe and, in the relatively mild British climate, caused so much damage to young trees that the growing of silver fir had to be abandoned.

For a long time the giant fir remained virtually immune to insect attack, but a tree represents a vast store of food, and wherever there is food, sooner or later something will come along to eat it. *Adelges piceae* then found its way to the United States and Canada. Its attacks first became noticeable in 1962, high in the Great Smoky Mountains, where it has killed vast numbers of trees. There the insect is known as the balsam woolly adelgid, and it feeds on the trunks and branches of balsam fir (*Abies balsamea*), red spruce (*Picea rubens*), Fraser fir (*Abies fraseri*), and yellow birch (*Betula lutea*). The noble fir (*Abies procera*) is also subject to severe attacks by species of *Adelges*.

These attacks can kill the tree. Other stem-feeding insects do little harm directly but may open the way to a much more serious fungal attack. Weymouth or white pine (*Pinus strobus*), for example, can become infested with *Pineus strobi,* which may be accompanied or followed by *Cronartium ribicola,* a rust fungus that causes much more harm than the insect alone.

There are some insects that attack the roots of trees as well as the bark. Pine weevils (*Hylobius abietis*) breed in the roots of coniferous trees and in the stumps of trees that have been felled. Their larvae, which are white, legless grubs up to 0.8 inch (2 cm) long, feed in the root or stump for between one and two years. They then turn into adults, climb into a young tree, and feed on its bark. In extreme cases they can girdle the tree and kill it. Weevils comprise the superfamily Cuculionoidea, with about 50,000 species of beetles, sometimes known as snout beetles, in which, to a greater or lesser extent, the head is extended into a kind of beak called a rostrum with the mouthparts at its tip. Many weevils are pests of crop plants or stored food.

Moths and Sawflies

Caterpillars, which are the larvae of moths and butterflies, feed on leaves and some specialize on conifer needles. Budworms are especially harmful. *Choristoneura pinus* is the pine budworm, feeding mainly on jack pine (*Pinus banksiana*), and its close relative *C. fumiferana* is the spruce budworm. Despite its name, the spruce budworm does not confine itself exclusively to spruce. Indeed, its outbreaks seem to begin where balsam fir (*Abies balsamea*) is abundant and summers are often warm and dry. Eastern Canada and Maine have suffered several large outbreaks each lasting several years, for example from 1807 to 1818, 1870 to 1880, and 1904 to 1914.

Moths that attack trees are sometimes very attractive. The pine beauty moth (*Panolis flammea*) is well named, with intricately patterned chestnut and white wings and long "fur" over the back of its head and the upper part of its thorax. Even its larva is pretty, with a chestnut-colored head and dark blue body with longitudinal chestnut and pale blue markings. Altogether it is the kind of insect a conservationist might want to protect until she or he learned that in northern Scotland it has destroyed entire forests of lodgepole pine (*Pinus contorta* var. *latifolia*).

The pine beauty moth is another immigrant to Britain and possibly a tramp species. Long known for its destructiveness in continental Europe, its entry to Britain was spectacular. In 1976 it completely defoliated and killed 445 acres (180 ha) of lodgepole pine in Sutherland in the far northwest of Scotland. Since then there have been several outbreaks covering large areas of forest. Outbreaks occur when the population of moths increases. The insects spend from August to the following March in the soil as pupae and emerge as adults in April. In April and May the females lay their eggs on pine needles, and from June to August the caterpillars devour the needles. Lodgepole pine is a native of northwestern North America that grows well on the poor soils of the Scottish moors. Before lodgepole pine was imported, the pine beauty moth fed on Scotch pine (*Pinus sylvestris*), but Scotch pine is little harmed by it. Perhaps the tree has adapted to deal with the caterpillars, or maybe Scotch pine harbors predators that feed on them. Perhaps the severity of pine-beauty-moth outbreaks illustrates what can happen when two introduced species meet for the first time, and it provides a clear warning to North American foresters of what they might expect should the moth find its way across the Atlantic, although fortunately that is unlikely.

Pine beauty moth is not the only moth with larvae that feed on lodgepole pine. There are also the pine looper moth (*Bupalus piniaria*) and larch bud moth (*Zeiraphera diniana*). Despite its name, the larch bud moth is not especially fond of larch, much preferring pine or spruce. The caterpillar that defoliates larch is the larch casebearer (*Coleophora laricella*).

Broad-leaved trees also attract moths, of course. *Operophthera brumata* is the winter moth, so called because the adult females emerge from their pupae in the soil between October and February. They have only vestiges of wings and crawl up trees to lay their eggs. These hatch as green caterpillars that feed on many broad-leaved trees, especially oak and also spruce. Adults of the northern winter moth (*Operophthera fagata*) emerge in October and November, and its caterpillars feed in May and June, mainly on birch leaves. It is called northern because it was first recorded in Britain, in 1848, in Cheshire in the northwest of England. It is now known to have a much wider distribution, and both of these winter moths occur throughout Europe and over most of Russia.

Caterpillars of leaf-roller moths live in little compartments that they construct with silk between leaves or flowers or in leaves that they roll up. The oak leaf-roller moth (*Tortrix viridana*), a small, pale green moth, lays its eggs on oak trees (*Quercus* species), sometimes in such vast numbers that its larvae can defoliate the tree.

Sawflies can also cause considerable harm. Their larvae look like caterpillars, but unlike most true caterpillars, they have a pair of true legs on each segment of their abdomens. Caterpillars have three pairs of true legs on the three segments of the thorax and usually five pairs of prolegs on abdominal segments 3–6 and 10. Prolegs are fleshy protuberances, each bearing at its tip a ring of tiny hooks that help the caterpillar keep a firm grip on the surface. Sawflies are closely related to bees, wasps, and ants (Hymenoptera), not to moths and butterflies (Lepidoptera).

Female sawflies have an ovipositor, the organ through which eggs are laid, with sharp teeth along its edges, like a saw. Using its ovipositor like a saw, the female cuts a hole in a leaf or twig and then lays her eggs in the hole. When the larvae hatch, they feed on the leaves, and just like true caterpillars, they have prodigious appetites. *Diprion pini* and *Neodiprion sertifer* are both pine sawflies that attack forest trees throughout Europe. *D. pini* causes only minor damage, but *N. sertifer* attacks lodgepole pine and can defoliate it in outbreaks that last several years and usually end when the insects succumb to a viral disease. The larch sawfly (*Lygaeonematus erichsonii*) is a European species that found its way to North America, where it became a highly destructive pest.

Just like insects that bite humans, many of the insect pests of forests are harmful more for the infections they transmit than for the amount they eat. Many of these infections are fungal, but certain fungi need no help from insects in order to destroy the trees they parasitize.

Fungi

In fall, temperate forests are good places to hunt for fungi, visible then as the fruiting bodies from which they release their spores (see the sidebar on page 178 "What Is a Fungus?"). Many fungal fruiting bodies are colorful, some are strangely shaped, some grow on the ground in forest clearings, and some grow on the trunks of trees. These include the bracket fungi and fungi that produce curious, reddish-brown sheets with a white border, all curled and twisted, at or close to ground level.

One of these species, *Heterobasidion annosum*, accounts for about 90 percent of all the decay in conifers grown in Britain, and it is an extremely serious enemy of coniferous trees throughout the temperate regions of the world. Once airborne fungal spores have entered the cut surface of the stump of a felled tree, the infection spreads through the stump, into the roots, and from there to the roots of neighboring trees, crossing where the roots touch below ground. The disease it causes is known as Fomes root rot (because the fungus used to be classified in the genus *Fomes*). Firs are not always killed by it, but pines are very susceptible. Curiously for so devastating a fungus, unless *H. annosum* establishes itself quickly, it fails

What Is a Fungus?

A fungus is a living organism that resembles a plant in certain respects. Like plants, fungi remain always in the same place, and their cells have walls. There the similarities end, however, and scientists class fungi as the kingdom Fungi, of equal taxonomic status to the kingdoms Plantae (plants) and Animalia (animals).

Fungi have eukaryotic cells with a rigid wall containing *chitin* (not cellulose). The body of most fungi is organized as a system of fine threads, called hyphae, comprising cells with many nuclei. Together the hyphae form a network called a mycelium that lies across the surface of the material on which the fungus feeds or that penetrates beneath its surface. Fungal mycelia sometimes extend over a very large area.

Fungi reproduce by means of spores that are produced sexually or asexually depending on the species. Basidiomycetes, which are fungi belonging to the subdivision Basidiomycotina, produce fruiting bodies that carry their spores. These are often conspicuous. Edible mushrooms, toadstools, and all the visible woodland fungi are of this type. Other types of fungus are less visible. These include yeast and *Penicillium,* originally the source of the antibiotic penicillin. Some of these fungi are parasites of plants or animals.

Fungal cells contain no chlorophyll, and no fungus is able to perform photosynthesis. Fungi feed on organic material. They excrete powerful digestive enzymes through their hyphae and absorb the soluble digestive products through their cell walls.

because it competes poorly with other fungi. One of these, *Peniophora gigantea,* is harmless, so painting the stumps of cut pines with spores of this species suspended in water can protect them against Fomes root rot.

True *Fomes* is *F. fomentarius,* and it also causes disease. It attacks the sapwood of broad-leaved trees and kills them, and it is quite common throughout Europe and North America, although in Britain it is confined to the Highlands of Scotland, where it grows only on birch trees. In the forests of central Europe, *F. fomentarius* prevents most beech trees (*Fagus sylvatica*) from reaching a very old age. It attacks the trees when they have grown almost to their full height. The fungus hollows out their trunks, then its fruiting bodies appear as projecting brackets all the way up the trunks. Branches fall and then the trunk splits.

Apart from the damage it does, *F. fomentarius*'s principal claim to fame is that it is the original tinder fungus, and people are believed to have been using it for this purpose 10,000 years ago. After suitable processing, it becomes highly flammable and is easily ignited by the spark made by striking flint. Since Roman times it has also been used to cauterize wounds, more recently by dentists, to clean and dry teeth cavities before filling them, and its suede-like texture has led to it being used to make pouches and items of clothing.

Artist's (or false tinder) fungus (*Ganoderma applanatum*) is a similar bracket fungus with which *F. fomentarius* used to be confused. Fairly common in the forests of Europe and North America and also found in parts of South America and islands of the Caribbean, *G. applanatum* also grows on beech trees, causing them to rot and eventually killing them. It is called artist's fungus not because artists have any use for it but because patterns drawn with a sharp point on the white surface where spores are produced turn brown and are fairly permanent. The related *G. lucidum* also causes trees to rot. It attacks most broad-leaved species but is found only in Europe.

Honey Fungus

Serious though these fungi are, the most destructive of all fungi, and probably of all tree parasites, are the 40 species of *Armillaria,* one of which, *A. mellea,* is the honey fungus. *Armillaria* fungi are common throughout the temperate regions, and different species of *Armillaria* attack different tree species with varying degrees of severity.

Infection begins in the stump of a fallen or felled tree and spreads outward through the soil as black rhizomorphs, which are strands that look like bootlaces and give the fungus its other common name, bootlace fungus. Sheltered below ground, the rhizomorphs can survive extremes of temperature and forest fires. The rhizomorphs usually infect young trees, entering near ground level, and broad-leaved species are more vulnerable than conifers. The rhizomorphs can penetrate healthy bark or enter through the tiniest crack in a root, and then the fungus spreads to rot the entire tree. It also attacks pit props in mines, so it may have caused mining accidents, and it is a serious parasite of all woody plants, not only trees. Its fruiting bodies are honey-colored toadstools, and some species are edible after they have been cooked (they are extremely bitter when raw).

Armillaria fungi possess the property of bioluminescence. That is to say, they glow in the dark with a green light, produced in specialized cells when a compound called luciferin is oxidized in the presence of the enzyme luciferase, and because its rhizomorphs can permeate apparently healthy wood, *Armillaria* infection makes the wood seem to glow. At one time, a piece of glowing root was believed to have magical powers. This may be the origin of the sorcerer's wand.

These fungi have yet another claim to fame: They may be the world's largest living organisms. In 1992 a mat of *A. bulbosa* rhizomorphs was discovered in a mixed oak forest near Crystal Falls, Michigan. The *A. bulbosa* fruiting bodies throughout the area were subjected to genetic testing, confirming that all of them grew from the same mycelium. That mycelium extends over more than 37 acres (15 ha), and its weight was estimated as 110 tons (100 t). Scientists thought that the fungus was probably at least 1,500 years old. Hardly had the Crystal Falls fungus entered the record books, however, when, later the same year, an even bigger specimen of *A. ostoyae* was identified on Mount Adams, Washington. That individual covers approximately 1,500 acres (607 ha) and is believed to be 400–1,000 years old. It probably weighs more than 4,400 tons (4,000 t).

Plant-Eating Mammals

Fungi that cause disease are described as parasites, but the distinction between a parasite and a *predator* is rather fuzzy. Both are species that feed on other species, often killing them in the process. That is what fungi do. Predators move from one prey individual to another, which seems to imply that they are animals, but in their own way so do fungi. Think of the "bootlaces" spreading out in search of

targets for *Armillaria* fungus. Predators are usually bigger than their prey. Cats are bigger than mice, for example, but it is hard to apply this rule when the prey is the size of a tree. Are defoliating insects parasites or predators? Perhaps it is a distinction without a difference because the result is the same no matter what people call it.

The question arises because there are also mammals that feed on and seriously harm trees. These are herbivores, so it sounds strange to call them predators, but it sounds even stranger to call them parasites.

Deer, rabbits, hares, sheep, and goats eat leaves and the tender young growing shoots of trees. This is browsing. It rarely kills the tree and, of course, browsers can feed only to the height that they can reach. Browsing is harmful only in plantations, where it can alter the shape and rate of growth of trees.

Attacks on bark are more serious because this can allow fungal infection to enter and may weaken the tree, and if the bark is stripped all the way around the trunk, the tree will die. Several species of deer strip bark from trees, and so do tree squirrels and voles. Deer also rub their antlers against trees to remove the velvet from them or as part of their territorial or mating behavior. This causes fraying, leaving the bark hanging in shreds. This can kill the tree, although in a forest, only small numbers of trees are attacked.

An ecosystem is a community of plants and animals that feed on one another. Temperate forests are no exception,

Unless their populations are controlled, deer can cause serious damage to forests. These are mule deer (*Odocoileus hemionus*), also called black-tailed deer, and in this case they are doing no harm because they are eating dwarf mistletoe (*Arceuthobium* species), which is a hemiparasite (a parasitic plant that takes nutrients from its host to augment those it makes by photosynthesis). *(The Village of Ruidoso, New Mexico)*

and the harm insects, fungi, and mammals do to individual trees rarely amounts to damage to the forest itself. The forest survives, and even if one species of trees is reduced in number or even eliminated, others take its place, and what was forest remains forest. No member of the community of organisms inhabiting a natural forest can be regarded as a pest or a weed. It is only in plantations, where the trees are a commercial crop, that fungi and animals feeding on them are seen as pests competing with the forester.

NATURAL FOREST AND PLANTATION FOREST

When a tree dies and eventually falls to the ground, it provides food for a hierarchy of animals, fungi, and bacteria. Together, these organisms break down the structure of the fallen tree. Then they set to work on its cells. Finally the bacteria feed on the large, organic molecules from which its cells are made, converting them into smaller, simpler, inorganic compounds that can dissolve in water present in the soil and enter the root systems of living plants (see "Decomposers" on pages 135–137). In other words, the chemical substances from which the forest plants are made are constantly recycled.

Suppose, though, that a farmer who owns the forest calculates that it is more profitable to grow corn than trees, so he has all the trees felled and removed, the old roots ripped from the ground, the land plowed, and a crop sown. What had been forest will have become arable fields, and when the corn crop ripens, it will be harvested. This breaks the nutrient cycle by removing most of the plants from the area, rather than allowing them to decompose on the spot.

A different landowner might make a different decision. Although it is true that corn is more profitable than trees, the owner might feel that since the forest supports many attractive animals and plants, as well as providing a popular amenity for local people, it should not be cleared. All the same, that same owner might decide that it could be managed in such a way as to bring in an income. As trees matured they could be felled and their timber sold, and young trees planted to replace them.

Trees as a Crop

So the forest remains, but the nutrient cycle has been broken, nevertheless. Removing the felled trees prevents them decomposing in exactly the same way that the harvesting and the removal of the corn crop prevents its decomposition. In both cases the removal of the crop removes from the soil the nutrients that would otherwise have been returned to it in the course of the cycle. In time this must lead to the

depletion of the soil and a decline in yields. The remedy is to apply plant nutrients, usually in the form of fertilizer. Like arable crops, managed forests need feeding (see "Nutrient cyclong" on pages 169–173).

Applying fertilizers can replace the plant nutrients that are removed by cropping, but the nutrient cycle remains broken at the point where dead plant material is converted into simple nutrient compounds. This conversion is accomplished by a hierarchy of organisms that feed on the material and on one another. Reducing the amount of dead matter available to the decomposers also reduces the size and possibly the complexity of their population. Ecologically, therefore, the repeated removal of timber alters the character of a forest.

Timber can be a crop just as much as corn or potatoes, but there are differences in the way it can be obtained. Removing selected mature trees while leaving the others untouched has little effect on the overall composition and on the appearance of the forest. Its ecological effects are local, confined to the area around each of the felled trees. Provided damage to other plants is kept to a minimum, selective felling can be beneficial because it allows light to penetrate, herbs to flourish, and young trees to grow to full size. Plantation forestry, on the other hand, is much more like arable farming. When the trees mature, the crop is harvested and a new tree crop is then sown on the same land.

Natural Forests and Plantations

The difference between a natural forest and a plantation is very much like that between a natural meadow and a field of wheat. Both types of forest begin on bare ground. Left to be colonized naturally, what will become the natural forest progresses slowly through a series of vegetation types (see "Succession and Climax" on pages 147–151), and it will be a century or more before the forest matures. During this time, species continually arrive, and those that can find the resources they need remain. Once established, this diverse climax community of plants, vertebrate and invertebrate animals, fungi, and microorganisms remains in existence for a long time, constantly adjusting to minor changes.

A plantation is very different. In most cases, the bare ground on which it begins will have been plowed and possibly drained. Early seral stages will be omitted, and the ecosystem will jump directly to the planting of tree seedlings belonging to the climax species chosen by the forester. The young trees will grow in the absence of the other species of herbs and shrubs that would accompany the colonizing trees of a natural forest. Indeed, invading plants that might inhibit the growth of the desired trees will be regarded as weeds and suppressed.

Since it is the aim of a plantation to produce a crop of timber for sale, the forester will grow those tree species most likely to achieve this. In practice this often favors coniferous species, which grow faster than broad-leaved species. It also means that each area within the plantation will contain the species that grows best on that site, and almost invariably in the case of conifers, this will be just one species. An area growing only one species is called a monoculture. A plantation of broad-leaves may contain two or three species. In Britain, however, many natural broad-leaved forests, but by no means all, contain six or more tree species.

The best trees from a commercial point of view are not necessarily those that occur naturally in the area chosen for a plantation. This is especially true in Britain, where for many years the principal need was for softwood timber from coniferous trees to make such things as telephone poles and pit props to support the roofs of the galleries in coal mines. The only coniferous trees native to Britain are Scotch pine (*Pinus sylvestris*), yew (*Taxus baccata*), and juniper (*Juniperus communis*). Scotch pine is the only species to produce the type of timber required, and none of the native species grow well on poor, upland soils. Forestry was encouraged in the uplands on soils where agriculture was impossible or unprofitable. Not surprisingly after two world wars during which ships bringing food to Britain were under constant attack with heavy losses, agriculture was considered more important than forestry. This forced the state forests onto upland soils that were marginal for agriculture. Private forest plantations also tended to be located in the uplands because it was only there that forestry could bring the landowner a better financial return than farming. Species imported in the 19th century were planted. These came mainly from North America, and they had been grown successfully as ornamentals, thus proving that they could thrive under British conditions. Consequently, British plantation forests are still dominated by Sitka spruce (*Picea sitchensis*), Douglas (or Oregon) fir (*Pseudotsuga menziesii*), and lodgepole pine (*Pinus contorta*), together with Norway spruce (*Picea abies*), Corsican pine (*Pinus nigra*), and Japanese larch (*Larix kaempferi*).

More recently, increasing demand for hardwoods has led to the establishment of more broad-leaved plantations, and these do grow predominantly native species. In North America, with a wider range of suitable species, plantations are stocked with species native to the continent, if not to the area in which they are being grown.

Altering Landscapes

Obviously, seedlings that are planted all at the same time will grow to form a block of trees all of the same age. In a natural forest, a similar area will often contain trees of all ages, although this is not always the case. Scientists have been able to plot the distribution of tree species in parts of the original forest that grew over much of Europe before people started managing or clearing parts of it and have discovered that it consisted of a mosaic of patches, each composed of even-aged stands. This pattern is not surprising, given the events that affect forests. Fires, floods, and similar disturbances are usually limited in scale, each time leaving an area cleared of all vegetation. This bare ground is then recolonized from the surrounding forest, producing an even-aged stand. Few areas of a forest are immune from disturbance, so the eventual result is likely to be the patchwork of even-aged stands that seems to have developed in the original, primeval forest.

Large blocks of coniferous trees, all of the same species and all the same size, often growing in straight rows, are monotonous and aesthetically unappealing. They are also unpopular with conservationists. Foresters have responded to the criticism. Nowadays plantations are designed to be more diverse. Areas adjoining roads, wet ground, and more difficult terrain are left unplanted, so they can be colonized naturally. Forest blocks are less rigidly geometrical in shape, and in some places trees have been cleared to make glades and to improve access to riversides. Change cannot be rapid, however. Trees, even fast-growing conifers, take about 50 years to reach a size at which a whole stand is clear-felled. This means that the shape and pattern of plantations can be changed only gradually. Desirable changes are now being made in areas that were planted half a century ago and have now completed their rotation.

Differences Can Be Exaggerated

There is considerable difference between a plantation and a natural forest, but this should not be exaggerated. It is true that within a plantation blocks are clear-felled and replanted, producing even-aged stands. Undisturbed forest also tends to form such stands, however, so in this respect the plantation is not quite so unnatural as it may seem. Nor is monoculture necessarily unnatural. There are areas in natural temperate forests in which there are only one or two tree species.

Conservationists sometimes point out that a plantation forest supports less wildlife than a similar area of natural forest. This may be true in some cases, but a large, mature plantation comprises many blocks at different stages of development, each supporting a range of species that together amount to a fairly rich community. If the plantation is established on land that was previously unforested and is a long way from other forests, it may take a long time, centuries perhaps, for forest species to colonize it. This may be the basis for much of the ecological criticism of plantations. Eventually, though, those species will arrive provided the plantation remains in being.

Meanwhile, the vast majority of forests in Europe and the United States are either plantations or, at the very least, managed. There are very few areas, and none in Britain, that have not been managed by humans in some way, even though they may have been forested continually for centuries. The distinction between natural and plantation forests is sharply defined only at the extremes. An unmanaged (or poorly managed) forest of native species, complete with dying, dead, and fallen trees in varying stages of decomposition, bears little resemblance to the neat rows of an exotic species in an even-aged plantation stand. The difference is less stark between an area that is natural forest but managed by the removal of certain trees, their replacement by planted seedlings, and the control of unwanted species of plants and animals, and a similarly managed area of plantation.

6

Biodiversity and Temperate Forests

A temperate forest typically contains several kinds of trees and many more species of other plants, fungi, and animals. This wide variety of organisms is an example of biological diversity, a term that nowadays is usually abbreviated to biodiversity. This chapter is about forest biodiversity. It begins by defining the term and advancing several reasons why biodiversity should be preserved.

The remainder of the chapter outlines some of that biodiversity. It does so by describing 14 of the most widespread gymnosperm trees of temperate forests and 36 of the broad-leaved trees.

■ WHAT IS BIODIVERSITY AND WHY DOES IT MATTER?

Plantation forestry is a type of farming and, although trees take much longer to grow than do other farm crops, foresters resemble farmers in many ways. The forester prepares the ground, plants seedlings raised from seed in a nursery (called an orchard), applies fertilizer, removes weeds, controls pests, thins the crop as necessary and, some decades later, gathers in the harvest.

The plantation supports much more wildlife than a field growing an arable crop, but this is mainly because it remains undisturbed for long periods. Other species have ample time to colonize the forest during the years it takes for the trees to grow to a marketable size, and once the blocks of young trees are established, they require little attention. All the same, a plantation is different from a natural forest. Ecologically, an area in which trees are felled when they reach about 50 years of age is not at all the same as one in which trees of the same species have grown, died, fallen, and regenerated from their own seed or roots for several centuries. The natural forest ecosystem will support a wider range of species than the plantation ecosystem, and when natural forest is cleared to make way for a plantation, some of that diversity of species is lost. The area is still forest, but the biodiversity of the area has been reduced.

Biodiversity is a word that has become fashionable in recent years, and the importance of the concept it describes was recognized internationally in June 1992 at the United Nations Conference on Environment and Development, held in Rio de Janeiro. It was there, at the so-called Rio Summit or Earth Summit, that representatives from more than 150 nations signed the Convention on Protecting Species and Habitats, now known as the Convention on Biological Diversity, or Biodiversity Convention. The convention came into force (and became a treaty) at the end of 1992. The nations that have signed and ratified the convention, and those like the United States that accept the principles of the treaty but are not certain of its details or implications and therefore feel unable to sign, have committed themselves to taking whatever steps they can to preserve biodiversity.

Defining Biodiversity

No one would urge nations to reduce biodiversity. It is a self-evident good that everyone can support, but there is a difficulty. Biodiversity is a contraction of biological diversity, and most people instinctively think that it means the number of different living organisms that there are. Unfortunately, it turns out to be rather more complicated when attempts are made to define it more precisely. Then, the expression seems to mean something like the "variety of life," but if that is what biodiversity means, the only way to protect it is probably to protect all living things. Many people would support that, but scientists faced with the task of applying the concept have to make choices, and they find that when biodiversity is defined in this way, it is extremely vague and so broad as to encompass just about everything.

The United Nations Environment Programme (UNEP) defines biological diversity as "the variability among living organisms from all sources, including, *inter alia*, terrestrial, marine and other aquatic ecosystems and the ecological complexes of which they are part. This includes diversity within species, between species, and of ecosystems." UN support for conserving biodiversity is not new. The World Charter for

The Biodiversity Convention

In June 1972 the United Nations sponsored the biggest international conference there had ever been up to that time. Called the United Nations Conference on the Human Environment, it was held in Stockholm, Sweden, and was known informally as the Stockholm Conference. Delegates to the conference resolved to establish a new United Nations program to be called the UN Environment Programme (UNEP). UNEP came into being in 1973. Its task was to collect and circulate information about the state of the global environment and to encourage and coordinate international efforts to reduce pollution and protect wildlife. In the UN family of specialized bodies, a program ranks lower than an organization. It is smaller and has fewer staff. UNEP and the WFP (World Food Program) are programs; the FAO (Food and Agriculture Organization) and WHO (World Health Organization) are organizations.

UNEP sponsored several other major conferences during the years, and in 1992, 20 years after the Stockholm Conference, it organized the UN Conference on Environment and Development, also known as the Earth Summit and the Rio Summit because it took place in Rio de Janeiro, Brazil. It was largest meeting of world leaders ever held. The aim of the 1992 conference was to relate environmental protection to economic development, and to this end the delegates agreed on the Convention on Climate Change and the Convention on Biological Diversity—also known as the Biodiversity Convention.

A convention is a binding agreement between governments. Government representatives sign the convention, and when their own parliaments have accepted it the governments ratify it, confirming their willingness to abide by its terms and passing laws to enforce those terms within their own jurisdictions. When a majority of signatory governments have ratified the convention, it becomes part of international law, and it is then known as a treaty. Although it is still called the Convention on Biological Diversity, technically the convention became a treaty in 1992. By May 2006, 188 countries had ratified the Biodiversity Convention.

The Biodiversity Convention reminds governments that natural resources are not infinite and promotes the principle of using resources in sustainable ways that ensure that future generations will also be able to enjoy them. The convention requires governments to develop national strategies and plans of action to measure, conserve, and promote the sustainable use of natural resources. National plans for environmental protection and economic development should incorporate these strategies and plans, especially in respect of forestry, agriculture, fisheries, energy, transportation, and city planning. As well as protecting existing areas of high biodiversity, governments should restore degraded areas. The convention strongly emphasizes the need to involve local communities in its projects and to raise public awareness of the value of a diverse natural environment.

Many countries have now taken positive steps to implement the convention.

Nature which the UN General Assembly adopted in 1984 states that all species warrant respect regardless of their usefulness to humanity.

Taking the UN definition as a starting point leads to the idea of biodiversity at three levels. "Within species" implies genetic diversity, the variation from one individual to another that is transmitted from parents to their offspring. "Between species" refers to the difference between one species and another, and the diversity of ecosystems refers to the features that make each community of organisms in some sense unique. The second definition, of the diversity of species, is the one to which the World Charter for Nature refers and that most people would prefer, but as the UNEP statement shows, it is not the only one. There is a hierarchy, and it is possible to divide its three levels into five: genes, populations, species, associations and communities (see, for example, "History of Ecological Ideas" on pages 126–129), and ecosystems or landscapes. Having gone this far, it is quite easy to subdivide the hierarchy further, into any number of levels.

Many scientists have attempted a definition, and while these mainly agree on one or another hierarchical arrangement for the "variety of life," they demonstrate another source of possible confusion over the level at which biodiversity is to be considered. Should this be local, confined within particular ecosystems, or global, as the total range of genes or species or ecosystems present in the world? If global, this total genetic range is not known and may never be known; scientists may never even know how many species there are on Earth.

All the members of this hierarchy, from genes to landscapes, are individual things, but some scientists prefer to look at biodiversity from a different perspective. They emphasize the importance of processes. Individual living things are dynamic. Genes are constantly being divided, reassembled, and shuffled in the course of cell division, and they are altered randomly. These altered (mutated) genes disseminate among populations; some vanish and others become fixed. Cell division, mutation, and dissemination are processes. Individual organisms require food, so the cycling of nutrient elements is important to them, and that is another process. Protecting a species, or a gene, or an ecosystem is impossible without safeguarding both the processes on which it depends and its own function, which is also a process, so the two approaches are complementary.

Measuring Biodiversity

Despite the difficulties about definition, there are now ways in which biodiversity can be measured. This is vital because if biodiversity is being lost and society wishes to reduce the rate of loss, people who make important decisions must know how much is being lost and how fast. Measurements of biodiversity are based on comparing chromosomes, proteins, other biochemical compounds, and DNA from cell nuclei, mitochondria, and chloroplasts.

One consequence of these studies has been to cast doubt on what used to be the conventional idea of a species. Boundaries between one species and another are now much harder to draw than they were. The old definition was based on the idea of a group of organisms whose members either do not mate with members of other groups or, if they do mate, produce hybrid offspring that are almost invariably infertile. This is now considered unsatisfactory because there are far too many exceptions to it. Scientists now rely more on differences among genotypes; *genotype* is the complete genetic constitution of an individual.

Once it can be measured, the distribution of biodiversity can be plotted on maps. Biologists have known for a long time that some areas support more biodiversity than others, but now they can know how much more. This knowledge leads to the identification of hot spots where there is a great deal of biodiversity. The hot spot is under threat, usually because the entire area is likely to be developed or put to another use. In the 1980s, the British ecologist Norman Myers identified 10 hot spots in the world as a whole, later increasing this to 20, and most of them are in the Tropics. Other scientists raised the number to 25.

In October 1996, the World Wide Fund for Nature (WWF) published a report listing 217 "ecoregions" that have high biodiversity. That number has now increased to 867 in the world as a whole. The Environmental Protection Agency recognizes 15 ecoregions within the United States.

Gap analysis has become a widely used technique. It was devised in 1978 by Michael Scott, an American ecologist who now works for the Biological Resources Division of the U.S. Geological Survey. First a number of important or endangered species are identified. Then the range of each of these species is plotted separately on a map, and all the maps are assembled as overlays above a map showing wildlife reserves and other protected areas. The result shows whether the protected areas coincide with the overlapping parts of the ranges and also exposes gaps where biodiversity is high but there is little protection. The gap analysis technique is only as accurate as the species range maps it uses, so these are being improved and updated.

Why Preserve Biodiversity?

Most people take it for granted that biodiversity should be preserved, but from time to time the reasons for preserving it have to be stated. Clearing an area of forest might provide land for farming, housing, mining, or any number of alternative uses that would bring obvious material benefit to people locally or even nationally, so anyone urging that the forest remain as it is must produce cogent reasons.

Usually, the answer given is a practical one. Essentially it amounts to saying that there would be more to gain by leaving the forest untouched than would be gained by clearing it. This is the reason most likely to be accepted because it compares one kind of utility with another.

Forests provide timber and wood pulp for making paper, but in temperate regions most of this is supplied from plantations, growing just a few species. A natural forest contains different tree species and probably more of them, and their wood may be of considerable commercial value if not now then at some time in the future. Wild trees may also possess heritable qualities, such as resistance to particular pests or diseases that could be transferred to cultivated species by genetic modification.

It is not only the wild trees that would be lost when the forest was cleared. So would most or all of the smaller organisms and with them species and substances of possibly immense value. These may well exist in familiar, temperate forests, unknown because until now no one has bothered to look for them. Just because a forest is located in a populated region of a highly developed country, it would be quite wrong to suppose that every species in that forest had been identified. Familiar forests can spring surprises.

One afternoon in the fall of 1994, a class of mycology students from Cornell University went on a field trip to Michigan Hollow State Forest at West Danby, New York. Mycology is the study of fungi, and the undergraduates were instructed simply to collect anything that looked interesting. Their finds were returned to Kathie T. Hodge, a graduate student of mycological classification, for identification. Among the specimens were the remains of two insect larvae with yellow, fingerlike, fungal fruiting bodies growing out of their backs. Hodge identified the fungus as the sexual state of *Tolypocladium inflatum.* This is the species from which the immunosuppressant drug cyclosporin is made. Its sexual state had been observed only five or six times before, although the fungus in its asexual state is fairly common in soils. Until then cyclosporin, the drug on which all transplant patients rely to prevent organ rejection, had been made from the asexual fungus collected from the wild. The discovery of the sexual form meant the fungus could be cultured. The Finger Lakes Land Trust, a conservation group in Ithaca, planned to establish a reserve in the forest devoted to chemical prospecting—the first reserve of this type outside the Tropics—and a major pharmaceutical company expressed interest in becoming a partner.

It is not only fungi that may prove useful. Very little is known about the bacterial populations of forest soils, and bacteria are now being genetically modified to break down industrial pollutants, recover valuable metals and other substances, and perform a wide range of other useful tasks. Plants themselves may produce substances with therapeutic properties. Taxol, found in the bark of the Pacific or western yew (*Taxus brevifolia*), inhibits ovarian, breast, and lung cancer. That tree grows along the Pacific coast of North America from California to British Columbia.

Pharmaceutical prospecting in temperate forests has barely begun, and it seems foolish to destroy what might turn out to be the source of treatments for many human ailments. In purely commercial terms, the retail value of drugs derived from plants runs to tens of billions of dollars a year and it is rising. Most of these drugs are from tropical plants, but the size of the market should be a consideration for anyone thinking of ways to turn a profit from land that supports nothing but a few square miles of apparently useless, unmanaged forest.

Aesthetics and Morality

Utility, then, is a powerful reason for maintaining biodiversity, but it is not the only reason. There are two others, heard less often because they do not directly involve money and because some conservationists mistakenly think them weak for that reason.

Both of these arguments begin by considering what would be entailed in restoring biodiversity once it had been lost. In the Blue Mountains, Oregon, for example, there are about 380 species of vertebrate animals, not counting fish, including salamanders, frogs, snakes, lizards, bats, squirrels, wolves, foxes, badgers, porcupines, and many birds, among them the golden eagle and the osprey. Britain cannot boast as much biodiversity as that, but at Monks Wood, Hertfordshire, there is an ancient forest covering 388 acres (157 hectares). Monks Wood contains 146 species of bryophytes (liverworts and mosses), 152 of lichens, more than

1,300 of fungi, and 305 of vascular plants (gymnosperms and angiosperms), as well as 149 species of vertebrate animals and 2,842 of invertebrates. In Poland, the Białowieża Forest, which is very much larger, has far more species, with nearly 1,000 species of vascular plants, 8,500 of insects, and 226 of birds. The Białowieża Forest supports such a large number of species because it consists of a patchwork of quite small areas of habitat, each favoring its own community of organisms.

Clear away the forest and those species will disappear. If, some years later, people decide that the clearance was a mistake, they can change their minds. They can plant trees or just leave the land fallow, and eventually it will probably develop into forest. The resulting forest will not be a precise replica of the forest that was lost, however, and may be very different from it, and the regeneration will take a long time. A natural pine forest might take up to 125 years to return, a forest of elm, oak, and hickory about 200 years, and a maple and beech forest longer than that. A natural forest cannot be restored within the lifetime of those who removed it.

People benefit from the presence of all those species. Vacationers and those living nearby are able to visit the forest, to walk in it, and to see and hear its wildlife and experience its atmosphere. Not everyone is able to visit, but there are many people who like to think that they might do so one day. Meanwhile, perhaps their children will go there with their schools or clubs. Even though they cannot directly experience the forest themselves, people like to know that it is there, and they are able to experience it indirectly through books, paintings, movies, and television programs.

These are aesthetic reasons. Aesthetics is a branch of philosophy concerned with beauty and works of art. A forest is not a work of art, but a book, a painting, or a movie depicting it is. Defending the forest on aesthetic grounds is entirely respectable, and the aesthetic argument is a powerful one. Many forests and other landscapes have survived because their owners thought them beautiful. The danger of this argument is that it favors the most beautiful areas or those with the greatest biodiversity and may be less concerned about other areas and tolerate their loss.

The moral argument is also powerful. It holds that nonhumans have a right to live without unnecessary interference. While people must eat and may defend their food supply and themselves, there are limits that they should not transgress. Applying this argument compels society to debate about where those limits should be located, what is or is not necessary, and how best to minimize harm to other species.

Biodiversity is difficult to define and complicated to measure. To some extent, it is a social and philosophical concept rather than a biological one. Nevertheless, it is proving extremely helpful in guiding the debate about the natural world and society's relationship with it.

Domesticating Trees

Trees that provide people with food or some other useful product, such as citrus fruits, olives, or rubber, have changed in the thousands of years during which they have been cultivated. Forest trees, grown in plantations for their timber, have not been altered to the same extent, but this is merely a consequence of the relatively short time during which they have been grown in this way. Domestication brings change, and forest trees have not yet been fully domesticated.

Historically, the change begins as soon as people gather seed from wild plants and sow it. They select seed from the plants that yield the biggest or best crop. When it is time to gather seeds from these, again the biggest or best are chosen. This is artificial selection, and, like natural selection, it can produce major modifications.

The domestication of forest trees has started with the selection of species to satisfy certain demands. Apart from the quality of their timber, these include their suitability for the soils and climates in which they are to be grown. Sitka spruce (*Picea sitchensis*) grows well in Britain, for example, but it is not just any Sitka spruce, grown from seeds gathered at random. In North America, this species grows from Alaska to Oregon over a 1,500-mile (2,400-kilometer) north-to-south range, and although all the trees belong to the same species, those growing in the north and those in the south have adapted to rather different conditions. Trees from the middle of the geographic range are best suited to the north and west of Britain and those from Washington to the southwest of Britain. Already selection has begun, and there are improved breeds of many commercially important species, some of them hybrids. These result from crossbreeding between closely related species to transfer desirable characteristics from one to the other. American and European poplars have been crossed in this way to produce varieties that are resistant to certain diseases, and European and Japanese larch hybrids are beginning to enter commercial cultivation.

Plant breeders have targets, precise ways in which they would like to improve the species with which they work. Some trees would be more valuable if they could be bred to grow straighter or if their branches produced wood of higher quality. If seedlings grew faster in the period after they were planted, they would compete more successfully with weeds and outgrow them sooner. This would reduce the need for weeding, and that would save money.

Genetic Engineering and Micropropagation

Traditionally, plant breeders selected on the basis of the physical appearance of plants, known as their phenotype. Today breeders are also geneticists and work from geno-types—the genetic composition of the plant. This approach allows them to identify the genes responsible for particular characteristics and transfer these to other species. This greatly accelerates the breeding process.

It is now fairly routine to produce trees by micropropagation. This is a technique for producing vast numbers of identical copies, or clones, of an especially desirable tree. Tissue samples are usually made up of meristem cells, which are capable of dividing indefinitely and produce new growth. Meristem cells are grown in cultures and develop into whole plants. In this way it is possible to copy a plant more than 100,000 times in a year.

Increasingly, trees grown in plantations are likely to differ genetically from their wild relatives and ancestors. This is what has happened to agricultural crop plants, but today it is happening very much faster. The cultivated varieties are greatly superior for the purpose for which they are grown, but they also resemble one another much more closely than would a stand of wild trees. Their genetic similarity makes them vulnerable to specialist parasites or changing environmental conditions. From time to time it will be necessary to improve them further by crossing them with, or incorporating genes from, a wild relative. This will then be seen as yet another reason for preserving the wild forests and their biodiversity.

■ GYMNOSPERM TREES

The gymnosperm trees of New World forests can be grouped as incense cedars, cedars, cypresses, swamp cypresses, firs, hemlocks, larches, pines, redwoods, and spruces. Trees are best known by their common names, but some trees have several common names, so the same species may go by different names in different places, and the common names are unreliable as guides to botanical relationships. Cedars and cypresses, for example, may or may not belong to the main genera *Cedrus* (cedars), which occurs naturally only in the Old World, and *Cupressus* (cypresses), and they may or may not be closely related to each other. For this reason the botanical name is more useful, and although these names are revised from time to time, until everyone is used to a new name, the old name is usually given as well.

In the Old World, the taiga is the mainly coniferous forest that forms a great swath across northern Europe and Siberia, effectively from the Atlantic to the Pacific. It is by far the biggest forest in the world. The taiga stretches from the shores of the Baltic Sea in the west to the Sea of Okhotsk in the east, and from the northern climatic limit for tree growth to the edge of the steppe, the southern shore of Lake Baikal, and the Chinese border in the south. It measures more than 6,200 miles (10,000 km) from west to east, and within the borders of the former Soviet Union, it covers some 2,446,520 square miles (6,333,750 km²). In addition, it covers 78,000 square miles (201,000 km²) in Finland, 118,150 square miles (306,000 km²) in Sweden, and 34,000 square miles (88,000 km²) in Norway. Overall, therefore, the Eurasian boreal forest covers approximately 2.7 million square miles (6.9 million km²).

Its composition varies greatly with latitude. In the north the forest becomes very open, like parkland, as it grades into tundra. In the south, for example to the south of Irkutsk, there are places where it becomes swamp forest, and on mountain slopes it extends into regions covered with mixed or broad-leaved forest at lower altitudes.

Around the Mediterranean, the natural vegetation is sclerophyllous forest (see "Forests in Lands Where Summers Are Hot and Dry" on page 6). This comprises broad-leaved evergreen trees as well as conifers. The gymnosperms in this region also include species belonging to the groups found elsewhere.

Arborvitae

Western red cedar (*Thuja plicata*) belongs to the genus of trees known as arborvitae, which means tree of life, and it is sometimes called the giant arborvitae. Western red cedar grows to a height of 100–200 feet (30–60 m). Its glossy, scale leaves have a fruity smell, which is very strong when they are bruised, and the cones are tiny, no more than 0.5 inch (1.2 cm) long. The tree is native to western North America, where it is an important source of timber, and it also occurs in China and Japan and at the tip of South America. The map shows its global distribution. Arborvitae wood is used in North America to make weatherboarding and roof shingles, and it is also the tree most often used by Native Americans to make totem poles.

In the east, the American or white cedar (*T. occidentalis*) is a smaller tree, up to 65 feet (20 m) tall, with cones 0.3–0.5 inch (0.8–1.2 cm) long.

Several species of arborvitae occur in China and Japan. Most are fairly small, the Korean arborvitae (*T. koraiensis*) often being no more than a shrub but sometimes growing into a conical tree about 30 feet (9 m) high. Chinese arborvitae (*T. orientalis*), of northern and western China, is of similar

The global distribution of arborvitae (*Thuja* species)

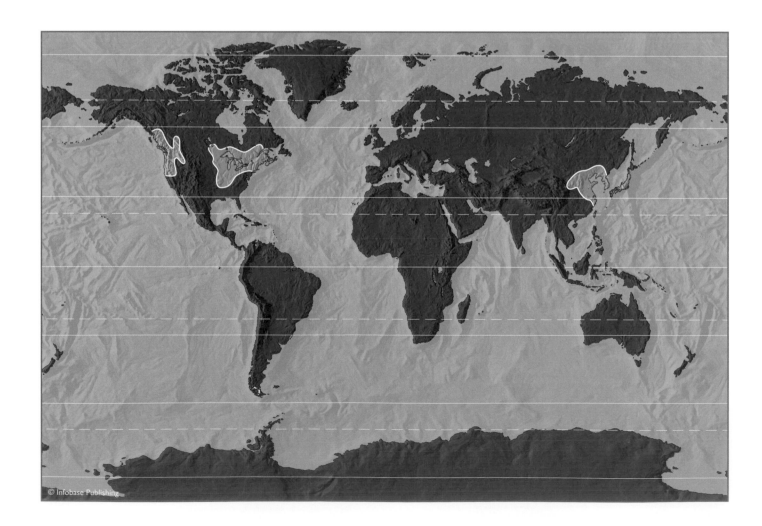

size. Japanese arborvitae (*T. standishii*) is rather bigger, growing into a handsome, conical tree about 65 feet (20 m) tall.

Hiba arborvitae differs from other arborvitae in having much broader branchlets (twigs), bearing larger scale leaves, and having much more rounded cones with thicker scales, and the tree grows much more densely near its base than other arborvitae. These differences have persuaded scientists to classify it in a genus of its own, as *Thujopsis dolobrata*. It is a handsome tree, shaped like a pyramid and growing to a height of about 50 feet (15 m), and there are many cultivated ornamental varieties. In Japan, where it occurs naturally, its wood is used for building.

Cedars

Cedars are grown widely for ornament, and cedar wood is valuable. Soft but durable, it is used in construction and for making furniture.

There are many trees with cedar in their names, but there are only four species of true cedars (*Cedrus*)—and some authorities consider them to be a single species. Cedars occur naturally in Algeria, the eastern Mediterranean, and in the western Himalayas. They are handsome trees with long and short shoots. The short shoots bear clusters of needles 0.25–two inches (0.5–5 cm) long. Male cones are up to 2 inches (5 cm) long and open between September and November. The female cones are up to 0.4 inch (1 cm) long. The fruiting cones are two–four inches (five–10 cm) long and take two–three years to develop.

The Atlas or Atlantic cedar (*C. atlantica*) occurs in Algeria. It is pyramid-shaped and up to 130 feet (40 m) tall. The Cyprus cedar (*C. brevifolia*) grows in the mountains of Cyprus. It has a dome shape and is up to 40 feet (12 m) tall. The cedar of Lebanon (*C. libani*) is represented

on the flag of Lebanon shown in the illustration. It grows in the Taurus Mountains of Syria as well as in Lebanon. Its branches form very distinctive tiers. The tree has needles about one inch (2.5 cm) long and grows to a height of 130 feet (40 m). The pyramid-shaped Himalayan or Indian cedar (*C. deodara*) is the tallest of all cedars, reaching a height of nearly 200 feet (60 m).

Douglas Firs

Douglas firs were formerly placed in the genus *Abies* but are now recognized as a genus of their own, *Pseudotsuga*, although they are not closely related to the hemlocks (*Tsuga*). There are two North American species. The large-coned Douglas fir (*P. macrocarpa*) is a tree up to 80 feet (25 m) tall, with cones four–seven inches (10–18 cm) long, that is native to southwestern California.

Oregon (or gray) Douglas fir (*P. mensiesii*) is the most commercially important timber tree in North America, although its quality is somewhat variable. It grows naturally from British Columbia and throughout the western states as far south as Mexico. About half of all the forest trees in this region are Oregon Douglas firs. It can grow to 330 feet (100 m), and its timber is used for just about everything that can be made from wood, from houses, boats, and railroad freight cars to furniture and flagpoles.

False Cypresses

American cypresses belong to the genus *Chamaecyparis* or false cypresses. Both *Chamaecyparis* and *Thuja* have scale leaves that hug the twigs so closely that no bark or leaf buds are visible beneath them. If the tip of a leaf-covered twig feels soft and fleshy to the touch, however, the tree is a *Thuja*, and if the tip is thin and hard, the tree is a *Chamaecyparis*.

Lawson cypress, or Port Orford cedar (*C. lawsoniana*), grows naturally in the Far West of the United States in California and Oregon, but it is a hardy, very attractive, spire-shaped tree and is widely cultivated, with more than 200 named varieties. It can grow 80–165 feet (25–50 m) tall, and its timber, which has a spicy, fragrant odor, is of high quality with many uses in building, furniture, and boat-building, and its strength and resistance to attack by fungi and insects makes it suitable for making railroad sleepers. It is grown mainly for ornament, however, because the tendency of its main stem to fork into two greatly reduces its value for timber.

Nootka cypress (*C. nootkatensis*), also known as yellow cypress and stinking cypress, is a very similar tree, growing to a height of 100–130 feet (30–40 m). Its scale leaves have a smell reminiscent of tomcats, hence the name stinking, but its timber is odorless. Nootka cypress grows from Oregon northward to Alaska and, like Lawson cypress, it is

The flag of Lebanon bears an image of the cedar of Lebanon (*Cedrus libani*)

cultivated for ornament and produces timber with similar qualities and uses.

Firs

Firs belong to the genus *Abies* and their leaves are needles, which in most species are flat, rather than the scales borne by the cedars and cypresses. The map shows their global distribution.

Perhaps the best-known and most widespread North American fir is *A. balsamea*, the balsam fir or balm of Gilead, which grows over much of the continent as far north as the Arctic Circle. Balsam fir and Fraser's balsam fir (*A. fraseri*) are the species most often used as Christmas trees. The balsam fir is a tree up to 80 feet (25 m) tall. The Colorado, or white fir (*A. concolor*) grows to 130 feet (40 m) in the mountains of Colorado and south as far as Mexico.

These trees are small compared with other North American firs. The red fir (*A. magnifica*), for example, grows as a very symmetrical tree 230 feet (70 m) or more tall. It is native to Oregon and California. Red silver fir (*A. amabilis*) grows to a height of 260 feet (80 m) in natural forest, although to less than half that size in cultivation. It occurs naturally in the mountains of British Columbia, Alberta, Washington, and Oregon. To its south, extending into California, the noble fir (*A. procera*) is of similar size. It is grown in plantations, where it usually reaches about 165 feet (50 m). In natural forest, the grand fir (*A. grandis*) or giant fir can grow to 330 feet (100 m). This species is widely grown in plantations because it tolerates a wide range of soil conditions and grows fast, but it rarely attains more than half the height of which it is capable. Its timber is used to make furniture, boxes, and in building, and is also pulped for paper.

In the Old World, the northern forest is dominated by fir, pine, larch, and spruce, often with juniper. Fir trees occur in the forests of western and central Siberia, throughout China as far south as the Himalayas, in east Asia and Japan, in Turkey and Syria, in central and southern Europe, and around the Mediterranean. The true firs, all of which belong to the genus *Abies*, are often called silver firs to distinguish them from species that have fir in their common names despite belonging to other genera.

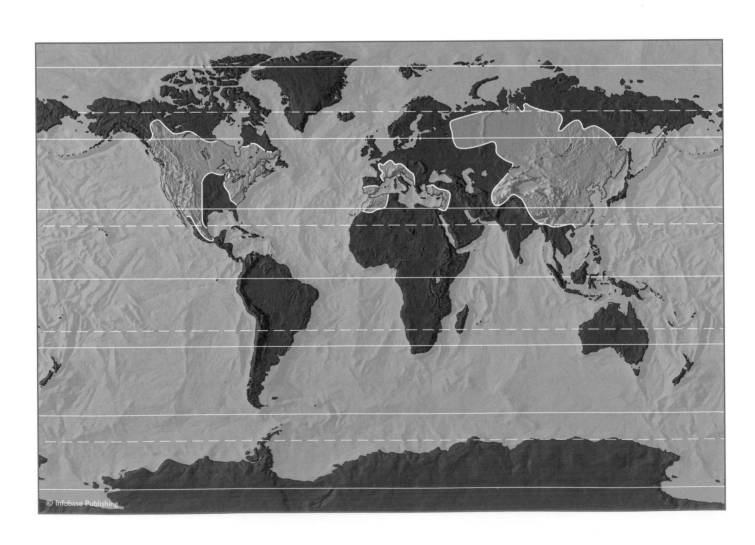

The global distribution of firs (*Abies* species)

As its common names suggest, the Amur or Kirghan fir (*Abies nephrolepis*) grows in Asia. The River Amur marks the border between Russia and China, and the land in which this tree is found is called Ussuriland, bounded to the west by the Ussuri River, to the east by the sea of Japan, and continuing to the south into China and Korea. Still farther east, the Kamchatka fir (*A. gracilis*) grows naturally only in the forests of the Kamchatka Peninsula and the Sakhalin fir (*A. sachalinensis*) only on the island of Sakhalin to the north of Japan. Manchurian or black fir (*A. holophyla*) is also a native of eastern Asia.

The three most important firs native to Japan itself are the momi and nikko species and Veitch's fir. The momi or Japanese fir (*A. firma*) is a conical tree occurring in southern Japan that grows to 100 feet (30 m). The nikko fir (*A. homolepis*) grows to a similar height and is found in the mountain forests of central Japan. Several species of firs are widely cultivated as ornamentals. Nikko fir grows in cities better than most firs because it is fairly tolerant of air pollution. Like the nikko fir, Veitch's fir (*A. veitchii*) occurs in the mountains of central Japan. It is rather smaller than the nikko fir, reaching about 80 feet (25 m), and although it shares the nikko fir's tolerance for airborne pollutants, it tends not to live as long.

Korea, to the south of Manchuria and in the same latitude as Japan, is mountainous, and the Korean fir (*A. koreana*) is native to the mountains of the south. It is a conical tree and fairly small, growing to no more than 60 feet (18 m) and often less, especially in cultivation.

Farther west, the Siberian fir (*A. sibirica*) is the dominant fir over much of the taiga in central Asia together with the Turkestan fir (*A. semenovi*). The Siberian fir is similar to, and the Old World equivalent of, the North American balsam fir. These give way in the northern part of the Caucasus to the Caucasian or Nordmann fir (*A. nordmanniana*), a column-shaped tree that grows to a height of 165 feet (50 m) or more. Continuing westward, the Cilician fir (*A. cilicica*) is native to the mountains of northern Syria and southeastern Turkey, part of the region known in ancient times as Cilicia. The tree grows to about 100 feet (30 m).

The common fir of the mountains of central and southern Europe is the silver fir (*A. alba*), sometimes called the common or European silver fir to distinguish it from other firs that are also called silver. The (common) silver fir grows to about 165 feet (50 m) and occurs naturally at altitudes of 2,600–6,000 feet (800–1,800 m). In many parts of Europe, this is the species most often used as a Christmas tree.

The sclerophyllous vegetation of the Mediterranean includes three species of fir, their common names identifying the countries in which they occur. In each case they grow naturally only in the mountains. Algerian fir (*A. numidica*) is rather rare in the wild, being confined to just a few areas of the coastal mountains. It is a handsome, almost columnar tree, growing to a height of 80 feet (25 m) and is a popu-

lar ornamental in gardens, growing well in urban areas. It grows to the same size and is closely related to the Spanish fir (*A. pinsapo*), which occurs naturally in the mountains of southern Spain. This tree is also known as the hedgehog fir because its short, rigid needles, about 0.8 inch (2 cm) long, grow all around the stems and stand out almost at right angles, like the prickles on a hedgehog. The Greek fir (*A. cephalonica*) is found in the Greek mountains and, despite its name, is not confined to the island of Cephalonia. A taller tree than the other Mediterranean species, it reaches 100–165 feet (30–50 m). Its needles are flattened, 0.8–1.2 inch (2–3 cm) long, and, like those of the Algerian and Spanish firs, are rigid and grow all around their stems.

Hemlocks

Hemlocks (*Tsuga* species) have a rather disjunct distribution. They occur naturally in eastern and western North America, in the Himalayas, and in eastern China, Taiwan, and Japan.

Hemlock trees (*Tsuga* species)   *(Vanderbilt University)*

The female cone of a Canada hemlock tree (*Tsuga canadensis*), also called the eastern hemlock *(The Arboretum at Penn State Behrend)*

They are handsome trees, roughly pyramid-shaped and with horizontal or slightly drooping branches.

Wood from the common, or Canada hemlock (*T. canadensis*) has many uses, but it is especially popular for making ladders, and its resin is sold as "Canada pitch." The tree is about 80–100 feet (25–30 m) tall and grows throughout eastern North America. Carolina hemlock (*T. caroliniana*) is smaller, rarely reaching 80 feet (25 m). It grows in the mountains of the southeastern United States.

In the west, mountain hemlock (*T. mertensiana*) is a tree 100 feet (30 m) tall, with some individuals in natural forest growing to 165 feet (50 m), that is native to the mountains from Alaska to California. Both it and western hemlock (*T. heterophylla*) are often grown as ornamentals, but western hemlock is also widely grown in plantations for its timber, used to make boxes, for building, and pulped to make paper. The tree occurs naturally near the Pacific coast, where it grows to a height of 100–200 feet (30–60 m). Western hemlock has distinctive needles which are of uneven lengths and crowded along the twigs in a very haphazard fashion.

In the Old World, hemlocks occur naturally only in southern China and Japan. Chinese hemlock (*T. chinensis*) is a tall tree, growing to 165 feet (50 m) in its natural habitat, but specimens grown for ornament in other parts of the world are usually very much smaller.

Two species of hemlock are found in Japan, one mainly in the center and north of the country, the other in the south. Like Chinese hemlock, both are much bigger where they grow in natural Japanese forests than they are when cultivated. Northern Japanese hemlock (*T. diversifolia*) reaches a height of about 80 feet (25 m) in Japanese forests. The south-

ern species, called Japanese or southern Japanese hemlock (*T. sieboldii*), grows to about 100 feet (30 m), but cultivated specimens of both species reach barely half these heights and are often no bigger than shrubs.

Incense Cedars

There are eight species of incense cedars, and none of them are true cedars (*Cedrus*). Of the eight species, five occur in New Zealand and New Caledonia, in the Pacific, one on the Pacific coast of North America, one in China, and one in Taiwan. At one time all eight were classified in the same genus (*Libocedrus*), but the North American, Chinese, and Taiwanese trees have now been placed in the genus *Calocedrus*. As though to confuse matters further, there is also a Chilean incense cedar, the only species in the genus *Austrocedrus*.

North American incense cedar (*Calocedrus decurrens*) can grow to a height of 148 feet (45 m) and is shaped like a narrow cone. Cultivated varieties, which are grown as ornamentals throughout temperate regions, are often almost cylindrical. The name *decurrens* refers to the tiny, scalelike leaves. These are markedly decurrent, which means that their bases extend down the stem, below the points where they join it, so they cover the stem completely. Beneath the leaves, the bark is reddish brown and deeply furrowed. Wood from the incense cedar is light, resistant to decay, and fragrant. It is used to make a range of products, including pencils, boxes, and fence posts.

Chilean incense cedar (*Austrocedrus chilensis*) is smaller, growing to about 82 feet (25 m) and less than that under cultivation. Its wood is scented and hard-wearing but is little used.

Junipers

Junipers are distributed throughout temperate regions of the Northern Hemisphere, with some species extending into the Tropics. There are about 60 species of them, all belonging to the genus *Juniperus*. Junipers have two kinds of leaves, although some species, including the common juniper (*J. communis*), bear only the young type. The young leaves are needles, sharply pointed and growing in groups of three or in opposite pairs. Adult leaves are scalelike, pressed closely to the stem and overlapping.

Most juniper species are small trees or shrubs, and some alpine and subarctic varieties of the common juniper are no more than 12 inches (30 cm) tall. There are also quite tall juniper trees, however, and just below the treeline on the mountains of Europe and Asia, there are forests dominated by juniper. Where it is obtainable, juniper wood is used in building, for fences and roof shingles, and for making furniture. Juniper also yields aromatic oils and tars with many pharmaceutical, perfumery, and other uses (cedarwood oil is obtained from a juniper, not a cedar). Juniper is also used as a flavoring for some meats, especially venison, and in gin. The word *gin* is a corruption of *geneva*, which has nothing to do with the Swiss city but is from *genévrier*, French for "juniper tree."

Chinese juniper (*J. chinensis*) can be a low shrub, but it also occurs as a tree shaped like a narrow cone and 65 feet (20 m) tall that occurs in Mongolia, China, in the Himalayas, and in Japan. It is often cultivated in gardens and parks, and there are many varieties. The needle or temple juniper (*J. rigida*) is one of the species that produces only juvenile, needlelike leaves, the feature that gives it one of its common names. Needle juniper grows naturally in Japan, Korea, and Manchuria, where it is sometimes a small tree, about 43 feet (13 m) tall with branches that hang down, and sometimes a shrub. In cultivation it is usually a shrub. Drooping juniper (*J. recurva*) also produces only needlelike leaves, growing in threes, but in this species they are crowded together, overlapping, and pressed close to the stem, and the plant itself is a shrub or a small tree about 33 feet (10 m) tall, with spreading, hanging branches. It grows in southwestern China, the Himalayas, and Burma (Myanmar), where the variety *coxii* is called the coffin juniper because its wood is used to make coffins.

Junipers also form part of the vegetation of the Mediterranean region and Near East. Syrian juniper (*J. drupacea*) is a narrow, conical tree 33–40 feet (10–12 m) tall, bearing only juvenile needle leaves, that grows in Syria and other parts of Asia Minor as well as in Greece. Prickly juniper (*J. oxycedrus*) also occurs in Syria and east as far as the Caucasus, as well as in Spain and North Africa. It has only needle leaves and is a small tree about 33 feet (10 m) tall.

Despite its name, Spanish juniper (*J. thurifera*) has a range extending from the Caucasus through Asia Minor, across North Africa, and into southwestern Europe. It bears adult, scale leaves and is a tree about 40 feet (12 m) tall. Phoenician juniper (*J. phoenicia*), also bearing scale leaves, grows all around the Mediterranean and in the Canary Islands. It is a very small tree, no more than 20 feet (6 m) tall, although those in the Canary Islands grow much taller. A little farther north, savin (*J. sabina*), also bearing scale leaves, is barely large enough to be called a tree, growing to only 15 feet (5 m). It occurs in central Europe.

Larches

Larches (*Larix* species) are among the most attractive of all trees. They are tall, with a graceful pyramid shape, but their beauty lies primarily in their deciduous foliage. This takes the form of bunches of needles that change color in the fall shortly before they are shed, producing patches of gold amid the otherwise dark green forest.

The map on page 194 shows their global distribution. Two species are native to North America. Tamarack (*L. laricina*), also known as hackmatack, eastern larch, and American larch, is not confined to the east of the continent. It grows naturally in Alaska, Canada, and the eastern United States as far south as Pennsylvania. It grows up to 65 feet (20 m) tall. Western larch or West American larch (*L. occidentalis*) is taller, reaching a height of 148–180 feet (45–55 m). It is a native of the west, growing naturally between British Columbia and northern Montana. Larch timber is tough and durable, and both species provide wood with many uses, including construction, telegraph poles, railroad sleepers, and pit props.

In the Old World, larches occur in the Alps, the mountains of eastern Europe, throughout most of the Asian taiga, and into Japan. They are one of the most common trees in the taiga, and they also occur farther south in the mixed forests. Over most of the taiga the Siberian larch (*L. sibirica*) is predominant, a tall tree, reaching a height of 100 feet (30 m). East of the Yenisei River, the Siberian larch first mixes and hybridizes with, and then gives way to the Daurian or Dahurian larch (*L. gmelinii* or *dahurica*). The Daurian or Dahurian region is a large area of plains and low hills with an extreme continental climate and areas of permafrost. It takes its name from the Dagur (Dahur is the Latinized form), a people living in northern Manchuria. The Daurian larch tolerates a wide variety of soils and has very shallow roots, which allow the tree to find water very efficiently in the upper soil. This feature enables Daurian larches to grow in permafrost areas, and their range extends farther north than that of any other larch. Farther east, the Korean Daurian larch (*L. olgensis*) and the Kamchatka larch (*L. kamschatica*), growing on the Kamchatka Peninsula, are very similar.

Larches also grow in the milder climate found in the Himalayan valleys and foothills. The Tibetan larch (*L. potaninii*), a tree some 60–70 feet (18–20 m) tall, is found in western China as well as Tibet, and the similar Sikkim larch (*L. griffithiana*) grows in Nepal and Tibet as well as in Sikkim.

Japanese larch (*L. kaempferi*) is a much bigger tree, reaching a height of 100 feet (30 m). It occurs naturally in the mountains of Japan. In 1861 it was introduced in Britain, and about 1910, foresters became interested in it because it grows faster than the native European larch (*L. decidua*) and seemed to thrive better than *L. decidua* (which is not native to Britain) in the British climate. Consequently, Japanese larch began to be grown widely in commercial plantations. About 1904, larches of both species were growing on the estate of the duke of Atholl at Dunkeld in Scotland, and female flowers of *L. kaempferi* were pollinated by male flowers of *L. decidua*. The resulting hybrid, known as Dunkeld larch (*L. × eurolepis*) or hybrid larch, exhibits heterosis, or hybrid vigor. This means that

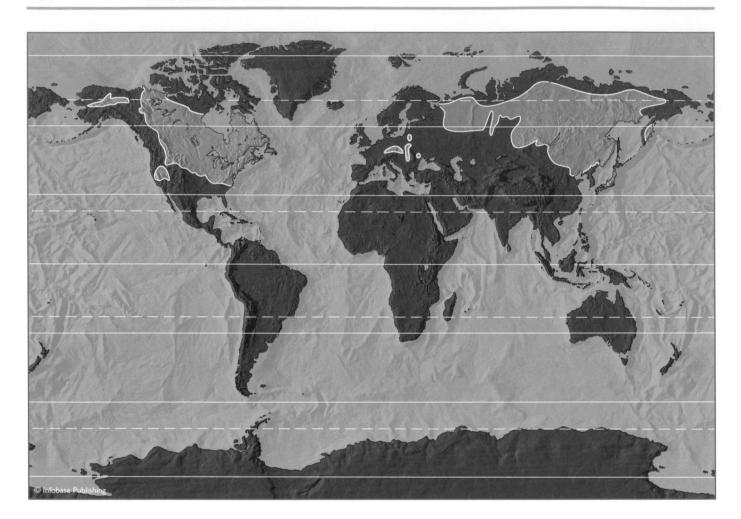

The global distribution of larches (*Larix* species)

the first-cross hybrids grow faster than either parent. They are also more resistant to pests and diseases and more tolerant of poor soils. Dunkeld larch is now grown widely in plantations from seed produced in nurseries where the two parent species are grown in alternate rows to increase the likelihood of cross-fertilization.

Pines

True pines belong to the genus *Pinus,* which includes nearly 100 species distributed throughout temperate regions of the Northern Hemisphere. The map on page 195 shows their distribution. A number of unrelated trees, some from the Southern Hemisphere, also have the common name *pine,* but all North American pines are true pines. Two species of bristlecone pine—Great Basin (*P. longaeva*) and mountain (*P. aristata*) pines—are famous for living so long that scientists have developed from them a chronology against which other dates, including radiocarbon dates, can be calibrated. The trees grow in California, in the mountains of Colorado,

Arizona, and in New Mexico. Some individuals are more than 4,600 years old. Correlating their growth rings with those of dead pines has produced a chronology that extends more than 8,000 years into the past.

Bristlecone pines grow in a climate that is too dry to support dense forest, but it is not the only North American species to be valued for something other than its timber. Monterey pine (*P. radiata*) grows to 65 feet (35 m) or more, but its value arises from its being broad and bushy, so it provides excellent shelter from the wind. In addition, it is able to remove salt from the air, which allows less salt-tolerant plants to grow nearby. Monterey pine occurs naturally only along the California coast near Monterey, but it has been planted in many other places, usually for ornament but sometimes for timber, especially in Australia and New Zealand.

Nine species of pines form important components of North American forests. Shore pine (*P. contorta*) is native to western coastal areas. The lodgepole pine is *P. contorta* variety *latifolia.* It grows in inland forests and was the tree preferred by Native Americans for making the frames of their lodges, hence its common name. It is a slender tree, often no more than 33 feet (10 m) tall, and is widely cultivated for timber because of its tolerance for poor soils. Its cones are said to be serotinous, meaning that they are stimulated to open and release their seeds only after they have been heated by fire, a feature lodgepole pine shares with several other species (see "Fire climax" on pages 151–155). Lodgepole pine grows in the west, from Alaska to California. In Canada almost to the Arctic Circle and in the northeastern United States, its place is taken by jack pine (*P. banksiana*), a tree about 65 feet (20 m) tall that also produces serotinous cones. Its timber is used to make railroad sleepers.

Jack pine is one of the "hard" pines. Although all coniferous trees produce softwood, compared with the hardwood of broad-leaved trees, pines are subdivided commercially into two groups, as soft or hard, depending on the ease with which they can be worked with carpentry tools. The groups are sometimes classified as the subgenera *Strobus* (soft) and *Pinus* (hard), so the botanic name of jack pine could be written as *Pinus Pinus banksiana*.

The soft and hard designation refers to the amount of resin in the wood. Soft pine has little resin; hard pine has much more. There are also anatomical differences that help

with identification. These center on three factors. Soft pines have one vascular bundle (see "Phloem" on page 105) in the leaves and hard pines have two. In soft pines the sheaths surrounding the shoots bearing clusters of needles fall away; those of hard pines do not. The scalelike bracts covering the longer shoots are decurrent in hard pines but not in soft pines. Bracts are modified leaves, and in decurrent bracts the base extends back down the stem. Surprisingly, perhaps, given its longevity, bristlecone pine is of the soft variety. Monterey and lodgepole pines are hard.

Longleaf, or pitch pine (*P. palustris*), another hard pine, also withstands fire well. A thick layer of leaves protects its terminal buds, from which new growth develops. These leaves prevent the temperature from rising high enough to kill the buds, so the tree survives while the surrounding vegetation is destroyed. On the coastal plain of the southeastern United States, the first European colonists found large areas covered by pure stands of longleaf pine. These probably developed because of fires that had been occurring naturally for thousands of years at intervals of 3–10 years, producing conditions only this species could tolerate. It is a big tree, growing to 130 feet (40 m), and where it is grown for timber or pulping to make paper, foresters

The global distribution of pines (*Pinus* species)

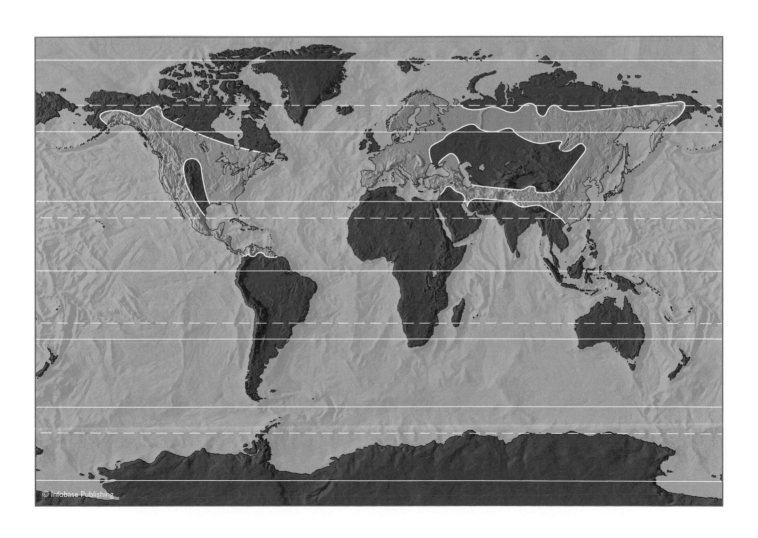

sometimes use fire to remove competing trees. The name *longleaf* refers to its bright green needles, which are up to 17.75 inches (45 cm) long.

Pitch is a tarlike substance made from resin that was used to caulk (seal) the planking of wooden ships. Until it became possible to manufacture synthetic versions more cheaply, pine resin was also the raw material from which turpentine and some types of glue were made. It was obtained from hard pines. Longleaf pine was a good source, which is why it is called pitch pine, but it was not the only one, so there are several species called pitch pine. Pitch pine itself (*P. rigida*), also known as northern pitch pine and easter pine, a hard pine up to 50 feet (15 m) tall, grows from Massachusetts south through the Appalachians and is one of the few pines that will thrive in salt marshes. The variety *serotina* is the pond pine.

Loblolly pine (*P. taeda*) is another fire-resistant species. It is a hard pine with abundant resin and a fragrance that gives it an alternative common name of frankincense pine. The resin is of too poor a quality to be of commercial value, however. Loblolly pine grows up to 100 feet (30 m) and occasionally to 165 feet (50 m) tall, and its needles, up to 10 inches (25 cm) long, are blue-green. It covers large areas of poor, sandy, or marshy land, called pine barrens, of the eastern and southeastern United States, and it often colonizes abandoned arable land where the soil is depleted of nutrients due to bad farming. There it is known as the old-field pine. Its timber has many uses. Loblolly pine often grows with shortleaf pine (*P. echinata*) as a codominant, especially where the pines establish themselves on abandoned fields. Slash pine (*P. caribaea*), or Cuban pine, occurs farther south, throughout Florida and on several islands in the Caribbean. This is a commercially important tree in tropical America. It is widely planted, mainly for pulp, there and also in South Africa.

The American pine with the most extensive range is the ponderosa pine (*P. ponderosa*), also known as the western yellow pine and the bull pine. It occurs naturally in mountainous areas from British Columbia south to the Black Hills of South Dakota and from there south to Texas and Mexico and also down the northern part of the Pacific coast where commercially it is second in importance only to the Douglas fir. The variety *arizonica* is the Arizona pine. Despite being a hard pine, its wood is easy to work, and it is an important plantation species. It is a big tree, growing to a height of 165–250 feet (50–75 m) with a trunk up to eight feet (2.5 m) in diameter. Farther south along the Pacific coast, its place in natural forests is taken by Jeffrey's pine (*P. jeffreyi*), which is very similar to the ponderosa. Timber from both species, produced mainly in California, Washington, Oregon, and Idaho, is known as ponderosa pine.

Coulter pine (*P. coulteri*) also grows in southern California and Mexico. It grows to about 80 feet (25 m) tall,

but its most remarkable feature is the size of its cones, which give the tree its alternative common name of big-cone pine. These are 10–14 inches (25–35 cm) long, about six inches (15 cm) in diameter, and can weigh four pounds (1.8 kg) or more, and their hard scales end in long, sharp, curved spines like claws. The digger pine (*P. sabiniana*) of California is closely related and very similar. Its cones also have scales with long, curved, spines, and the cones are often six inches (15 cm) across, although they are only up to nine inches (23 cm) long. The seeds of both are large and resemble nuts. Native Americans used to gather them for food.

Timber from soft pines can also be valuable. American soft pines are further divided into the following groups. White pines produce very pale wood. Stone pines grow mainly near the timberline and are important mainly as protectors of watersheds. Foxtail pines have needles that live for 15 years or more, and as more and more accumulate, branches acquire a bushy appearance like a fox's tail. Bristlecone pines are foxtail pines, and there is also a tree known as the foxtail pine (*P. balfouriana*) that is found in the Sierra Nevada. Nut pines produce the seeds sold as pine nuts or pine kernels. Most of the nuts come from the nut pine (*P. edulis*).

One of the most valuable of all North American timber trees is a soft pine, the eastern white pine (*P. strobus*), also known as the Weymouth pine. It grows to a height of 80–165 feet (25–50 m), and its wood is white with a very even grain. At one time it dominated forests that stretched from the Atlantic coast westward to Minnesota and Manitoba and south to Georgia, but so many trees have been felled that today few mature specimens remain. It is the only soft pine to occur naturally east of the Great Plains. There is an equivalent species, no less valuable, on the other side of the continent. The western white pine (*P. monticola*) occurs in the northwestern United States and British Columbia. It is much the same size as the eastern white pine, but its wood is slightly darker. Its young shoots are covered in down, a feature that distinguishes it from *P. strobus*.

A third North American soft pine that has also been extensively harvested for its timber, which commands one of the highest prices of all U.S. timbers, has the distinction of being probably the biggest pine in the world. The sugar pine (*P. lambertiana*), also a white pine, grows to a height of 165–330 feet (50–100 m) with a trunk that can be 12 feet (3.7 m) across. It grows in California, which is where almost all the timber from this species is produced, the small remainder (about 5 percent) coming from Oregon. Its seeds are edible, but it is called the sugar pine because, when damaged, its heartwood exudes a sweet sap with laxative properties.

Pines are probably the most widespread coniferous trees in temperate regions of the Old World. Except for the far north and the steppe lands and deserts of the continental interior, they occur throughout Europe and Asia north of the Himalayas. Pines grow over such a vast area because they tolerate a wide variety of soil and climatic conditions. In the taiga, forests of Scotch pine (*P. sylvestris*), stone pine (*P. cembra* or *sibirica*), and larch occupy more than half of the total area.

Scotch pine, the only conifer native to Britain (and the subspecies *scotica* grows naturally only in Scotland), has needles growing in pairs. The tree reaches a height of 65–130 feet (20–40 m) with a straight trunk up to 4 feet (1.2 m) in diameter. It is one of the hard pines with a range extending from the Atlantic to Kamchatka. The Norsemen of old knew it as the *fur,* and it is often called "fir" in Britain, Germany, and Scandinavia, although it is a true pine (*Pinus*), not a fir (*Abies*). It forms an important component of the forests of Germany and Poland as well as the taiga proper.

Swiss and German woodcarvings are famous. The best of them are made with the wood of the stone pine, a soft pine also known as the Swiss stone pine, Siberian stone pine, arolla pine, and, in Russia, the cedar. It grows to about 80 feet (25 m), sometimes more. As its names suggest, its range is wide. It occurs in the Alps at altitudes of 4,000–6,000 feet (1,200–1,800 m) and through the whole of the west Siberian plain and the basin of the Yenisei River (and it also occurs in America). Its seeds yield oil that is used locally for cooking and formerly for lamps.

In the Far East, the Scotch and stone pines are joined by several species with a very local distribution. These include the mourning pine (*P. funebris*), an endangered species, the Korean cedar pine (*P. korainensis*), and the Japanese dwarf pine (*P. mugo* or *pumilia*), a stunted tree that occurs on mountain ridges. The same hard pine also grows in the mountains of central Europe, where it is known as the mountain pine. It closely resembles Scotch pine but seldom grows larger than a bush with long stems that sometimes take root where they touch the ground. In German-speaking regions, it is called *Krummholz* (literally, "crooked wood") or *Knieholz* ("knee-[high] wood"). There is another tree that is also known as the mountain pine. This is *P. uncinata,* a much bigger tree, up to 80 feet (25 m) tall, that grows in the mountains of southern Europe from the Pyrenees to the Alps.

A little to the east, the Austrian pine (*P. nigra*) grows in Austria and the Balkans. A fine tree, often planted elsewhere for ornament, it grows to a height of 130 feet (40 m) or more. Its straight, fairly rigid needles grow in pairs and are 3.5–6.3 inches (9–16 cm) long. It has no value as a timber tree, but it is sometimes used in shelter belts and, because it is very bushy when young, as a nurse tree to protect seedlings of a more valuable species that is removed, once they have grown big enough to fend for themselves. The subspecies *P. n. maritima* is the Corsican pine, needles of which are 4.7–7.0 inches (12–18 cm) long and immediately recognizable because they are twisted. Corsican

pine is an important timber tree. *P. n. caramanica* is the Crimean pine. Its needles are the same length as those of the Corsican pine, but they are straight. As their names indicate, these two subspecies occur in Corsica and the Crimea respectively. Other species are native to the Balkans, including Bosnian pine (*P. leucodermis*), a tree up to about 100 feet (30 m) tall that also grows naturally in Italy, and Macedonian pine (*P. peuce*), a smaller tree growing to a height of about 65 feet (20 m).

At altitudes of 7,000–12,000 feet (2,100–3,700 m) in Bhutan, parts of Nepal, and Afghanistan, the Himalayan, Bhutan, or blue pine (*P. wallichiana,* a soft pine also known as *P. griffithii* and *P. excelsa*) usually grows about 115 feet (35 m) tall but can reach 165 feet (50 m). Its name *blue* refers to the bloom on its needles. These grow in bunches of five and are 4.7–8.0 inches (12–20 cm) long. The Chinese pine (*P. tabuliformis*), a tree about 80 feet (25 m) tall, grows in southern China and Korea, and in northwestern China there is the lace-bark pine (*P. bungeana*), a tree of similar size. Its common name refers to the way patches of its bark fall away, exposing the brightly colored wood beneath.

Austrian pine needles are very dark in color and earn the tree its alternative common name of black pine. There is also a Japanese black pine, or kuromatsu (*P. thunbergii*). Much cultivated for ornament, it grows to about 100 feet (30 m). There is also a Japanese white pine (*P. parviflora*), grown in many gardens. In the wild it grows to about 80 feet (25 m), but most cultivated specimens are less than half that size.

Aleppo pine (*P. halepensis*) is a fairly small tree, up to 50 feet (15 m) tall, that grows around the shores of the Mediterranean and in parts of the Near East including, of course, the area around the town of Aleppo in Syria. In Italy and elsewhere in the Mediterranean region, the Italian stone pine (*P. pinea*) is also known as the umbrella pine because of its shape. Since the days of ancient Rome and still today, it has been grown for its edible seeds, called pignons. These ripen when they are nearly four years old and are kept in their cones until needed to prevent the oils they contain from oxidizing and becoming rancid. At the western end of the Mediterranean and in North Africa, the cluster or maritime pine (*P. pinaster*) is a tree about 100 feet (30 m) tall that grows vigorously in coastal sand. It has been used extensively to stabilize sand dunes on the coasts of the Bay of Biscay and, to a lesser extent, in many other parts of the region and even in southern England, where it has become naturalized. On some sandy soils in France, cluster pines have provided protection for other trees, so whole forests have developed. Not only do these trees produce forests where there had been bare sand, but they also provide a useful source of income for local people because the cluster is a hard pine, rich in resin, from which high-quality turpentine is obtained.

Redwoods

Redwoods have been planted in many parts of the world, but they are native only to the Pacific coast of North America. There are only two living species, nowadays classified in separate genera in the same family (Taxodiaceae) as the swamp cypresses. The dawn redwood (*Metasequoia glyptostroboides*), native to southern China, is probably related, although some scientists believe it should be placed in a family of its own.

The coast redwood (*Sequoia sempervirens*) grows naturally in the "fog belt" along the coast from southern Oregon to Monterey, California. It is rarely found growing naturally more than 25 miles (40 km) from the coast or at an altitude higher than 3,500 feet (1,000 m). It produces seeds prolifically, but barely 25 percent of them germinate, and the trees reproduce mainly vegetatively by producing suckers. New trees grow readily from the stumps of fallen or felled trees.

The Sierra redwood (*Sequoiadendron giganteum*) grows on the slopes of the Sierra Nevada at altitudes of 3,000–8,500 feet (900–2,600 m). This is a tree with several common names. As well as Sierra redwood, it is known as the big tree, giant sequoia, mammoth tree, and wellingtonia, and it is so impressive that some individuals have names of their own. The General Sherman tree, for example, stands in the Sequoia National Park and has been measured in every dimension. It is 272.4 feet (83 m) tall, 27 feet (8 m) in diameter measured 8 feet (2 m) above ground level, and its circumference at the base measures 101.5 feet (31 m). It is estimated to weigh about 6,167 tons (5,595 tonnes).

It is a huge tree, but both redwoods are huge. Coast redwoods can grow to a height of 400 feet (120 m) and Sierra redwoods to 330 feet (100 m). They are also long-lived. Coast redwoods commonly live between 400 and 800 years or even longer, and Sierra redwoods were thought to be the most long-lived of all trees until the age of bristlecone pines was measured. One specimen of bristlecone pine was found to be 3,200 years old.

Both redwoods have a thick, spongy bark. The bark is almost impossible to ignite and thus protects the tree from fire.

Timber from the coast redwood is soft, fine-grained, and easy to work, and the size of mature trees means that the timber can be produced as long planks up to six feet (2 m) wide. Its popularity led to extensive cutting after about 1860, but through the efforts of conservationists, the rate of felling declined from the 1940s. Large forests remain, some of primary stands and others from regrowth. Sierra redwoods have survived better because their timber is more brittle, and so it is less in demand.

The name *sequoia* honors Sequoya (ca. 1760–1843). The son of a British trader and a Cherokee mother, Sequoya was born in the Cherokee town of Taskigi, Tennessee. His mother brought him up, and he never learned English. He became a skilled silver craftsman and mechanic, but after a hunting accident crippled him for life, he spent his days studying and thinking. Believing that the settlers derived their power from being able to write information, he devised an alphabet of 86 characters, representing every syllable in the Cherokee language. His alphabet was easy to learn and proved popular. Eventually it allowed books to be published in Cherokee, and in 1828 *The Cherokee Phoenix* newspaper appeared in Cherokee and English.

Spruces

Spruces (*Picea* species) are tall, mainly conical trees with horizontal or hanging branches at irregular intervals along the stem. When their needles fall (or are pulled away gently), a small peg is left behind on the stem. This distinguishes spruces from firs, which have round, flat scars where needles have been detached. Spruce needles grow singly, distinguishing them from pines in which the needles grow in bunches. Spruce cones hang down, fir cones are upright or at right angles to the stem, and pine cones are hanging, upright, or at right angles to the stem, depending on the species.

Spruces in which the branches hang and the smaller branches bearing the needles hang down below them have a distinctly weeping habit. Brewer's spruce (*P. breweriana*), also known as weeping spruce and siskiyou spruce, is a striking example, with branches that often touch the ground. A tree up to 130 feet (40 m) tall, it grows in the mountains of Oregon and northwestern California.

Brewer's spruce is one of seven North American species, four of which occur in the west. The map on page 199 shows the global distribution of this genus. At lower altitudes and along the coast from Alaska to California, Sitka (*P. sitchensis*) is the most common spruce. It is named after the town of Sitka in southern Alaska. It grows to 200 feet (60 m) tall and sometimes more, making it the largest North American spruce. It is commercially important, but probably less so than Engelmann spruce (*P. engelmannii*), a smaller tree, growing to a height of 65–165 feet (20–50 m). Engelmann spruce is very common in British Columbia, but it has a range extending from there and western Alberta to northern California and Montana at altitudes of about 1,500–12,000 feet (455–3,660 m). It lives to a good age, often for 500 years or more.

One variety of Engelmann spruce (called blue Engelmann spruce) has distinctly blue needles, a feature it shares with a variety of the Colorado spruce (*P. pungens*). Both blue varieties are called *glauca* (the Latin *glaucus* means bluish-green or gray). The blue Colorado spruce is called the blue spruce, and there is also a variety with pendulous branches called Kosteriana spruce. Colorado spruce grows up to 165 feet (50 m) tall, at altitudes rather lower than those preferred by Engelmann spruce, from the Yellowstone National Park to Idaho, Utah, and Arizona.

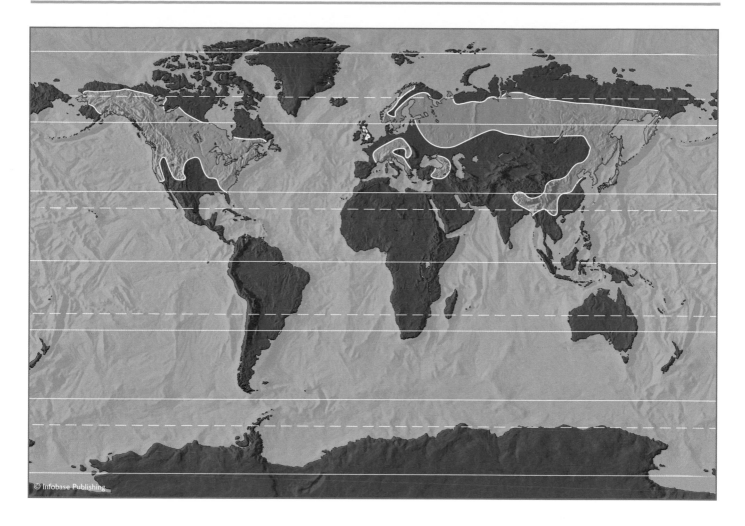

The global distribution of spruces (*Picea* species)

The eastern spruces are the red, the white, and the black, and all three are of great commercial importance. Red spruce (*P. rubens*) grows from Nova Scotia south to the Appalachians and Georgia to a height of 60–100 feet (18–30 m) or more, with trunks 60 cm (24 inches) in diameter. It is grown on plantations for its timber. As its botanical name suggests, white, or Canadian, spruce (*P. glauca*) has bluish needles. Growing to a height of 60–70 feet (18–21 m) and sometimes to 120 feet (37 m), it forms pure stands over large areas; elsewhere it forms forests with other coniferous trees, birch, aspen, or balsam poplar. It grows naturally in Alaska and across Canada, up to the northern limit for tree growth, in the east in Labrador, New York, and around the Great Lakes. It shares this range with black spruce (*P. mariana*), a rather larger tree, growing 65–105 feet (20–30 m) tall, which also has rather bluish needles. Spruce gum is obtained from black spruce.

All the North American spruces produce soft, light-colored, straight-grained, rather lustrous wood, containing very little resin and having no taste or odor. Spruce wood has

many uses, and because sounds make it resonate, sounding boards in pianos and the bodies of violins, guitars, and other stringed instruments are often made from it. It is also used to make chipboard, hardboard, and other types of board, and it is the best wood for making rayon and cellophane. Most spruce, however, is pulped to make paper.

Spruce occurs throughout temperate regions of the Northern Hemisphere and on mountainsides in lower latitudes. Like firs, these trees have dense foliage and cast a deep shade, making natural spruce and fir forests dark, gloomy places. Often they are eerily quiet, for their shelter means that there is little air movement near ground level.

Norway spruce (*P. abies*) is the most common species in much of the western part of the taiga, and it is recognizable by its thick, blunt needles, red-brown shoots, and long cones. This is the tree that is most widely used in northern Europe as a Christmas tree, especially the big trees erected in public open places, although in Germany the common silver fir (*Tannenbaum* in German) is preferred. The Christmas tree that stands every year in Trafalgar Square, London, is a Norway spruce donated by the people of Norway to the people of London. Norway spruce is a slender, conical tree that grows to a height of 165 feet (50 m). It yields timber that is widely used as "white wood" or "white deal." Purified, its resin is known as Burgundy tar, which is used in high-quality varnishes, and its leaves and shoots are processed to make Swiss turpentine or, processed in a different way, to make a spicy alcoholic drink known as spruce beer. Its bark is used in tanning leather.

East of the Urals, Siberian spruce (*P. obovata*) is more common. It is very similar to Norway spruce, and its timber is also widely used, but it has much smaller cones.

As its name suggests, Serbian spruce (*P. omorika*) occurs naturally in Serbia and Bosnia-Herzegovina, especially on limestone soils near the River Drina and less commonly in other parts of southern Europe. It is a narrow, almost columnar tree that reaches a height of about 100 feet (30 m). It is cultivated for ornament and grown for timber in northern Europe. Its branches are more or less horizontal, and its leaves are flattened with pointed ends, distinguishing the tree from most spruces, which have needles that are diamond-shaped in cross section. The needles are 0.5–0.7 inch (1.2–1.8 cm) long and in two rows on either side of the stem and almost at right angles to it, like the teeth of a comb (the technical term for this arrangement is *pectinate*).

Schrenk's spruce (*P. schrenkiana*) and oriental, or Caucasian, spruce (*P. orientalis*) replace Norway and Siberian spruce farther south in central Asia and the Caucasus. Both grow to 100–115 feet (30–35 m). Oriental spruce is cultivated in Europe and America as an ornamental and sometimes for timber.

South of the Kirghiz Steppe of central Asia, the Tian Shan mountain range extends for about 1,520 miles (2,450 km) from west to east. Its lower slopes are covered with broad-leaved forest, but above about 5,600 feet (1,700 m) this gives way to spruce forest that is dominated by dense pure stands of Tian Shan spruce (*P. tianshanica*).

To the south of the Tian Shan range, Himalayan spruce (*P. smithiana*) grows in the western Himalayas. Also known as west Himalayan spruce, morinda spruce, and khutrow, the Smith of its botanical name is Sir James Edward Smith (1759–1828). Carolus Linnaeus (Carl von Linné 1707–78) was the Swedish botanist who introduced the system of using two names to classify plants and animals. After he died, Sir James Smith bought his manuscripts and natural history collection and took them to London. When the Linnean Society was founded in 1788, Smith became its first president, and after his death the material from Linnaeus was bought by the society, which still owns it. Himalayan spruce is a large tree, usually growing to a height of 100–165 feet (30–50 m), but some grow much taller, and it is an important timber tree in forests at around 10,000 feet (3,000 m). It has a weeping shape, much like that of Brewer's spruce. An even taller tree grows at the opposite end of the Himalayas. Sikkim, or east Himalayan spruce (*P. spinulosa*), can attain a height of 200 feet (60 m).

To the east of the mountains, several spruces occur in western China. Chinese or dragon spruce (*P. asperata*) is an attractive tree, with rather blue foliage, that is often grown as an ornamental. In natural forests it can be 100 feet (30 m) tall, but most cultivated specimens are smaller. Likiany, or Likiang spruce (*P. likiangensis*) was first discovered near the town of Lichiang in Yunnan Province of western China, which gives it its name, and the tree also occurs in Tibet. It grows to a height of about 100 feet (30 m), and the outermost two rows of needles on each stem overlap like tiles (the technical term for this arrangement is *imbricate*) and point forward. Sargent spruce (*P. brachytyla*) also occurs there as well as in central China. A tree about 80 feet (25 m) tall, its needles are flattened (like those of Serbian spruce) and are bright green on the upper side and grayish white below.

Sakhalin is a large island off the eastern coast of Russia and immediately north of Hokkaido, the northernmost island of Japan. There, and in Japan, is where Sakhalin spruce (*P. glehnii*) occurs naturally, a tree that reaches 130 feet (40 m) in height.

Several species of spruce are native to Japan itself. Alcock spruce (*P. bicolor*) is similar to Sakhalin spruce, although smaller, reaching only about 80 feet (25 m). Both species, and also Likiany spruce, have needles that are diamond-shaped in cross section but somewhat flattened so that the diamonds are wider than they are high. Tiger-tail spruce (*P. polita*) can grow to 130 feet (40 m) and is planted for ornament in Europe and America. Its shiny, deep green, four-sided needles are 0.6–0.8 inch (1.5–2.0 cm) long, curved, stiff, and with a very sharp point rather like the claws of a cat. Yeddo spruce (*P. jezoensis*) grows to 165 feet (50 m) and occurs over much of northeastern Asia as well as Japan. Its needles, 0.4–0.8 inch (1–2 cm) long, are pointed but not sharp like those of the tiger-tail spruce. Hondo spruce is a variety (*hondoensis*) of this species with shorter needles that thrives better in cultivation.

Swamp Cypresses

Swamp cypresses comprise three, closely related species in the genus *Taxodium*, all of them growing naturally in the southern and southeastern United States and Mexico. Related to the redwoods (family Taxodiaceae), they are deciduous or evergreen in part of their range, with pale green leaves, often flattened, growing either in one plane to either side of their shoots or all around them. As the name suggests, they often grow in swamps or shallow rivers, but they will also grow in well-drained soil. One specimen of Mexican (or Montezuma) cypress (*T. mucronatum*), called El Gigante, is said to be still growing in the village of Santa Maria del Tule, where it was seen by Hernán Cortés, who wrote about it more than 450 years ago. This species grows to about 165 feet (50 m). It is evergreen in Mexico but deciduous in cooler regions.

Pond cypress (*T. ascendens*), also known as upland cypress, grows to about 80 feet (25 m) and occurs in the southeastern United States as far to the west as Alabama. The most common of the three is the swamp cypress (*T. distichum*), with pale green, feathery leaves. It occurs as far to the west as Illinois, grows to 165 feet (50 m), and its timber is used for outdoor products, such as garden furniture, because it resists attack by insects and damp conditions do not affect it. Both the pond and swamp cypresses are also known as bald cypresses.

Yews

Yews (*Taxus* species) differ from coniferous trees in not producing seeds in cones and in lacking resin canals in their wood and leaves. Instead of cones, their seeds are contained in a fleshy, scarlet covering called an aril. This is the only part of the plant that is not extremely poisonous to all mammals, and fallen leaves and twigs remain poisonous even when they are brown and withered. Because they are so dangerous, farmers clear yews from their land. They are most often found in graveyards, where they are not disturbed and can reach a great age (1,000 years or more), and in gardens, where they are grown for ornament. Cultivated varieties are often cut and trained into exotic shapes, and they make excellent tall, dense hedges. In places, however, they survive in forests. Their hard, durable wood is valuable but not easily obtained, and it is used for floor blocks, panels, and veneers in cabinetmaking. Traditionally yew was the preferred material for making longbows.

The common or English yew (*T. baccata*) occurs naturally throughout Europe, North Africa, and the Near and Middle East. It grows to a height of 40–65 feet (12–20 m) and has flat leaves, narrow but not quite needles, up to 1 inch (2.5 cm) long and usually in two rows, one on either side of the stem.

Japanese yew (*T. cuspidata*) occurs in the forests of eastern Siberia and China as well as in Japan. Its leaves are similar to those of the common yew but are not arranged in two rows. The tree grows to about the same size as the common yew and is sometimes grown in gardens. Chinese yew (*T. celebica*) reaches no more than about 40 feet (12 m), often less, and is barely more than a shrub. It occurs widely in China and Taiwan, and its range extends almost to the equator in the Philippines and in Sulawesi, Indonesia.

■ ANGIOSPERM TREES

The broad-leaved forests of temperate regions occur in latitudes immediately below those of the boreal forest, or taiga, with which they merge across wide belts of mixed forest comprising both broad-leaved and coniferous trees. A mixed forest is defined as one in which the majority of the trees are of one type, either coniferous or broad-leaved, but at least 20 percent are of another type. Mixed forests are extensive, especially in Russia, where they occupy a much larger area than those of the purely broad-leaved type.

In higher latitudes, the broad-leaved trees are deciduous, as an adaptation to the low temperature and dry conditions of winter when water in the upper soil is likely to be frozen. In lower latitudes many broad-leaved trees are evergreen, and in Mediterranean climates they are often sclerophyllous. Many of the broad-leaved trees in tropical and equatorial forests are close relatives of those found in temperate forests.

Alder

Not only do alders (*Alnus* species) tolerate cool weather and wet ground, they positively prefer these conditions. That is why they are found beside rivers and lakes where the soil is wet but their roots are able to remain above the water table. In some places alders form pure stands, called alderwoods, that are either isolated from other types of forest or enclosed by them to form distinct regions beside water and on wet, peaty soils in low-lying areas within the main forest. In Britain, alderwoods are sometimes known as carrs. Alders are small trees and shrubs with simple, alternate leaves. Their flowers are catkins, and their fruits are woody and look very much like small pine cones.

Alder trees are found in North America from the northern limit for tree growth in Alaska and Canada throughout the continent except for desert regions, through Central America, and down the western side of South America. In the Old World they occur from northern Norway, throughout all but the most extreme north of the taiga, and as far south as the Himalayas.

Probably the most widespread American species are the American green alder (*A. crispa*) and the Sitka alder (*A. sinuata*). The American green alder is a small tree, about 10 feet (3 m) tall, that grows in the mountains of the East from Labrador to North Carolina. Sitka alder is a taller tree, growing to about 43 feet (13 m). It is found in the West from Alaska—Sitka is an Alaskan town—to northern California. Red or Oregon alder (*A. rubra*) also occurs naturally in the West from Alaska to California, Idaho, and Oregon. It can attain a height of 80 feet (25 m). The "red" in its name refers to its bright red shoots.

Black or common alder (*A. glutinosa*) is the most widespread Old World species. It can reach a height of up to 115 feet (35 m), although it is usually smaller. It has rather rounded leaves that make it easy to distinguish from the gray, speckled, or European alder (*A. incana*), which has pointed leaves. As one of its common names suggests, the leaves are rather gray in color. Gray alder grows to about 65 feet (20 m) tall.

The Italian alder (*A. cordata*) occurs farther south in southern Italy and Corsica and is smaller still. It grows to about 50 feet (15 m). Like all alders, it is deciduous, but its leaves are glossy, which probably helps conserve moisture.

Some Asian species are big. *A. cremastogyne* and *A. lanata,* which grow in China, can reach 100 feet (30 m), as can *A. nitida,* of the western Himalayas.

Alder bark is sometimes used in tanning leather, and Native Americans used to hollow out the main stems of the red or Oregon alder to make canoes. Today the wood is used to make furniture and paper.

Ash

Ashes (*Fraxinus* species) are small trees and shrubs occurring from the Great Lakes to as far south as Mexico. They are also found throughout Europe and Asia, including the Indian subcontinent and Vietnam, and as far north as southern Scandinavia and the southern edge of the taiga. They also grow naturally in North Africa. Their leaves are opposite and their flowers usually have no petals. The fruit is a samara, which is a seed with a single wing—the common name for this type of fruit is a key—that is produced in bunches.

White ash (*F. americana*) is a tree up to 130 feet (40 m) tall that grows in eastern North America. It is cultivated as an ornamental, and, like most ashes, its wood is valuable for furniture and interior carpentry. Other species occurring naturally in the east are also known by their colors. Red or green ash (*F. pennsylvanica*) is a smaller tree, reaching no more than 65 feet (20 m). Blue ash (*F. quadrangulata*) is much the same size, and black ash (*F. nigra*) grows to about 100 feet (30 m). All four species are commercial sources of timber. Small amounts of wood are also obtained from the pumpkin ash (*F. profunda*), a small tree that grows in the coastal swamps of the southeastern states.

Oregon ash (*F. latifolia*), which grows in the northwestern United States, is also exploited locally for timber. It reaches a height of about 80 feet (25 m). Arizona or velvet ash (*F. velutina*), another western species, is a tree 25–40 feet (8–12 m) tall with twigs and leaves that are densely covered with fine hairs.

Two-petal ash (*F. dipetala*) is uncommon in producing flowers with two showy petals. It is a shrub that is about 13 feet (4 m) tall that grows in California. Fragrant ash (*F. cuspidata*) also has flowers with petals, and in this case the flowers are heavily scented. It grows in Texas and New Mexico. Single-leaf ash (*F. anomaea*), native to the drier parts of the southwestern United States, has leaves that are reduced to single leaflets.

The ash is one of the most important trees in European folklore (see "Forests in Folklore and History" on pages 236–241). According to some myths, humans are made from ash wood. The cultural significance of ash may be due, at least partly, to its usefulness. Its wood is light, hard but springy, warps very little, and is durable. Spears, tool handles, and wagons are made from it, and it also burns well. The traditional yule log is of ash.

Common or European ash (*F. excelsior*) is probably the most famous Old World representative of the genus. It is a big, handsome tree, up to 150 feet (45 m) tall that grows naturally throughout Europe, including Britain. It is easily recognizable in winter by its black leaf buds, which stand out clearly against the pale gray bark. Later in the year, like all ashes, it produces bunches of winged seeds—keys. The Mediterranean and North African species is the narrow-leaved ash (*F. angustifolia*). This species is smaller than the common ash, growing to about 80 feet (25 m), and it has darker, rougher bark.

Manna is the yellow, sweet-tasting sap of the flowering ash or manna ash (*F. ornus*), a tree up to about 65 feet (20 m) tall that grows around the Mediterranean, including North Africa. In Italy and Sicily it has traditionally been used as a mild laxative, given to children. Himalayan manna ash (*F. floribunda*) is a tree twice the size of the Mediterranean species that also yields manna. This is only one of the medicinal uses of ash. At one time the bark of the common ash was steeped in both hot water and the liquor used to treat liver complaints such as jaundice, and the ashes from the fire were used for scalp infections.

Aspen, Cottonwood, Balsam Poplar, Poplar

Aspens, cottonwoods, balsam poplars, and poplars all belong to the genus *Populus,* and they occur throughout the temperate regions of the Northern Hemisphere from the Mediterranean and the Himalayas to the Arctic Circle. They grow quickly into medium-sized or large trees. Their flowers are catkins that appear before the leaves, and the fruits are capsules with long silky hairs at their bases. These look rather like cotton bolls, which is why some species are known in America as cottonwoods.

Populus trees are divided botanically into four types or subgenera of which one occurs naturally only in China. Those found in North America are the white and gray poplars and aspens (subgenus *Populus*); the balsam poplars (subgenus *Tacamahaca*); and the black poplars and cottonwoods (subgenus *Aegiros*). *Populus* trees hybridize readily within their own groups but less commonly with members of other groups. Wood from poplars is light, strong, odorless, and fairly nonflammable. It is used to make food containers, brake blocks for railroad cars, and matches, and long ago it was the best wood for making arrows.

Quaking or American aspen (*P. tremuloides*) has the largest range of any North American tree, occurring from Alaska in the West and Labrador and Newfoundland in the East, south to Pennsylvania, Missouri, and Nebraska, and extends into the Sierra Nevada and Mexico. Its leaves grow on flat stems, which makes them tremble in the slightest breeze. It is a large tree, growing to a height of 100 feet (30 m). Bigtooth aspen (*P. grandidentata*) is a slender tree growing about 55 feet (17 m) tall with a name that refers to its leaves, which feel firm to the touch and have a row of large serrations along both edges. The tree is native to the northeastern United States and southeastern Canada.

In the second group, the balsam poplar or tacamahac (*P. balsamifera*) occurs naturally in the north from Labrador to Alaska and south across the northern United States. It is a big tree, up to 100 feet (30 m) tall. Balsam is a resinous gum, sometimes used as an ointment, which has a strong fragrance. The balsam poplars exude the gum from their leaf buds, releasing a pleasant odor in spring.

The balm of Gilead, which grows wild in the northeastern United States and southeastern Canada and was once very popular as an ornamental, is a natural hybrid (*P. × candicans*) of uncertain origin but possibly between the balsam poplar and the eastern cottonwood. On the western side of the continent, from Alaska south to California, the western balsam poplar or black cottonwood (*P. trichocarpa*) is a quick-growing tree, up to 120 feet (37 m) tall or sometimes to as much as 200 feet (60 m). Its timber is used for veneering and pulping.

Cottonwoods proper, comprising the third group, are trees with more or less triangular leaves. The group includes the eastern cottonwood or necklace poplar (*P. deltoides*), a large, broad tree, growing to a height of 150 feet (45 m). Although native to the east, its range extends from Québec to Montana and south to Texas and Florida. It is an important source of timber and is planted extensively along city streets. Many ornamental varieties and hybrids are cultivated.

In the Mississippi valley and the coastal swamps of the southeast, swamp cottonwood (*P. heterophylla*) is a tree growing up to 90 feet (27 m) tall. Farther west, in Texas and New Mexico, Fremont cottonwood (*P. fremonti*) grows along riverbanks and around water holes. The plains cottonwood (*P. sargenti*) grows in the central United States; farther west and into the Rocky Mountains, its place is taken by the narrowleaf cottonwood (*P. angustifolia*).

Leucoides, the fourth subgenus, comprises four species of trees with rough, scaly bark and round or square petioles (leaf stalks). *P. lasiocarpa* is typical. This is the Chinese necklace poplar, native to central and western China and cultivated elsewhere as a novelty. It grows up to 70 feet (22 m) tall and has gray–green bark. Its branches are rather widely spaced, giving the tree the appearance of being sparsely covered. The Chinese necklace poplar is the only poplar to bear male and female flowers on the same tree. The male catkins are red, and the females yellowish green; sometimes both are on the same catkin, the male at the base and the female, or a bisexual flower, at the top.

Lombardy poplar (*P. nigra* var. *italica*), a tree that can reach a height of 100 feet (30 m) but looks taller because its branches all point upward, is instantly recognizable by its shape, which makes it possibly the most famous of poplars. This growth habit, called *fastigiate,* is as a freak in many trees but is difficult to propagate. In the case of the Lombardy poplar, propagation is easy because the tree grows well from cuttings, and that, in fact, is how Lombardy poplars are grown; almost all of them are male. The tree originated in northern Italy, and Lord Rochford brought the first cuttings to England in 1758. The tree is now grown widely but only for ornament or screening, for the timber is of little value.

Black poplar (*P. nigra*), from which the Lombardy poplar is derived, grows naturally throughout Europe, including Britain, and Southwest Asia. It is a big, rather cylindrical tree up to 115 feet (35 m) tall. Its bark is brown and rather rough, but the tree is called black to contrast it with the white poplar, or abele (*P. alba*), a tree up to 65 feet (20 m) tall that has pale, silvery bark with darker diamond-shaped marks. Gray poplar (*P. canescens*) has ridged, dark gray bark and grayish white leaves. It grows vigorously, reaching nearly 130 feet (40 m), and is found through most of Europe.

Some botanists believe the gray poplar to be a hybrid between the white poplar and the aspen (*P. tremula*), which is also a European species. Aspens reach a height of about 65 feet (20 m) and grow readily from suckers, so they often form dense stands. Their oval or almost round leaves with wavy edges are attached by long, slender, flattened stalks that allow the leaves to move in the slightest breeze, making the entire tree look as though it is trembling; that is how it earns the name *tremula.*

Basswood, Lime, Linden

Basswoods, limes, and lindens all belong to the genus *Tilia,* about 30 species of fairly tall trees that are distributed throughout the temperate regions of the Northern Hemisphere, except for the drier parts of central Asia. They are recognizable by their leaf buds and flowers. Each bud is protected by two scales, one much bigger than the other, like a finger and a thumb, and each flower grows on a long stalk that is attached to its bract for about half the length of both bract and stalk. They have broad, toothed leaves borne on long stalks. *Tilia* trees thrive on moist soil that is neither very acid nor alkaline. Their wood has many uses and is especially popular for making piano keys.

Basswood, or American basswood, American lime, or American linden (*T. americana*) occurs naturally in the northeastern United States and southeastern Canada but has been planted in the streets and squares of cities over most of America. It grows as a conical tree to a height of about 70 feet (21 m) and then broadens to produce a dome-shaped crown with big leaves and about 4.7 × 4.0 inches (12 × 10 cm). White basswood (*T. heterophylla*) also grows in the East. It may be a variety of *T. americana*, but its leaves are somewhat smaller.

Small-leaved lime (*T. cordata*), a tree up to 115 feet (35 m) tall, occurs throughout Europe as far east as the Caucasus, and as its name suggests, its leaves are smaller than those of the broad-leaved or large-leaved lime (*T. platyphyllos*), a slightly smaller tree occurring over a similar range. In Britain, the two have hybridized to produce the common lime or linden (*T.* × *europea*). This is a very big tree, growing to 145 feet (45 m) or more, with a roughly conical shape and rather open, domed crown.

Caucasian linden, Caucasian lime, or Crimean lime (*T. euchlora*) is a smaller tree, reaching about 65 feet (20 m). It is closely related to *T. cordata* but has bigger leaves; some botanists suspect that it may be a hybrid between *T. cordata* and another Caucasian species. Silver pendent lime or weeping silver lime or weeping white linden (*T. petiolaris*) is often grown in parks and gardens and may come originally from the Caucasus, although no one can be sure. It grows to more than 100 feet (30 m). Silver lime (*T. tomentosa*) is a tree of similar size that is native to southeastern Europe and south-western Asia, although it is planted far from these areas. Mongolian lime (*T. mongolica*), from northern China and Mongolia, and Oliver's lime (*T. oliveri*), from central China, both fairly tall trees, are also grown as ornamentals.

Beech

Beeches (*Fagus* species) occur naturally in the eastern United States and part of southeastern Canada. In the Old World they occur naturally only in Europe and the Near East and in China and Japan. The map shows the global distribution of beeches. Where they do occur, beeches are often the dominant or codominant species in Old World forests.

The global distribution of beeches (*Fagus* species)

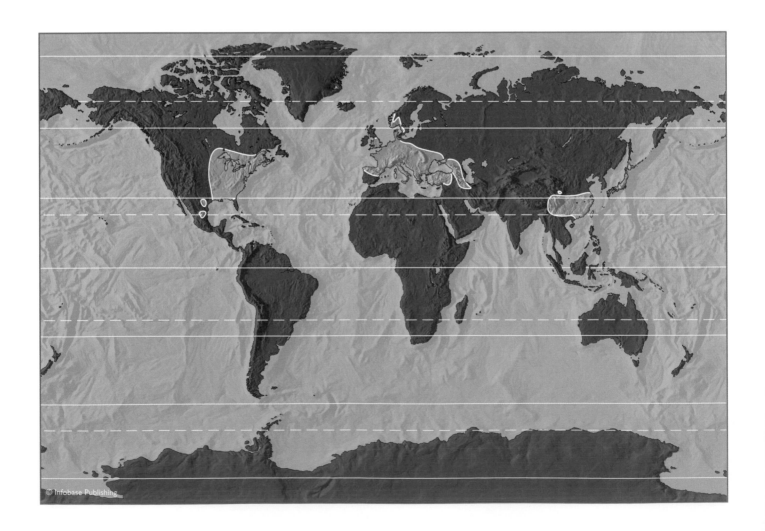

There is only one species native to North America. American beech (*F. grandifolia*) grows to about 80 feet (25 m) tall. Its leaves are larger than those of most *Fagus* species and have clearly marked serrations around the edges. The leaves help in the identification of American beech where this species grows near cultivated beeches from other parts of the world.

Julius Caesar wrote that beech trees did not grow in Britain. In fact they do, and their pollen has been found in soils dated to pre-Roman times, so scientists believe that Caesar was referring not to beech but to the sweet chestnut, which the Romans introduced. The Romans may have called the sweet chestnut *Fagus* rather than *Castanea,* the name by which botanists know it now. The most widespread European species is the common or European beech (*F. sylvatica*). It is a handsome tree, up to 100 feet (30 m) tall, that yields a valuable timber as well as nuts that can be fed to pigs and from which oil can be extracted. In the Balkans and eastward to the Caucasus and Asia Minor, the common beech gives way to the oriental beech (*F. orientalis*). The two are very similar, but the oriental beech grows only on sites that are more sheltered than those the common beech will tolerate.

Chinese beech (*F. englerana*) is a small tree, up to about 50 feet (15 m) tall, that occurs in central China, together with *F. lucida* and *F. longipetiolata,* the commonest species in the east and west respectively. Siebold's beech (*F. crenata*), a tree up to 100 feet (30 m) tall, grows in Japan, where it is an important timber tree. Japanese beech (*F. japonica*) is somewhat smaller.

Birch

Birches (*Betula* species) are deciduous trees that grow naturally throughout the Northern Hemisphere, the ranges of some species extending into the Arctic and southern Greenland (Kalaallit Nunaat). There are about 50 species, and all of them have pale, almost white bark that becomes rough and develops dark patches in older trees. The flowers are wind-pollinated catkins (sometimes called lambs' tails) that emerge at the same time as the leaves. The seed is a little nutlet with two wings. Birches grow readily from seed and are often the first trees to colonize disturbed ground. In northern parts of its range, birch forms extensive stands and is an important component of many forests.

Many birches have very attractive bark, especially the paper birch or canoe birch (*B. papyrifera*), which has bark that peels off in layers like sheets of paper. It is the most widespread of all North American birches, occurring throughout the northern parts of the continent from Labrador to Alaska and to New England, Nebraska, and Washington. Its height varies from about 50 feet (15 m) to 100 feet (30 m), and it is widely planted for ornament, in the Old World as well as the New. Its bark is waterproof and was used by Native Americans for covering canoes. The wood is also used for roofing, for fuel, and for making a variety of small articles.

Sweet birch, also called black birch and cherry birch (*B. lenta*), grows in eastern North America and is a source of timber, used mainly for flooring and furniture in Pennsylvania and West Virginia. The name *sweet birch* refers to the sweet-smelling oil that is obtained from its bark and wood. Its bark is very dark, which is why it is called the black birch, and reminiscent of the bark of a cherry tree. It grows to about 80 feet (25 m).

As a timber tree, yellow birch (*B. alleghaniensis,* formerly known as *B. lutea*) is much more important. Its wood has many uses, including flooring, furniture, and boxes. During World War II, the frames of the wings and fuselage of the Mosquito fighter bomber, which remained in service with the Royal Air Force for many years, were made from it. The tree grows to 100 feet (30 m) and grows naturally from Newfoundland to western Minnesota and south to Georgia. River birch (*B. nigra*), a tree of similar size, grows in the central and eastern United States and is the only birch to grow at low altitudes in the southern states, where it is found on wet ground and riverbanks. White or gray birch (*B. populifolia*) is a small tree, reaching 20–40 feet (6–12 m), that grows in northeastern North America.

Dwarf birch (*B. nana*) is the most northerly species. No more than a shrub, up to 40 inches (100 cm) tall, dwarf birch grows as part of the tundra vegetation in northern Scandinavia and Siberia and on mountainsides as far south as Scotland.

Through most of the Old World range, the most widespread birch trees are the silver birch and downy birch. Silver birch (*B. pendula*), also called European birch, common birch, and warty birch, grows throughout Europe, northern Asia, and North Africa. The *pendula* of its name refers to way the tips of its branches hang down, unlike those of downy birch (*B. pubescens*), and this is the easiest way to distinguish between the two species at a distance. *Silver* refers to the color of the bark and *warty* to the small, pale, wartlike bumps on its branches. These provide one of the two other ways to tell one species from another because the branches of the downy birch have no warts. In addition, the young twigs of silver birch are smooth, and those of downy birch (sometimes called the hairy birch) are covered in the hairs that give the species its common name. Both are handsome trees. Silver birch is the larger of the two, growing to about 80 feet (25 m); downy birch is about 65 feet (20 m) tall. An oil obtained from the bark of downy birch is used in Russia in the tanning of leather and gives Russian leather a characteristic smell.

Himalayan birch (*B. utilis*), native to China and the Himalayas, is about the same size as downy birch and has much hairier twigs. All three of these species produce useful timber.

Box

There are about 30 species of boxes (*Buxus*) distributed around the Northern Hemisphere in regions with a Mediterranean climate and through the southern fringes of the broad-leaved deciduous forest. Boxes grow naturally across central and southern Europe and on dry soils as far north as southern England, most famously as a grove on Box Hill in Surrey, a place named for it. Boxes also grow on hills in eastern France and on limestone soils in the Jura region on the border between France and Switzerland. *Buxus* species grow naturally in Central America and the West Indies but not in North America.

Box is interesting because it is a broad-leaved evergreen with small, shiny, leathery leaves that are adapted to dry conditions. Left to grow, the common box (*B. sempervirens*) is a small tree 20–30 feet (6–9 m) tall. More usually, however, it is found in the form of a hedge or a small tree cut to an ornamental shape. Its tolerance of close clipping has allowed it to be used for the dense, square-edged hedges, sometimes less than 40 inches (100 cm) high, that are used to produce the geometrical shapes of formal gardens. The wide range of leaf shapes and colors provided by many cultivated species and varieties enhances the appearance of these hedges. Farther south, Balearic box (*B. balearica*) takes the place of common box. It is very similar to common box but is a little taller and has bigger, less shiny leaves.

Box produces a very heavy hardwood that can be polished to give a fine finish. It has many uses and is especially popular with woodturners and engravers. Traditionally, boxwood was used to make woodblocks for printing finely detailed engravings by a technique introduced in about 1800 by the English artist and naturalist Thomas Bewick (1753–1828).

Buckeye, Horse Chestnut

Buckeyes or horse chestnuts (*Aesculus* species) are deciduous trees and shrubs that occur naturally throughout the United States. They also grow naturally in southeastern Europe and parts of southern Asia and Japan. The *buckeye* of the American name is the seed, which is large and shiny. It is known in Britain as a "conker" and is used in a traditional children's game that involves threading the seeds on string and striking one conker with another to see which breaks first.

Red buckeye (*A. pavia*) grows to about 13 feet (4 m). Native to the southeastern United States, it is cultivated widely as an ornamental, partly for its flowers and partly for its foliage, which turns a brilliant red at the end of the summer. Sweet buckeye or yellow buckeye (*A. flava* or *A. octandra*) grows in the same region but is a more substantial tree, growing to 100 feet (30 m) tall. Ohio buckeye (*A. glabra*) is a tree that grows naturally up to 26 feet (8 m) tall in the central and southeastern United States. In the West, the California buckeye (*A. californica*) is a tree that grows up to 40 feet (12 m) tall.

The European horse chestnut or conker tree (*A. hippocastanum*), a handsome tree that grows up to 115 feet (35 m) tall or more, is recognizable by its large leaves composed of seven or, less commonly, five leaflets. These lack individual stalks, but the leaf as a whole has a single, long stalk. A native of Albania and Greece, it has been planted widely elsewhere. It was introduced to Britain in 1616 and is now naturalized. The name *horse chestnut* is believed to refer to the fact that in Turkey the seeds, closely resembling those of the sweet chestnut, were used to treat ailments in horses; other livestock will not eat them.

Moving eastward from Greece, horse chestnuts next appear again naturally in the northwestern Himalayas, where the Indian horse chestnut (*A. indica*) is found. This is also a big tree, growing to 100 feet (30 m) or taller. The Chinese horse chestnut (*A. chinensis*), up to 90 feet (28 m) tall, grows in northern China. The Japanese horse chestnut (*A. turbinata*) is also tall, reaching a height of 100 feet (30 m). All species of horse chestnuts are grown for ornament, but their wood is of little commercial value.

Cherry

Cherries belong to the genus *Prunus,* as do plums, apricots, peaches, and almonds, and there are about 400 species in the world as a whole. The pin cherry or wild red cherry (*P. pennsylvanica*) grows naturally in North American forests. Despite its name, it has a range extending from the Atlantic to the Rockies. It is a tree that grows up to 40 feet (12 m) tall.

Most cultivated cherries are descended from the wild cherry, also known as the gean or mazzard (*P. avium*). This is a tree that reaches up to 65 feet (20 m) tall or more and is found in Europe, parts of Russia, and North Africa as a component of broad-leaved forests. It is also grown for its valuable timber, as is the bird cherry (*P. padus*), though to a lesser extent. In the Balkans, the cherry plum or myrobalan (*P. cerasifera*) is a tall shrub cultivated for its fruits that now occurs more widely, having escaped from cultivation.

Chestnut

Chestnuts (*Castanea* species) grow naturally in the eastern and southeastern United States, but the American chestnut (*C. dentata*), once an important component of many forests, is now uncommon, having been almost destroyed by chestnut blight (see "Tree Predators and Parasites" on pages 173–180). Chestnuts also grow naturally in southeastern Europe and western Asia, in eastern China, Korea, and Japan. They are deciduous trees that grow quickly, often to a large size, and live for a very long time.

The sweet, or Spanish, chestnut (*C. sativa*) comes originally from the eastern Mediterranean and was introduced to northern Europe, including Britain, by the Romans. The name *Spanish* may refer to the chestnuts that were being imported to Britain from Spain at about the time the tree was being introduced, but the true origin of the name is lost. The tree itself grows up to 130 feet (40 m) tall with a trunk that can be 40 feet (12 m) in circumference.

Several species of chestnuts occur in China, the most important being the Chinese chestnut (*C. molissima*). It is a tree that grows up to 65 feet (20 m) tall and is found on high ground at altitudes up to 8,200 feet (2,500 m). The Japanese chestnut (*C. crenata*), also found on high ground but not usually above about 3,000 feet (900 m), is a tree that reaches up to 33 feet (10 m) tall. Both of these produce edible nuts, although they are smaller than those of *C. sativa*.

Crab Apple

Apple trees (*Malus* species) grow naturally in temperate forests of the Northern Hemisphere. Cultivated apples are grown to either side of this geographic belt and also in the Southern Hemisphere. There are probably less than 30 species in all, but they have been hybridized and selected for cultivation for so long that those few species have given rise to about 1,000 cultivated varieties. Some of these have escaped, established themselves in the wild, and then crossbred among themselves and with wild apples, so the position now is somewhat confused. Wild apples and escaped cultivars are both commonly known as crab apples.

The wild crab apple (*M. sylvestris*), a tree that stretches up to about 30 feet (9 m) in height, grows throughout Europe and is common in Britain. Its fruits are very acid but are used to make conserves, and the wild crab provides the rootstock for most cultivated apples. In Siberia, northern China, and eastern Asia the wild apple that occurs naturally is the Siberian crab (*M. baccata*). It is taller than the European apple, sometimes reaching about 50 feet (15 m). There are many varieties of both these species, and both of them are ancestors of domesticated apples.

In northern Italy there is the hawthorn-leaved crab apple (*M. forentina*), a small tree with leaves resembling those of the hawthorn. In the Himalayas, Assam, China, and Japan, the Hupei apple (*M. hupehensis*) has fragrant flowers and leaves from which a tisane (herbal tea) is made.

Dogwood

Dogwoods (*Cornus* species) occur naturally over the eastern half of North America south of the Great Lakes and in the west from northern California to British Columbia. In the Old World they grow naturally in western Europe and in central and eastern Asia.

Flowering dogwood (*C. florida*) grows up to 23 feet (7 m) tall in the understory of forests in the east of North America. Nuttall's or Pacific dogwood (*C. nuttalli*) grows in the west. Most specimens are up to 50 feet (16 m) tall, but some reach 100 feet (30 m).

Common dogwood (*C. sanguinea*) is a shrub or a small tree that grows up to about 13 feet (4 m) tall in the understory or shrub layer of broad-leaved forests throughout Europe, including southern England. The *sanguinea* in its name refers to the blood-red color of its leaves in the fall. Cornelian cherry (*C. mas*) bears attractive yellow flowers that appear before the leaves in early spring, and scarlet, oval-shaped fruits resembling cherries that can be made into jam and syrup. It is a small tree, up to about 25 feet (8 m) tall, which grows in central and southern Europe but is cultivated more widely for ornament.

Bentham's cornel (*C. capitata*) also produces attractive fruits, in this case resembling strawberries rather than cherries. It is a tree up to 45 feet (14 m) tall that is found in the Himalayas and in China. Over much the same range but extending into Japan, the table dogwood (*C. controversa*) is a little bigger. It produces attractive foliage, sometimes purple, in the fall. *C. kousa*, growing about 20 feet (6 m) tall, produces attractive flowers as well as strawberrylike fruits. It occurs in China, Korea, and Japan.

Elder

Elders (*Sambucus* species) are small trees or shrubs that occur naturally over Europe south of Scandinavia, in North Africa, and in a belt running from the eastern Mediterranean roughly northwestward across Asia north of the Himalayas to eastern Siberia north of Kamchatka. They grow as understory species in floodplain forests and pine forests.

The most widespread European species is the common elder (*S. nigra*), which grows to about 25 feet (8 m). Cordials as well as a white country wine are made from the flowers and a red cordial and wine from the berries. Made well, these can be of high quality, and the red wine was once used to dilute imported red wines. Elder wood is soft and easily hollowed to make a flute, and it was being used in this way thousands of years ago.

Elm

Elms have occurred naturally throughout the temperate regions of the Northern Hemisphere for at least the last 65 million years, and during that time they have changed very little. The last of the dinosaurs may have gazed up on them!

One way to tell whether a tree grows naturally in an area is to look for it among place names, and to do that it is first necessary to find its name in the local language. The wych elm or Scotch elm or mountain elm (*U. glabra*) is a true native of Britain. In Gaelic (the native language of the Scottish Highlands), the elm is *leamhan* (pronounced "leven"). There is a Loch Leven, with Kinlockleven beside it, two towns called Leven (one of them in northern England), as well as Levencorroch on the Isle of Arran off the Ayrshire coast. In fact, wych elm grows right across Europe and Asia, as far as Korea and on mountainsides farther south. It is a big, spreading tree, up to 130 feet (40 m) tall. The name *wych* means "supple" or "pliant" and is used in this sense in the name of witch (wych) hazel.

Elms can be recognized by their leaves. These are lopsided, so either the stalk, the tip, or both seem bigger on one side than the other. The edges of the leaves have serrations resembling saw teeth. This is a common feature, but the teeth of elm leaves are double (the technical term is *bidentate*).

Elms (*Ulmus* species) have suffered badly from Dutch elm disease (see "Dutch Elm Disease" on page 174), although this may have affected North America less severely than it did Britain. Nevertheless, American elm, or white elm (*U. americana*), is no longer so familiar a shade tree as it once was on American streets, squares, and campuses. It is a big, handsome tree, reaching a height of 130 feet (40 m), native in southeastern Canada and throughout the United States to the east of the Rockies. Many individual trees have been named for famous people. The American elm is an important source of timber, used for many purposes.

Slippery elm (*U. rubra*) is about half the size of American elm, but its timber is also used. It grows naturally within an area defined by a line drawn from Quebec to North Dakota and from there to Texas. The *slippery* of its name refers to a fragrant, mucilaginous substance with therapeutic properties obtained from its inner bark. Branches and twigs of the cork or rock elm (*U. thomasii*) have flanges or "wings" of cork. This tree grows to about 100 feet (30 m) and occurs naturally in the northeastern United States and eastern Canada. It yields hard, dense wood.

Other American elms occur naturally in the forests of the southeastern states and Mexico, extending into northern Colombia, and all have branches with wings of cork. These are smaller trees than those of the north. The wahoo or winged elm (*U. alata*) grows to about 50 feet (15 m) and is usually found on dry, upland sites. The cedar elm (*U. crassifolia*) and September elm (*U. serotina*) prefer moist soils. Both trees grow to about 65–80 feet (20–25 m). Most elms flower in spring, but these two both flower in the fall, which explains the *September* in one name. The Mexican elm (*U. mexicana*) is a tree that grows up to 65 feet (20 m) tall, found throughout Central America and into northern Colombia.

Until they were devastated by Dutch elm disease, elms were among the most common of European trees, largely because they grew in hedgerows as well as in forests, so they were visually prominent in the countryside. The elm that is most often seen in hedgerows or standing alone in a field is the English elm (*U. procera*), a tree about the same size as the wych elm.

Elms are very variable, and the best way to distinguish one species from another is by the overall shape of the full-grown tree. The English elm produces foliage on branches that grow almost all the way down its trunk so that the shape is rather cylindrical. Wych elm is more spreading and is shaped a little like a mushroom. The European field elm or smooth-leaved elm (*U. minor*) is rather shorter, reaching about 100 feet (30 m), and it is difficult to identify because there are very many local varieties. These grow naturally from England to central and southern Europe, North Africa, and the Near East, and where they grow they are often the most common elm, so there are few others with which to compare them. There is also a very common elm hybrid, being a cross between the wych elm and European field elm, with many varieties. The Dutch elms belong to this group.

In eastern Europe and as far north as Finland, there is the European white elm (*U. laevis*), a tree that grows to 100 feet (30 m) tall and very variable in shape. It is also known as the fluttering elm because its flowers are borne on very long stalks and tend to flutter in the wind. As with all elms, the flowers open in February or March, long before the leaves. In central and eastern Asia, the Siberian elm (*U. pumila*) is a smaller tree, up to 80 feet (25 m) tall, which bears its leaves on very long stalks.

Elms may have appeared first in China and spread to Japan, Korea, and then westward from there. More species occur naturally in China and the Himalayas than elsewhere. Most have no English common names, although there is a Chinese elm (*U. parvifolia*), which grows in southeastern Asia, Korea, and Japan as well as in China. It differs from most elms by flowering in the fall. In the south of its range it is almost evergreen, keeping its leaves through the winter and losing them in the spring.

Caucasian elms (*Zelkova* species) belong to the same plant family as the elms (Ulmaceae), and like the *Ulmus* elms, the bigger ones are important timber trees, producing very hard wood. They occur naturally in two areas, from the eastern Mediterranean eastward to the south of the Caspian Sea and in eastern Asia from Kamchatka and eastern Siberia through Japan and eastern China, but they are grown for ornament in parks in many other parts of the world. The Caucasian elm or Siberian elm (*Z. carpinifolia*), a native of the Caucasus Mountains, is rather cylindrical in shape with a very dense mass of branches and reaches a height of 80 feet (25 m). Cretan zelkova (*Z. abelicea*), native to Crete, is smaller, growing no more than about 50 feet (15 m) tall, and the cut-leaf zelkova (*Z. verschaffeltii*), also from the Caucasus, is also barely taller than a shrub.

The biggest of the Asian species is the keaki or keyaki, or Japanese zelkova (*Z. serrata*), a Japanese tree that reaches a height of 130 feet (40 m) in the wild but much less than that in cultivation. Its timber is used for building, and the tree is also popular for bonsai cultivation. Chinese zelkova (*Z. sinica*) grows to no more than about 50 feet (15 m).

Eucalyptus

Eucalyptus (*Eucalyptus* species) are broad-leaved evergreen trees and shrubs that are native to Australia, with a few species occurring in New Guinea, Indonesia, and the Philippines. This makes them predominantly tropical and subtropical, but the genus comprises about 450 species, and some are grown in temperate regions for ornament or timber. They are the principal timber trees and the most characteristic trees and shrubs of Australia, but they do not grow naturally in New Zealand. Eucalyptus range in size from shrubs about three feet (90 cm) tall to the mountain ash (*E. regnans*) of Victoria, which is the tallest of all flowering plants. Mountain ash can grow to a height of more than 330 feet (100 m).

Eucalyptus are the dominant trees in most natural forests in Australia from the tropical forests of the north to the temperate forests of the south and Tasmania. There is no other region of the world where a single plant genus dominates the forest canopy to this degree.

This large genus is divided into six main groups that are recognizable by the appearance of their bark. Gums have smooth bark that peels away from the trunk. Bloodwoods have rough, scaly, and flaky bark. Boxes have rough, fibrous bark. Peppermints have bark consisting of fine fibers. Stringybarks have bark consisting of long fibers. Ironbarks have hard bark, dark in color and with deep fissures. These names are not used consistently, however, and unrelated species may have the same common name in different parts of Australia.

Eucalyptus are adapted to the fires that from time to time sweep through the Australian forests. In some species fire destroys the trees, but their seeds survive in the ground. Most species, however, survive by growing new shoots from resting buds in the bark once the fire has died down.

Jarrah, messmate, and blackbutt are especially important as sources of timber in Australia, and they are big trees. Jarrah (*E. marginata*), native to Western Australia, is probably Australia's most widely exploited species, and large amounts of its timber are exported. Jarrah is a big tree, growing to a height of 80–115 feet (25–35 m). Both messmate (*E. obliqua*) from southeastern Australia and blackbutt (*E. pilularis*) from eastern Australia are even taller, growing to about 200 feet (60 m). Mountain ash is also commercially important for its timber and as a source of wood pulp.

There are other less important timber species, some of which are even bigger. Alpine ash (*E. delegatensis*) grows to 165 feet (50 m) or more, and karri (*E. diversicolor*) from western Australia can reach 245 feet (75 m). Tasmanian blue gum (*E. globulus*) from southeastern Australia and Tasmania grows to about 150 feet (45 m) and is cultivated for ornament in many places around the Mediterranean and in California.

Gum

Gums are deciduous trees that form an important component of forests in the southern United States and Central America. There are two distinct types. Sweet gum (*Liquidambar styraciflua*), found in the southeastern states, Mexico, and Guatemala, is a deciduous tree growing to about 150 feet (45 m) in the wild but barely one-third that size when cultivated, which it is often is for its magnificent fall colors. Its valuable timber is sometimes called satin walnut.

The other gums, also known as tupelos, belong to the genus *Nyssa*. The best known is the black gum, tupelo, or pepperidge (*N. sylvatica*), a tree somewhat similar to an oak. It grows to about 100 feet (30 m) in swamps or moist soil on hillsides and is much cultivated for its fall foliage. Water gum, twin-flowered nyssa, or tupelo (*N. biflora*) is similar but only about half the size, and *N. ursina*, the third member of the group in which the flowers are borne in groups of two or more, is a shrub.

The second group, in which flowers are borne singly, contains three species. The cotton gum or water tupelo (*N. aquatica*) grows to about 100 feet (30 m), and the sour tupelo, ogeche lime, or ogeche plum (*N. ogeche*) grows to about 30 feet (9 m) on riverbanks. As its common names suggest, its fruits are edible. *N. acuminata* is a shrub, that grows to about 16 feet (5 m) tall in pine swamps. All the tupelos yield a valuable wood.

Hawthorn

In all, there are probably about 280 species of hawthorn (*Crataegus* species). Most are North American, but around 35 to 55 grow in the Old World, distributed throughout the temperate regions of the Northern Hemisphere.

The fruits of hawthorns are called haws, and those of black haw (*C. douglasii*), native to western North America, are used to make jellies. Washington thorn (*C. phaenopyrum*) is cultivated as an ornamental for its red fall foliage and its orange or red fruits that remain on the tree until well into the winter. All hawthorns bear thorns (although there are thornless cultivated varieties). The cockspur hawthorn (*C. crus-galli*) has the biggest thorns, which are up to three inches (8 cm) long.

In the eastern United States there are several species that grow in thickets. These include green (*C. viridis*), May (*C. opaca*), Allegheny (*C. intricata*), dotted (*C. puntata*), and fleshy (*C. succulenta*) hawthorns.

The common hawthorn (*C. monogyna*) is a shrub or a small tree that grows to about 33 feet (10 m) tall throughout Europe and western Asia. In parts of Britain it is planted to make dense, stockproof hedges. This is the species that is linked to May Day traditions, and the variety *biflora* is the Glastonbury thorn, which flowers a second time in December in mild winters. According to legend, the thorn arose from the staff that Joseph of Arimathea drove into the ground when he visited Glastonbury as a trader, accompanied by Jesus. May is the Midland hawthorn (*C. laevigata*), occurring throughout Europe and North Africa (and a different species from the North American May hawthorn). It is smaller than the common hawthorn, with smaller leaves, and, as its name indicates, it flowers in May.

Around the shores of the Mediterranean, the Mediterranean medlar or azarole (*C. azarolus*) is a small, spreading tree, about 30 feet (9 m) tall, the fruits of which taste like apples and are used in conserves and to flavor liquors. Hungarian thorn (*C. nigra*), a tree about 20 feet (6 m) tall, grows in southeastern Europe. Tansy-leaved thorn (*C. tanacetifolia*) grows naturally as a tree, up to 33 feet (10 m) tall, in Asia Minor.

Chinese hawthorn (*C. pentagyna*), stretching up to about 20 feet (6 m) tall, is either native to a region extending from southeastern Europe to Iran and the Caucasus or has been introduced to this region from China, its real home, and has become naturalized. Oriental thorn (*C. laciniata*), a tree of similar size, has also been introduced from China and now grows naturally in Spain and southeastern Europe.

Hazel

Hazels (*Corylus* species) are approximately 10 species of shrubs and small trees that grow throughout the temperate regions of the Northern Hemisphere. They are very common in the understory of European broad-leaved forests. Their Old World range covers Europe south of Scandinavia and eastward to the Black Sea, and they also grow in eastern China and Japan. They were once grown extensively for their wood (see "Coppicing" on page 291) and for their nuts, called cobnuts, filberts, or hazelnuts. The American hazel or filbert (*C. americana*) of eastern North America is widely grown for ornament as well as for its nuts.

European hazel or cobnut (*C. avellana*) is a shrub or tree that grows up to 23 feet (7 m) tall and can form dense thickets. It was valued for its wood, but the more reliable source of nuts was the filbert (*C. maxima*), a tree of similar size that is native to southern Europe.

Turkish hazel (*C. colurna*), from eastern Europe and Asia Minor, is also cultivated for its nuts. It is a much bigger tree, reaching 80 feet (25 m) in height. Chinese hazel (*C. chinensis*) are somewhat larger, growing up to 100 feet (30 m) tall, but are otherwise so similar to Turkish hazel that they used to be regarded as two varieties of the same species. They occur in China. Tibetan hazel (*C. tibetica*) grows up to about 23 feet (7 m) tall, in Tibet, and Japanese hazel (*C. sieboldiana*) is a shrub, about 16 feet (5 m) tall, that grows in Japan.

Hickory

Hickories (*Carya* species) are 17 species of deciduous trees, belonging to the same family (Juglandaceae) as walnuts, that grow naturally throughout North America to the east of the Great Lakes. A few species are native to eastern Asia, but hickories do not grow naturally in Europe. Hickories grow fast and yield valuable timber, and their fruits are nuts, which in some species are edible. The trees are often cultivated for their nuts and also for ornament.

The most important edible hickory nut is the pecan, obtained from the tree of the same name (*C. illinoensis*). It grows in the Mississippi Basin (its fruits were once known as Mississippi nuts), rapidly reaching a height of 150 feet (45 m). Bitternut or swamp hickory (*C. cordiformis*) is smaller, growing to about 90 feet (27 m). The coating to its nuts contains so much tannin that the nuts are inedible. The nuts of water hickory, or bitter pecan (*C. aquatica*), are also inedible for the same reason. This is a small tree, about 50 feet (15 m) tall, that grows in swamps and rice fields.

In addition to the pecan, shagbark (*C. ovata*), big shellbark (*C. laciniosa*), and Carolina (*C. carolinae-septentrionalis*) hickories are the most important sources of edible nuts. Shagbark, or little shellbark hickory, yielding hickory nuts, is a tree that grows up to 118 feet (36 m) tall and sheds its bark in narrow strips. The big shellbark hickory or kingnut is very similar, but unlike white hickory nuts, kingnuts are yellow-brown. Pignut, or smoothbark hickory (*C. glabra*), is a tree that grows up to 80 feet (24 m) tall on wet ground as far north as Ontario. The sweet pignut, or false shagbark or red hickory (*C. ovalis*), is very similar but tends to shed its bark. Both produce edible nuts. Mockernut, or big bud hickory or white heart hickory (*C. tomentosa*), a handsome tree that grows naturally up to 100 feet (30 m) tall in the northeastern states and is often planted as an ornamental, produces nuts with thick, hard shells and tiny, but edible, kernels. All the hickories produce useful wood, used traditionally for making the handles of tools.

Holly

There are about 400 species of holly (*Ilex* species), growing naturally in warm-temperate, subtropical, and tropical regions of both Northern and Southern Hemispheres. Their popularity as ornamental plants has produced innumerable cultivated varieties. Most species are evergreen but not all. About 15 species grow naturally in the New World in the southeastern United States and Central America.

Holly prefers a moist, mild climate. Under suitable conditions it forms part of the understory of broad-leaved deciduous forests, and holly can also form stands by itself, called (obviously) hollywoods.

The most widespread American holly (*I. opaca*) is a small tree, growing up to 50 feet (15 m) tall. It produces red berries. The inkberry (*I. glabra*) produces black ones. It is a shrub no more than 6.5 feet (2 m) tall. Another bearer of red berries is emetic holly (*I. vomitoria*), also known as Carolina tea, cassena, Indian black drink, and yaupon, a tree that grows up to 26 feet (8 m) tall. It was used medicinally by Native Americans in the southeast and on some Caribbean islands, where it also grows naturally. Dahoon holly (*I. cassine*), which grows to a height of 80 feet (25 m), has a similar range.

Holly provides the foliage and berries that Europeans have been using to decorate their homes in midwinter at least since Roman times, and this was a Celtic as well as Roman tradition. European hollies are evergreen, but some American, Chinese, and Japanese species are deciduous.

Red holly berries are used for Christmas decoration, usually together with their leaves, but not in the case of the winterberry, or dogberry or black alder (*I. verticillata*), a shrub that grows up to 10 feet (3 m) tall in swamps. It is deciduous, its leaves turning black before they fall, but the berries remain. Smooth winterberry, or hoopwood (*I. laevigata*), is very similar and also grows in swamps, as does the possum haw (*I. decidua*), a shrub or small tree, up to 33 feet (10 m) tall.

The most widespread Old World species is the common or European holly (*I. aquifolium*), a bushy tree that grows up to 80 feet (25 m) tall, throughout western Europe, North Africa, and western Asia. Farther east, its place is taken by species native to the Himalayas, China, and Japan. Most are shrubs, but Himalayan holly (*I. dipyrena*) is a tree that grows up to 50 feet (15 m) tall in the eastern part of the mountain range.

Hornbeam

Hornbeams make up 35 species (genus *Carpinus*) of deciduous trees that grow in North America, Europe, the Near East, and, east of the Himalayas, in China and Japan. The map shows their distribution.

The global distribution of hornbeams (*Carpinus* species)

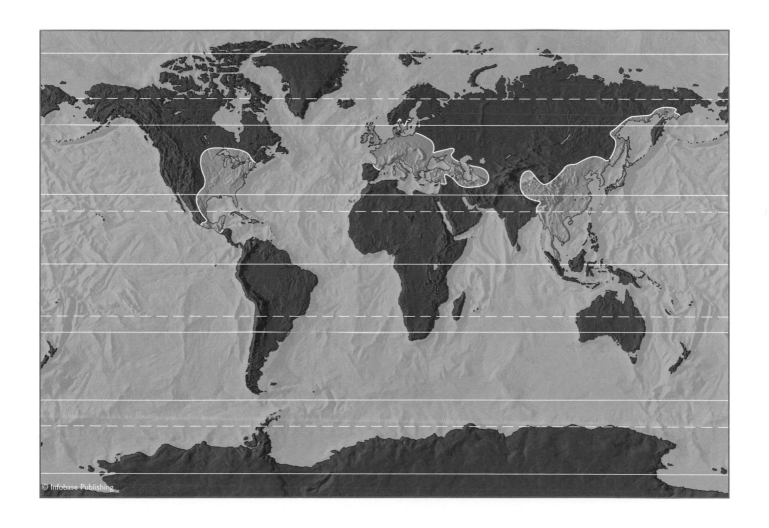

Only one species grows naturally in North America. This is the American hornbeam, or American muscle tree or blue beech (*C. caroliniana*), and as the map shows, it is distributed over much of the western part of the United States, Central America, and the islands of the Caribbean. American hornbeam is a small tree, growing slowly to about 40 feet (12 m). It is often planted for ornament because of the splendid fall colors of its leaves, and its wood is used mainly to make tool handles. Hornbeams produce very hard, dense wood that was once used to make the yokes with which oxen were harnessed to plows and wagons.

The most widespread Old World species is the common hornbeam (*C. betulus*), growing throughout Europe and Asia. It grows to about 80 feet (25 m) tall and can be mistaken for beech. Common hornbeam can be distinguished in late spring by its catkins, which are longer than those of beech, and in spring and summer by its leaves, which are more markedly toothed and veined than those of beech. In winter its leaf buds resemble those of beech but hug the twigs rather than pointing outward.

Oriental hornbeam (*C. orientalis*) is a small tree, or sometimes shrub, of southeastern Europe. Japanese hornbeam (*C. japonica*), native to Japan, is a tree that grows up to about 60 feet (18 m) tall and that is sometimes cultivated in Europe.

Magnolia

Magnolias (*Magnolia* species) are famous for their large, showy flowers and are cultivated widely. There are approximately 125 species, some evergreen and some deciduous. They grow naturally in the southeastern United States,

extending into Central America and southern and eastern Asia. The genus is named for Pierre Magnol (1638–1715), a French professor of botany and medicine.

The most important American forest species are the Fraser and southern magnolias. The Fraser magnolia, or ear-leaved umbrella tree (*M. fraseri*), is deciduous, grows to about 40 feet (12 m), and produces big, fragrant flowers. It occurs in the southern Appalachians. Southern, evergreen, or laurel magnolia, also called bull bay (*M. grandiflora*), is a pyramid-shaped, evergreen tree that grows up to 100 feet (30 m) tall, with fragrant, cream-colored flowers that measure up to 12 inches (30 cm) across, and grows on the southeastern coastal plain. The umbrella tree (*M. tripetala*), about 40 feet (12 m) tall, produces flowers with a strong and unpleasant smell. It is closely related to the southern magnolia, as is bigleaf magnolia (*M. macrophylla*), a rather larger tree that produces leaves 12–40 inches (30–100 cm) long and flowers as wide as 14 inches (35 cm) across. The cucumber tree (*M. acuminata*), which produces fruits that resemble cucumbers while they are green and unripe, is a deciduous, pyramid-shaped forest tree that reaches about 65–100 feet (20–30 m) tall. Its timber is used for flooring.

Maple

Maples, comprising 111 species of the genus *Acer*, are distributed throughout the temperate regions of the Northern Hemisphere, and their range extends into some parts of the Asian Tropics.

As a commercial source of timber, the Oregon or bigleaf maple (*A. macrophyllum*) is probably the most important North American species. It appears from Alaska to California and is a tall tree, growing to 100 feet (30 m). The vine maple (*A. circinatum*), growing up to 40 feet (12 m) tall, is the only other maple native to the Pacific coast. It grows on riverbanks from British Columbia to California. Farther inland, rock or dwarf maple (*A. glabrum*) and bigtooth maple (*A. grandidentatum*) grow near mountain streams. Both are small trees, up to 40 feet (12 m) tall.

Apart from timber, the best-known maple product is the syrup, obtained from the sap by a process devised by Native Americans living around the Great Lakes and near the St. Lawrence. Syrup is produced mainly from the sugar maple (*A. saccharum*), a tree that grows up to 80 feet (25 m) tall and, to a lesser extent, from the very similar black maple (*A. nigrum*). The leaf of the sugar maple in its red, fall color is the national emblem of Canada. Box elder or ash-leaved maple (*A. negundo*) has also been used in the past as a source of maple syrup. A tree that grows up to 50 feet (15 m) tall, beside rivers and in swamps in most states.

Red maple (*A. rubrum*) grows to about 100 feet (30 m) and is found in most natural woods on low-lying sites in the east. It is often cultivated for its spectacular fall colors,

Magnolias are widely cultivated for their large, showy flowers. Magnolias are among the most ancient of flowering plants, known from the early Cretaceous (about 130 million years ago), and the first flowers may have looked very much like those of magnolias. This is bigleaf magnolia (*Magnolia macrophylla*). *(Vanderbilt University)*

as is the silver maple (*A. saccharinum*), which grows naturally from Quebec to Florida. Striped maple, also called moosebark or moosewood (*A. pennsylvanicum*), is a tree that grows up to 40 feet (12 m) tall, often in the understory of eastern maple forests. As its name indicates, its smooth bark is striped green and white. It also has large leaves that turn bright red in the fall. On higher ground its place may be taken by mountain maple (*A. spicatum*), which grows to about 25 feet (7.5 m).

In the Old World, more species of maple occur in Asia than in Europe, where only 14 species grow naturally. The most widespread European species is the field, hedge, or common maple (*A. campestre*), a tree growing up to 80 feet (25 m) tall and recognizable by its small, dark, deeply lobed leaves, which turn yellow or red in the fall. Field maple occurs throughout Europe, North Africa, and western Asia. In eastern Europe its identification can be more difficult because it sometimes hybridizes with the Cretan maple (*A. sempervirens*), a smaller tree that grows up to 33 feet (10 m) tall, and with the Montpellier maple (*A. monspessulanum*), which grows up to 50 feet (15 m) tall. Both of these have leaves with wavy edges rather than lobes. Cappadocian maple (*A. cappadocicum*) grows naturally from the Caucasus to China but is widely cultivated elsewhere. It grows quickly, reaching a height of about 80 feet (25 m). Heldreich's maple (*A. heldreichii*), a tree that grows up to about 65 feet (20 m) tall, with leaves lobed almost to their bases, occurs naturally in the Balkans but is also widely cultivated. The Balkan (*A. hyrcanum*) and Italian (*A. opalus*) maples are both similar but smaller. Lobel's maple (*A. lobelii*), from Italy, has deeply lobed leaves that are twisted at the tips, and almost all its branches rise vertically. Norway maple (*A. platanoides*), occurring naturally over most of Europe, grows to about 80 feet (25 m) or more and is grown extensively for ornament.

The Old World sycamore (*A. pseudoplatanus*) is sometimes called the great maple. An attractive tree growing to 80 feet (25 m) or more, it is native to central and southern Europe. It was introduced to Britain in the Middle Ages and is mistaken by some people for the mulberry fig, or sycamorus, of Palestine and by others for the plane tree, which is how it acquired its names. It is now fully naturalized and readily invades woodland. Trautvetter's maple (*A. trautvetteri*) of the Caucasus is generally similar but somewhat smaller.

Oriental maples are generally smaller than those of Europe. Many species are popular garden and park trees, perhaps the best known, with countless varieties, being the Japanese maple (*A. palmatum*). In the wild this tree can grow to 50 feet (15 m), but it reaches barely half that in cultivation and is grown for the red of its leaves in the fall. The trident maple (*A. buergeranum*), from China and Japan, is grown for its summer foliage and the gray-budded snakebark maple (*A. rufinerve*) from Japan for its attractive bark as well as its foliage colors.

A grove of silver maple (*Acer saccharinum*) displaying the rich fall color for which maples are renowned   *(VTweb.com)*

Oak

Oaks (*Quercus* species) belong to the same family (Fagaceae) as beeches and southern beeches. Most are trees but some are shrubs, some are deciduous and others are evergreen or half-evergreen, keeping their leaves through the winter and shedding them in the spring. There are about 600 species of oaks and, as the map shows, they grow naturally throughout most of the temperate Northern Hemisphere, extending through Central America and the Caribbean islands into the northern Andes and in Asia reaching through the Malayan Peninsula and Indonesia. Their range extends across the equator in southern Asia but does not reach Australia. Outside the temperate regions, oaks are found only in mountain ranges where temperatures are cool. They grow from sea level to an altitude of 13,000 feet (4,000 m). The widest variety of oaks is found in North America, with more than 85 native species, where oaks form an important component of many forests, the greatest concentration of species occurring in the Sierra Madre highlands of Mexico.

Commercially, oaks are of great importance. Their timber is one of the strongest, most durable of hardwoods.

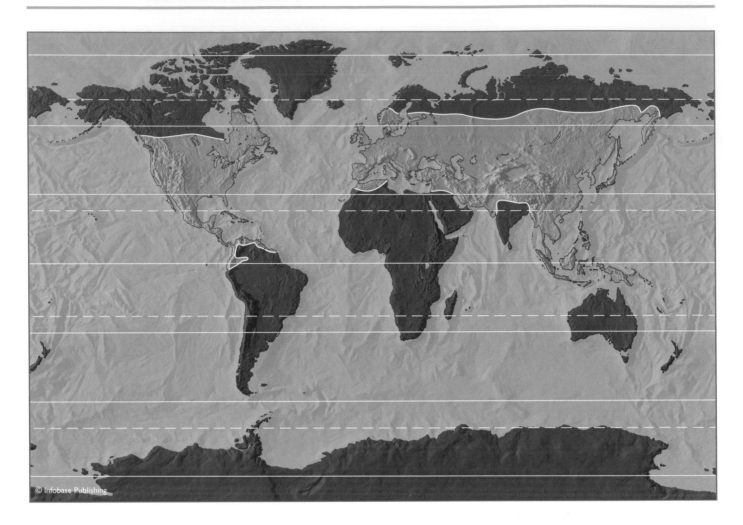

The global distribution of oaks (*Quercus* species)

Timber from different species are often classified as red or white oak. White oak is the harder and more durable, but both types are used for the same purposes.

Most of the oaks native to North America are large trees, and many provide valuable timber. Those growing in the eastern United States yield some of the finest timber in the world for structural use and cabinetmaking, but oaks have other uses. Some supply fuel, others produce acorns that are edible for humans or farm livestock. Oaks also yield tannins that are used in tanning leather and formerly for making blue-black ink from oak marble galls induced by the gall wasp *Andricus kollari*. Bark from the black oak is the source of quercitrin, which is processed to make yellow flavine and red flavine, the basis of dyes that are used to color wool. Many species are grown for ornament, some of them producing spectacular colors in fall. Oaks also feature strongly in folklore and literature (see "Forests in Folklore and History" on pages 236–241).

Evergreen oaks are often known as live oaks. Most North American oaks are deciduous, but the live oak (*Q.*

virginiana) is said to produce the most durable of all oak timbers. It used to have many uses, including shipbuilding, but is now scarce. It grows to about 65 feet (20 m) in the southern states and Mexico.

Californian live oak, or encina (*Q. agrifolia*), found in California, is somewhat larger, reaching 80 feet (25 m). Along the coastal ranges of California and southern Oregon, on mountainsides up to about 9,000 feet (2,800 m), there is the canyon live oak, or maul oak (*Q. chrysolepis*), a tree only about 60 feet (18 m) tall, but more than twice as wide as it is high. Its strong wood was formerly used to make farm tools. The interior live oak (*Q. wislizenii*) has leaves resembling those of holly.

Laurel oak (*Q. laurifolia*), found along the Gulf Coast and in Florida, is half-evergreen, which means it retains some of its leaves throughout the year. It is a tall tree, growing to 100 feet (30 m). The laurel oak growing in the uplands is *Q. haemispherica*. Willow oak (*Q. phellos*), native to the coastal plains of the southeastern states, is half-evergreen in the southern part of its range and deciduous farther north. It is a tall tree, growing to 65–100 feet (20–30 m) with leaves resembling those of a willow, which give it its name. These are yellow when first they open, then they turn green, and in fall they turn yellow again. Willow oak is grown as a shade tree along streets and is also an important source of red oak timber.

Other oaks are deciduous. The white oak (*Q. alba*) is the source of some of the most valuable timber, used for shipbuilding, to make railroad ties, furniture, and barrels. It is a huge tree, growing to a height of 150–165 feet (45–50 m), with a trunk up to 5 feet (1.5 m) across, that occurs on the eastern side of the continent. Burr oak, or mossy cup oak (*Q. macrocarpa*), another eastern species, is even bigger, sometimes reaching 180 feet (55 m), with a trunk as much as 7 feet (2 m) in diameter. The name mossy cup refers to a fringe of scales, resembling hairs, around the rim of the acorn cup. Post oak (*Q. stellata*) also provides valuable timber. It is a smaller tree that grows up to 65 feet (20 m) tall on fairly dry ground on the eastern side of the continent. Many

Black or quercitron oak (*Quercus velutina*) is the source of quercitrin, a yellow dye (its name made from *quer* and *citron*) invented by Edward Bancroft (1744–1821); it is also used in tanning leather.   *(Vanderbilt University)*

of its leaves have at least one very wide lobe, giving them the approximate shape of a cross.

Chestnut, basket, or rock oak (*Q. prinus*), with edible acorns, and swamp chestnut oak (*Q. michauxii*) both have leaves resembling those of chestnuts. They are big trees, growing up to 100 feet (30 m) tall in the eastern United States, chestnut oak growing on dry ground, mainly on upland sites, and swamp chestnut oak being confined to permanently wet ground on the coastal plain. Overcup oak (*Q. lyrata*) is also a tree of the coastal wetlands. Its common name refers to its acorn cups, which are thin and scaly and completely enclose the nut. Swamp white oak (*Q. bicolor*), another tree of wet ground in the southeast, grows to about 80 feet (30 m) and yields dense, hard wood that has been used in construction and to make railroad ties but also for cabinetmaking. The same area is also the habitat of the water oak (*Q. nigra*), a tree of similar size.

The most valuable oak of the Mississippi valley is the Shumard oak (*Q. shumardii*). It grows to 100 feet (30 m), sometimes more, and its leaves turn a magnificent red in the fall. The leaves bear tufts of pale hairs on their undersides, which distinguishes them from those of the scarlet and red oaks.

Black or quercitron oak (*Q. velutina*), found in forests on the eastern side of the continent, is also a tall tree, growing to a height of 100–165 feet (30–50 m). It is one of the red oaks, as are the northern and southern red oaks. The northern red oak (*Q. rubra*) grows faster than other oaks native to the region, sometimes at a rate of 8 feet (2.5 m) a year, to reach a height of about 80 feet (25 m). It occurs in forests from Canada to Florida and Texas. The southern red oak, or cherrybark oak (*Q. falcata*), grows to a similar size on dry land in the southeastern United States. It has dark, gray-brown bark with narrow ridges, somewhat reminiscent of the bark of a cherry, hence the name. Blackjack oak (*Q. marilandica*) grows in the same region but on poor soils, and it is a smaller tree, reaching no more than about 33 feet (10 m), although its wood is useful as fuel.

Shingle oak (*Q. imbricaria*) has leaves resembling those of a willow, like the willow oak, although in this case the tree is deciduous. It is fairly small, reaching no more than about 50 feet (15 m) in natural woods, but often growing taller when cultivated, and it occurs widely in the central and eastern United States in forests growing on rich, moist soil on hillsides and riverbanks. Its common name refers to the fact that early settlers used its wood to make roof shingles and clapboard.

Scarlet oak (*Q. coccinea*) occurs on sandy soils in the northeastern United States and is widely cultivated for its brilliant fall foliage. It is a handsome, spreading tree that grows to a height of about 80 feet (25 m). Pin oak (*Q. palustris*), also native to forests in the northeastern states, grows to about 100 feet (30 m). Its timber is used in construction and for making clapboard.

Black oak (*Quercus velutina*) is a tall, handsome tree that is native to eastern North America. *(Vanderbilt University)*

The most important western timber tree is the Oregon white oak (*Q. garryana*), used for flooring and making furniture. It grows from British Columbia to California. California white oak (*Q. lobata*) and California black oak (*Q. kelloggii*), trees that grow about 80 feet (25 m) tall in California and Oregon, are less important commercially, although acorns from the California black oak were once the staple food of Native Americans living in the area. Wood from this tree was their principal fuel.

The cork oak (*Q. suber*), a native of southern Europe and North Africa, is the tree which supplies all of the world's genuine cork. It is one of the evergreen oaks and occurs over much the same range as another evergreen, the holm or holly oak (*Q. ilex*), named for its hollylike leaves (*ilex* is holly). Macedonian oak (*Q. trojana*), which grows in southern Italy and the Balkans, is also evergreen. All three species grow to about 65 feet (20 m). A Chinese evergreen oak, the bamboo-leaved oak (*Q. myrsinifolia*) is rather smaller; its smooth-edged, narrow leaves are only faintly reminiscent of bamboo leaves.

Algerian or Mirbeck's oak (*Q. canariensis*) is half-evergreen, not losing its leaves until early spring. It is a big tree, up to 115 feet (35 m) tall, that grows naturally in North Africa and Spain.

Deciduous oaks extend into higher latitudes, but they also thrive in southern Europe and Asia Minor. Perhaps the most famous is the pedunculate, or English, oak (*Q. robur*), a magnificent tree spreading up to 150 feet (45 m) tall. It shares its range with the durmast, or sessile, oak (*Q. petraea*). The two species are very similar but can be distinguished by their acorns. Pedunculate oak has acorns that are attached by long stalks; those of sessile oak have almost no stalks and sit pressed against the twig. Both oaks are important sources of timber. Turkey oak (*Q. cerris*), from southern Europe but planted elsewhere, is very similar but has narrower leaves than the others, and its acorn cups are covered in woolly hairs. It is grown mainly for ornament, its timber being inferior to that of the other species.

Daimio oak (*Q. dentata*) grows in China, Korea, and Japan. A deciduous tree that grows to about 60 feet (18 m) tall, it has much bigger leaves than most oaks, some of them 12 inches (30 cm) long.

Olive

There are about 20 species of olives (genus *Olea*), centered in Africa and distributed throughout much of the warm-temperate and tropical regions of the Old World, but only one species is of major importance. The olive (*O. europaea*) has been cultivated in southern Europe since ancient times. There is evidence that it was being grown in Asia Minor in about 3700 B.C.E., and it is of great importance in all Mediterranean cultures. The tree itself is a broad-leaved evergreen 10–40 feet (4–16 m) tall that can live for 1,500 years. It will not grow where winter temperatures average less than 37°F (3°C), and it dies if the temperature falls to 16°F (–9°C).

Pear

Pears (*Pyrus* species) occur naturally throughout the temperate regions of the Northern Hemisphere. There are about 20 species.

The common, or European, pear (*P. communis*) is a tree that grows to about 50 feet (15 m) tall and is found throughout Europe and western Asia. This is the species from which many cultivated varieties have been bred, but others have made important contributions. The wild pear (*P. pyraster*) of central Europe is one of these, and the fruit of the snow pear (*P. nivalis*) of Switzerland and southern Europe is sometimes used to make the fermented drink, perry. Caucasian pear (*P. caucasia*) trees are vigorous colonizers of open ground and grow in forests. They have also been used in developing cultivated varieties.

In China and Japan the oriental, sand, Chinese, or Japanese pear (*P. pyrifolia*) is widely cultivated. The "sand" in one of its common names refers to the stone cells that give most pears their slightly gritty texture.

Plane, Sycamore

Plane trees (*Platanus* species) have a very discontinuous distribution. In all, there are six or seven species. One occurs in southeastern Europe and eastward as far as Iran, one in southeast Asia, and all the others in North America. The sycamore (*P. occidentalis*), also known as the American plane, buttonwood, and buttonball, may well be the largest broad-leaved tree in North America. It grows to a height of 165 feet (50 m), with a diameter (always measured at chest height) of as much as 11.5 feet (3.5 m). It grows on rich, alluvial, lowland soils in the eastern and southeastern United States, sometimes in pure stands, but more often in association with a number of other broad-leaved species.

Its western equivalent is the California sycamore, or California plane (*P. racemosa*), which grows in the Californian coastal ranges and Sierra Nevada. Not quite so large as the eastern species, it reaches a height of about 130 feet (40 m). The Arizona sycamore (*P. wrightii*), which grows in the southern states and Mexico, is smaller, reaching about 80 feet (25 m). American sycamores are not closely related to the European sycamore (which is a maple).

Rhododendron

Rhododendron (*Rhododendron* species) grows naturally throughout most of North America, except for the Great Plains. North American rhododendrons are shrubs or small trees, but in some places they are important components of the understory in broad-leaved forest and can form dense thickets. The most common species are the catawba (or mountain rosebay) rhododendron (*R. catawbiense*), which grows to a height of about 10 feet (3 m), and rosebay (or great laurel) rhododendron (*R. maximum*), which reaches more than 40 feet (12 m). Both are evergreen. Catawba rhododendrons grow in the southeastern United States, and their flowers produce a spectacular display in June in the Great Smoky Mountains National Park. The rosebay rhododendron shares a similar range, but it extends farther to the northeast. The rhodora rhododendron (*R. canadense*) grows in northeastern North America.

Of the 850 species comprising the genus *Rhododendron*, six are native to Europe. Most rhododendrons grow naturally in the Himalayas, where there are forests dominated by tree-sized rhododendrons, as they do in the mountains

of Southeast Asia and the Malayan Peninsula. They produce large, colorful flowers, and beginning in the 17th century, but especially during the 18th and 19th centuries, European plant hunters brought back many species, which they then hybridized to produce ornamental plants.

In 1763 *R. ponticum,* native to the Mediterranean region, was brought to England from Gibraltar—no one knows by whom—and by early in the 19th century it was being sold in London as a pot plant. It thrived in the British climate and was used as the rootstock for many park and garden hybrids. Unfortunately, this introduced species escaped from cultivation and is now a widespread and aggressive weed in light, acid soils throughout Britain, invading many forests and forestry plantations. Once established, *R. ponticum* can be controlled, but it is extremely difficult to remove completely.

Rowan, Whitebeam, Service Tree

Rowans, whitebeams, and service trees belong to a genus (*Sorbus*) of about 85 species of small trees and shrubs that are found throughout the temperate regions of the Northern Hemisphere. Rowan (*S. aucuparia*), also known as mountain ash and quickbeam, grows naturally throughout Europe and western Asia. In Scotland it grows at altitudes up to 2,000 feet (600 m), often standing alone but at lower levels forming part of the understory of natural forest. Its name *mountain ash* refers to its *pinnate* leaves, which resemble those of the ash (*Fraxinus* species).

A tree that grows to about 50 feet (15 m) tall, the rowan is the subject of many ancient beliefs. It was believed to give protection against witches, and to this day there are people who will not injure a rowan, far less fell one, for fear of the bad luck this would bring. It appears as though by magic, germinating readily from seeds dropped by birds and then growing rapidly. Its bright red berries are used to make a jelly to accompany meats such as mutton and venison, and its hard wood is used to make tool handles and other small items.

It is not the only Old World rowan. Japanese rowan (*S. commixta*) is a very similar tree, also bearing tight clusters of white flowers and red berries. It is found in Sakhalin and Korea as well as Japan. Sargent's rowan (*S. sargentiana*) is a shrub, about 16 feet (5 m) tall, that grows throughout China, and Vilmorin's rowan (*S. vilmorinii*), with rather paler berries, is a somewhat smaller shrub native to western China. Also in western China there is the Hupeh rowan (*S. hupehensis*), a tree about 50 feet (15 m) tall with berries that ripen to pale pink or white. It is grown for ornament in some European parks and gardens.

The common whitebeam (*S. aria*) grows naturally on chalk and limestone soils throughout Europe, and as one or other of its many cultivated varieties, it is also grown in parks, gardens, and streets. It is a small tree, growing to about 50 feet (15 m) tall, with berries that are sometimes used to flavor drinks (if they can be collected before the birds take them). In parts of northern England the tree is known as the sea owler, *owler* probably being a corruption of *alder*, and the fruits are known as chess apples; eaten fresh they are unpleasant, but once they start to decay (or "blet") they are quite edible. Swedish whitebeam (*S. intermedia*) is smaller, growing to only about 33 feet (10 m). It grows around the Baltic Sea and in northern Germany, as well as in northeastern Scotland, where it is presumed to grow from seeds dropped by migrating birds. Like the common whitebeam, it is grown in many parks and gardens. In southwestern Europe there is also a Pyrennean whitebeam (*S. mougeotii*), a shrub or small tree that grows high on mountainsides.

The service tree (*S. domestica*) grows throughout southern Europe, North Africa, and western Asia. It can reach a height of more than 50 feet (15 m), and its fruits look rather like small apples or pears, about one inch (2.5 cm) long. They are edible, more so when bletted, and are used to flavor certain kinds of beer. The tree is cultivated in parts of Europe for ornament but also for its fruit. *Service* comes from the Latin *cerevisia*, which means "beer."

The wild service (*S. torminalis*) is a larger tree, growing to more than 65 feet (20 m), although it is often smaller. Unlike most *Sorbus* species, its leaves are lobed like those of a maple, for which it is easily mistaken. The bark cracks to produce shapes that give the tree its other name, checkers, and the fruits are sometimes called checkers berries. Its range is similar to that of the service tree but extends into southern England and Wales. Its fruits have been eaten and used medicinally, but the tree is not cultivated. It can develop from seed but more often spreads from suckers, establishing itself very slowly.

In several parts of Britain, the fact that the wild service has never been planted deliberately means that its presence is taken to indicate that woodland in which it occurs is ancient or that what is now open ground was once forested. Ancient woodland occupies land that has been forested continuously at least since 1600, which is when plantation forestry began.

Southern Beech

Southern beeches (*Nothofagus* species) are close relatives of beeches and interesting because of their discontinuous distribution (see "Trees That Grow on Either Side of a Vast Ocean" on page 11). They occur only in the Southern Hemisphere in South America and Australasia. In the New World they are found only in Chile and southern Argentina, but there are several species.

Some *Nothofagus* trees are large. Coigüe, or Domey's southern beech (*N. dombeyi*), of central and southern Chile can grow to a height of 165 feet (50 m). It is a broad-leaved evergreen, as are the coigüe de Magallanes (*N. betuloides*) and roble de Chiloe (*N. nitida*). Both occur in Chile and Argentina, coigüe de Magallanes in the south of the region and roble de Chiloe to its north, and both are trees that grow to 100 feet (30 m) tall. *Roble* is the Spanish word for "oak tree." Southern beeches are not oaks, of course, but their timber is very similar, which is how some of the commercially important species have earned this name.

All the other South American species are deciduous. Ruil (*N. alessandri*) and roble de maule (*N. glauca*) are trees that grow to 130 feet (40 m) tall, and rauli (*N. alpina*) reaches about 100 feet (30 m). These species are found in central Chile. *N. procera*, also known as rauli or raoul, is an important, fast-growing timber tree that stretches up to 100 feet (30 m) tall and that grows naturally in the Andes in Argentina and Chile. Roble pellin, also called coyan or hualle (*N. obliqua*), once formed extensive forests in central and southern Argentina and Chile, and it is still very important as a source of timber. It grows to 155 feet (35 m). Antarctic beech, also known as Ñire or guindo (*N. antarctica*), is smaller, reaching no more than 60 feet (18 m), and grows as a low shrub on exposed mountainsides in southern Chile and Argentina.

The South American species produce more valuable timber, but some of the southern beeches native to temperate Australasia are big, impressive trees, all of them evergreens. Native to southeastern Australia, the myrtle beech (*N. cunninghamii*) reaches a height of 165 feet (50 m), as does the Australian beech (*N. moorei*) of eastern Australia.

Species native to New Zealand include the silver beech (*N. menziesii*), red beech (*N. fusca*), and hard beech (*N. truncata*), all of which grow to 100 feet (30 m), and the smaller mountain beech (*N. cliffortioides*), reaching 50 feet (15 m), and black beech (*N. solandri*), growing to 80 feet (25 m).

Spindle

Spindle trees comprise a genus (*Euonymus*) of about 175 species of shrubs and small trees, some deciduous and some evergreen, which is centered on the Himalayan region and Asia, but spindle trees occur naturally throughout the temperate regions of the Northern Hemisphere.

Some spindle trees produce brilliant flowers that give them the popular name of burning bush. Two of these grow naturally in North America. *E. atropurpurea*, known also as Indian arrow wood and wahoo, as well as burning bush, is a shrub or small tree that grows naturally up to 26 feet (8 m) tall in the eastern and north-central United States. The western burning bush (*E. occidentalis*) grows along the U.S. western coast. It is smaller, reaching only about 18 feet (5.5 m) high.

The common spindle tree (*E. europaeus*) of Europe and western Asia can reach a height of about 30 feet (9 m) in cultivation, but in the wild it is usually smaller. Its young shoots are square in cross section, and its bright red fruits have three to five lobes, making them very distinctive. The wood is very hard and was once used to make such items as skewers and spindles for spinning, which explains its name. Its charcoal is excellent for drawing and has also been used in making gunpowder. The fruits are poisonous to sheep, and the tree harbors the eggs of an aphid that attacks bean crops, so many farmers destroy spindle trees when they find them. Despite this, the tree is grown in some parks and gardens, as are several Asian *Euonymus* species.

Strawberry Tree

Strawberry trees (*Arbutus* species) are named for their edible fruits, which resemble strawberries in appearance but not in flavor (which not everyone finds pleasant). There are 14 species, of which 12 are American and two grow around the Mediterranean region. All of them are evergreen shrubs or small trees.

Madroña, madrono, or Pacific madroña, also known as laurelwood and Oregon laurel, is a tree (*A. menziesii*) that grows up to about 75 feet (23 m) tall and that occurs naturally in western North America from British Columbia to California. It is grown for its timber, and its bark is used in tanning.

One Old World strawberry tree (*A. unedo*) has an interesting extension of its Mediterranean range into southwestern Ireland, where it forms part of what is known as the Lusitanian flora, a relic of a former interglacial period. This tree grows to about 30 feet (9 m) in height and is a popular garden and park ornamental, cultivated for its foliage, clusters of delicately colored white, pink, or pale green flowers, and fruits. Its fruits can be preserved and are used as a flavoring in certain liqueurs, and where the tree grows naturally its bark is used in tanning. The Cyprus strawberry tree, or eastern strawberry tree (*A. andrachne*), from the eastern Mediterranean grows up to 33 feet (10 m) tall and has serrated leaves.

Tulip Tree

If plane trees have a discontinuous distribution, tulip trees (*Liriodendron* species) have an even more extreme one. They are deciduous trees related to the magnolias (family Magnoliaceae), and today there are only two species. One grows naturally in eastern North America and the other in a few locations in central China.

The American species is the tulip tree (*L. tulipifera*), also known as the tulip poplar, yellow poplar, and whitewood. It grows naturally from Nova Scotia to Florida and as far west as Michigan and is cultivated widely elsewhere. It forms an important component of broad-leaved forests, where it grows

rapidly on moist, fertile soils to fill gaps caused by the death of other species, and it can live for 500 years. Apart from its large, tuliplike flowers, which open in summer, its unusual four-lobed leaves turn a splendid orange and yellow in the fall. The tree itself is tall and stately, with dense foliage for most of its height. It grows to 200 feet (60 m). Its timber, called whitewood, is valuable and is used for house interiors.

Walnut

Walnut trees (*Juglans* species), which are related to hickories, grow throughout North America south of the Great Lakes, their range extending through Central America and into tropical South America. All walnuts produce nuts, but not all of these are edible, and some walnuts are better known for their timber. Because of its beauty, walnut wood is used to make valuable items, such as furniture, pianos, and gunstocks. Some nuts that are not eaten provide oils used in soaps and paints, and some yield dyes.

Most walnuts are big trees that often live for several centuries. The black walnut (*J. nigra*) grows to a height of 115 feet (35 m) or more. It grows naturally from south-

Black walnut (*Juglans nigra*) is an American species cultivated mainly for ornament and for its timber. *(The Natural Resources Conservation Council)*

ern Ontario, New England, and Michigan to Georgia and Texas, and this is the species most famed for its timber. Oils from its nuts are used in food. The range of the butternut, or white, walnut (*J. cinerea*) extends farther north than that of the black walnut. It grows to about 65 feet (20 m), and its wood is also used to make furniture.

There are several western species. The Texas walnut (*J. microcarpa*) grows naturally in the southwestern United States and Mexico. It is a small tree, growing to only about 33 feet (10 m) tall. The Arizona walnut (*J. major*), growing on alluvial soils on upland plains and in mountain valleys at altitudes of 1,500–6,000 feet (450–1,800 m) from Arizona into New Mexico and Mexico, is slightly bigger. It reaches a height of about 50 feet (15 m). The California black walnut (*J. californica*) is about the same size and grows in southern California. Although short, it often produces many branches low down, giving it a very bushy appearance. In northern California, the Hinds black walnut (*J. hindsii*) produces edible nuts, which are marketed, and wood that takes a high polish. It is a larger tree, reaching a height of about 65 feet (20 m), and is cultivated in California and Oregon as a roadside tree and for ornament.

Walnuts grow in the Old World in warm temperate regions from the eastern Mediterranean to southern Asia and in eastern China, Korea, and Japan. They were introduced throughout Europe, including Britain, in Roman times, and the most widespread species, the common walnut (*J. regia*), which grows naturally from southeastern Europe to China, is also known as the English walnut and Persian walnut. The name *walnut* is from the Old English *walhhnutu* and means "foreign nut," so in England, where the tree is not a native, *J. regia* must be the English foreign nut, a name not quite so peculiar as it may sound. Its crushed leaves release a dark brown pigment that some people once used to darken their complexions. The tree itself grows to about 100 feet (30 m), and its timber is used to make high-quality furniture and for veneers.

Somewhat smaller trees that grow to about 65 feet (20 m) tall occur in eastern Siberia, China, and Japan. All produce edible nuts, and Siebold's walnut (*J. sieboldiana*) from China and Japan is valued for its timber.

Wayfaring Tree

Viburnums comprise about 150 species in the genus *Viburnum*, some evergreen and some deciduous, that are distributed throughout temperate regions of the Northern Hemisphere, with some species in the Asian Tropics. They are shrubs or small trees. *V. acerifolium* is the arrowwood of eastern North America. Its bark has medicinal properties and, as its name suggests, its wood was once used for making arrows. The mooseberry, moosewood, or hobblebush, also of eastern North America, is *V. alnifolium*.

One of the best known Old World species is the wayfaring tree (*V. lantana*), which grows throughout Europe and western Asia as a small tree up to about 16 feet (5 m) tall with attractive flowers and splendidly colored foliage in the fall. No one really knows how it acquired its name, but in the 16th century, writers were drawing attention to the frequency with which it was found in roadside hedges, where it cheered wayfarers.

Another attractive species is the guelder rose (*V. opulus*). It grows no more than about 13 feet (4 m) tall. Its name derives from a cultivated variety, in which all the flowers are sterile, that was first grown in a Dutch province called Gelderland, or Guelders.

Willow

Willows (*Salix* species) grow throughout the temperate and tropical regions of the world, except for New Guinea and Australasia, although they are rare in the Tropics and more common in the Northern than the Southern Hemisphere. There are about 400 species in all, some small shrubs, but others large trees. Willows are pollinated by insects, and their flowers are catkins, with male and female flowers borne on separate plants. Most willows prefer damp conditions and grow best along riverbanks and in marshes and swamps. Many produce valuable wood.

The black willow (*S. nigra*) is the largest North American species. Growing to the east of the Rockies, it is a tree that grows to 100 feet (30 m) tall, sometimes taller, and its wood is used mainly to make boxes. Peach-leaved willow (*S. amygdaloides*), another eastern species, grows to 65 feet (20 m). In the Pacific states, red willow (*S. laevigata*) can reach 50 feet (15 m) and western black willow (*S. lasiandra*) 45 feet (14 m) tall.

White willow (*S. alba*), in lowland areas from Europe to central Asia, is a large tree that grows to 80 feet (25 m). It is called white because of the way its leaves reflect light.

The only willow that is grown commercially in Britain for its wood is the one used to make cricket bats. The cricket-bat willow is a variety of the white willow (*S. alba* var. *coerulea*) that is grown from cuttings, often in rows along riverbanks.

Crack willow (*S. fragilis*) is so called because its shoots and twigs break off easily with a loud crack if they are struck sharply. This is not the disadvantage to the tree that it may seem because each broken fragment is capable of growing into an entire tree if it falls on a suitable spot. The tree grows to about the same size as the white willow, and hybrids of the two are fairly common.

Weeping willow (*S. babylonica*) is a tree that grows to about 50 feet (15 m) tall. It probably originated in Iran. Although much cultivated for its weeping habit, it is not very hardy in northern climates, where most willows of this form are a variety of the white willow (*S. alba* var. *tristis*).

Sallow, or goat or pussy, willow (*S. caprea*) and gray sallow (*S. cinerea*) are very similar, grow in similar places, and hybridize readily. These trees, found throughout Europe and western Asia, are about 33 feet (10 m) tall. The male flowers (catkins) are collected for decoration, especially on Palm Sunday. Bay-leaved, or laurel-leaved, willow (*S. pentandra*), a tree that grows to about 65 feet (20 m) tall, shares this range. Its dark, shiny leaves are broader than those of most willows and resemble laurel or bay leaves.

7

History of Temperate Forests

When humans first entered the temperate forests in pursuit of game, plants they could eat, and materials to build their homes, they became part of the forest ecosystem, and like any other member of the community, they began to modify their surroundings. From that time, the story of the forests was intricately intertwined with the histories of the people who lived in and beside them. This chapter is about the history of temperate forests, which is the history of the relationships between people and the forests. It begins, as stories should, at the beginning, and that beginning involved substantial changes in the forests. Those changes accelerated when people began to tend domesticated animals and to cultivate crops. The chapter describes the effect of farming on temperate forests and explains the techniques scientists use to unravel the clues to the past that lie buried in forest soils.

To the people living in or close to them, forests have always been more than mere stores of food and materials. Forests are sometimes dark, mysterious, and potentially dangerous places that provide the background for traditional tales of fairies, witches, and animals that talk. Particular trees have acquired a special significance that, in many cases, survives to this day in local beliefs and customs.

Yet the history of the forests was mainly the story of forest clearance. By medieval times it was evident that timber was being extracted faster than it could regenerate, and unless steps were taken to replace trees, this vital resource might be lost. That realization marked the start of forest conservation and of forestry—the deliberate planting of trees for profit.

■ FOREST CLEARANCE IN PREHISTORY

Starting about 10,000 years ago, as the last ice age drew to a close and climates grew warmer, little by little the edges of the glaciers and ice sheets retreated to higher latitudes and higher altitudes. Meltwater flowed, gradually cutting river channels, and seeds dropped by the wind and by birds began to germinate on the thawed ground.

Plants colonized the newly exposed land, and in time what had once been barren ice or sparse tundra became forest. Forests covered most of what are now the temperate regions. At first they were forests of the kind that today grow only in the far north and south, along the edge of the tundra. Then, as the climate continued to grow warmer, dense coniferous forests were able to survive until they were replaced by broad-leaved forests (see "Holocene Recolonization" on pages 91–95).

Origin of the Prairie

Not everywhere was forested. By the time the first European explorers and settlers arrived, grasses and herbs had become the dominant type of vegetation over large areas in the continental interior of North America. French explorers called these vast grasslands *meadow*, for which the French word is *prairie*.

Trees can and do grow in the prairie, not only in the tall-grass prairie but even in the mixed- and short-grass prairie where the climate is semiarid. Trees grow naturally on the sides of ridges and valleys where the ground is not level, and tree plantations and tree belts that are planted to provide shelter survive for many years. Studies of pollen found in lake sediments (see "Using Pollen to Study the Past" on pages 228–231) show that as recently as 5,000 years ago what is now a prairie landscape was more like open forest or parkland, with grasses and scattered groves of trees often dominated by pines.

No one is quite sure why the trees disappeared, but the most likely sequence of events began between 5,000 and 6,000 years ago. At that time the climate became markedly drier, favoring grasses rather than trees. Trees became more widely scattered, and tall grasses became dominant. This established the tall-grass prairie from eastern Iowa to western Michigan, with shorter grasses in the still drier regions to the west and south. After about 4,000 years ago, the climate changed again, this time becoming cooler and moister. Trees grew well under these conditions. In the north, the boundary between the boreal and mixed forests moved south in response to the cooler conditions, and the boundary between forest and prairie moved west in response to increased precipitation. This did not lead to the recreation of open forest, however, because humans now began to influence events.

Grassland Fires

In those days, most of the Native Americans living in the interior of the continent lived by gathering plant foods and hunting game. Game was plentiful, with large herds of grazing animals, especially bison, of course, and a good way to hunt them was by lighting a fire to drive a herd into an ambush. In any case, prairie fires were not uncommon. There were huge storms then, just as there are now, and lightning readily ignited dry grass. After the fire died down, fresh grass and herbs would grow vigorously in soil fertilized by the ashes. The overall effect was to drive game in a way that made it easier to kill, and at the same time, to improve the pasture so that there was more food for more animals. The herds increased in size and the people ate well.

Driving game and improving the pasture were not the only effects, however. A more profound change was also taking place. On the plains, where the ground is open, the wind blows freely. Fires are driven by strong winds, and there was nothing to halt their spread. They traveled fast. As they did so, the flames caught first those plants that were standing above ground level. Trees and shrubs burned. So, too, did the grass, of course, but grass grows from a point at or just below ground level where it is protected from all but the most intense heat. The fires did much more damage to woody plants than they did to prairie grasses. Indeed, on natural grassland a fire will burn off a layer of dead, dry grass that would otherwise suppress the growth of fresh, new grass.

Some tree seedlings would have survived the fire, and tree seeds present in the soil would germinate. As long as fires happened only occasionally, the forest would have recovered and adapted (see "Fire Climax" on pages 151–155) to the conditions. Deliberately setting fires increased their frequency, however. This gave the woody plants a briefer interval to grow back between fires. The interval was too brief for many species and they disappeared.

At the same time, there was another factor that favored grasses at the expense of trees and shrubs. When the new flush of grass emerged in the aftermath of each fire the grazing herds would return. As they grazed, the bison and other large animals would bite off the growing tips of some of the emerging woody plants and trample others. In time and with repeated fires, fewer and fewer trees would mature to produce seeds, so the soil would store fewer and fewer viable tree seeds, and the grassland would become permanent.

Much later, hunting and employing fires to drive game was replaced by farming. The natural grasses were cleared, the land plowed, and cereals, which are domesticated grasses, were grown instead. Cultivation also prevented trees from colonizing the land, and between them the Native American hunters and the farmers who succeeded them produced the landscape that is seen today.

Scientists strongly suspect that not only large parts of the North American prairie but also of the South American pampas and the steppe grasslands of Europe and Asia were formed in this way. In Africa, the savanna grassland is still maintained by being deliberately fired by the people who live there.

Long before people began to record their history, they were drastically altering the environment in which they lived. They were clearing forests.

The First European Farmers

In northwestern Europe, forest is still the most common natural vegetation type. When farmland is abandoned, usually scrub and then trees colonize it. Ecologists believe that if the land were to be abandoned and left undisturbed, in time much of Europe would once again be forested. In Europe it was not hunters who removed the forest but farmers. They needed land to grow crops and pasture for their livestock. The forest clearance began about 5,500 years ago in Britain and it accelerated about 2,000 years ago during the Roman occupation (see "European Forests of the Middle Ages on pages 233–236).

In temperate Europe, early farmers kept pigs and cattle and possibly sheep and goats as well. Although aurochs (*Bos primigenius*), the ancestor of domesticated cattle, lived in Britain, it is believed that migrants brought cattle with them rather than attempting to domesticate the wild ones already present, and cattle were certainly taken in small boats to Ireland, where there were no aurochs.

Pigs are forest animals, feeding on the forest floor, and they would have had no effect on the forest. Cattle, on the other hand, would have eaten grass in clearings, but probably they fed mainly by browsing, eating such tree leaves and shoots as they could reach. Their herders would have augmented this resource by climbing into the trees to cut down browse from higher levels. As they fed, the cattle would have trampled tree seedlings and damaged tree bark, sometimes badly enough to kill the tree. Cattle enlarged the clearings in which they lived. Sheep are not browsers. If farmers wanted to keep sheep they had to find pasture for them, and this meant clearing forest. Goats are adaptable, but around the Mediterranean they caused serious harm by eating growing shoots from vegetation that was already becoming sparse.

The Elm and Lime Decline

Browsing cattle may have been responsible for what paleo-botanists call the elm decline. The evidence for this is a sharp reduction in the amount of elm pollen present in the soil at levels dated to approximately 5,000 years ago. The reduction in elm pollen is found over most of northwestern Europe but not in North America. The decrease at that time is not probably to have been due to a change in the climate because such a change would have affected elm in some places more than in others, and almost certainly it would also have affected elms in North America. Perhaps the decline may have been caused by a severe disease epidemic, Dutch elm disease (see "Dutch Elm Disease" on page 174) being the most likely culprit.

Alternatively, a new farming technique may have been responsible. Farmers at that time may have started to keep their cattle in pens. Penning the animals would protect them from wild predators, such as wolves, and make them easier to control, but they would no longer have been able to find their own food. Farmers would have had to bring their food to them—as they do today to cattle raised in feedlots. Wild cattle are forest-dwellers that prefer to feed on the leaves and soft twigs of shrubs and those tree branches that are within their reach. This food is called browse, and cattle are naturally browsers. Early farmers would have gathered browse to feed their stock, and they would probably have preferred leaves and twigs from elm to those of other tree species because elm is especially nutritious. If people were removing elm browse each year before the trees flowered, less pollen would have been produced. In this case, the elm decline would have been a reduction in the amount of pollen but not necessarily of elm trees. Elms in northern Europe do not rely on flower fertilization and seed production; they propagate veg-etatively. Preventing the trees from flowering would not reduce the number of trees.

One possible cause might have led to the other. *Pollarding* (removing the growing shoots) increases the vulnerability of elms to Dutch elm disease. By cutting the trees for browse, farmers may have exposed them to disease, and that may have killed them.

Much later there was also a lime decline. This, however, was definitely associated with climatic change. About 2,500 years ago the climate became cooler, and the warmth-loving lime (*Tilia* species) declined throughout the northern part of its range. At the same time European beech (*Fagus sylvatica*) expanded northward to fill the gap.

Swidden Farming

Field crops such as wheat and barley were also being grown at the time of the elm decline, and trees had to be removed to make room for them. The first fields were forest clearings made by farmers.

Early farmers practiced *swidden farming*. This is a form of shifting cultivation very similar to the slash-and-burn type of cultivation that is still practiced in parts of the Tropics. An area of land is cleared of trees and shrubs, and crops are grown in it for several seasons. After a few years, yields start to decrease because nutrients removed from the soil in the form of crops have not been replaced, and as crop yields decline, weeds start to overtake and smother the crop plants. At this stage it is no longer profitable to continue cultivating the plot, which is abandoned; the process is repeated elsewhere. Wild plants colonize the abandoned plot, and in a few years the fertility of the soil recovers. The system works as a cycle in which each plot is revisited every so many years. It is a type of farming that can be sustained indefinitely, but it requires a large area to feed each person.

Clearing forest with hand tools is less difficult than it seems. Early European farmers used axes with blades made from chert, the stone that is called flint when it is found surrounded by chalk, and also used handles of ashwood. Some years ago, Danish scientists fitted a handle to a genuine chert axe blade that was 4,000 years old and found that they could fell 100 trees with it before it needed sharpening. Using axes of this type with blades of polished stone, it took three men just four hours to clear about 720 square yards (600 m²) of birch forest. As trees fall they bring down others, especially on sloping ground, speeding the process, and clearance ends with a fire to burn off shrubs and wood for which there is no use.

The First Mines

Chert and flint were important industrial materials in those days, and where they occur naturally they were mined on an industrial scale by workers who used picks made from antlers and the bones of aurochs.

In Norfolk, England, there is an area of deep pits and underground galleries called Grimes Graves from where flint was being mined about 4,000 years ago and made into tools that were traded. The mines are surrounded by spoil heaps and sites where a substantial community lived and grew food. Grimes Graves is possibly the earliest industrial site and it is very large. The mined area alone covers more than 4,300 acres (1,750 ha). Clearing the forest presented those early farmers with no serious difficulty.

■ FARMING SINCE ROMAN TIMES

At its peak in about 400 B.C.E., the Roman Empire extended from Turkey to Spain and from North Africa to the north of Britain—beyond which the land that is now Scotland

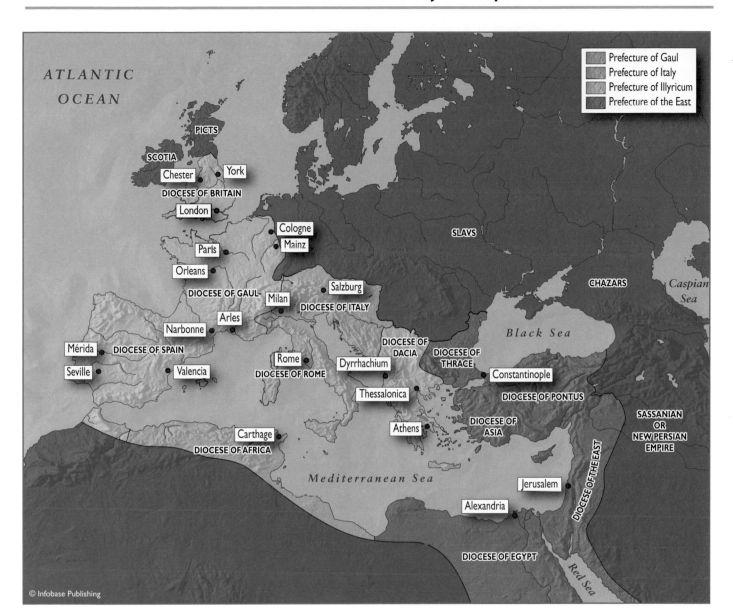

ATLANTIC
OCEAN

PICTS

SCOTIA

Chester · York

DIOCESE OF BRITAIN

London

Cologne

Mainz

Paris

Orleans

DIOCESE OF GAUL · Milan

Salzburg

DIOCESE OF ITALY

Arles

Narbonne

Mérida DIOCESE OF SPAIN

Seville · Valencia

Rome

DIOCESE OF ROME

Dyrrhachium

DIOCESE OF DACIA

DIOCESE OF THRACE

Thessalonica

Athens

Carthage

DIOCESE OF AFRICA

Mediterranean Sea

SLAVS

CHAZARS

Caspian Sea

Black Sea

Constantinople

DIOCESE OF PONTUS

DIOCESE OF ASIA

SASSANIAN OR NEW PERSIAN EMPIRE

DIOCESE OF THE EAST

Jerusalem

Alexandria

DIOCESE OF EGYPT

Red Sea

Prefecture of Gaul
Prefecture of Italy
Prefecture of Illyricum
Prefecture of the East

© Infobase Publishing

The Roman Empire at its greatest extent, in 400 C.E.

was inhabited by the Picts. The map shows the area under Roman rule at that time.

Britain was very important to the Romans (see "Roman Britain: Forest to Factory" on pages 232–233). It held rich mineral deposits. Tin from Cornwall, on the southwestern peninsula of the country, was exported to the farthest regions of the Roman world. Still more valuable, perhaps, the British climate was ideal for cereal farming. The Roman Empire existed during a climatic optimum, which is a time when the climate is relatively warm. Britain was farmed intensively. Roman estates were large and the farmers often wealthy. Most of their produce was exported.

Britain had been farmed since long before Roman times. Iron Age people lived in circular houses made from wood and mud and with thatched roofs. Their chieftains usually lived in defensible locations in hill forts. When the Roman legions arrived, the chieftains traded with them, but it was some time before their lifestyle changed. When it did, they moved down into the Roman towns, built rectangular houses from stone rather than wood, and adopted fashionable Roman customs, but the Iron Age farmers continued farming as they had done for centuries.

Farms were often managed on an in-field out-field system. Around the farmhouse and its outbuildings there were small, irregular fields where vegetables and herbs were grown. These were the in-fields. Cattle and sheep were brought into the central area at night and in winter, and their manure was dumped on the in-fields, which were consequently very fertile. Surrounding the in-fields were the out-fields. These were larger and were used to grow crops, including wheat, barley, rye, oats, flax, and hemp. Beyond the out-fields, there were still larger fields, called parks in many parts of the country, where livestock grazed. These outermost fields marked the limits of the farm, but farmers shared the use of the land between farms, known as the waste or wilderness, to graze their stock and gather fuel and other materials. "Waste" simply described uncultivated land, and a "wilderness" was where wild plants grew.

Beyond the limits of the empire, tribes hostile to Rome were on the move. Roman garrisons were frequently attacked, and sometimes they were overwhelmed. Their opponents, whom the Romans called barbarians, invaded the Empire, and eventually they fatally eroded its military and political power. The Visigoths and Ostrogoths came from Scandia (Scandinavia), the Franks and Vandals from northern Germany, and the Huns from central Asia; the Angles and Saxons, from what is now Denmark, invaded Britain, conquered it, and settled.

The Angles and Saxons were farmers. They occupied the Romano-British farms. By then these farms varied widely in their design with fields of many different sizes and shapes. What changed Roman and Saxon farming methods was not so much the nationality of the farmers as the introduction of plows suitable for cultivating different types of soil (see "Plows and Ditches" on pages 233–234).

Medieval Farming

In 1066 armies led by William of Normandy invaded and conquered Britain. William the Conqueror became King William I, and the Middle Ages began. Under Norman rule, farming changed because the political structure changed. Around every village there were large, open fields. Each field was divided into strips. The strips were owned or rented by individuals, but they were farmed in common. Although this was the general pattern, it was extremely flexible. Villagers adapted it to suit local circumstances.

The most widespread arrangement was that each village family would own or rent a number of strips in the big arable field. Their strips were not in a block but scattered randomly, reflecting the fact that farmers acquired strips as they fell vacant. Each family farmed its own strips, but after the harvest all of them turned their livestock onto the arable field to feed on the residue of the crop and manure the field as they did so. The farmers also shared the outlying meadows for grazing, and every few years each arable strip was left fallow, and livestock were allowed free access to fallow land. Beyond the farm, everyone could graze livestock on the waste.

Medieval farming worked because it was very strictly regulated. Farmers met regularly to agree on the rules. These determined which crops could be grown on each strip and the number of animals each farmer could release onto the arable field and into the wasteland.

The Agricultural Revolution

Most of the large common fields had disappeared by about 1500, and by 1600 the medieval strip fields survived only in certain places. The forests and other areas of uncultivated land remained largely unchanged until the 18th century.

The 17th and 18th centuries were a time of change throughout Europe. During this period, often called the Enlightenment, old ideas were challenged, science flourished, and philosophical, political, social, and economic theories and reforms were being debated. Farmers were often reluctant to change their methods, but they came under pressure to do so. Between about 1600 and 1700, the population of England increased from approximately 4.5 million to 6 million, and the populations of other countries were growing at comparable rates. More food was needed, which meant that either more land had to be brought into cultivation or that existing farms had to become more productive.

Farming changed, but the new techniques spread fairly slowly, and it was not until the end of the 18th century that they were being applied widely. As early as the late 16th century, many farmers were in the habit of sowing grass on land that had previously grown arable crops and leaving the land for several years to recover its fertility. At the same time, they plowed up grassland to grow arable crops. This technique became more popular. It increased the productivity of the land because, as well as giving the soil time to recover lost nutrients, farmers were able to sow new and more digestible grass varieties as well as crops such as clover and alfalfa that added nitrogen to the soil. They introduced other crops, especially turnips, for feeding their livestock. These improvements allowed farmers to keep more livestock, and the livestock manured the land. Soil fertility improved. Vegetable crops, such as various kinds of cabbages, carrots, parsnips, and potatoes that had once been grown only in gardens, became farm crops, grown in arable fields.

New machinery began to appear. After experimenting for several years, in 1701 the English lawyer and farmer Jethro Tull (1674–1741) invented a seed drill—although it was not in widespread use until after 1750. This device, which could be drawn by a single horse, sowed seeds at regular intervals in rows, allowing the farmer to destroy weeds between the rows. It was not the first seed drill—the ancient Egyptians had used them—but it was much more efficient than its rivals. The drawing on page 227 shows Tull's drill. Its use increased yields and prevented weeds from depleting the soil by consuming plant nutrients. In 1714 Tull invented a horse-drawn hoe to mechanize weed control between rows.

Plows were also improving. Medieval plows varied widely in design. Some were light and others heavy, some had wheels, but all of them were made from wood and iron by local carpenters and blacksmiths who followed traditional patterns. By 1700, however, agricultural writers were

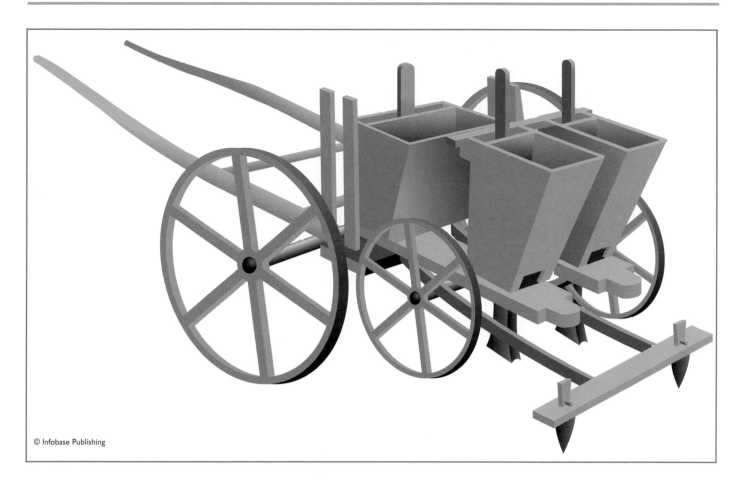

© Infobase Publishing

The seed drill invented by Jethro Tull. This is a model made in about 1730.

criticizing all the existing plows as being too heavy and difficult to control. Gradually, improved designs came into use that turned a better furrow and demanded less work from the plowman and the draft oxen or horses.

These changes, in crops, farming techniques, and farm machines, greatly increased agricultural productivity. Together they amounted to what came to be called the agricultural revolution, which established the general patterns of farming throughout Europe and North America until the even greater changes of the 20th century. The more recent changes came with the introduction of farm chemicals, new and more productive crop varieties, and machines driven by internal combustion engines that replaced draft animals, freeing for other uses the land that had formerly been needed to feed them.

Peasant Farming and Industrial Farming

Farm machines are expensive. They also need space in which to maneuver. Harvesting machines, especially, are able to work efficiently only in large fields. Consequently, the mechanization of farming favors large farms that have the space to utilize the machines and that generate an income big enough to justify the capital investment.

Agricultural chemicals, like seeds, are purchased as a cost of production. The farmer must apply them long before the crop is ready to harvest. Either the farmer must have sufficient funds to meet these costs in advance and wait to recoup them until the crop is sold, or a bank or credit facility must make a loan. In both cases, the owners or tenants of large farms are more likely to have funds or access to credit than are farmers with smaller holdings.

Highly mechanized farms making effective use of agricultural chemicals are large farms, and their size brings other advantages. Compared with small farms, large enterprises can market their produce efficiently and profitably and, because of the quantities involved, buy the materials they need at the lowest prices. Economies of scale mean they can buy cheap and sell dear.

In some countries, however, the system of land tenure makes it difficult for farms to expand and grow to a large size. When a landowner dies, instead of being sold or passing to a single heir, the property is divided among members of an extended family. The heirs may prefer not to retain and work the land, but they are entitled to do so if they choose. The result is that holdings tend to become smaller, and it is

A fairly typical village in central Europe, with the church dominating the countryside from a hilltop, dwellings clustered in the valley, vineyards providing the main economic activity, and forested mountains (extinct volcanoes) in the background   *(Fogstock.com)*

usually impossible for the families owning them to accumulate enough capital or obtain credit to expand. This is peasant farming, and its alternative, based on large, mechanized farms, is sometimes called industrial farming.

The two types of farming produce quite different landscapes. Its high level of mechanization means that industrial farming employs few workers. Those who do work on the farms often live in nearby towns and commute. The large size of the farms, in the case of arable farms with large fields, produces open landscapes with few houses and small villages.

Peasant farming produces much richer landscapes. More people live in the countryside. They require services, so there are larger villages and more small towns to supply them and to accommodate those employed in the shops, workshops, small factories, schools, and other facilities. Agriculturally, however, this type of farming is less efficient at producing low-cost food.

■ USING POLLEN TO STUDY THE PAST

Seed plants reproduce by releasing pollen from the male organs (see the sidebar on page 229 "What Is Pollen?"). This is transported to the female organs, where it releases sperm for fertilization. Pollen grains themselves are extremely small and light. In some species they are transported by the wind, in others by animals such as insects and birds, and they are produced in vast quantities to allow for considerable wastage.

Inevitably, most pollen grains are lost. Of these, some fall into lakes or bogs and become incorporated into mud that accumulates as sediments. Many years later, those sediments can be examined. By then, the environment will probably have changed greatly. What was once a lake or wet ground may be dry and hard. Nevertheless, provided that it has not been disturbed, the soil can be identified as having once been an accumulation of sediment. The date the sediment formed can then be calculated, either by comparing it with other material nearby, the date of which is known, or by measuring the decay of radioactive carbon-14 in once-living material trapped in it.

Often, scientists are able to recover pollen from the former sediment. This is possible because each pollen grain is

What Is Pollen?

Pollen is the mass of minute grains—pollen grains—that develops within the anthers (the male organ) of a flower or on the pollen cone of a gymnosperm. A pollen grain holds the male sex cell, which contains three haploid nuclei (nuclei that carry only one set of chromosomes). One is the tube nucleus and the other two are sperm nuclei. When the pollen grain arrives on the stigma of the female flower a tube grows from it, containing the tube nucleus. The pollen tube penetrates the stigma and extends down the style and into the ovary. There it grows toward an ovule and eventually reaches the embryo sac where the tip of the tube ruptures. The two sperm cells travel down the pollen tube and enter the embryo sac where one sperm cell fuses with the nucleus of the ovum (egg) and the other sperm cell fuses with the two polar nuclei present in the embryo sac, giving rise to the triploid (three chromosomes) endosperm.

Every pollen grain is protected by a tough outer coat called the exine. Although most pollen grains remain viable for only a very short time, the exine resists decay and under suitable conditions can survive for thousands of years. The size and shape of the exine varies according to the family, genus, and sometimes even the species of plant. These external characteristics allow scientists to identify the types, and sometimes the species, of plants that once grew in soil that can be dated.

encased in an outer coat, called an exine. Exines are so tough that they are virtually indestructible if they become trapped in acidic, airless conditions, such as those in the wet mud of a bog or on the bed of a stagnant lake. The outer coat of a pollen grain can survive unaltered in such places for many thousands of years.

Pollen Analysis

In the early years of the 20th century a Swedish geologist, Lennart von Post (1884–1951), realized that preserved pollen might be used to reconstruct past patterns of vegetation. Von Post was particularly interested in forests, and it was with forests that his work, first published in 1916, mainly dealt. Pollen studies were already being made but for a quite different purpose. Many people suffer from hay fever, a distressing complaint caused by an allergy, commonly

to pollen, and medical researchers were studying airborne pollen to identify the plants that released it. Von Post was among the first scientists to study pollen preserved from the fairly distant past. In 1929 he was appointed a professor at Stockholm University, a post he held until 1950.

Pollen grains evolved along with the rest of the plant that produces them. Just as plants can be differentiated into families, genera, and species, so can their pollen grains, at least up to a point. Pollen grains from different plants have different shapes, and their exines are marked with grooves and pits that form characteristic patterns. Grasses and herbs can usually be identified from their pollen down to the taxonomic level of the family. Trees and shrubs can be identified to the level of genus. A very few plants can be identified at the species level. Spores from plants that do not produce seeds are sometimes preserved in the same way and can also be used for identification.

The study of such ancient pollen and spores is called pollen analysis, and it has also led to another scientific discipline, palynology (although some people use the two names interchangeably). Palynology involves classifying and plotting the distribution of pollen, spores, and other microfossils.

Material for pollen analysis is obtained by drilling cores and then removing samples from carefully measured depths. The pollen grains must be separated, which involves treating the sample with reagents and centrifuging it (spinning it very fast). Then the pollen must be mounted on a microscope slide. Pollen grains are very small and can be studied only under a microscope with a magnification of about ×400 or higher.

Having extracted pollen grains and mounted them, the scientist identifies them by comparing them with grains that are labeled. To do this, she or he may work from published pictures of pollen grains or by referring to a library of actual grains mounted on slides. Scientists who specialize in pollen analysis accumulate libraries of their own slides for this kind of reference use.

What Does Pollen Tell Us?

Someone looking through a microscope and identifying a pollen grain might suppose that the identification proved that a particular plant once grew at the place from which the pollen sample came. The abundance of particular pollen might also reflect the abundance of that plant. Unfortunately, it is not so simple, and pollen has to be interpreted with great care.

In the first place, some plants produce much more pollen than do others. Alder (*Alnus*) and birch (*Betula*), for example produce twice as much pollen as elm (*Ulmus*) and spruce (*Picea*) and eight times more pollen than either lime (*Tilia*) or maple (*Acer*). When the grains on the slide are counted, allowance must be made for this difference in output.

The method by which the grains were transported also affects their distribution. Pollen grains from coniferous trees have small sacs that help to keep them airborne, so if the pollen was carried high into the air and there was a wind blowing constantly from a particular direction, conifer pollen would travel farther before falling than would pollen from broad-leaved trees. This would make less difference if the wind carrying the pollen was blowing at a low level because friction with plants would make the airflow turbulent, and eddies would bring the pollen grains to the ground. Pollen can also be carried by water, and this, too, will tend to sort one type of pollen from another.

The pollen retrieved from a sample of sediment represents what is known as a death assemblage. That is to say, the pollen has been transported, possibly over a long distance, and sorted at the same time. The proportions of the plant species that are present as pollen may be very different from those an observer would have seen at the site when the plants were alive. Suppose, for example, that the sediment containing the pollen came from what had been a lake bed and that on one side of the lake the forest was mainly elm, propagating itself vegetatively, and the forest on the other side was predominantly oak. What no one has any way of knowing is that in those days, but not now, the prevailing wind blew from the elm toward the oak. The wind would have carried elm pollen across the lake, with some falling into the water, but much of the oak pollen would have been carried away from the lake. Consequently, elm would be greatly overrepresented in the sample. The aim of pollen analysis is to convert the death assemblage revealed by the sample into a life assemblage, accurately representing the plants that really grew at that place.

Pollen Frequency

There are two ways in which scientists can report the abundance of pollen. Often, the amount of pollen from each family or genus is counted as a proportion of the total amount of pollen present and reported as a percentage. This is called the relative pollen frequency (RPF). For example, the report might state that the sample contained 50 percent pine, 30 percent oak, and 20 percent alder. Some scientists find this less satisfactory than reporting the absolute pollen frequency (APF), which is the total number of pollen grains for each plant type.

It was Lennart von Post who pointed out that the RPF is misleading unless the relative pollen productivity of each plant is known, as well as the way the pollen is dispersed and the effect of the form of dispersal on the distribution. By reporting percentages, the RPF suggests the proportions of the trees in the original forest, which it is not entitled to do. That is why many scientists prefer to use the APF, although the difficulties von Post mentioned are much less severe

today than they were then because to a large extent information now exists on pollen productivity and the means and effects of dispersal.

Obviously, once the pollen grains have been mounted on slides, they must be counted, one type of grain at a time. Each sample must contain at least 200 identifiable grains and preferably more because numbers smaller than 200 cannot be interpreted reliably. After all, a few grains could have been blown for hundreds of miles and bear no relation whatever to the vegetation among which they fall; it is much less likely that hundreds of grains will arrive in that way. Since the purpose of the study is often not simply to determine the original vegetation pattern but to observe changes in it over a period of time, there will be several samples to examine collected from different soil levels. Once the grains have been counted, the resulting numbers, the raw data, are subjected to various statistical procedures to standardize the way they are interpreted.

Finally, the results are often presented as a pollen diagram. This shows a vertical cross section through the site, with the soil depth marked. Each plant then occupies its own column as a block varying in width according to the percentage of the total pollen it represents at each level (the RPF). The illustration on page 231 shows an imaginary pollen diagram showing what the result looks like; a real diagram would be headed with the name of the site from which the samples were taken.

History of Plants and History of Climate

Dates can often be assigned to depths in a sedimentary profile, so a pollen diagram shows the history of the vegetation at a particular site. Data from a number of sites can then be combined to give a vegetation history for a wide area. These have shown, for example, that 10,250 years ago the forest in England and Wales was dominated by birch (*Betula*), pine (*Pinus*), and juniper (*Juniperus*). At 9,798 years ago ±200 years (there is always a plus-or-minus margin of error in dating), the forest was of birch, pine, and hazel (*Corylus*). At 8,880 ± 170 years ago, birch had disappeared, leaving a hazel and pine forest. At 8,196 ± 150 years ago, the forest was of pine, hazel, and elm (*Ulmus*). At 7,107 ± 120 years ago, the composition had changed yet again, this time to oak (*Quercus*), elm, and alder (*Alnus*). Finally, at 5,010 ± 80 years ago, the forest was of oak and alder.

A history of this kind is interesting, but a pollen diagram has an even more interesting story to tell. Tree species have climatic preferences. If one species replaces another over a wide area, it is probably because the climate changed. Birch, pine, and juniper suggest a cold climate, like that of the taiga or the Canadian boreal forest. The arrival of hazel indicates rather warmer conditions. The warming seems to

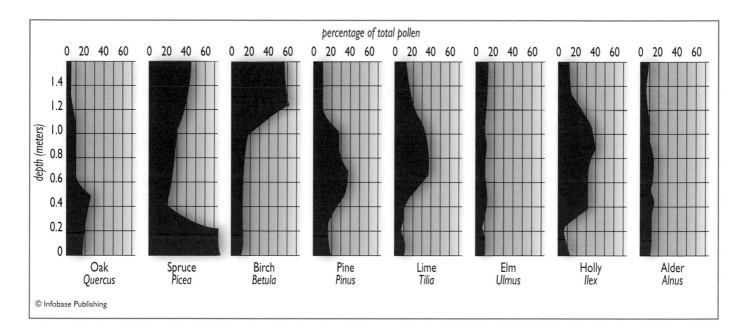

percentage of total pollen

depth (meters)

Oak
Quercus

Spruce
Picea

Birch
Betula

Pine
Pinus

Lime
Tilia

Elm
Ulmus

Holly
Ilex

Alder
Alnus

© Infobase Publishing

A pollen diagram

have continued, until the climate was mild and wet, which are the conditions that suit oak and alder.

During the 80 or so years since they were introduced, the techniques of pollen analysis have been greatly refined and have been applied at many sites. They have allowed scientists to compile a history of vegetation for many areas and, from that, a history of the climate and of the environment in which animals and humans lived.

FORESTS IN CLASSICAL TIMES

Dionysus was the Greek god of the vine. The Romans knew him as Bacchus, and the drunken rituals performed by his worshippers were known as Bacchanalia. Drunken revelries are still described as *bacchanalian*. As with most Greek gods, there was much more to Dionysus than that. For one thing, Dionysian festivals gradually developed into theatrical performances for which plays were written to be performed on a platform, called a *skene,* and accompanied by a commentary and dances performed by a *khoros.* A modern performance of a play in a theater or the showing of a movie is an event that can be traced all the way back to the worship of Dionysus. *Scene* and *chorus* are Greek words that are still in use from those days so long ago.

Dionysus was often accompanied by pans, satyrs (the Romans called them fauns), and similar spirits. These were partly human and partly goats, and Dionysus himself was

sometimes represented as a goat. Throughout Europe, spirits in the shape of goats were woodland deities, perhaps because goats like to wander in the forests where they browse on the leaves and nibble the bark. All of them, and Dionysus himself, were associated with the cultivated crops, which they could encourage, but especially with the forests.

Gods do not appear suddenly out of nowhere. Dionysus was probably introduced to Greece from Thrace, a country occupying most of what is now Bulgaria, or from Phrygia in what is now Turkey. Wherever he came from, the Greeks seem to have known of him by about 1200 B.C.E., and he is mentioned, though as only a minor deity, in the poems attributed to Homer and probably dating to about 850 B.C.E.

Ancient Greece

People would not have worshipped woodland gods and spirits unless they were familiar with forests. At one time, therefore, a substantial part of Greece must have been forested. There are references in the writings of Homer to "wooded Samothrace" and "wooded Zacynthos." Trees and forests were often used in metaphor. For example, Homer told in the *Iliad* how Sarpedon, a Lycian leader and ally of the Trojans, fell in battle "as falls an oak or silver poplar, or a slim pine tree, that on the hills the shipwrights felled with whetted axes, to be timber for shipbuilding." Elsewhere in the *Iliad* Homer described a battle as raging like a forest fire: "Through the deep glens the fierce fire rages on some parched mountainside, and the deep forest burns, and the driving wind whirls the flame every way."

Deforestation began early, however, mainly to free land for farming. In *Works and Days,* the Greek poet Hesiod (ca. 800 B.C.E.), who lived in Boeotia in southern Greece, described the simple life and work of a farmer and extolled its virtues to his brother, Perses. Apparently Perses had made off with the bulk of the family inheritance and opted for a life of luxury and ease. Unlike the first people to settle in Greece whose tools were of stone, these farmers had metal tools with which they could clear woodland much more quickly. The mountains were deforested first and then, as immigration and natural increase led to a rise in the size of the population, the lowlands were cleared. The Greek historian Thucydides (ca. 465–400 B.C.E.) reported in his *History of the Peloponnesian War* that ". . . the productiveness of the land increased the power of individuals; and in turn was a source of quarrels by which communities were ruined, whilst at the same time they were exposed to attacks from without." Strife forced people to migrate, and many sought asylum in Athens, but the poor soils of southern Greece were unable to feed the burgeoning population, and eventually Athenians were compelled to establish colonies elsewhere.

When Athens found itself at war with neighboring states, it needed a navy. The age of the Greek wars, leading to the establishment of the Greek Empire, reached their climax during the lifetime of Pericles (ca. 495–429 B.C.E.). By that time, timber for shipbuilding had to be imported. Athenian expansion led to the greatness of Athenian civilization but also to the loss of most of the Greek forests.

Eventually the rate of deforestation slowed. The historian Polybius (ca. 200–some time after 118 B.C.E.) wrote of forests in the mountains of his home country of Arcadia in the Peloponnese. The clearance was most thorough in Attica, the southernmost part of mainland Greece. Plato (428 or 427–348 or 347 B.C.E.) stated in the *Critias* that the trees of Attica had been felled to provide building materials for Athens, leaving the uplands bare.

Other Mediterranean lands fared better. In his *Geography*, the geographer Strabo (born ca. 63 B.C.E.; it is not known when he died) reported that the mountains of Spain were densely forested and that there were large forests in many other parts of the Mediterranean region. However, Strabo also reported that the lowland forests of Cyprus had been cleared long ago to provide timber for shipbuilding and fuel for smelting copper and silver.

Today, more than half the land area of Greece supports scrub vegetation that has replaced the original forest. There are four distinct types of scrub, called maquis, pseudomaquis, phrygana, and shiblyak. Maquis, found in the south, includes such trees as strawberry tree (*Arbutus unedo*), holm oak (*Quercus ilex*), Judas tree (*Cercis siliquastrum*), and Aleppo pine (*Pinus halepensis*). Pseudomaquis includes some oak, box (*Buxus*), and juniper (*Juniperus*). The other types of scrub consist of shrubs. Forests still survive in the hills. There are oak and chestnut forests on the lower slopes and coniferous forests on higher ground.

Rome and Its Empire

Polybius worked for the Romans at a time when they, too, were expanding their empire. Eventually, this became by far the largest empire of the ancient world. In 400 B.C.E. it extended from Jordan in the east to the Atlantic and from North Africa, including Egypt, in the south to Britain in the north. (See the map in "Farming Since Roman Times" on pages 224–228).

Many Romans became extremely wealthy from war or trade and took to investing their money in land, producing an agricultural system in some respects very much like that of the United States today. Land was bought and sold by speculators hoping to make a fast profit, and farms became very large. Small farmers found life difficult.

Most Roman landowners did not farm their estates themselves but entrusted the management to bailiffs. Agriculture was not mechanized, of course, and slaves did the actual work. The empire thrived on trade. This included trade in food, and the demand for food was huge. Roman Britain, comprising what are now England and Wales, but not Scotland or Ireland, became one of the most important food-producing regions of the empire and a major food exporter.

When the Romans invaded Britain in 43 C.E. with a well-equipped army of 40,000 soldiers, they found a land that had been inhabited and much of it managed for a very long time. It is impossible to know for certain what Britain was like then, but some archaeologists believe that clearance of the original-natural forest began in about 1400 B.C.E. during the Bronze Age and that by 1000 B.C.E. the total area of forest in Britain was smaller than it is today. Other scientists think that about half of the original forest had disappeared by 500 B.C.E. and that the major clearance came later during Iron Age and Roman times.

Roman Britain: Forest to Factory

There is no doubt that the Romans, with their large estates and superior technologies, greatly intensified British farming. Agriculture expanded over most of the lowlands, large quantities of wheat and wool were exported, and landowners became very prosperous.

The Roman conquest also necessitated a great deal of building. The invasion and occupation was proceeded by a series of military advances and the establishment of forts, each garrisoned by up to 1,000 soldiers. These forts, with accommodation for the troops and the civilians supporting them, were made from wood. Once an area had been subdued, colonists followed and built houses, also from wood. There was a great demand for timber.

At the same time, Britain was also becoming industrialized. The country possessed valuable metal ores and fuel was needed to smelt them. Iron and lead were exported, and some tin and silver. Fuel was also needed to make pottery, bricks, and tiles and to refine salt. Before the Romans arrived, the Britons obtained their salt by heating stones in a fire, then throwing them into the water of a saline spring and scraping off the crust of salt that formed on the stones. The Romans taught them to produce salt by boiling brine slowly in a shallow pan—over a wood fire.

In the British climate, grain often needs drying after harvest to prevent it from germinating. This is especially important if the grain is to be exported because germination during transport will greatly reduce its value on arrival. Exporting farms needed grain driers. Romans liked their homes centrally heated, and they enjoyed hot baths. It all needed fuel.

Wood was the only fuel, used directly or in the form of charcoal (see "Making Charcoal" on page 248), but huge trees are not ideal. With no power tools to help, felling and then cutting them into small pieces is hard, slow work. It is more probable that each furnace, villa, and town obtained its fuel from areas nearby where small trees were *coppiced* (see "Coppicing" on page 291) on a permanent basis.

In much of the country, it is likely that when an opening appeared naturally in the forest, livestock were allowed in to graze. The animals destroyed young seedlings and prevented the clearing from closing again. Little by little, what had been closed-canopy forest gave way to a more open countryside, with large areas of pasture. Trees that survived grew to a much larger size because they were no longer surrounded and shaded by other trees. Then, as grain crops became more important, the grasslands were plowed.

Gaul (France) was conquered by the Teutons in the fifth century, and Britain was then isolated from the rest of the empire. Many Roman troops had already been withdrawn, but for a long time the administration continued independent of Rome itself. There was no time when the Roman occupation of Britain can be said to have ended. It simply decayed, and Celtic influences became much stronger. By this time, most of the British original-natural forest had disappeared and, with no plantation forests to take its place, Britain was no more forested than it is now.

■ EUROPEAN FORESTS OF THE MIDDLE AGES

The Roman occupation of Britain lasted from the year 43 C.E., when an army of about 40,000 men led by Aulus Plautius landed in Kent, until early in the fifth century, when the Teutons conquered Gaul, leaving Britain cut off from the main part of the empire. After that, the central government in Rome ceased to send senior officers and

governors, and the rulers of Roman Britain were increasingly left to their own devices. Many troops were withdrawn before the route south was finally closed, but not all of them, and there was no sudden end to the occupation. The empire did not depart, authority was not passed from one government to another, no flags were lowered or raised. Rome simply became irrelevant. By the sixth century, the remaining Romano-British forces and government were retreating as Saxons invaded from the east.

During the four centuries of direct Roman rule, Britain was made highly productive. Its agriculture and industries served Rome, and to provide the land and fuel they needed, forests were cleared. What happened in Britain was similar to the way that many parts of Roman-occupied Europe were developed. Where land was to be settled and farmed on behalf of the imperial government, the area was laid out according to a careful plan, with field and estate boundaries and roads usually arranged in a grid pattern. It was not Roman organization and planning that made the agricultural expansion possible, however, but Roman technology.

Plows and Ditches

Before the Romans arrived, only the lighter soils could be cultivated, because farmers had only a simple plow that scratched a shallow rut in the ground, throwing the soil to either side. To plow their land thoroughly, they had to do the job twice, with the second set of furrows at right angles to the first. Even then the plows were useless on heavy, clay soils, nor were the farmers able to drain low-lying wet ground.

The Romans knew how to remove surplus water by digging ditches. This greatly increased the area suitable for cultivation, often bringing into production fine-textured, silty soils that were highly fertile once they were no longer waterlogged. The Romans also brought with them a variety of plows, each suitable for a different type of land. The *romanicum* was used on heavy soils, for example, the *companicum* on rather lighter soils, and the *ard* on the lightest soils. Some plows had a coulter, which is a knife fixed to the front of the share that cuts through the soil, and some had moldboards which are winglike attachments that turn over the soil to one side, burying weeds.

These technological improvements—and the Romans also introduced a range of other agricultural tools—made it possible to expand farming onto the more difficult soils. To achieve this, it was first necessary to clear the land, and that meant removing the forest. As new land was brought into cultivation, the farmers needed housing, of course, so some timber was felled for immediate use. Wood was also needed as fuel, for providing the hot water and central heating wealthy Romans expected in their homes, and for the public bathhouses used by the soldiers and poorer citizens, as well as for heating grain-dryers, kilns for making bricks, mortars, and pottery, and furnaces for working metals. It was agricultural expansion that drove the forest clearance, however, rather than the demand for timber.

What historians know about land use during this period is based mainly on archaeological detective work. No one at the time bothered to record what was happening, but much later a detailed record was compiled. In 1066, the most famous date in English history, William of Normandy defeated the army of the Saxon King Harold at Hastings, and Britain came under Norman rule.

Domesday

William wished to know more about the land he now possessed, and in 1085 he ordered a survey of his realm. Commissioners were sent to every corner of England (although some counties were not included because at that time they were not part of England) and took evidence, on oath, concerning every community. Everywhere they went, the commissioners noted the size of the population as well as detailing such matters as how much land the community had and who owned it, how the land was used, and numbers and species of livestock. A second team of commissioners followed behind to verify their findings. The task was completed in 1086, having taken less than one year, and the resulting documents were copied into two volumes. The survey was exceedingly thorough, so much so that the people compared it to the interrogation that they expected to receive on the Day of Judgment. They called the completed survey the Domesday Book.

Along with much else, Domesday records the area of forest or woodland that was available to each community, and the result is startling. Many settlements had no woodland at all, and others had very little. A place called Polroad in Cornwall, for example, had a population of eight (including one slave), 17 acres (6.9 ha) of pasture, but only three acres (1.2 ha) of woodland. Some villages had 100 acres (40 ha) of woodland or more, but when the total area is added together it becomes evident that England was no longer the forested country it once had been. Some parts of the country had more forest than other regions, but over large areas the landscape was one of cultivated fields and widely scattered, small woods. On average, it seems that forests covered about 7 percent of the land area. This figure refers only to England (Wales and Scotland were separate countries then). Today almost 12 percent of the land area of the United Kingdom (England, Wales, Scotland, and Northern Ireland) is forested. This comparison is slightly misleading, however, because forests cover a much higher proportion of the land area in Scotland than they do in England, so the proportion of forested land in England is likely to be about the same now as it was 900 years ago. By the time of Domesday, the original forest had largely been cleared. Such

forest as remained would have comprised mainly native species, but there may also have been some of those species that had been introduced during Roman times, such as the sweet chestnut (*Castanea sativa*).

Elsewhere in Europe, forests at that time were more extensive. To this day, a little more than 28 percent of the land area of Poland is forested, and the Białowieża Forest, covering 482.5 square miles (1,250 km²), is largely virgin forest, unaffected by human intervention. In medieval times, European forests no doubt covered a larger area than they do today, but probably not a much larger area.

Landscapes of Medieval Europe

Lowland England has been fairly densely populated for much of its history, and even without the clearance in pre-Roman and Roman times, much forest would have been lost to agriculture. The effect of this can be seen when the English countryside is compared with that in Scotland, which is much more sparsely populated. In the 16th century, most of Highland Scotland below the tree line was covered with forest of pine, birch, and oak. In 1618 a poet called John Taylor walked from London to Edinburgh, a distance of 380 miles (611 km) on modern roads, and then continued northward. On his way to Braemar, inland from Aberdeen, he walked for 12 days without seeing a single sign of human habitation of any kind nor a single cultivated field, and he began to fear that he would never again set eyes on a human dwelling. Braemar

---

Scotland, showing Edinburgh and Braemar. John Taylor would have crossed the Firth of Forth by ferry; today it is spanned by a rail bridge and a road bridge.

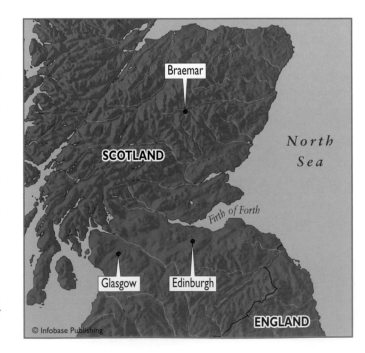

is still surrounded by forest, although the countryside between Edinburgh and Braemar is mainly farmland. The map on page 234 shows the locations of Edinburgh and Braemar.

The picture that emerges over the lowland regions of Europe as a whole is of a generally forested landscape, but with areas around cities, towns, and villages cleared for farming. The size and number of such cleared areas obviously depended on the size of the human population farmers had to support, so in densely populated regions such as England, Flanders, and the Netherlands the countryside was one of cultivated fields, open pasture land, and isolated, usually small, woods. This is the kind of landscape shown in paintings by European masters from the late medieval period and a little later, for example in "The Harvesters," by Pieter Brueghel the Elder, painted in about 1565.

Forests for Food and Raw Materials

Vast tracts of forest were set aside for hunting. This was a sport for the wealthy, of course, but it was also an important means of procuring food. The game that was killed was eaten, and to ensure a plentiful supply, hunting forests were carefully protected. That still left large areas, remote from towns and farms, where the forest was what today we would call wilderness.

In inhabited areas, however, what remained of the forest was an important resource. It was often called waste or wasteland, suggesting it was land for which no one had any use, but that is not what the word *waste* meant. The waste was simply the uncultivated land beyond the better pasture and the fields where crops were grown. Although some areas of forest were privately owned, usually the waste was land used by the community as a whole rather than being reserved for the exclusive use of its owner.

Not only were the forests used, they were used fairly intensively. Except on the moors, where peat was obtainable and in those places where coal could be dug from shallow mines, wood was the only fuel for cooking and heating (see "Forests as Sources of Fuel" on pages 246–248). Ordinary homes, sheds, and barns were built wholly or partly from wood. Wood was needed for making tools, boats, wheeled vehicles, boxes, furniture, fences, and a vast array of everyday items. People used metals, of course, but the fuel that was used to smelt and forge them was charcoal, made from wood.

These different needs called for different kinds of wood, and this led to the development of different kinds of forest. Big machines and power tools had not been invented. There was no way that people could take just any kind of wood, chop it into chips, and then reassemble the chips into any shape they chose. Nor could they easily cut large timber into particular shapes. Instead, they had to choose the tree to fit the purpose. Timber was felled as the local carpenters and builders needed it, and it was used at once without being seasoned. The bent and warped timbers that can still be seen in surviving medieval buildings show clearly that these were installed while the wood was still green.

Needs varied. One village might be expanding, increasing the demand for structural timber, or it might be decided to build a new bridge over a river, while another village changed little from one generation to the next. Sometimes it might be necessary to buy timber from outside the community if no local tree was large enough. For instance, the post around which a windmill is built is immense and might have to be carried, although its weight made that difficult, so it would be bought as close to the building site as possible.

Over the years this led to specialization in types of forest, and forests came to vary greatly. A forest from which people cut small wood (see "Small Wood" on pages 250–254), consisting of poles to make tool handles, fences, and other small articles, was not the same as one supplying large timber for building, and both of these were different from wood pasture—that is, open woodland where the trees are too far apart for their crowns to form a closed canopy, allowing grasses to grow between them. Wood pasture supplied wood and also sheltered grazing for cattle and sheep.

Managing the Forests

Exploitation of the forests had to be regulated in order to ensure a continuity of supply. It was that regulation and the techniques associated with it that led to the emergence of different categories of forest. Powerful landowners jealously guarded their forests against overexploitation and dealt harshly with those who took more than the permitted amount of wood.

Concern to conserve the forest resource seems to have been widespread. In Germany, the amount of wood that people were allowed to take was strictly rationed. A letter has survived, written in the middle of the 12th century by a bishop of Norwich, England, rebuking his woodward (the official in charge of the forest) for giving away wood. "I appointed you the custodian of the Wood," he wrote, "not the rooter up of it . . . Guard the Wood of the Holy Trinity, as you wish to be guarded by the Holy Trinity, and to continue in my favor."

Ensuring a continuing supply of wood and timber meant that no tree should be felled unless another tree would grow up to replace it. Young saplings had to be fenced to protect them from grazing animals, and when trees were felled the damage they caused to others as they fell had to be minimized. Small wood could be obtained without killing the trees supplying it. Coppicing (see "Coppicing" on page 291) was a common practice by the middle of the 13th century.

Careful assessments were made of the amount of wood and timber particular areas of forest could yield, and permitted harvests were calculated accordingly. Those who broke the rules were fined.

Medieval forests in populated areas were managed, but they were not planted. They were what remained of the original, primeval forest, and although their composition may have changed, such change was not planned. Unlike plantations, they were permanent features of the landscape where trees did not grow in rows or in blocks where all the trees were of similar age. They were scattered irregularly with clearings here and there and paths and rivers winding through them. Commercial forestry, based on orderly plantations, came later.

FORESTS IN FOLKLORE AND LITERATURE

Little Red Riding Hood (her original German name was *Rotkäppchen*, Little Red-Cap) was a little girl who succeeded in outwitting a wolf. She had set out to deliver meat and wine to her grandmother, who lived a half-hour's walk away, deep in the forest, but when she arrived, her grandmother had disappeared and the wolf had taken her place. Like all such tales, this one has several meanings, and it is not really about genuine wolves and forests. Nevertheless, stories must

be set somewhere, and to the German country people from whom the Brothers Grimm (see the sidebar on page 237) collected their stories, the word *forest* would immediately evoke images of darkness and danger. Wolves lived in the forest, and although most of the time they kept away from people and were harmless, they could never be trusted. Occasionally they attacked for no obvious reason.

Wolves were but one of many hazards facing those who wandered in the forest, and they were far from the worst. In *Hansel and Gretel* the two children became lost in the forest and fell into the hands of a witch, who waylaid children and then cooked and ate them.

Poor Woodcutters and Abandoned Children

Unlike the woodland immediately surrounding farms and villages, the more distant forest was of little material value to ordinary people. Kings and nobles might hunt game in it, but it was no place to try to earn a living. Hansel and his sister Gretel were the children of a woodcutter who was so poor that he could barely feed the family even when times were good, and when famine afflicted the whole country, they faced starvation. Again, the story has deeper meanings. It was the woodcutter's second wife, the children's stepmother, who was determined to save her husband and herself by allowing Hansel and Gretel to perish from hun-

A glade in a forest in Hampshire, in southern England, from the book *Memorials of Hampshire*, edited by G. E. Jeans and published in 1906. This shows the kind of countryside in which many folktales are set. *(Frederick Golden Short)*

The Brothers Grimm

Jacob Ludwig Carl Grimm (1785–1863) was a grammarian (an expert on grammar), and his brother Wilhelm Carl Grimm (1786–1859) was a literary scholar. They were born at Hanau, near Frankfurt, Germany, Jacob on January 4, 1785, and Wilhelm on February 24, 1786. They attended the Friedrichs Gymnasium (high school) in Kassel, and then they both studied law at the University of Marburg. It was while they were university students that both brothers first became fascinated by the study of medieval German literature.

The brothers were very close. They were inseparable as children, and in 1805, during a visit to Paris, Jacob wrote to his brother that he hoped they would never again be separated. They were not, and they spent their entire lives together. They even worked in the same room, and it was they who called themselves Die Gebrüder Grimm, The Brothers Grimm.

Following their graduation, the brothers worked in the Kassel library until 1830 when Jacob was appointed professor and librarian at the University of Göttingen and Wilhelm was made underlibrarian. Wilhelm was made a professor in 1835. In 1837, however, they were dismissed for having signed a petition protesting against the repeal of the constitution by King Ernst August I of Hanover. Banned from the kingdom of Hanover, the brothers returned for a time to Kassel. In 1841 they moved to Berlin, where both of them were professors and both were elected members of the Academy of Sciences. Their principal project was a dictionary of the German language.

The task was so huge that despite recruiting many other scholars to help, it was not finally completed until 1960.

Their linguistic researches took the brothers to every part of German-speaking Europe. They needed to hear local people speaking in their own accents and dialects in order to trace the way sounds change in time and from place to place, and they found that the best way to achieve this was to persuade people to tell them stories. These were traditional tales, and the brothers wrote them down as they heard them. Jacob was more interested in the language itself, and it was Wilhelm who became entranced by the stories. The first edition of their collections appeared in 1812 in a book called *Kinder- und Hausmärchen* (Children's and household tales). The book was immediately popular. A second volume appeared in 1814, and several more were published in later years. The version most widely read today in English is a translation of the seventh edition, originally published in 1857. One of the storytellers the brothers used as a source was of French descent, which is why some of the Grimm stories are of French origin. It is their stories for which the brothers are loved today, but these were a by-product of their principal life's work, which was their linguistic research.

Jacob was never ill, but Wilhelm was never robust. Wilhelm died in Berlin on December 16, 1859. Jacob died, also in Berlin, on September 20, 1863. They are buried side by side in the St. Matthäus Kirchhoff Cemetery in Schöneberg, Berlin.

ger, lost in the forest, and it was the children's own cleverness, courage, and loyalty to each other that saved them. People connect images automatically, and it is as though "dwelling in the forest" was synonymous with "living in dire poverty." *The Woodcutter's Child* tells of a woodcutter who was so poor that it was all the family could do to place food on the table every day of the week. Snow White, another child abandoned by a wicked stepmother, finds refuge with a party of dwarfs, who protect her. They live, of course, deep in the forest.

Poverty was very real, and until modern times there were famines in which poor people died from hunger. Desperate people really did abandon infants in the forest because they were unable to feed them. *Babes in the Wood* is a children's story based on *The Children in the Wood*, a ballad believed to have been written in 1595, and several similar tales were written in England at about that time.

Tales That Are Told Everywhere

French folktales view forests in much the same way. "Beauty and the Beast" (*La Belle et la Bête*) teaches us that beauty is only skin-deep and that a fearsome exterior may conceal a noble heart. Probably every culture in the world has its own version of this story. The beast dwells in a palace surrounded by barely penetrable briars and brambles and set deep inside a forest. A merchant, on his way home, loses his way in the forest, comes across the apparently unoccupied palace, and plucks a rose as a gift for his daughter (La Belle). The penalty for this theft is that La Belle must live in the palace as a companion to the beast, whom she comes to love.

Like "Beauty and the Beast," most cultures have their own versions of all of the classic fairy and folk tales. We learn many of them from those compiled by Jacob and Wilhelm Grimm (1785–1863 and 1786–1859 respectively) in Germany,

What appear to be ancient, primeval forests often turn out to be plantations. This is Needwood Forest in Tutbury, Staffordshire, in the English North Midlands (from *The Story of Some English Shires* by Mandell Creighton, published in 1897). In genuinely ancient forest the trees grow fairly close together and are straight, tall, and thin. These mighty specimens, gnarled with age, were planted and allowed to grow; the picture is of parkland.

Charles Perrault (1628–1703) in France, and Andrew Lang (1844–1912) in Britain. Names and minor details change, but the story remains always the same, and so does its location. Where this involves a forest, the forest is mysterious, full of concealed hazards, a metaphor for life and the ease with which people may lose their way and fall into spiritual danger. There is always hope, of course. Woodcutters, most of whom seem to have beautiful daughters, are decent, hardworking folk living in poverty, but it is a poverty that can be ended by a stroke of good fortune, which must always be earned.

There are exceptions, however. Russian fairy tales and folktales seem to ignore the forest that covers such a vast area of their country. They set stories in countries far away on the far side of a huge forest as a literary device to show that what follows will be remote from events that happen in the everyday, familiar world, the equivalent of the "Once upon a time, in a far distant land . . ." opening. The forest is an obstacle but no more. Nor are Russian stories peopled with fairies or other spirits. There are benign forces and evil monsters, the direct descendants of pre-Christian gods and demons, but they do not lurk in the darkness among the trees, waiting to leap out at travelers. The stories have their share of magic, but it is the magic of nature, involving birds and mammals that talk and involve themselves in human affairs.

The Forest Wilderness

Fear and dislike of the forest were not confined to folktales. They were part of the way most people saw the world around them. To Shakespeare, the forest was a wilderness, mysterious, secret, and dangerous. It was land that was unused and unusable, an environment inherently hostile to humans. Like other poets and playwrights, Shakespeare thought of forests in much the same way that authors nowadays might think of deserts.

As You Like It is a comedy that takes place in the Forest of Arden, where the duke, who has been usurped by his brother, lives in exile with his supporters. The Forest of Arden was real, many Elizabethan authors referred to it, and Shakespeare described it as "this desert inaccessible, Under the shade of melancholy boughs" (*As You Like It*, Act II Scene 7). Clearly, it was a place where a large number of people could hide. In *A Midsummer Night's Dream*, the forest, this time located somewhere not far from Athens, is inhabited by fairies and other spirits. They may be charming, and their accidental involvement with humans leads to great confusion and much merriment but to no serious harm. Yet that involvement arises from the quarrel between Oberon and Titania, the king and queen of the fairies, over a changeling that Titania has taken and refuses to give to Oberon, who wants the stolen human child as a page.

Dangers Lurking in the Shadows

Metaphors succeed by using what is familiar to illuminate what is not, and many authors used the general idea of a journey through a forest to stand for the journey through life. As in life, it is difficult to find a true way through the forest among many paths in some places, but no signposts, and in others none. There are terrifying wild beasts, witches, mischievous sprites, and malevolent demons, dangers that lie in wait in every dark corner. If the parallels were so obvious, it must be because this is something like what people really believed about forests. Had it not been part of their received wisdom that forests were best avoided, the stories could not have communicated their messages.

Forests really were dangerous places. Even today, it is surprisingly easy to become lost in one, even when there are paths. Without a compass or some other means of determining direction, there are no reference points, no landmarks to show the way, and for most people, being lost in a large forest still means serious trouble. Modern people are not trained to find food, drink, and shelter in a forest, far less to deal with any medical emergency that might arise—and neither were the villagers who strayed from home centuries ago.

Outlaws and Brigands

In earlier times people could come to more serious harm than finding themselves lost or twisting an ankle. Robin Hood is a folk hero who may or may not be based on a real person, but he is portrayed as an outlaw, and forests really did harbor bands of outlaws. Few if any of them gave to the poor what they took from the rich. The Robin Hood stories probably date from the 13th century, and Robin was not originally a contemporary of Richard I. Robin Hood personified the discontent of the poor in the north of England. They were unhappy with the way they were being governed, and that discontent eventually spread to the whole of England and culminated in the Peasant's Revolt of 1381.

Robin Hood was a political rebel and he was not the only one. Eadric the Wild, or Eadric of the Woodland, was a Saxon rebel who fought a guerrilla war against the Normans from forests in the west of England, near the Welsh border. In 1069, Eadric and his followers led an assault on a Norman castle and burned it to the ground and then overcame the garrison in Shrewsbury castle and for a time occupied the castle himself.

Other forest outlaws were brigands. They operated in bands and were violent. Malcolm Musard, for example, who lived in the early 14th century, seems to have started as the leader of a gang of poachers in the forests of Worcestershire and Gloucestershire. After a time, he commanded about 300 followers and took to extortion, demanding money with threats. He was capable of carrying these out. When one landowner refused to pay up, the Musard gang felled about 200 of his trees and grazed cattle on his land. Eventually Musard was captured, but he was acquitted of all charges and appointed Chief Forester, protecting the royal forests mainly by the means he had always used and profiting handsomely in his new profession.

There were gangs of this kind in most forests, led by men who gave themselves fancy titles. Near York there was one who called himself Lionell, King of the Rout of Raveners. There were outlaw bands that set up their own forms of government and legal systems, and some of these were formidably large. William Beckwith, whose band formed a parliament and appointed local officials near Knaresborough, Yorkshire, in about 1390, commanded 500 or more followers. There were times when large tracts of the forested English countryside would be under the complete control of a gang and remain so for several years. When gang leaders were caught, they were often allowed to buy a pardon, which also involved their appointment to some official position. A brigand was thus transformed into a loyal servant of the crown and the peace was restored, but the resulting officials could not shake off their background quite so easily. Local people paid heavily.

Not all of those who found refuge in the forest were especially dangerous. Some, probably petty offenders and individuals whose political or religious beliefs rendered them unpopular with the authorities, lived as best they could and as inconspicuously as possible. They might trap game and take some wood. Both activities were unlawful, but they took only what they needed for their own use. Some worked for the charcoal makers, who asked no questions. Any traveler encountering them would find some of these ragged, filthy, folk sullen and probably verbally aggressive. Others, though, would simply be evasive. They

Robin Hood, the popular hero, shown here surrounded by his "merry men" and, unusually, with a full beard. *(Painting by Daniel Maclise, courtesy of Nottingham Castle Museum and Galleries)*

were not dangerous robbers, but meeting them may well have been an alarming experience.

People had some reason to fear the forest, and sometimes parts of forests were cleared for reasons of security. In 1712, Viscount Weymouth cleared part of Selwood Forest in Somerset and built a church on the land in order to drive out the bandits and forgers who had been operating there. Many forests harbored criminal bands, not all of them engaging in robbery. Even the trees growing beside roads outside the forest could conceal highwaymen.

The number of robbers may have been exaggerated, but the stories about them and the others who hid in the forests accorded with the general opinion, held right up until the last century, that our ancestors, as people who once lived in the forest, were necessarily barbarous, uncivilized, and rough. Our word *savage* comes from the Old French *sauvage,* from the Romanic *salvaticus,* which is from the Latin *silvaticus,* a wood-dweller (*silva* means "wood").

Wild Beasts

Forests also harbored wolves, and, rightly or wrongly, these were much feared. When royal lands were granted to tenants in medieval times, it was often a condition that the occupant must keep dogs for hunting wolves and must attempt to destroy all wolves on those lands. Still earlier, during the reign of the Saxon English king Athelstan (reigned 924–939), there were so many wolves in Yorkshire that near Filey a person called Acehorn built a retreat where travelers could shelter if they were being attacked.

Athelstan intervened in a quarrel between two Welsh princes, and as his price for resolving the dispute, he imposed an annual fine on Constantine, the Welsh king. Athelstan's successor Edgar remitted the fine on condition that Ludwall, who had succeeded Constantine as king of Wales, pay him an annual tribute in the form of the carcasses of 300 wolves. Ludwall paid the tribute for two or three years but then ceased because no more wolves could be found. At one time historians believed that the penalty imposed by King Edgar might have brought about the extinction of the wolf in Wales. It is now thought that the wolf finally disappeared from the country about a half-century later. The wolf became extinct in England some time in the late 15th or early 16th century. In Scotland, the first of several laws requiring the destruction of wolves was passed in 1427, but Scotland was much more densely forested than England, and wolves survived until the end of the 18th century. The last wolf was killed in Ireland in the late 18th century. In April 1644, the English writer John Evelyn saw the heads of wolves nailed to the gates of the castle at Blois on the banks of the River Loire, France.

Nowadays there are conservation groups and government agencies that try to preserve natural forest, but until about the middle of the last century, the "wild wood" was a fearful place, and its clearance was to be encouraged. Drive back the forest, most people believed, and many benefits will follow. Land is freed for farming, reducing the risk of famine. Visibility is improved so that approaching dangers can be seen well in advance of their arrival. Travel, and the trade associated with it, becomes easier and safer. Wild animals live farther away from farms and habitations, so they pose less risk to people and livestock. Outlawed criminals also can no longer hide in the forest and prey on communities near the forest edge. The rule of law can be extended and people can live in peace. Throughout most of history, civilization expanded by clearing the forest.

The first European settlers in North America took these attitudes with them. Behind the coastal strip and in places reaching all the way to the coast, they faced the American Forest, in their eyes a vast, desolate place haunted by wild beasts and wild men. They set about clearing the forest, not simply to win ground for cultivation, but to make the land fit for civilized people. As settlements expanded, bounties were paid for killing wolves. As recently as 1868, the state of Minnesota paid $10 apiece for the scalps of wolves, and in that year it paid wolf hunters a total of $11,300.

Bears also inhabit forests, and the bear of European forests is *Ursus arctos,* known in Europe as the brown bear and in North America as the grizzly. It is a big animal and a dangerous one, but at one time it was hunted extensively for its meat and fur and simply for sport. In Europe bears were also trapped and used for bearbaiting, an entertainment in which the bear was tethered and set upon by dogs, and captive bears were made to dance on their hind legs at fairs. People could afford to laugh at a bear that was securely fastened by a collar and a stout chain, and they could gamble on the outcome of a bearbaiting competition (the competition was for the fiercest and strongest dogs), but an encounter with a wild bear was to be dreaded.

Brown bears still survive in Russian and some continental European forests, but their numbers are few. No one can be certain, but they were probably extinct in Britain by the 10th century, although a few scattered populations may have survived for longer in remote areas. Bearbaiting and dancing bears continued because captive bears were imported for the purpose. The North American grizzly is also endangered.

The black bear (*U. americanus*) resembles the grizzly but is smaller. It occurs only in North America.

Worshipping Trees

There was a great difference between the untamed forest and those parts of the forest from which people obtained wood and fuel and where their pigs were allowed to roam. This was safe and useful forest, a valuable and valued resource, and within it people recognized trees as individuals.

Trees were once worshipped, not only in Europe and North America but everywhere in the world. Like all natural objects, trees were believed to possess spirits, but theirs were big, important spirits, reflecting the stature of trees themselves. Consequently, trees had to be treated with respect. In some places, however, only certain trees had spirits. For example, in parts of Dalmatia, a province of Croatia, trees without spirits could be cut down with impunity but not those that had spirits. A woodsman who felled a tree that had a spirit would either die quickly or be crippled for life unless he used the same ax with which he cut down the tree to behead a chicken on the stump of the felled tree. Everywhere, trees were thought to be sensitive, and they had to be treated with great care.

The major religions displaced most of these old beliefs, but they never quite banished them completely. May Day celebrations are the most widespread surviving relics of tree worship. Either on May 1 or Midsummer Day, people used to cut down a tree in the nearest wood and set it up, garlanded with flowers, in the center of the village, and they would decorate their houses with branches. Details of the customs vary from place to place, but in one way or another May Day is celebrated with trees, branches, flowers, and a great feast throughout all of Europe and Russia.

Ash and Oak

Each species of tree has its own particular qualities, and these qualities were believed to derive from the characters of the spirits inhabiting them. Ash was one of the most important. In one of the Norse creation myths, in the beginning there was Yggdrasil, the great ash tree. One of its three roots reached to the domain of the god Odin, the second to the land of the Frost Giants, and the third to the kingdom of the goddess Hel, the land of the dead. One day, the gods Odin, Hoenir, and Lódurr were walking along the seashore when they came across two trees. From the ash tree (Askr) they made the first man, and from the elm tree (Embla) they made the first woman. The gods then brought the humans to life.

Farther south, the oak was the most revered of trees. Both the Greeks and the Romans associated it with Zeus, known to the Romans as Jupiter, the highest of the gods and directly responsible for the sky, rain, and thunder. Zeus dwelt in the Greek mountains where the oak trees grow, and in Italy every oak was sacred to Jupiter.

The ancient Germans, too, held the oak to be the most sacred of all trees, and they linked it to the thunder-god Thunar, or Donar, whom we remember still every Donnerstag, Thunar's day, or Thursday. Oaks were also sacred to the Slav thunder god, known in Russian as Pyerun and to the Lithuanians as Perkunas. Perpetual fires, kindled with oak wood, burned in his honor, and to ensure good crops, men made sacrifices to oak trees and women to lime trees. Among

Celtic peoples, the mistletoe was sacred as well as the oak on which it grows.

Elder and Rowan

Lesser trees also had personality. Elder would keep away evil spirits, and a sprig of elder leaves worn in the hair or on the coat would give protection outdoors. Elder trees were grown near homes for protection, and the spirits were paid with offerings of cakes and milk.

Rowan linked this world with the next. It was inhabited by fairies, which are believed by some to be the spirits of the dead and able to fend off supernatural harm. It is lucky to have rowan growing near the home. In one Scottish legend it even saved the life of a fox. The fox was chased to the edge of a cliff by men and dogs and escaped by dropping from rowan to rowan down the cliff face.

It is not surprising that so many beliefs should be associated with forests and with the trees that grow in them. Until very recently, the wild wood really was wild and dangerous, a place to be feared. At the same time, trees supplied so many needs that they became very familiar as sensitive, spiritual beings, each possessing its own individual characteristics.

■ HISTORY OF FORESTRY

On November 5, 1662, John Evelyn recorded in his diary that the council of the Royal Society met at Gresham College, "where was a discourse suggested by me, concerning planting His Majesty's Forest of Dean with oak, now so much exhausted of the choicest ship-timber in the world." The discussion Evelyn describes led in 1668 to a program of planting in the Forest of Dean, today covering 27,000 acres (almost 11,000 ha) in western Gloucestershire, England.

John Evelyn (1620–1706) was a man of letters, the author of about 30 books, and he kept a diary from the age of 11. With his reputation as an author already established, in 1661 he joined the scientific society that the following year became the Royal Society of London. He was appointed to its council and remained active in its affairs for the rest of his life.

It was in 1662 that the commissioners of the navy asked the newly formed society to comment on the depletion of timber suitable for shipbuilding, and the matter was passed on to John Evelyn. As part of his broad education, Evelyn had studied the cultivation and uses of plants in Paris, and he loved trees. His reply to the navy came in the form of a book, published in 1664, called *Sylva, or a Discourse of Forest-Trees, and the Propagation of Timber in His Majesties Dominions.* The book was written for estate owners and was in three parts. One part gave instructions for growing apples for cider, and a second part was a general gardening manual, but the main part of the book described the cultivation of different species of trees and the uses of their timber, all written in plain but excellent English. The book was a huge success. An enlarged second edition was published in 1670, a third with an added section about soils in 1679, and a fourth enlarged still further this time by the addition of a section on salad crops in 1706, soon after Evelyn's death. More editions appeared in subsequent years, some consisting of the forestry section by itself, the last edition appearing in 1825.

Conserving Forests

For about 150 years *Sylva* was the standard work on forestry, and its publication is often taken to mark the beginning of the deliberate planting of trees as a crop. Obviously, it was not quite that simple. Evelyn did not invent the idea of organized tree-planting. Indeed, he included in *Sylva* examples of landowners who had already established timber plantations. He was not even the first propagandist for tree-growing, but he was certainly the most famous and, perhaps, the most persuasive.

Penalties for the unlawful felling of trees were being exacted in the seventh century, and forest management had been practiced since Norman times. When trees were felled in the royal forests, the area that had been cleared was fenced to allow young trees to grow up and replace them. During the 15th and 16th centuries, a number of laws were enacted to prevent the overexploitation of forests.

Areas of land were not being deliberately planted with trees, however. Farm leases often required trees to be planted to replace those that were felled or had died. Farmers were expected to gather seeds for this purpose, and some writers of the time gave instructions for transplanting, but the policy was mainly to control felling and allow natural regeneration rather than to plant. Plantation planting began during the 16th century but patchily and on a small scale. Often it involved planting a mixture of seeds collected locally and then managing the resulting trees as coppice (see "Coppicing" on page 291). This method imitated natural woodland.

The first recorded instance in England of planting a block of trees all of one species was in 1580 when Lord Burghley (formerly William Cecil) had 13 acres (5.26 ha) of oaks planted in Windsor Park (they are still there). William Cecil (1520–98) was a statesman who became a principal adviser to Elizabeth I. It was not until the reign of Charles II that forestry became more widespread, starting with the planting scheme in the Forest of Dean that John Evelyn had recommended.

Trees for Profit

From about 1600, both the Royal Navy and the British merchant fleet began to increase in size. Those wooden ships were built almost entirely from British timber, apart from the masts, for which tall, straight, conifer trunks were imported. This expansion of the fleets led to a widespread fear that the demands of shipbuilding would deplete the forests. That is the fear that John Evelyn examined on behalf of the commissioners of the navy.

It is very doubtful that there was a material shortage of timber. Despite its rapid expansion, the military and merchant fleets were quite small, and so were the vessels that comprised them. The major expansion in shipping occurred in the late 18th and early 19th centuries, and since suitable timber was available at that time, it must also have been available earlier because of the rate at which oak trees grow. What is more likely is that the commissioners of the navy were offering much too low a price for timber, landowners were refusing to sell to them, and so there appeared to be a shortage.

The purpose of forestry was to make a profit. The propagandists for tree planting, including Evelyn, made much of the amount by which tree plantations could increase the value of estates. One estimate was that after 18 years, white poplars that cost £30 ($54 at the 2006 exchange rate, although it is difficult to compare 17th-century monetary values with those of today) to plant would be worth £10,000 ($18,000). Adam Smith (1723–90), the famous author of *The Wealth of Nations,* believed that timber was as profitable a crop as wheat or pasture.

Clearing Forests to Improve the Land

Despite evidence that growing timber could be profitable, the English tradition of clearing forests persisted for a surprisingly long time. Many landowners believed that the removal of trees improved the land. Early in the 19th century, the Board of Agriculture, a government agency, commissioned its secretary, the agricultural writer Arthur Young (1741–1820) to assess the benefits that might accrue from enclosing and cultivating open land. In his *General Report on Enclosures* published in 1808, Young included forests as wasteland suitable for improvement. "When it is considered that some of the Royal Forests are situated upon soils which would be productive in all the usual crops raised by the common agriculture of the kingdom," he wrote, "they will, without question, appear to be an object which merits no slight attention."

Young appears to be saying that felling trees contributed to progress. What is more, there was biblical authority for this point of view. Psalm 74 states (verse 5) that "A man was famous according as he had lifted up axes upon the thick trees." William Ewart Gladstone (1809–98), a major landowner as well as four times British prime minister, certainly believed this, and Gladstone's passion for tree-felling was much publicized. That is why, when Gladstone visited the German chancellor in 1895, Bismarck presented him with an oak sapling to plant back home in Britain. Germans had been cultivating coniferous forest plantations for a very long time. The earliest record of one, at Nuremberg, dates from 1368. The Black Forest, covering about 2,300 square miles (almost 6,000 km²) in Baden Württemberg, consists mainly of Norway spruce, but that is only because beech and oak are kept confined to the valleys and are not allowed to colonize the higher ground. The Black Forest is plantation forest.

Meanwhile, aristocratic owners of large estates were planting trees for ornamental as much as commercial reasons on land that had been cleared of forest long ago. Landscape designers encouraged their clients to create parks, with trees scattered informally. This was a highly ostentatious display of wealth because the scale of planting was huge. Historians have calculated that between 1760 and 1835, private landowners planted at least 50 million timber trees in Britain. Many of these were imported, exotic trees, the taste for which grew to such an extent that by about 1840, it was only exotic trees or greatly improved native ones that were considered fashionable. Attitudes toward trees became thoroughly sentimentalized, and in England individual trees and groves came to be cherished. In France, on the other hand, mature trees were harvested, and it was rare to see an old tree.

The inadequacy of the British approach became evident during World War I, when timber had to be imported through a partial sea blockade. Timber was an essential raw material, not least for making the pit props that supported the roofs of galleries in coal mines at a time when coal was Britain's primary fuel. After the war, in 1919, the Forestry Commission was established. A government agency, its task was to acquire land and develop a state-owned forest large enough to provide a strategic reserve of timber. Today forests cover about 10 percent of the land area of Britain. In 1999 (the latest year for which figures are available), the total area of national forest amounted to 6,570,389 acres (2,659,000 ha) distributed throughout England, Wales, and Scotland, of which 2,201,661 acres (891,000 ha) was state forest managed by the Forestry Commission. Of this total, 60 percent consisted of conifer plantations.

Russian Forestry

Russian forestry is a mixture of management and planting. Interest in protecting the forests began in the 17th century. Areas were set aside, inside which felling was strictly controlled or forbidden. This was done partly to protect watersheds. Then, in 1888, all forests, both private and state-owned, were made subject to conservation laws.

Following the revolution, private ownership of forests ceased. Cutting continued according to the demand for timber, but management improved and large areas were planted, including about 15 million acres (6 million ha) of shelter belts around farmed land.

Forestry in North America

Between them, Canada and the United States have 15 percent of the world's forests. The Canadian forests cover 1 billion acres (417.6 million ha), amounting to 45 percent of the country's land area. Forests cover 33 percent of the land area of the United States, a total of 736 million acres (298 million ha). Approximately 94 percent of Canadian forests and 28 percent of the commercial forests in the United States are publicly owned. The U.S. Forest Service, an agency within the Department of Agriculture, manages the public forests in the United States.

As North American settlers moved westward during the 19th century, they cleared large areas of the apparently endless forest. Trees were so abundant that there seemed no need to cultivate them, and it was not until 1877 that Carl Schurz, secretary of the interior, persuaded Congress to take seriously the risk that if felling continued, the entire American forest might disappear. In 1891 and 1897, legislation was passed authorizing the president to take land into public ownership in order to protect it from exploitation, and in 1921 and 1924, Congress authorized the federal government to buy forested land. At first, protection of the American forests was intended to safeguard watersheds and only secondarily to produce timber. Canada is much more sparsely populated, the northern climate is severe, and Canadian forests came under less pressure. Canada still retains more than 90 percent of its original forest cover.

In both the United States and Canada, plantation forestry of the kind that is practiced in Europe is probably less important than management of the natural forest with controls on the amounts of timber to be taken and the time allowed for regeneration.

At one time, forests covered much of temperate Europe and North America, but people came to realize that if they went on taking timber as they needed it, in time the forest would disappear. They had to find a way to produce timber on a sustainable basis so that the resource would remain available indefinitely. The method that they developed led to modern forestry.

8

Economics of Temperate Forests

Properly managed, a forest is a renewable resource. For each tree that is felled, another grows to replace it, so that the supply of timber can continue indefinitely. This chapter, about the economics of temperate forests, begins by explaining the concepts of renewable and nonrenewable resources and their implications.

The remainder of the chapter describes the materials that are obtained from forests and the ways that they are prepared for use. It describes the use of wood and charcoal as fuel and the preparation of timber and small wood. Each tree species produces wood with individual characteristics that make it suitable for certain uses but not for others. The chapter lists some of the most widely used trees and their uses.

■ RENEWABLE AND NONRENEWABLE RESOURCES

Trees grow. This means that felling a tree in order to use its timber does not necessarily reduce the amount of timber available for future use. A new tree may grow to replace the one that was felled. If trees are compared to money in a bank account, the forester is living on the interest while the capital, which is the total number of trees, remains unchanged.

Builders use a large amount of timber in the construction of private houses. Some houses are made entirely from wood, of course, but even those with walls of brick or stone and roofs of slates or tiles contain a great deal of wood. It is timber that provides the framework for the roof and, very often, a timber frame that supports the whole house. The brick, stone, slates, and tiles form exterior cladding for protection and appearance.

Perhaps at the same time as new houses are being built, not far away other builders are hard at work erecting an office block or shopping mall. These larger buildings also have frames, in this case made from steel. Steel and timber both perform the same function in construction, steel in large buildings and timber in smaller ones. Both are materials on which society depends, and to some extent they are interchangeable.

Buildings, both large and small, are essential, and therefore the materials from which to construct and maintain them must be readily available so that people can take what they need when they need it, like withdrawing money from a bank account. Money, stored in a bank until it is needed, is sometimes called an asset or a resource. When someone wants to start a business, an adviser may ask her whether she has the resources to do so, meaning the money the start-up will cost. In the same way, timber and steel are resources. Without them, planning a building would be a waste of time.

Renewable and Nonrenewable

Timber and steel are both resources, but there is an important difference between them. Steel consists mainly of iron, with added carbon and small, measured amounts of other metals to strengthen it and give it particular qualities. The iron and other metals are obtained from ores, which are minerals that are separated from rocks taken from the ground. Timber comes from trees, and different qualities of timber are obtained by growing different species of trees. Take a ton of mineral ore from the ground and it is gone. There is a hole where it used to be. It is possible to fill in the hole but not to fill it with newly made ore that is similar to the ore that was removed. Cut down a tree and it will leave a gap in the forest, but in this case, not only can the gap be filled, but it can also be filled with a new tree just like the one that was felled. The metal ore is a resource that cannot be replaced once it has been used. It is what is often called a nonrenewable resource. Timber, on the other hand, is a renewable resource. Like a farm crop, it can be replaced.

Explained like this, the distinction between renewable and nonrenewable resources is very clear, and at one time there were fears that the world would one day face shortages of some nonrenewable resources. Obviously, if people continue to use them, eventually they will be gone. That is what *nonrenewable* means. Renewable resources were seen as much less of a problem simply because they are easily replaced. In fact, the difference between the two kinds of resources is not quite so clear as it seems.

The first qualification concerns time scales. It is not literally true that certain resources are nonrenewable. Eventually, metals return to the ground as junk, the part of the Earth's crust on which they lie is subducted into the mantle, and that mantle rock reemerges somewhere else to form new crust. The process takes millions of years, but metals do not leave the planet. They are called nonrenewable because in the modern world, they are being used much faster than they can be replaced naturally. Timber is replaced more rapidly, of course, in tens of years rather than millions, but it is still possible for it to be used faster than it can regrow. Modern forestry avoids this risk (see "Plantation Forestry" on pages 294–300), but modern forestry is not practiced everywhere in the world.

Resources and Reserves

A more serious complication arises from the very idea of a resource. This is a substance for which people have a practical use, and where people have a use for something, other people will do their best to provide it. If there is a demand for timber, foresters will fell more trees, and if the foresters believe that the demand will be sustained, they will plant more forests. If there is a demand for a particular metal, mining companies will extract more of the ore.

These companies know how much ore their mines contain, so they can calculate how long it will be before the mine is worked out and their operations must move elsewhere. Preparing for that day, they conduct surveys to locate the places where the next mines must be opened, and they identify the most likely sources of ore to which to move after that. Ores that they have identified are known as reserves. If the amount of ore has actually been measured, it is called a proven reserve. If geologists have taken samples and calculated the amount for a mine that has not yet opened, it is an indicated reserve. If the amount is calculated on the basis of the rock structures and past experience but without making measurements, it is an inferred reserve. If the presence of ore is suspected because of the types of rock, it is a potential reserve.

Making measurements and using the word *reserve* make it sound as though the quantities involved are precise and fixed, but it is not like that. Eventually, the highest-quality ore will have been taken from a mine. There will be plenty of ore left but of lower quality, which means that it will be more expensive to extract metal from it. Other companies that have access to higher-grade ore may be able to produce the metal more cheaply, so the mine becomes uneconomic and must close. It can happen, however, that a new technology is developed that makes it practicable to mine poorer ores. It can also happen that demand for the metal increases

rapidly and prices rise. In both cases it becomes worthwhile to work the poorer ore so that the mine stays in business. At the same time, increasing demand compels companies to search more vigorously for new sources and to identify new reserves. The resource may be nonrenewable, but if the demand for it increases, the size of the reserves also increases. For example, between 1950 and 1970, because of increased demand, reserves of the principal ore for tin increased 10 percent, for copper 179 percent, for aluminum 279 percent, and for chromium 675 percent.

Changing Technology

Changes in the way materials are used can also be dramatic, and rising prices can produce major economies. Present CAFE (Corporate Average Fuel Economy) standards in the United States call for a passenger car to travel 27.5 miles on one gallon of gasoline (11.7 km/l). Most small family cars will now achieve this. About 30 years ago, most European family cars achieved no more than 21 miles per gallon (9 km/l); the average car now travels about 37.5 miles on one gallon (16 km/l). The improvement was the result of better car and engine design driven by higher fuel prices. If the high oil prices of 2006 are maintained, they will likely lead to further improvements.

Technological change also affects the way resources are used. Copper is used mainly for electrical wiring, and until recently, huge amounts were needed for the cables that carry telephone lines. Today, copper telephone wires are being replaced with optic fibers, made from a special kind of glass, and an increasing amount of long-distance communication travels as radio waves to and from orbiting satellites rather than by submarine cable. Copper is much less important than it was.

Optic fibers and satellites are replacing copper wires because they are more efficient, not because there is a shortage of copper. If there were a shortage of copper or anything else, technologists would find alternatives.

This does not mean that society can afford to be careless in its use of resources. For one thing, any use of materials affects the environment, and the less that is used, the less environmental damage its use is likely to cause. It does mean, however, that hoarding a supposedly nonrenewable resource for the benefit of future generations is not necessarily wise. The resource may be more abundant than it seems, and in years to come, technological advances may render it obsolete.

Caring for Renewables

Substitutes can also be found for most renewable resources, but doing so is more difficult. People have been using timber for building ever since they started making shelters for themselves, and builders are still using it. Wood continues to be the most widely used material for making tables and chairs, cupboards and shelves, tool handles, and a vast range of other household items. There are substitutes: Metals and plastics are just as good. Many people prefer wood, but the real reason its rivals have so far failed to supplant it has nothing to do with sentiment. It is simply that planting more trees is easy and cheap, at least in temperate regions, and wood requires little processing. It does not have to be smelted or refined or synthesized in a factory but merely stored while it matures and then cut to the desired shape. Some commodities may be as good as wood, but wood is always likely to be cheaper.

Its renewability is what gives wood its main economic advantage, but this conceals a danger. Planting more trees is easy and cheap, but only provided that there is suitable land on which to plant them. In many parts of the temperate regions, there are competing demands for land, and most forms of land use are more profitable than forestry. A field of corn is worth less than a field of houses, but a field of trees, especially young ones that will not be ready to harvest for some years, is worth less than either. Consequently, communities should take special care that their forests continue to occupy an area large enough to supply timber and wood (see "Forest Timber" and "Small Wood" on pages 248 and 250 for an explanation of the difference) at a sufficient rate to match their use of them.

FORESTS AS SOURCES OF FUEL

About 10,000 years ago people in Southwest Asia began to cultivate wheat and barley and to tend domesticated livestock. In parts of China they had already been growing rice and millet for a long time. Until they domesticated crop plants and animals, people lived by gathering wild plants, hunting animals, and scavenging, stealing what meat they could from the bodies of large animals that more powerful predators had killed.

When they obtained these foods, at first they ate them raw, but later people learned how to use fire to cook their food. This greatly extended the range of foods that they could eat because many plant and animal parts are inedible or unpleasant to eat raw but are delicious cooked. The possession of fire also meant, of course, that people could move into higher latitudes, where the climate was cooler but large game was abundant.

Many generations must have passed before the users of fire became makers of fire. Fires occur naturally. After a spell of dry weather, lightning readily ignites withered vegetation, and once a fire starts, it is not difficult to keep it burning. Brands would be taken to the campsite, and the fire there kept fueled. When people left on hunting or food-gathering expeditions, they would leave behind a fire damped down so

that it would still be alight when they returned, and a little gentle blowing would be enough to fan it into flame. When they moved permanently, they took fire with them as slow-burning brands that would be used to light a cooking fire at each overnight stopping place. Scientists believe that this way of life began some hundreds of thousands of years ago and that the first people to use fire were not modern humans (*Homo sapiens*), but *H. erectus,* an earlier species.

Wood as Fuel

The first fuel was wood, and for many people in developing countries today, wood is still an important fuel. According to figures from the World Energy Council, in 1999 wood supplied approximately 70 percent of the primary energy used in Kenya, and in rural areas the contribution rose to more than 93 percent. In Africa overall, wood provided 60–86 percent of primary energy. In Mexico fuelwood provided about 36 percent of the energy used in private homes. In 1998, Vietnam derived about 22 percent of its primary energy from wood. Primary energy is energy that is derived from a fuel that is used in its natural form, such as coal, petroleum, gas, and wood. Even in the industrialized countries of Europe and North America, wood is widely used as fuel, especially in rural areas. In Austria, Finland, and Sweden, wood provides 12–18 percent of primary energy.

Those who depend on fuelwood may have to make do with whatever is available locally, but not all trees produce wood that burns equally well. Elder, for example, spits hot cinders, and country folk will not use it on their fires. Softwoods, such as pine and spruce, contain resin. They are easier to ignite and burn hotter than hardwoods, such as oak and elm.

An open log fire is a cheerful sight, and wood is the most obvious of all fuels because it is plentiful in many places and it burns readily. Unfortunately, open fires are extremely inefficient because most of the heat is lost in the rising hot air and gases. The first person to propose a remedy was the American statesman, scientist, and inventor Benjamin Franklin (1706–90). Franklin invented a closed stove that stood in the center of the room (see the sidebar on page 247), so all of its heat was retained in the room. The Franklin stove was the ancestor of all subsequent wood-burning stoves.

Compared with a coal fire, however, even an efficient wood fire feels cool, and an ordinary coal fire gives out less heat than an anthracite fire. Coal, including anthracite, is a form of fossilized wood, and both wood and coal are made mainly from carbon and hydrogen, but coal has been tightly compressed, which explains why it burns at a higher temperature: Its carbon and hydrogen, the combustible ingredients, are packed more closely together than they are in wood. One ton of wood, dried to 20 percent moisture content, has a volume of about 100 cubic feet (about 3.2 m³/t); one ton of coal occupies about eight cubic feet (about 0.25 m³/t).

Benjamin Franklin and His Wood-Burning Stove

Benjamin Franklin (1706–1790) was one of the greatest statesmen of his day, and at the same time, he was the first truly great American scientist. As a scientist he is most famous for his studies of electricity, but he was also a talented inventor. He invented—and wore—bifocal spectacles and also invented a wood-burning stove called the Franklin stove or the Pennsylvania stove.

In Franklin's day, people heated their homes with open fires. These burned on hearths that were recessed into the wall to allow the smoke to escape up a chimney. Franklin knew that the fire radiated heat in all directions and that the air it heated transmitted heat by convection. Some of the radiant heat and almost all of the convective heat disappeared up the chimney. Open fires are notoriously inefficient.

Franklin's solution was to place the fire in the middle of the room. The fire was contained in an egg-shaped iron stove that was closed by doors. Air was drawn into the stove from the top, passed out at the bottom, and was conducted to the chimney through pipes beneath the floor. The metal of the stove became very hot and radiated heat, air warmed by contact with the stove, transported heat by convection, and the pipes warmed the floor. The stove achieved almost total combustion of its fuel. The stove was not simple to light, however, because unless it warmed slowly, it failed to establish a correct flow of air and filled the room with smoke. Apparently Franklin said that his servants were too stupid to manage it properly.

Franklin invented the stove in 1742 and it was an immediate success. Typical of the man, he refused to patent it. A London manufacturer made a great deal of money from a very similar stove. When Franklin learned of this, he wrote in his diary "Not the first instance." For him, it was reward enough to be imitated.

Coal has lost most of its water. Freshly cut green wood can contain as much as two parts of water to every part of combustible fiber by weight, and most of the smoke from a wood fire is water that has been vaporized by the heat and then condensed again into minute droplets. Wood can be dried, but unless it is dried in kilns, the dry wood still holds about 15–20 percent of water.

Composition of Fuels

Combustion is the oxidation of carbon to carbon dioxide ($C + O_2 \rightarrow CO_2$) and of hydrogen to hydrogen oxide (water: $4H + O_2 \rightarrow 2H_2O$). Both reactions release energy. Burning one pound of carbon fully to carbon dioxide releases about 14,500 BTU of energy (33.7 megajoules [MJ] per kg), and one pound of hydrogen releases about 62,000 BTU (164 MJ/kg). Clearly, the higher the proportion of carbon and hydrogen the fuel contains, the more energy it will deliver. The table shows how wood that has been dried completely compares with coal. Wood contains an average 48.5 percent of carbon by weight compared with 77 percent in the kind of shiny, black coal that people burn in domestic fires (bituminous coal) and 90 percent in anthracite. Brown coal (lignite) is a low-grade fuel, rather like peat.

Early fire-users and fire-makers would have preferred the driest wood they could find. It gives the hottest flame and it weighs least, which is an important consideration to someone who has to carry it back to the village. Wood was perfectly adequate for providing light on dark evenings for security and a flame to drive away hungry predators. Its heat was sufficient to keep everyone warm in winter and to cook food.

Smelting Ores

Then a difficulty arose because people began to use metals. Copper melts at 1,982°F (1,083°C). The heat from a well-

Composition of Fuels (percentage of total)

FUEL	CARBON	HYDROGEN	OXYGEN	NITROGEN	SULFUR	NONCOMBUSTIBLE SOLIDS
Wood	48.5	6.0	43.5	0.5		1.5
Brown coal (lignite)	67.0	5.1	19.5	1.1	1.0	6.3
Bituminous coal	77.0	5.0	7.0	1.5	1.5	8.0
Anthracite	90.0	2.5	2.5	0.5	0.5	4.0

made and managed wood fire is enough to melt copper from its ore. Tin melts at only 449°F (232°C), so it was perfectly feasible to alloy copper and tin to make bronze. Iron was more difficult, however. It melts at 2,795°F (1,535°C). The problem was not that wood cannot produce enough heat, but that it is too bulky, so its heat is insufficiently concentrated to achieve the necessary temperature.

The solution that people devised was ingenious, and not at all obvious. They found a way to heat a mass of wood under airless conditions. This dries the wood and drives off its oxygen, but without allowing the wood to catch fire. As the table on page 247 shows, the higher the quality of the fuel, the greater its content of carbon and the lower its content of oxygen. Anthracite contains only 2.5 percent oxygen, compared with 43.5 percent in dry wood. Drive off the oxygen and the volume of the remaining wood will be greatly reduced and the proportion that is carbon will increase.

They had invented charcoal, the substance artists use (in a purified form) for drawing and that is still used in industrialized countries as a fuel for barbeques (it also has uses in the chemical industries). Charcoal has a carbon content similar to that of anthracite, so charcoal can be described as an impure form of carbon.

Charcoal is much less dense than anthracite, but burning it produces at least 50 percent more heat than burning the same volume of wood. It is enough for smelting iron, and it is the fuel that was used in the metal industries until coal replaced it in the early years of the Industrial Revolution. Until well into this century, charcoal was being made in Europe by the traditional method, and small amounts still are, although mainly out of historical interest.

Making Charcoal

Small sticks and branches were used to make charcoal. Often these were coppice poles grown for the purpose (see "Coppicing" on page 291), and oak was the preferred tree. All the bark was removed, partly because oak bark could be sold for use in the tanning of leather and partly because burning oak bark releases large amounts of choking, sulfurous gases.

Charcoal was made in a shallow, circular pit, about 15 feet (4.5 m) in diameter. The floor was of earth or ash, and there was a wooden pole about 6.5 feet (2 m) tall at the center. Short lengths of wood were stacked around the pole, with longer sticks outside them, sloping inward, and stacking continued until a dome-shaped heap had been made, reaching almost to the top of the central pole. The stack was then covered with bracken, leaves, and turf, all packed down tightly and sealed with a coating of mud, made from a mixture of earth and ash. The central pole was withdrawn, leaving a hole down which a small amount of charcoal was dropped. Burning charcoal was then dropped into the hole.

This ignited the charcoal and wood at the center, and the top of the hole was sealed as soon as flames were seen.

After a short time, smoke would emerge from the stack. At first it was white, but later, after a day or sometimes several days, it would turn blue. Blue smoke meant that the process was completed. The stack was allowed to cool, and the charcoal was then ready for use.

Making charcoal was hard and very dirty work. While a stack was smoldering, it had to be checked at regular intervals. Holes in the coating had to be sealed, and the top of the central hole had to be opened to check that the wood inside had not caught fire. If it had, water was used to extinguish the flames. Meanwhile, as one stack burned, another was being prepared, and charcoal was being removed and bagged from a stack that had cooled.

Today, very few people practice the craft of the charcoalmaker. Other fuels and other technologies are used to smelt and work metals, but these are quite recent innovations. For most of history, charcoal was the most widely used industrial fuel, and wood was the fuel with which most people warmed their homes and cooked their food.

■ FOREST TIMBER

Timber is a word with a long history that has been traced back to Indo-European, the language ancestral to almost all modern European and western Asian languages. More recently, it was closely related to the Latin word *domus,* a dwelling, so timber is something connected with building, in fact a building material. At one time the word meant any kind of building material. Then it came to mean wood as a building material for ships as well as for buildings, and from there the meaning broadened to include any kind of wood.

Today, people often use *timber* and *wood* as though they were synonyms, but strictly speaking they are not. Wood is the material from which small articles are made (see "Small Wood" on pages 250–254). Timber is used to make large objects.

Apart from this general sense, timber can also be used specifically. Timbers support roofs and floors and provide the main structural components of ships, as in "ships' timbers." In an American sawmill, timber is wood not less than five inches wide and five inches thick (12.7 × 12.7 cm), although this is a rather vague designation.

Timber and Lumber

Felled trees are stripped of their branches and then cut into logs of a length that can be transported conveniently. In North America, at this stage the material is often called lumber, a term that is not much used in Britain. Then each log is measured and sold to the sawmills by solid volume. The volume is calculated by measuring the length and the diameter at mid-length and then using $V = (\pi d^2 L) - 40,000$, where V is the volume in cubic meters, d is the diameter in centimeters, and L is the length in meters (to convert cubic meters to cubic feet, multiply by 35.31). Sawmills then cut the logs to popular sizes, and customers specify the sizes they need.

There are exceptions. If uncommonly large sizes are needed, special arrangements must be made. Masts for a ship, for example, may be bought as logs that are longer than those usually sent to the sawmills, and the buyer may select in advance the individual trees that will supply them.

Seasoning the Wood

Water accounts for a substantial part of the weight of wood (see "Wood as Fuel" on pages 246–248). Timber is not sold by weight, so customers are not paying for water, but its presence greatly affects the behavior of the wood. All timber starts wet because of the water being transported through it. In this condition, it is known as "green" timber. As wood dries, its volume decreases, and it may shrink unevenly, acquiring irregular twists and bend, called warping, and splits may appear. Such flaws are inconvenient at best and can be catastrophic in construction timbers.

To avoid this, timber is dried, or "seasoned," before it is sold. Seasoning takes place at the sawmill after the logs have been cut into planks, boards, and timbers. Traditionally, it was done by stacking the timber in layers outdoors. Thin boards are laid crosswise across each layer, with spaces between them to allow air to circulate freely through the stack, and the stack is covered with a roof to allow rainwater to drain away without wetting the timber. How long the stack must be left like this varies from one tree species to another and with the climate. Generally, the process takes 6–18 months. The stacks must be inspected at intervals.

A modern lumberyard, with stacks of logs and cut wood, and planks stacked for seasoning *(Missouri Natural Resources Conservation Service)*

In practice, most timber is kiln-dried, a technique that reduces to about one week a process that otherwise can take more than a year. Kilning also improves the quality of the seasoned timber. Often, outdoor air-drying and kiln-drying are combined, with the green timber starting to season in an outdoor stack and being finished in a kiln. Timber for kilning is stacked in a building resembling a large shed, heat is supplied by steam flowing through pipes, fans ensure a full and even circulation of air, and the temperature and humidity are controlled within fine limits. Alternatively, timber can be seasoned in 12 hours or less by heating it strongly in a chamber from which the air is evacuated. This causes the water to vaporize rapidly. Precisely how long seasoning takes by either of these processes, and the temperature at which the timber must be held, vary according to the tree species.

Grading and Sorting

Timber is graded while it is still green. Once the bark has been removed, knots become visible. These can determine the maximum length into which the timber can be cut because sections with too many knots must be removed, and a standard calculation based on the number of knots and other imperfections is what determines the grade.

After seasoning, the timber is sorted. This time each piece is marked according to its tree species, dimensions, and grade. Then it is ready for finishing, in which two sets of sharp, rapidly revolving blades shave both sides, smoothing and slightly polishing them.

Hardwood and Softwood

Wood is classified as softwood and hardwood. These names refer to the types of trees producing the wood, softwoods coming from gymnosperms (coniferous trees), hardwoods from angiosperms (see "Seed plants, Conifers, Flowering Plants" on pages 101–103), which are trees that are flowering plants rather than cone-bearing or spore-producing ones. The adjectives *soft* and *hard* refer to differences in the structure of the wood, but the names are unfortunate because they are misleading guides to how hard the wood is. They will not tell the carpenter which woods blunt tools fastest. Balsa, the wood used mainly for model-making, is technically a hardwood because the tree producing it (*Ochroma* species, mainly *O. lagopus*) is a flowering plant (angiosperm), but a cubic foot of balsa wood weighs no more than about 10 pounds (150 kg/m³). Balsa will not blunt tools, but despite being a hardwood it is so soft that unless very sharp tools are used it tends to crumble. It is easy to hammer nails through it, but it will not hold them firmly, and gluing is the best way to fix together two piece of balsa. Wood from the kauri pine (*Agathis australis*) of North Island, New Zealand, on the other hand, has a density of about 38 pounds per cubic foot (610 kg/m³), yet it is classed as a softwood.

Suitability for a particular use depends on many factors, of course. Hardness is important, but what matters is whether the timber will survive the treatment for which it is intended.

Appearance is also important for timber that will be used to make furniture and other articles that will be displayed in homes or in public buildings. Timber from species with an attractive color or grain, which is usually very expensive, is often cut into very thin sheets that are glued to the exposed surfaces of articles made from cheaper wood. Such a covering is called a veneer. A softwood cabinet with a walnut veneer will look like walnut, but it will cost far less.

Testing Timber

When the Forestry Commission was established in Britain in 1919, the principal need was for pit props, the timbers that support the roofs of the galleries in coal mines. Timber was preferred to steel because it is slightly elastic and, no less important, it makes loud creaking and groaning noises before it breaks, often providing enough warning for miners to escape a roof fall. Props in coal mines or anywhere else must withstand pressure applied on their ends, acting parallel to the grain. This is called the compressive strength of the timber.

Timbers may also be loaded at right angles to the grain like a shelf held at both ends and loaded with books. This loading will cause the timber to bend, and it is important to know by how much and the point at which it will bend no further and the wood will break. The maximum bending strength is the stress that a timber can bear when loaded slowly and continuously. This test also measures the stiffness of the timber, or to give it its technical name, its "modulus of elasticity." The stiffer the timber, the less it will bend under a given load. Loading is not always applied at the center of timber supported at its ends. Rail ties carry the rails and the weight of passing trains at their ends.

Loads are not always applied slowly. Picture the floor of a dance studio, for example, where a class of dancers all jump into the air and land simultaneously. Resistance to sudden loads is measured by dropping a standard weight onto the timber repeatedly, increasing the height of the drop each time until the timber breaks.

For some purposes, wood has to be bent. This is achieved by steaming the wood while applying pressure, and the bending property is measured by continuing this until the timber snaps. The result is given as an S/R ratio, where S is the thickness of the timber and R is the maximum radius to which it can be bent without breaking. Beech, for example, has an S/R ratio of about 1/2, which means that a piece one inch thick can be bent to a radius of two inches, or a piece two inches thick to a radius of four inches.

Timbers must often be fastened to one another by joints or bolts. Fastenings will tend to depress the surface, slowly making hollows. These will loosen the fastening with a risk that eventually the structure will fail. It is important, therefore, to measure the resistance of timber to indentation of this kind. It is done by striking the timber with a steel ball, exerting a known force, and measuring the indentation caused. Another test measures the amount of stress timbers bolted to a wall or other structure will tolerate before they shear, breaking parallel to the grain.

It is also important to know how easy it is to work with the timber. How many hours of working are possible before saws, planes, and other edge tools need resharpening affects the cost of using a particular timber. Sharpening tools takes time, during which the tools cannot be used. People who are planning to work with a species unfamiliar to them also need to be warned of the ease with which the material splits, especially if they plan to nail or bolt it.

Most timber is exposed to attack by beetles and other insects, but some species are more resistant than others to insect attack. Again, this is a quality that must be known, and it must include the insect species to which the material is vulnerable. If timber is to be used outdoors, its weather resistance is important as a measure of its durability. Preservatives are commonly used to protect timber, but species vary in their permeability to preservatives, and this must also be known.

These properties are measured on samples of timber by standard tests performed in standard ways. The results are published so that those working with timber can easily learn the species most suited to their particular requirements.

■ SMALL WOOD

Almost all paper is now made from wood. In 2004, a total of 389.9 million tons (354.5 million tonnes) of paper and cardboard was made from wood in the world as a whole. As with any industry, the totals vary somewhat from one year to another. In 2003, the world produced 374.1 million tons (340.1 million tonnes), and in 2002, total production was 363.9 million tons (330.8 million tonnes).

Paper consists of fibers lying across and adhering to one another. Until the last century, the fibers were obtained from straw, various grasses, and the bark of certain trees, but the best paper was made from cloth fibers, mainly in the form of rags that were soaked, bleached, and shredded. Paper was made one sheet at a time, and a really good worker could produce about 750 sheets in a day. Not surprisingly, paper was expensive.

In 1863, however, a pulp mill in Maine, owned by I. Augustus Stanwood and William Tower, began to use wood as a source of fiber, pulping it mechanically. Within a few years, several other mills were pulping wood. Prices began to fall because once machines were available to pulp it, the temperate forests of America, Europe, and Asia were able to supply good quality fibers in vast quantities. At the same time, paper manufacture became a continuous process. Some paper of the highest quality and with special uses, such as the paper used for printing bank notes, is still made from rags. Some straw is still used, and some high-quality writing and printing paper is made from esparto grass. Paper for watercolor painting is often made by hand. Synthetic fibers are used in some papers, but more than 90 percent of all the paper in use today is made from wood pulp by huge machines.

Wood for Pulping and Chipboard

Wood destined for pulping is grown for the purpose in plantation forests (see "Plantation Forestry" on pages 294–300). Sometimes people talk of economizing in their use of paper to save the tropical forests or even the temperate forests, but although there may be good reasons for saving paper, its use threatens no forests. In Europe and North America the wood is all grown specially, and felled trees are replaced. This is not the case everywhere in Russia, but in time it will become so because this is the most economical way to produce pulping lumber.

Most pulp is made from softwoods, but not all. Ash (*Fraxinus excelsior*), beech (*Fagus sylvatica*), and poplar (*Populus* species) make excellent pulp. Coniferous species supply most pulp, however, because these are the trees that grow fastest. Sitka spruce (*Picea sitchensis*) is the most extensively grown tree in Britain and is also the principal source of wood for pulping in that country. Sitka spruce is particularly suitable because its wood contains little resin. Norway spruce (*P. abies*) and Douglas fir (*Pseudotsuga menziesii*) are among the other conifers used for pulping. Poplars are also among the species with wood that can be cut into soft shavings and used for stuffing, called excelsior.

Chipboard manufacture also consumes a large amount of forest wood. Chipboard has many uses in the construction industry, where panels made from it screen work sites and are made into internal partition walls. Packing cases, furniture, and a range of other articles are also made from it. Because it is made from wood chips, pressed and sealed together with resin, chipboard is strong and does not warp. It is more durable than the plywood it has largely replaced. Plywood is made from several thin sheets of board glued together, and it tends to separate when it becomes wet and the glue dissolves. Attractive hardwoods are also used as veneers.

Veneers apart, these uses destroy the individual character of the wood. No one can tell which tree species were used to make the pages of any book, nor is it easy to identify the trees from which the chips in a piece of chipboard came. Yet although these account for a very large proportion of the output from temperate forests, there are also many items made from wood for which the type of wood matters and can be recognized.

Alder and Elm

Clogs are the wooden shoes traditionally worn in parts of Europe, including the Netherlands and northwestern England, that have become fashionable in recent years. The people who wore them originally lived in a region with a moist climate, and if their feet were to remain dry, the clogs had to have waterproof soles. Alder (*Alnus glutinosa*) is the tree from which they were usually made. It is very waterproof, extremely durable whether it is wet or dry, and it is easy to carve, so an entire shoe made from it can be decorated with a carved pattern. The bark, stripped before the wood is sold, can be used for tanning, and any waste chippings or ends of wood can be made into excellent charcoal.

Elm (*Ulmus* species) wood is also very resistant to water. Many years ago water pipes were made from it, and they survived intact for more than two centuries. It was also used to make clogs, wheelbarrows, farm carts, and weatherboarding, but elm is especially durable if it remains permanently immersed in water; where it is alternately wet and dry, it does not last so well. For this reason it was used to make waterwheels, the piles on which jetties were supported, sea defenses, and various structures at seaports.

Oak and Cherry

Oak (*Quercus* species) also yields very durable wood that withstands prolonged immersion in freshwater or saltwater. It is the wood from which ships were once made, and it has also been used for sea defenses. Wood from live oak (*Q. virginiana*) is greatly prized in North America, as are the red oak (*Q. rubra*) and white oak (*Q. alba*). In Europe the most widely used species are the pedunculate oak (*Q. robur*) and durmast (or sessile) oak (*Q. petraea*). Turkey oak (*Q. cerris*) is the exception. Although this tree is attractive, its wood is not durable.

Many articles can be made from oak, and it is also used for veneering. Small poles make excellent stakes for outdoor fences, gates can be made from it, and indoors it once was used to make wall paneling that was sometimes carved. Oak bends well, which means it can be used to make chairs, and it also makes other fine furniture. With a slight bend, oak staves make the best barrels for storing and maturing wines, beers, and ciders.

Wall paneling and veneers can also be made from the wood of the wild cherry (*Prunus avium*), also known as the gean and mazzard, a species that is nowadays grown commercially for its wood. The tree itself grows to about 60 feet (18 m) tall, and its heartwood is a rich, reddish brown and rarely cracked (the technical term is *shaken*), with a very straight grain. Quality furniture and ornamental items are also made from it, and cherry wood is especially useful where its straight grain is needed. Woodwind instruments are sometimes made from cherry.

Beech and Birch

Beech (*Fagus* species) produces wood that is hard and strongly resists compression, so the European (*F. sylvatica*) and American beeches (*F. grandifolia*) are both very valuable, but beech wood deteriorates rapidly out of doors. The wood makes excellent blocks for flooring, the wedges that support ships while they are being built, high heels for shoes, and furniture. Stocks for rifles and shotguns are often made from beech, as were the wooden blocks of old-fashioned carpentry planes. Beech also bends well when steamed, so it is used to make bentwood chairs. Kitchen utensils, such as wooden spoons and breadboards, are often of beech.

Birch (*Betula* species) also perishes rapidly if it remains wet, so it is useless for outdoor applications. Kept dry, though, it has many indoor uses. In Britain, the silver (*B. pendula*) and downy birches (*B. pubescens*) seldom grow tall and straight enough to yield timber, but their wood is made into the backs of brushes, bowls, toys, and similar small items, and it is also used to make chipboard. In the past it was used to make barrels, especially those in which herring were stored. Farther north, in Scandinavia, large, thin sheets are peeled off the trunks of the same species to make plywood. In North America, the cherry (*B. lenta*) and yellow birches (*B. lutea*) yield valuable wood that is used to make furniture, as does the paper birch (*B. papyrifera*). This tree is also known as the canoe birch because its waterproof bark used to be peeled off and used to cover the frames of canoes.

Birch twigs are used to make besoms, which are yard brooms made from long twigs fastened around the end of a handle. Identical besoms are also used to beat out fires in forestry plantations, which is odd because birch twigs are the best kindling there is for lighting a fire.

Basswood, Lime, and Willow

American basswood (*Tilia americana*) is especially prized for its long, straight grain. Before wood was replaced by plastic, the slats of venetian blinds were made from American basswood, and, covered nowadays with plastic but at one time with ivory obtained from elephant tusks, it is the wood from which piano keys are made. Many beautiful European carvings are made from the soft, firm wood of the large-leaved lime (*T. platyphyllos*).

Wood from lime (also called linden) is also used to make boxes and matches, but the wood most often used to make matches comes from poplars (*Populus* species), various groups of which are known as cottonwoods or aspens. The European aspen (*P. tremula*) and the quaking aspen (*P. tremuloides*) of North America have light, soft, yet tough wood with a straight grain, making it easy to cut into straight sticks. In medieval times arrows were made from it.

This quality apart, aspen may seem to be a curious choice for making matches because it is nonflammable. That is why it was formerly used to make brake blocks that refused to smolder, far less burst into flames, when they were jammed hard against a fast-turning wheel. Nor does it splinter, which makes it valuable for boxes. Pallets on which goods are stacked so that they can be moved around by forklift trucks are usually made from poplar. The nonflammability of poplar is a drawback to its use in matches but is countered by impregnating the wood with paraffin wax.

Willow was also used to make brake blocks. Willows belong to the genus *Salix* of which there are approximately 400 species, not all of them large enough to supply wood that can be worked. The most important of those that grow to the size of trees in Europe are the white willow (*S. alba*) and the crack willow (*S. fragilis*). White willow is used for pulp. The black willow (*S. nigra*) of North America is also tree-sized, and its wood is used to make boxes. Willow wood is light, pale, strong, and does not splinter readily, which is why the variety *caerulea* of white willow is the traditional wood for making cricket bats and is known as cricket-bat willow. Willow is also used to make the sides of the sieves that gardeners use, and it is cut into thin strips that are woven to make garden baskets, sometimes called trugs, for carrying cut flowers and other produce.

Slender willow stems, called osiers, withes, or withies, are used to make wickerwork items, such as furniture and baskets. The preferred species are *S. viminalis, S. fragilis, S. triandra,* and *S. purpurea*.

Willow and poplar bark were once the sources of salicin, a drug used to reduce inflammation and relieve pain. Originally a country remedy, salicin was later processed to make acetylsalicylic acid, marketed as aspirin. Aspirin is now made from chemical feedstocks that are not derived from tree bark.

Willow is finding a new use as an environmentally friendly fuel. It grows quickly from cuttings and can be harvested for use in industrial furnaces. Burning biomass fuels, a name often shortened to biofuels, releases carbon dioxide into the air, but it is carbon dioxide that the plants themselves absorbed during photosynthesis, so its release does not increase the atmospheric concentration of this greenhouse gas.

Ash, Hornbeam, and Hickory

European ash (*Fraxinus excelsior*) yields a tough, pliable wood that resists shock, so it is used for anything requiring a material that will absorb shocks without breaking. The list includes a range of sports equipment, such as hockey sticks and oars and the handles of tools. The American white ash (*F. americana*) and the Oregon ash (*F. latifolia*) are valuable for their timber, which is used to make furniture and for other indoor purposes. Ash is not durable, however, and is unsuitable for fencing or other articles that are in permanent contact with the ground, but it does make excellent firewood.

Hornbeam (*Carpinus betulus*) derives its name from its hard, tough wood, and at one time its name was *hard-beam*. It is the toughest of all British woods. This quality, combined with often being cross-grained, also means that hornbeam is difficult to work, which restricts its use. Mallet heads are made from it, as are articles such as cogwheels and pulley blocks. Nowadays these are usually made from metal, but wooden ones were used in old windmills and watermills where grain was ground into flour.

American hornbeam, or blue beech (*C. caroliniana*), just as hard, is also used to make tool handles, but the most popular North American trees for this purpose are hickories (*Carya* species). Many hickories are grown for their nuts (see "Hickory" on page 210), but the wood is tough and absorbs shock well. Shagbark (*C. ovata*), shellbark (*C. laciniosa*), Carolina (*C. carolinae-septentrionalis*), mockernut (*C. tomentosa*), pignut (*C. glabra*), and sweet, or red, hickory (*C. ovalis*) are the trees most often used.

Spruce

Conifers are usually thought of as sources of large timber or pulp, but smaller items are also made from them. Thin sheets of pine, bent to the proper shape, are used for the belly of violins and string instruments related to it, the back and sides (called ribs) of the instruments being of hardwood.

Wood from the spruces (*Picea* species) is known as whitewood or white deal. It, too, can be used to make sounding boards for string instruments. To be suitable for work of this quality and indeed for all high-quality work, the wood is taken from trees that have grown slowly because of being shaded by larger trees. Slow growth results in many annual growth rings and, therefore, a fine grain (see "How Wood Forms" on pages 113–116). The wood must also be free from knots, a fault to which spruce is very prone. Knots are dark in color, hard, and resinous, and if the wood is cut into thin sheets, then the knots tend to fall out, leaving holes.

Removing the lower branches (called *brashing*) and pruning the higher ones reduces the number of knots. Sitka spruce (*P. sitchensis*) usually yields fairly coarse wood, but it can be of a quality good enough to be cut very precisely and used for better-quality furniture.

More commonly, spruce wood is used to make kitchen furniture, boxes, and packing cases. Matchboxes are usually made from cardboard nowadays, but years ago they used to be made from spruce wood, sliced very thin. Rather thicker slices are a major ingredient of plywood, provided they are free from knots. Until scaffolding came to be made from metal tubing and ladders from lightweight metals, these were usually made from Norway spruce (*P. abies*). Norway spruce also supplies the wood to make flagpoles and the masts of small boats, although metal masts have largely replaced wooden ones.

Maple, Box, and Pear

The hardwood used in string instruments is often European sycamore (*Acer pseudoplatanus*) or some other species of maple (*Acer*). In North America, the wood of the sugar maple (*A. saccharum*) is often used. Pale pink in color, this wood is also used to make sports equipment and, if it has an attractive grain, for high-quality furniture. Kitchen utensils, bowls, and chopping boards can be made from sycamore, and it is also used to make rollers for machinery and, in the days before spin dryers, for the mangles operated by turning a handle that squeezed water out of the laundry by passing it between two rollers.

Obviously, boxes can be made from almost any kind of wood, but there is one genus of trees (*Buxus*) that is called "box." Boxes are grown mainly as ornamental hedges in formal gardens, where one of their advantages is that they grow very slowly. Slow growth also means that the annual growth rings are very narrow and, consequently, that the grain is very fine. The wood is an attractive yellow color, and the boxes made from it are of the finest quality. The wood can be sculpted and carved with very delicate designs. It is one of the best woods for marquetry and can be polished to a high gloss. Boxwood boxes are luxury items. Originally the common box (*B. sempervirens*) was used, an Old World species native to Britain, although the wood was more often imported from Turkey.

Boxwood is harder and heavier than wood from any other tree native to Britain, and it is used for articles where hardness matters. Drawing instruments used to be made from it before plastic replaced wood, and some school rulers still are. Musical instruments have been made from it since Roman times. It is the wood from which the English printmaker and illustrator Thomas Bewick (1753–1828) engraved the woodblocks that made him famous. Bewick was responsible for reviving the art of woodblock engraving.

Modern woodblock engravers use a different wood, however, from *Casearia praecox,* the Venezuelan box.

The English woodcarver Grinling Gibbons (1648–1721) worked with pearwood from the common pear (*Pyrus communis*), a tree that grows naturally up to about 40 feet (12 m) tall throughout most of Europe. In the past it has been used for a variety of ornamental work and drawing instruments such as set-squares, protractors, and the handles for cutlery, and colored piano keys were sometimes made from it.

Cedar, Sweet Chestnut, and Hazel

Wood from the western red cedar (*Thuja plicata*) is light and durable. It can be used outdoors for the frames of greenhouses, gates, roof shingles, and cladding on buildings. Small boats can be made from it, and in North America these trees used to be hollowed out to make light, robust canoes.

Sweet-chestnut (*Castanea sativa*) wood is very durable, making it ideal for outdoor use. Fences and the posts to which they are fastened are often made from chestnut. The tree itself grows to a large size, but the wood usually contains deep cracks (shakes), which limits its value as timber. Smaller pieces of chestnut wood are used to make furniture and coffins.

Not all fences are meant to stand permanently in the same place. Farmers use temporary fences, called hurdles, to contain livestock. By moving the hurdles, the animals can be made to graze one section of pasture at a time. Gardeners also use temporary fences to provide shelter. Traditionally, hurdles were made from sticks of hazel (*Corylus avellana*) in Europe and the very similar American hazel (*C. americana*) in North America. As with so many everyday articles, farm hurdles are now more often made from metal, but weaving them from hazel was once a highly skilled job.

Even today, in this age of metal and plastic, our homes, schools, and workplaces contain countless small articles made from wood, and in the past they would have contained more. Each article is made from the wood of those plants best suited for the purpose, and there are few plants for which no use has been found. For small items it is the quality of the material that matters rather than the quantity, and even those trees and shrubs that produce only thin stems and branches can be used. Wood from some species of spindle trees, for example, including the common European *Euonymus europaea,* was once used to make skewers and, as its name suggests, spindles for spinning.

9

Health of Temperate Forests

Many people are concerned that forests are disappearing at an alarming rate. This concern is justified in the case of tropical forests, which are being cleared in many regions, but temperate forests are expanding, not contracting. This chapter on the health of temperate forests begins with an assessment of the total area of the forests and how this is changing.

Although the total area of forest is not decreasing, it is possible that the forests are not in good health. The chapter recounts the history of concern about the effect on forests of acid rain and explains it. It then describes the climatic effects of clearing forests. There is widespread anxiety about the present rise in average temperatures and the contribution carbon dioxide and other greenhouse gases are making to this rise. Trees are made mainly from carbon. Temperate forests are affected by rising levels of atmospheric carbon dioxide and, at the same time, play an important part in reducing the greenhouse warming effect by absorbing carbon dioxide. The chapter explains this. It also describes the way clearing forests can lead to the erosion of soils. The chapter then summarizes the present state of each of the principal types of temperate forest and ends with an outline of the need for conservation and of managing forests in ways that can be sustained, with commercial interests accommodating the needs to preserve biodiversity.

■ MODERN FORESTS

Are forests disappearing from the world at an alarming and possibly accelerating rate? Is air pollution killing some trees and damaging many more? What will happen to temperate forests as the world's climates continue to grow warmer?

These are among the questions that concern many people today. They concern the health of the forests now and in the future. This chapter discusses the issues surrounding forest health, first at a general level and then with regard to particular types of forest.

As the glaciers retreated at the ending of the last ice age, the land they had covered was left bare. Gradually, plants colonized a belt along the southern margin of the newly exposed land. The vegetated belt expanded northward, and all the time the composition of its plant communities was changing. Eventually, after thousands of years of ecological succession (see "Succession and Climax on pages 147–151) the temperate regions were blanketed by forests.

It is wrong to suppose that the first modern humans to explore and hunt in the ice-free landscape roamed through a vast, primeval forest. This forest did develop, but slowly, and the first human hunters stalked their prey through a countryside that was mainly tundra, a land of bare rock interspersed with patches of grasses and sedges and scattered shrubs and, here and there, a few small, stunted trees. Landscapes are constantly changing, and they would change even without our intervention.

Humans did intervene, of course, by using fire to drive game and by clearing forests to provide land for farming (see "Forest Clearance in Prehistory" on pages 222–224). The original forest once covered about 95 percent of the area of western and central Europe, but by the 16th century 80 percent of that forest had been cleared. Domesday records show that most of the English forest had disappeared by the 11th century (see "Domesday" on page 234). What happened to European forests was repeated in temperate forests throughout the world. North America was affected later, of course, as the area colonized and farmed by European settlers expanded. In the United States, nearly one-third of the total deforestation had been completed by 1850 and the remaining two-thirds by 1920. Overall, probably no more than about one-tenth of the original temperate forests remain in the world as a whole.

How Much Forest Is There?

This makes it sound as though temperate forests have almost vanished. Combine this with the widespread concern about the rate at which tropical forests are being cleared, and it is easy to assume that, if present trends continue, before long no forests will remain, temperate or tropical. All types of

Despite appearances, clearcutting does not mean the forest as a whole is disappearing. This is a management technique, and new trees will replace those that have been felled. *(U.S. Forest Service)*

forest tend to be lumped together. During attempts in April 1997 to win agreement at the United Nations for an international convention to control logging, the Canadian Natural Resources Minister Anne McLellan warned that the world was losing its forests at an alarming rate. She seemed to make no distinction between one type of forest or geographical region and another. If she really did believe all types of forest to be under threat, she was not alone. Many environmentalists fear that the temperate forests are disappearing, due mainly to logging.

Despite the past clearances, in 2005 temperate forests covered about 11,702 square miles (30,310 km²) of the United States, 11,974 square miles (31,013 km²) of Canada, and 38,664 square miles (100,139 km²) of Europe, including 31,227 square miles (80,879 km²) of the Russian Federation. The forested area of the United States is increasing by about 15 square miles (40 km²) a year, mainly through the reclassification of woodland as forest. Between 2000 and 2005, European forests expanded by 34 square miles (88 km²). These are small increases, but they demonstrate that temperate forests are not disappearing.

Logging and Farming

It is very unlikely that so large an area will be cleared. Indeed, it is doubtful whether it could be cleared without a major international effort. Certainly, commercial logging is an improbable cause. Forests are cleared when land is deliber-

ately put to a different use. It is not logging that intentionally alters the use of the land but farming, and the main purpose of clearing forests has always been the provision of land for pasture and cropping. This is still the principal cause of forest clearance in the Tropics, but in temperate latitudes the process was completed long ago. Agricultural surpluses and the high cost of supporting their production and then storing or disposing of them are causing most countries in the temperate regions to take land out of agricultural production, not to increase the area of farmed land. This frees land that had been farmed for conversion to other uses, one of which is afforestation. In other words, the change in land use from forests to farms has now been reversed.

Even so, forests might disappear if the demand for timber were so high that for a long enough period the rate of logging exceeded the rate at which forests are able to regenerate. In British Columbia, for example, large quantities of aspen (*Populus*) are used to make disposable chopsticks for the Japanese market, and some environmentalists fear this will lead to the progressive deforestation of a large area. After all, if the chopsticks are disposable, the demand should be ceaseless and the market unlimited. In fact, the area involved is a very small proportion of the total forested area, and the forest is either replanted or allowed to regenerate naturally. The loggers have no choice but to move out once an area has been cleared of timber, and if the forest is left to itself it will grow back. Forests disappear only if the land on which they grow is converted to another use.

Temperate forests are nowadays recognized as economic and environmental assets, and most governments enforce policies that at least maintain the total forested area and more often seek to increase it. New forests are being planted, and although almost 75 percent of the timber produced commercially throughout the world is taken from the temperate forests of the Northern Hemisphere, much of it is grown in plantations, and the area of most of the natural forests is increasing.

Replanting and Natural Regrowth

After the period of deforestation in the United States, which ended around 1920, there was a period of about 40 years during which the total area of forest increased. Then the area began to decrease again. Since 1990 it has stabilized and is now increasing once more. European forests are also expected to continue increasing in size. These increases are due mainly to planting, but the situation in temperate regions differs from that in the Tropics by involving no loss of natural forest. In the United States and Europe, including the densely forested countries of Scandinavia, almost all the timber produced is from plantations or second-growth forest.

Cleared blocks of forest are almost always replanted, but land that was originally forested will often revert to forest if it is allowed to do so. When European settlers arrived to colonize New England, they converted much of the natural forest into a countryside that was characterized by fields and pasture with scattered woods from which they obtained the timber they needed for building and wood for smaller articles and fuel. Now, some 350 years later, the remains of those fields and settlements can still be found as broken walls, holes that once were cellars, and old dirt tracks. Farming was abandoned and the forest has regenerated. As the forest returned, so did the plants that naturally accompany the trees and the animals associated with them. Ecologically and floristically, the regenerated New England forest is very similar to the original forest that was cleared.

Disease

Appearances can deceive, however. Diseases such as chestnut blight and Dutch elm disease (see "Tree Predators and Parasites" on pages 173–180) have altered the composition of the forests, and the parasites that caused them were introduced by humans. Elsewhere in the world, there are economic pressures to weaken the protection of large forests by allowing uncontrolled logging. This will not destroy the forests, but it may alter them, and some species of plants and animals may become rarer as a result.

Timber is traded internationally on a large scale, and the transport of logs from one continent to another greatly increases the risk of transmitting organisms that cause disease. It can also transport insect pests, traveling without the predators that keep their populations in check in the regions where they originate.

It is important to remember, therefore, that although the temperate forests occupy a vast area and there is no realistic possibility that they might be cleared, there are risks to them that should be taken seriously. The forests will remain, but they consist of trees, other plants, and animals that can be harmed.

Forests and the World Around Them

The temperate forests, and especially the boreal forests, are so large that is it tempting to think of them as complete in themselves, enclosed and self-sufficient, as though they had no close links to the world beyond their boundaries. This is an illusion, of course, and a dangerous one. Plants and animals need water, for example, and they need air. Should the water flowing into them be contaminated, forest plants may be poisoned. The air may deposit harmful substances on leaves and the by-products of industry and transport may wash down in the rain (see "Acid Rain" below).

Still more serious, perhaps, the climate itself may continue to change. All plants have distinct climatic preferences and do not thrive if temperatures and precipitation amounts remain for too long outside certain limits. Average temperatures throughout the world as a whole have risen slightly during the past century or so, and many scientists fear that climates may continue to grow warmer (see "Forests and the Greenhouse Effect" on pages 268–272). If this happens, the composition of forests would change, and it is not certain that all species could adapt quickly enough.

Crop cultivation and the tending of domesticated livestock were introduced to the temperate regions long ago. As the farms and pastures expanded, forests were cleared to make fields. That process has long passed its peak, however, and the area of land devoted to temperate agriculture is no longer expanding. Forest clearance stabilized, and in most countries the process has now been reversed. New forests are being planted, and the total forested area is increasing. This is a situation very different from that in the Tropics, where agriculture is still contending strongly for fertile land. Although temperate forests are much more secure than tropical forests, however, there are threats facing them.

ACID RAIN

At a meeting of the Manchester Literary and Philosophical Society in 1852, Robert Angus Smith (1817–84), a Scottish chemist, read a paper called "On the air and rain of Manchester." The paper was subsequently published in the *Memoirs and Proceedings of the Manchester Literary and*

Philosophical Society and it became a classic. Smith had described how rain falling downwind from Manchester—a large industrial city in northwestern England—was very acid, the acidity decreasing with distance from the city. The cause of this acidity, Smith reported, was a mixture of hydrochloric acid and sulfur compounds from the factories of Manchester. This was the first report of the phenomenon now known as acid rain. In 1872, Smith brought his findings to the attention of a wider audience with the publication (by Longmans, Green, London) of a book, *Air and Rain: The Beginnings of a Chemical Climatology.* R. A. Smith became Britain's first Alkali Inspector, the head of the government agency called the Alkali Inspectorate that was responsible for regulating industrial air pollution, and he did much to reduce emissions of hydrochloric acid from the alkali works producing soda for the soap and other industries.

Routine monitoring of the acidity of rainfall began the following year at Rothamsted in southern England, but it did not cover the whole of western Europe until the 1950s. In the United States, there were scientific studies in 1938 of the toxicity to plants of sulfur dioxide, and in 1944 damage to plants was reported in two areas along the Delaware River. Scientists from Rutgers College of Agriculture investigated, discovering that a wide range of wild and cultivated plants showed signs of injury. Air pollution was found to be the cause.

Then, in the 1960s, acid rain emerged as an environmental and political issue, and now it was taking a different form. The problem Smith had identified was fairly simple. Pollutants emitted from sources that could be identified were falling close to those sources. A century later, however, the acidity was caused not by the pollutants themselves but by the compounds into which complex chains of chemical reactions had altered them, and damage was occurring far from any identifiable source.

Forest Damage and Factory Emissions

Acid rain that has been reported since the 1960s has affected temperate forests, first in Scandinavia, then in Germany, central and eastern Europe, and North America. At first, many environmentalists believed the culprit to be sulfur dioxide emitted from power plants burning oil or, more commonly, coal with a high sulfur content. The idea was that in the air sulfur dioxide (SO_2) is oxidized by hydroxyl radicals (OH) to particles onto which water vapor condenses to form sulfuric acid (H_2SO_4). The reactions are:

$$SO_2 + 2OH \rightarrow SO_3 + H_2O \qquad (1)$$

$$SO_3 + H_2O \rightarrow H_2SO_4 \qquad (2)$$

The measures that had been taken to deal with the original problem—local pollution—were responsible for a new problem—long-range pollution. Poisoning is a matter of the dose to which a victim is exposed. A minute dose may cause no injury at all, a rather larger dose may cause illness, and a still larger one may be fatal. In the same way, the amount of harm caused by air pollution depends on the dose of pollutants received, and this depends in turn on the concentration of pollutants in the air. That is why pollution levels are always reported as concentrations, for example in parts per million, or as an amount of a substance per unit volume of the air or water containing it, such as micrograms per cubic meter ($\mu g/m^3$).

Pollution damage can be reduced, therefore, by diluting the pollutant, and until the acid rain reports of the 1960s, it was believed that if sulfur dioxide were diluted to about 0.000000025 ounce per cubic foot (25 $\mu g/m^3$), plants would not be damaged at all. Factory emissions were diluted by redesigning smokestacks. These were made very much taller than they had been previously, and they were no longer simple chimneys—basically vertical tubes—but more sophisticated devices that drew the fumes upward very efficiently. Gases left the stacks at a considerable speed, rising through the overlying air and rapidly mixing with it. The diluted mixture then moved away from the source, and air quality at ground level improved greatly.

All was not well, however, for two reasons. The first is that the pollutants that accumulate to high concentrations near to their source form only a small proportion of the total amount of those pollutants. Regardless of chimney height, most sulfur dioxide is diluted and carried away by the air, and altering the height of factory chimneys has no effect at all on pollution levels more than about 100 miles (160 km) downwind from the source. The second problem was that dilution did not always happen. It is possible for a cloud of gas from a factory chimney to remain concentrated long enough for it to travel a considerable distance. Air quality improved greatly in industrial regions close to the factories, but damage farther away continued unabated, caused by substances that were the end products of chains of chemical reactions. It came to be recognized that the only satisfactory way to reduce air pollution and damage from acid rain was to reduce the amounts of pollutants emitted. Emission controls were the result.

Sulfur and Nitrogen

High concentrations of sulfuric acid certainly harm trees. At one time, the culprit was sulfur dioxide from industrial Lancashire in northwestern England, which made it impossible to grow trees in the Pennine hills to the east. A similar case in Canada was documented in detail. In 1896, a copper smelter opened at Trail, British Columbia, located in the gorge of the Columbia River close to the U.S. border. From then until 1930, it emitted sulfur dioxide, eventually at a rate of almost 10,000 tons (9,100 t) a month. At the peak of the emissions, 30 percent of the trees were dead or dying for 52 miles (84 km) southward along the gorge, and 60 percent were dead or severely damaged for 33 miles (53 km). Sulfur dioxide from a copper smelter at Anaconda, Montana, also caused serious harm in the first decade of the 20th century.

Sulfur is still a problem in some parts of Europe, but not everywhere. Most lichens, especially the shrubby species, are very sensitive to sulfur, and quite small concentrations will kill them. A scale has been devised linking lichen species present at a site with the atmospheric concentration of sulfur dioxide, and the presence of tar spot, a fungal infection producing circular black marks on leaves, is also related to sulfur dioxide concentration. Sulfur dioxide kills the tar spot fungus, so the more tar spot there is, the cleaner the air. In Germany, where tree damage came to be called *Waldsterben*, "forest death," lichens grew in abundance, and analyses of needles from coniferous trees confirmed that sulfur levels were very low. Something was harming the trees, but it was not sulfur.

Scientists now recognize that the phenomenon is much more complex than was at first supposed. In the first place, rain is naturally acid. Acidity and alkalinity are measured on a pH scale where pH 7.0 is neutral, values below 7.0 are acid, and values higher than 7.0 are alkaline (see the sidebar "Acidity and the pH Scale"). Carbon dioxide (CO_2) dissolves in rain droplets in a sufficient amount to give ordinary rain falling from clean air a pH of 5.6. Nitrogen oxides, from the oxidation of nitrogen gas around lightning sparks and from forest fires, and sulfur dioxide, from natural fires and volcanoes, make rain still more acid so that the average pH for clean rain is about 5.0. Acid rain is defined as precipitation with a pH below 5.0.

Airborne sulfur is not an obvious suspect. At low concentrations, sulfur dioxide is either harmless or even beneficial. Sulfur is an essential plant nutrient, so plants benefit from a modest airborne supply. The reaction by which airborne sulfur dioxide becomes sulfate takes place fairly slowly, and sulfate is less harmful to plants than sulfur dioxide. When sulfate dissolves in cloud droplets, the resulting sulfuric acid is often too dilute to harm plants directly.

Buffering

What happens when very dilute acid reaches the ground depends on the soil. It is the positively charged hydrogen in acids (sulfuric H_2SO_4, nitric HNO_3, hydrochloric HCl, for example) that gives acids their chemical properties (see the sidebar). Dissolved in water, sulfate becomes sulfuric acid but with the sulfate and hydrogen as separate ions. With values and signs of their charges shown, the components of the acid can be written as $H^+ + H^+ + SO_4^{2-}$. A charged particle

Acidity and the pH Scale

According to the theory published in 1923 by both the Danish physical chemist Johannes Nicolaus Brønsted and the British chemist Thomas Lowry who were working independently of each other, acidity is a measure of the extent to which a substance releases hydrogen ions (protons) when it is dissolved in water. Also in 1923, the American theoretical chemist Gilbert Newton Lewis defined acidity as the extent to which a substance acts as receptor for a pair of electrons from a base. The two theories describe different ways of looking at the same thing and do not contradict each other. A substance possessing an excess of hydroxide ions (OH^{-1}) is said to be alkaline.

Acidity is measured on a scale of 0–14 that was introduced in 1909 by the Danish chemist Søren Peter Lauritz Sørensen. The acidity of a solution, measured at 77°F (25°C), is equal to the negative logarithm of c ($-\log_{10}c$), where c is the concentration of hydrogen ions in moles per liter. The scale measures the "potential of hydrogen," which is abbreviated to pH, so it is known as the pH scale.

Pure water is neutral—neither acid nor alkaline. A neutral solution has a hydrogen-ion concentration of 10^{-7} mol/l, so it has a pH of 7. A pH lower than 7 indicates an acid solution and one higher than 7 an alkaline solution. The scale is logarithmic, so a difference of one whole number in pH values indicates a 10-fold difference in acidity. A carbonated soft drink has an acidity of about pH 3, making it 100,000 times more acid than distilled water (pH 7), and ammonia (pH 12) is 100,000 times more alkaline than distilled water.

is called an ion; one with a negative charge is an anion, and one with a positive charge is a cation.

Particles of clay and decomposed organic material carry a negative charge, and as an acid solution flows through the soil, the hydrogen ions are adsorbed onto them by the electrostatic attraction of positive-to-negative charge. Gradually, the soil becomes more acid, as does the water flowing through it with a load of free hydrogen ions. This is the case, however, only where the soil lies above an igneous rock such as granite and there is a high rainfall. In this situation, rain falling even from the cleanest air will produce acid soil and water, because of the acidity of ordinary rain. Nor is rain the only source of hydrogen ions in soils.

Natural decomposition processes produce organic acids, and the uptake of nutrients with a positive charge can leave an excess of hydrogen.

If the underlying bedrock is sedimentary, there is a good chance the soil will be fairly rich in calcium and magnesium. If the parent material (see "Soil Formation and Development" on pages 15–18) is chalk or limestone, the rock will be made from calcium and magnesium carbonates, and the soil will have an abundance of these two metals. Calcium and magnesium atoms also carry a positive charge, so they are readily adsorbed onto clay and humus particles.

Hydrogen cations will compete for adsorption sites, and if there are enough of them, they will overwhelm and displace other cations by a process called mass action. The resulting cation exchange leaves the acid molecules adsorbed onto soil particles and releases the calcium and magnesium cations to be carried away in the water. The water leaving the soil into groundwater, rivers, or lakes remains neutral because when these cations dissolve, they bond to hydroxyl (OH⁻) ions, carrying a negative charge. In this neutral form, they are sometimes known as base cations. The soil will not become acid for as long as calcium or magnesium cations occupy most of its negative adsorption sites, and in soils derived from rocks such as chalk and limestone, the supply of these cations is limitless. The process by which a medium neutralizes an acid in this way is called buffering.

Much depends, therefore, on the rock from which soils are derived, and in southern Scandinavia where acid rain damage to forests was first noted, the underlying rocks are granitic. The soils are poorly buffered, and both soil and soil water are acidified fairly easily. Acid water flowed into lakes, and aquatic organisms were harmed as well as trees. On chalk and limestone soils, with much stronger buffering, the risks are much lower.

Ozone and Nitrogen

As the German forests demonstrated, sulfur is not the only cause of tree damage. Ozone came under suspicion. This is an extremely reactive gas that forms by a series of reactions driven by the energy of intense sunlight in air containing fairly high concentrations of nitrogen oxides, mainly from vehicle exhausts. Experiments found that ozone can injure plants. The first of these experiments were conducted in the 1950s to discover whether ozone from the Los Angeles basin was causing needle damage to the ponderosa pines growing in the San Bernardino Mountains of southern California. Ozone can enter stomata and damage the membranes of cells containing chloroplasts, thus reducing the rate of photosynthesis (see "Photosynthesis" on pages 113–120). Experiments in Germany found that Norway spruce trees were also damaged when exposed to ozone at 100–300 parts per billion mixed with acid droplets to give a pH lower than

3.0. Although the experiments seemed convincing, the damage scientists observed experimentally did not entirely match what was seen in the forests, and the contribution of ozone remains uncertain.

Excessive nitrogen seems a more likely cause, at least for some of the damage. Gaseous nitrogen is chemically inert (see "Nitrogen Fixation and Denitrification" on pages 122–125), but ammonia (NH_3) is an available form of nitrogen that can enter the air, dissolve in rainwater, and be washed to the ground miles from its original source. That source is most likely to be the urine of farm animals. Many forests grow on relatively poor land where at least some plant nutrients are scarce, and it is often the availability of nitrogen that limits tree growth. If nitrogen is added in large quantities, tree growth is vigorously stimulated, but it is unbalanced growth because plant tissues are short of other, mineral nutrients. This nutrient imbalance inhibits the hardening processes that prepare trees for winter, increasing the likelihood that they will be damaged by cold or desiccation.

Contaminants do not always act in isolation from one another. It is possible for the effects of two or more to combine in such a way that the damage they cause together exceeds the sum of the damage each would cause separately. The phenomenon is called synergism and is well known in several forms of pollution. No one knows whether synergistic interactions are involved in the injuries we attribute to acid rain, but it is possible.

Acid air pollution does not travel only as rain. Indeed, taking the term literally, acid rain is but a minor risk to foliage because rain strikes only the upper surface of leaves and branches, then quickly runs off them, and falls to the ground. Fine mist, on the other hand, consists of minutely small droplets that adhere to surfaces. They cling to both upper and lower surfaces of leaves and stay there while their water evaporates, concentrating the acid. As anyone who has walked through a mist will know, these droplets, so small that they hang almost motionless in the air, can make a person wetter than a shower of rain. Acid can also be deposited on surfaces directly from dry air. Water droplets are not always needed. Acid mist and dry acid deposition cause more serious damage to foliage than does acid rain.

How Acid Rain Damages the Soil

Acid rain damage, as opposed to damage from other forms of acid deposition, occurs mainly belowground. As well as helping to buffer the soil water, magnesium and calcium are also plant nutrients. Where these cations are displaced by hydrogen, they enter water and are carried down into groundwater or away by rivers. If magnesium-hydrogen and calcium-hydrogen cation exchange takes place at a rate faster than magnesium and if calcium can be replaced by the weathering of the underlying soil parent material, what remains may be insufficient for the needs of plants. Then trees, with a big demand because of their size, may suffer a magnesium or calcium deficiency.

Cation exchange may also involve exchangeable aluminum. Aluminum is very abundant in rocks and especially in clays, but almost all of it is securely bound in the chemical compounds from which these minerals are made. Aluminum never occurs naturally as the pure metal because it is so powerfully reactive. (Aluminum kitchen utensils are covered with a very thin layer of aluminum oxide that forms the instant aluminum is exposed to air.) In soils, however, there are some free aluminum cations, and these adhere to exchange sites. At a pH of 5.5 or below, some are dislodged by overwhelmingly large numbers of hydrogen cations. Liberated into the soil water, aluminum cations form groups in which an aluminum atom is surrounded by six molecules of water or by hydroxyl radicals (OH). At a pH of about 5.0, one of the water molecules will lose a hydrogen ($H_2O \rightarrow OH + H$). The release of a hydrogen cation increases the acidity, exacerbating any adverse effect already arising from the low pH. One of those effects is a slowing in the rate at which organic matter decomposes. Since decomposition is the mechanism by which nutrients are recycled, this can lead to nutrient depletion.

Aluminum cations can also enter root hairs. They are absorbed in the same way as calcium and magnesium and can take their place, blocking further uptake of calcium and magnesium. This leads to a nutrient deficiency in the plant and, in addition, aluminum can interfere with the transport of water through the plant, which can increase the plant's susceptibility to drought.

Water-borne aluminum is also poisonous to fish. The permeability of gill membranes is regulated by calcium cations, which can be displaced by aluminum cations at a pH of 5.0–5.5. At the same time, aluminum causes the release of mucus, which clogs the gills. Older fish are more susceptible to aluminum poisoning than young ones, but the effect in lakes can be serious. It is made worse by the fact that the lower the pH of water, the less phytoplankton (small, mainly single-celled, aquatic plantlike organisms that float near the surface) it contains, and the phytoplankton forms the base of the aquatic food chain. Acidified lakes become very clear. They look clean, and so they are, but this is because they support relatively few living organisms.

Drought and Disease

It was in the early 1980s when the amount of tree damage was increasing, apparently rapidly, that there were fears of what became known in Germany as *neuartige Waldschäden,* "new forms of forest damage." Pollution is not the only cause. Trees may suffer from drought, an entirely natural phenomenon, and symptoms of drought damage may not appear

until a few years after the rains have returned. The forest damage observed in Europe in about 1980, for example, followed a severe drought that affected all of western Europe in 1976. Beech (*Fagus sylvatica*) has shallow roots and is especially susceptible to drought. In 1985, Friends of the Earth surveyed beech trees in Britain and discovered that their leaves were changing color and falling some weeks earlier than usual. This was more probably due to stress induced by drought than to pollution, and the phenomenon was not repeated in subsequent years.

Trees may also be attacked by infestations of pests or disease organisms, and there were some outbreaks during the period of the worst damage. Studies of German soils in the 1980s found viruses and viruslike particles in soils and water from forest sites and in the needles of spruces, pines, and firs that were showing signs of ill health, although the studies did not suggest that viral infection was the primary cause of the observed damage.

Finally, these causes are likely to interact. Just as a severely malnourished or dehydrated person will be more susceptible to infectious disease, so trees suffering a nutrient deficiency will be more vulnerable to frost, drought, pest attacks, and disease. A tree that has been weakened by drought, pest, or disease may be killed by a concentration of acid that it would otherwise have survived.

During the 1980s acid rain became a hotly debated and controversial topic because it turned out to be very difficult to unravel just what was harming forests.

Measuring the Damage

The Scandinavian worries, starting in the 1960s, centered on the acidification of lakes. It was in Germany that forest trees were the main cause for concern, and Norway or common spruce (*Picea abies*) was the species identified as the first victim of *neuartige Waldschäden*. It is the most widely grown tree, making up about 40 percent of the German forests, and commercially it is the most important.

Damage was defined in terms of the proportion of its needles that a tree had lost, and at first Norway spruce was believed injured if it had lost more than 10 percent of its needles. On this basis, one-third of all spruce trees were classed as damaged. Further studies showed that the diagnosis was incorrect, however. All coniferous trees shed needles, and the number they are without at any one time varies from one individual tree to another. Some perfectly healthy trees lose more than 10 percent of their needles.

Today, foresters accept an international scheme for classifying damage. Up to 10 percent needle loss indicates no damage (class 0), 11–25 percent means slight damage (class 1), 26–60 percent is moderate damage (class 2), 61–98 percent is severe damage (class 3), and with more than 99 percent loss the tree is dead (class 4). Applied to all tree species, this classification reduced the scale of damage from more than half to less than 20 percent of the total forest area. In Europe, forest damage is assessed annually across plots in 2.5 × 2.5 mile (4 × 4 km) grids. The number of trees in each of the damage classes is then converted to the area of the national forest that those trees would occupy if they were all together in a single stand. This technique was first used in Germany, and it has now been adopted by the United Nations Economic Commission for Europe and by the European Commission.

Different Damage, Different Causes

There are several visible signs of ill health, now often called forest decline, and these signs vary from one species to another. For example, five different kinds of damage are known in Norway spruce (*Picea abies*), each with a different cause. At high elevations in the mountains of central Germany, and also in Austria, France, Belgium, and the Netherlands, there is a yellowing and then dropping of needles that are exposed directly to sunlight. This is due to magnesium deficiency and may lead to the death of the tree if there is also injury from another cause, such as frost, drought, insect infestation, or disease. Where Norway spruce grows alongside other species, such as silver fir (*Abies alba*), Douglas fir (*Pseudotsuga menziesii*), Scotch pine (*Pinus sylvestris*), and beech (*Fagus sylvatica*), these species may also be affected. At middle elevations in the German mountains, mainly on soils that are poor in nutrients and that are about 1,300–2,000 feet (400–600 meters) above sea level, there is a thinning of the tree crowns, sometimes accompanied by a yellowing of the needles. This is associated with fairly high sulfur-dioxide concentrations and wet deposition of hydrogen cations, combined with nutrient deficiency, especially of calcium and magnesium, but in some places also of phosphorus and potassium. In southern Germany, needles were seen to turn orange-yellow in September and then red and finally brown. The brown needles may remain on the tree for some months, but most fall around the end of October, causing a marked thinning of the crown but in the absence of other stresses rarely leading to the death of trees. This damage is due to infection by various fungal species. There has been yellowing of needles and thinning of crowns at elevations above about 3,300 feet (1,000 meters) on shallow, calcareous (calcium-rich) soils in parts of the Alps where the underlying rock is limestone. This has been observed in Austria, Switzerland, Italy, and Germany and is linked to nutrient deficiencies, especially of potassium. Crown thinning has also been observed in coastal areas. In 1983, aerial surveys found that one-third of the Norway spruces more than 60 years old in German coastal forests showed signs of damage in classes 2–4 and that tree growth was very slow. Similar damage occurred in Belgium and the Netherlands. The cause is uncertain.

Forests in the industrialized regions of central Europe continue to suffer damage from sulfur-dioxide pollution originating in the lignite-mining areas along the border between Germany and the Czech Republic. Large areas of forest have been affected on both sides of the Ohre Mountains and extending into the Harz and Fichtel Mountains of Germany. Trees have died only at elevations above 3,000 feet (900 m) in the Ohre Mountains, however, where strong winds exacerbate the effect. According to Otto Kandler, emeritus professor of botany at the Institute of Botany, Ludwig-Maximilian University, Munich, photographs of damaged trees in the Ohre Mountains were often shown as examples of *Waldsterben,* and many people assumed that the damage was much more widespread than in fact it was.

European forests have also experienced needle yellowing, needle death, crown thinning, and other injury to silver fir (*Abies alba*). Beech (*Fagus sylvatica*) has shown signs of damage due to a variety of causes. Insect infestations and fungal infections have caused some damage, there has been an excessive supply of nitrogen associated with deficiencies of other nutrients, and soil acidification has been found in affected areas.

In Britain, routine monitoring of the health of trees was initiated in 1984 and gradually expanded in subsequent years. It now covers Sitka spruce (*Picea sitchensis*), Norway spruce (*P. abies*), Scotch pine (*Pinus sylvestris*), oak (*Quercus* species), beech (*Fagus sylvatica*), and mixed broad-leaved trees. There are more than 350 level 1 sites, at each of which 24 trees are studied, and 20 level 2 sites, where conditions are monitored in more detail. The Forest Condition Survey, a monitoring program required by European Union legislation, examines trees at about 6,000 level 1 sites and 860 level 2 sites on a 10 × 10 mile (16 × 16 km) grid across Europe. In Britain there are 90 of these sites, monitoring five species (oak, beech, Scotch pine, Norway spruce, and Sitka spruce). Once data from different countries are adjusted to ensure that they are all applying the same standards, British forests emerge as little different than those elsewhere in Europe. There is no conclusive evidence, however, of a clear relationship between patterns of air pollution and the health of trees. In fact, trees in areas of high air pollution appear healthier than those growing in clean air. Probably this is because the climate in the polluted regions of England favors trees with dense crowns. There is evidence of damage to lakes due to acidification in areas underlain by igneous rocks.

Pollution Damage in North America

Particular pollution episodes, such as those at Trail and Anaconda, have affected forests in the western United States. Ponderosa pines (*Pinus ponderosa*) and Jeffrey pines (*P. jeffreyi*) growing in mixed conifer forests in southern California were found in the 1970s to have been suffering from ozone damage since the 1950s. In descending order of susceptibility, white fir (*Abies concolor*), California black oak (*Quercus kellogii*), incense cedar (*Libocedrus decurrens*), and sugar pine (*P. lambertiana*) were also affected.

In general, however, forest damage due to acid rain in North America is primarily a phenomenon of the eastern side of the continent, and it has been observed over a large area. South of the Great Lakes, a line drawn from the western tip of Lake Superior to the Gulf coast just west of New Orleans marks a boundary to the east of which the average pH of rainfall is below 5.0. The acidity of precipitation increases to a maximum, averaging pH 4.4, in the northeastern states and southeastern Canada, and the distribution of acidity closely matches the deposition of sulfate.

A study conducted at Hubbard Brook Experimental Forest near North Woodstock, New Hampshire, by Timothy Fahey and Stephanie Juice from Cornell University and published in the journal *Ecology* in May 2006 found clear evidence that acid deposition had reduced the number of sugar maples (*Acer saccharum*). There are detailed records of the chemical composition of soil at Hubbard Brook Forest since the 1950s. Acid rain can cause soils to become depleted in calcium while manganese levels increase, so the researchers added a source of calcium to one 25-acre (10-ha) plot while using another plot of similar size as a control. After 5–10 years the soil in the experimental plot became less acidic, making it more suitable for maples, and the concentration of calcium in maple leaves increased. By the fourth year previously high manganese levels in maple leaves had fallen to healthier levels. Seed production and the size and density of seedlings increased in the years following treatment. The study showed the effect of acid deposition on sugar maple, but it also showed how quickly the maples recovered when soil acidity was returned to normal levels.

Appalachian Spruce and Fir

In the 1980s scientists began to observe a sharp deterioration in the health of spruce and fir stands along the crest of the Appalachian Mountain chain from eastern Canada to the Mount Mitchell National Park in North Carolina. The situation has been monitored closely on Mount Mitchell in the Black Mountain range of the southern Appalachians, where it seems especially severe.

The deterioration was rapid. A study of an admittedly very small sample—272 red spruce (*Picea rubens*) and 213 Fraser fir (*Abies fraseri*) between 25 and 100 years old in 16 plots in the Black Mountains—found all the trees healthy in 1984. Spruce were the healthier of the two species with 79 percent of the trees having lost less than 10 percent of their needles (class 1) and 19 percent less than 50 percent (class 2). Of the firs, 60 percent were in class 1, 15 percent in class 2, and 25 percent had lost up to 99 percent of their needles

(class 3), although there were no dead trees in the sample stands. By 1986, however, 9 percent of red spruce and 16 percent of the firs had died, and by the spring of 1987, 41 percent of the spruce and 49 percent of the firs were dead.

The Forest Service of the U.S. Department of Agriculture studied spruce–fir forests in 1985 using aerial photography and ground inventories. Their examination of a much larger area than the Black Mountains study found that in 24 percent of the total area (15,614 acres; 6,319 ha), more than 70 percent of trees were dead (class 3). Between 30 percent and 70 percent of trees had suffered severe damage (class 2) in 6 percent of the area (4,245 acres; 1,718 ha). There were fewer than 30 percent of dead trees (class 1) in 70 percent of the area (45,894 acres; 18,573 ha). The severity of the damage increased with elevation, and the investigators noted that although the tops of the mountains occupied only a very small proportion of the total area, the dead trees at these high elevations were more visible than trees at lower levels. They also reported that fir, with 44–91 percent mortality, was more susceptible than spruce, with 3–14 percent of trees dead.

Balsam woolly adelgid is the principal cause of damage (see "Aphids and Adelgids" on page 176). A native of Europe, this insect was first reported in Maine in 1908, and by 1957 it had reached Mount Mitchell, although it had probably been established in the southern Appalachians since about 1940. Damage inflicted by the pests weakened trees, increasing the damage caused by a prolonged summer drought in 1986, followed by heavy rime icing in December of the same year. Acid precipitation is likely to have exacerbated the situation. Montane forests are more exposed than those at lower elevations because cloud droplets are significantly more acid than rain or snow falling from the clouds. Mountain forests are exposed to mists with a pH of 3.0 to 4.0 for between 30 and 80 days each year in the northern Appalachians and for 200 to 280 days in the southern Appalachians.

Adirondack and Appalachian Red Spruce

Red spruce also declined during the 1980s at elevations above 3,000 feet (900 m) in the northern Appalachians and the Adirondacks. Balsam fir is the dominant species above about 3,300 feet (1,000 m), and sugar maple (*Acer saccharum*) is that below 2,300 feet (700 m), red spruce being a minor species between about 2,600 feet and 3,300 feet (800–1,000 m) and rare above about 4,000 feet (1,200 m). These forests support a number of species, so changes in the abundance of red spruce might pass unnoticed were it not being monitored. Between 1982 and 1987, red spruce above 3,000 feet (900 m) were dying at a rate of about 4 percent each year compared with a rate of 0.5 percent at lower elevations. After 1987 it seemed probable that the remaining severely damaged trees (class 3) would die, but the healthy trees (classes 1 and 2) were showing no sign of deterioration.

Trees growing at high altitudes are subject to more climatic stress than those at lower elevations, and spruces did suffer from pests and fungal attacks, although at high altitudes these are less severe. It is possible that air pollution contributed to the decline in red spruce, although this still remains unproven.

Could Pollution Controls Make Things Worse?

By the late 1980s, acid-rain damage appeared to have stabilized. Governments had set limits to industrial emissions of sulfur dioxide and nitrogen oxides, atmospheric levels of which were starting to fall. There are now fears that acid-rain problems may reappear as a direct consequence of pollution controls. It is true that sulfur emissions have fallen dramatically over Europe and eastern North America, but so have dust emissions. Dust particles can be harmful to people with respiratory illnesses: They cause haze, which reduces visibility, so there would seem to be no good reason not to prevent their release where that is possible, and most governments have taken measures to reduce particulate pollution. At the same time, technological advances in industrial processes have reduced particle emissions. Not all fine particles are chemically inert, however. Some are alkaline (pH greater than 7), and when they dissolve in cloud droplets that are acidic, they reduce the acidity. This reaction releases base cations, especially calcium, magnesium, and potassium. Southern beech (*Nothofagus* species) trees in Chilean forests obtain almost all their calcium from atmospheric particles. When base cations fall to the ground in raindrops, they help reduce acidity in the water moving through the soil, thus mitigating the harmful effects of acid rain.

Reducing particle emissions has reduced the base-cation contribution they make to the soil. At the Hubbard Brook Experimental Forest in New Hampshire, the concentration of atmospheric base cations has fallen by 49 percent since 1965. In the forests of Sjoangen in southern Sweden, there has been a 74 percent drop since 1971. Norwegian forests have lost between 56 percent and 74 percent of their base cations since the 1940s. Similar reductions have occurred over most of Europe and North America. Some scientists believe the extent of the loss of atmospheric base cations is large enough to have offset between 54 percent and 68 percent of the reduction in atmospheric sulfur in Sweden and up to 100 percent of the sulfur reduction at some places in eastern North America. Spruce forests in the Fichtel Mountains of Germany may have been affected, and the reduction in atmospheric base cations may have contributed to the damage to red spruce in the southern Appalachians. This implies that although sulfur dioxide emissions have fallen, the reduction in base cations resulting from particulate pollution control has increased plant susceptibility to acid damage.

When Will Forests Recover?

When acid deposition ceases, soil and water recovery may be slow. Acidification involves the loss of base cations, and once those are gone, it may take a long time for them to be replaced. Until this happens, the pH will remain low.

Acid precipitation has undoubtedly contributed to the decline of forests in Europe and North America, but the extent of the European decline was sometimes exaggerated, and acid precipitation is not the only cause. Forests sometimes suffer serious damage and many trees die. During the last 200 years, there have been five forest declines in different parts of Europe, and there were 13 in North America during the 20th century, with airborne pollutants being implicated in six. It is very misleading, therefore, to attribute all damage to forest trees to acid rain. Pests and diseases have also caused harm, as have episodes of severe weather.

Emissions of sulfur and nitrogen oxides have fallen in response to the implementation of pollution controls. Provided the resulting improvement is not offset by the reduction in atmospheric base cations due to reductions in particulate emissions, in time most acidified soils and waters will recover and tree damage from this cause will cease.

■ CLIMATIC EFFECT OF FOREST CLEARANCE

At those times of year when the ground is bare, between the harvesting of one arable crop and the emergence of the next, a spell of dry weather can reduce a finely textured soil to the consistency of a powder that will trickle through the fingers. In this condition, a spell of windy weather can blow tons of it away in a dust storm that clogs ditches, reduces visibility, and darkens the sky. In the worst cases, the wind carries away expensive fertilizer and seed along with the soil. This is not a new problem. At one time, such storms happened fairly often in parts of eastern England, Denmark, Norway, and on the wide plains of Russia. They also occurred in the southern part of the Great Plains of North America, which is where so many farms failed tragically in the Dust Bowl years of the 1930s.

The Dust Bowl region has always had a climate subject to periodic episodes of severe drought, and it was drought that caused the damage of the 1930s. Only a drastic change in farming methods could make that land reliably cultivable, but elsewhere, in less arid climates, there is no mystery about how such dust storms can be prevented. European farmers found a solution as long ago as the 18th century. They planted trees in lines or as small, isolated clumps in certain, strategically chosen places. The trees absorb much of the energy of the wind, sheltering the land downwind of them. Today it is standard farming practice in all temperate regions to plant trees to form shelterbelts, and there are experts to advise on the location and configuration that will produce the optimum effect. In some places, shelterbelt trees have a powerful effect on the landscape. In the Rhône valley in southern France, cypress trees are used for this purpose and contribute much to the appearance of the countryside, as do the tall Lombardy poplars planted to provide shelter in the Netherlands. Shelterbelts alter the climate on their downwind side over a distance proportional to the height of the trees, and the effect can be increased by planting two or more belts parallel to one another some hundreds of meters (yards) apart.

If planting trees can reduce the speed of the wind, it is reasonable to suppose that removing groups of trees will also affect climate. As the original forests of temperate Europe and North America were cleared mainly to provide land for cultivation, the climate must have changed locally, and a change in wind speed will have been only one of the ways in which it did so.

Wind and the Forest

On a day when the wind in the open is blowing at 20 MPH (32 km/h), 100 feet (30 m) from the edge of a European forest, depending on its composition and density, the wind speed will be 12–16 MPH (19–26 km/h). Regardless of the composition of the forest, the wind speed will be reduced to about 1.2 MPH (2 km/h) 400 feet (120 m) from the edge. Obviously, American forests have a similar effect. A study found that inside a broad-leaved deciduous forest in Tennessee, the wind speed in January was 12 percent of that in adjacent open country, and in August, when the trees were in full leaf, it was only 2 percent. Trees absorb the energy of the wind, but that energy is sometimes powerful enough to cause them serious harm. Many beech trees were blown down when severe gales struck southern England in October 1987. The beeches were vulnerable because they were still in full leaf, so they absorbed a great deal of the force of the wind.

Remove the forest, and the climate becomes windier, not only over the area that was formerly covered by trees but also in a belt about 0.6 mile (1 km) wide surrounding it. It is not only inside the forest that wind speed is reduced but on the downwind side as well, where the forest provides shelter. Most open, level land is windy. In East Anglia, the bulge on the map in eastern England north of the River Thames, there

is usually a wind. The region is flat, low-lying, and exposed directly to easterly winds blowing off the North Sea. At one time, much of the land was forested and the remainder was swamp. In those days the climate inland would have been less windy. There is a similar story to be told in many other parts of the temperate latitudes.

Dust and Fog

Once the forest is gone, the air also becomes dustier, and fog, rolling gently forward on a light wind, travels farther. This is especially important near coasts, where sea fogs may penetrate much farther inland than they did before the forest was cleared. It happens because a forest is like a filter. Trees and shrubs trap water droplets, which adhere to leaves and bark, so the air leaving the forest on the downwind side is drier than the air entering it.

Some solid particles are trapped in the same way. Others fall because the amount of solid material that can be transported through the air depends on the amount of energy that the wind possesses. Friction, as the wind flows into and around trees and over the uneven ground surface, absorbs much of that energy. It is why the wind speed decreases with distance into the forest. The loss of energy also reduces the capacity of the wind to transport material that falls to the ground or onto plant surfaces from where rain soon washes it to the ground. There it remains because the wind speed inside the forest is never high enough to raise it again.

Frost and Windchill

Wind also affects the temperature in several ways. Regardless of the air temperature, people will feel colder outdoors on a windy day than they will on a still day because the wind carries away the thin layer of air surrounding and warmed by the body, which expends its own warmth warming this air. The result is that the wind can actually reduce the body's temperature. This is called windchill, and at low air temperatures it can be dangerous (see the sidebar on page 266). Fell the trees and everyone may need to dress more warmly in winter.

Sometimes, usually in spring or fall, a TV weather forecaster predicts that although the temperature overnight will fall close to freezing, frost is unlikely because of the wind. Given that the same forecast may also have warned of windchill, the absence of frost may seem paradoxical, but it is not. Ground frosts decrease when forests are cleared and wind speeds increase. During the day, the ground and plants close to it are warmed by the Sun. They also radiate the heat they receive, but during most of the day, they receive more energy than they radiate away, so they grow warmer. In the late afternoon, the balance shifts, and radiation from the surface exceeds the energy received from the Sun so that the plants and the ground surface cool. This chills the layer of air in contact with them. If this air is still and moist, its water vapor condenses onto cold surfaces as dew or, if the surfaces are cold enough, as ice crystals. If the air is moving, however, the air at ground level is constantly mixing with warmer air above, and there is no opportunity for a layer of still air to form and be chilled, so frost and dew are much less common.

This effect is not as beneficial as it may seem. Dew forms when water vapor condenses onto surfaces out of humid air. Condensation releases about 17,000 calories of latent heat per ounce weight (2.5 MJ/kg) of water that condenses, and the energy so released is absorbed by the surface onto which the water condenses, warming it (see "Latent Heat" on pages 43–44). Frost, formed by the sublimation of water vapor directly into ice, releases even more warmth: 19,000 calories per ounce (2.83 MJ/kg). Wind has no effect whatever on the rate at which plants and the ground radiate away their heat at night, but preventing the formation of dew and frost allows the ground surface temperature to fall further than it would in still air. The ground inside a forest freezes later to a shallower depth than ground outside. Clearing the forest reduces the formation of dew and frost, but the loss of latent heat allows the ground to freeze earlier in winter and the frost to be harder.

Changing Color

Removing the forest produces a still more dramatic change in the amount of solar energy that reaches the ground surface and is absorbed by it. When electromagnetic radiation, such as radiant heat and visible light, strikes a surface, some of the radiation is reflected and some is absorbed. The proportion that is reflected varies from one type of surface to another and is measured as the albedo of the surface (see "Why Surface Color Matters" on pages 34–35), expressed as a percentage of the total.

A coniferous forest, dark in color because of the dark green of its needles, has an albedo between about 8 percent and 14 percent, meaning that between 86 percent and 92 percent of the solar energy falling on it is absorbed. A broad-leaved deciduous forest, generally with somewhat paler leaves and, of course, with no leaves at all in winter, is more reflective. It has an albedo between 12 percent and 18 percent, so it absorbs between 82 percent and 88 percent of the energy it receives. The energy is absorbed by the tree foliage, so it is that the leaves that experience any warming effect and that it is used to drive photosynthesis (see "Photosynthesis" on pages 116–122).

Clear the forest and this complicated pattern of radiation reflection and absorption vanishes. The albedo of the surface changes. If fields growing wheat replace the forest, their albedo is between about 18 percent and 25 percent. This is much higher than the albedo of a forest, meaning that less energy is absorbed at ground level. If the crop is corn or potatoes, on the other hand, the albedo will be between 3 percent and 15 percent. Between crops, when the ground is bare, the albedo is between 5 percent and 25 percent depending on the color of the soil. It is only when the ground is bare, of course, that it absorbs all the unreflected energy. At other times, crops absorb most of the energy.

Although the albedo becomes highly variable once the forest has gone, the surface where energy is being reflected or absorbed is at a much lower level. More light and warmth reaches the ground, and its temperature rises higher during the day and through the summer, falling again at night and in the fall.

Shading the Ground

Energy that is absorbed by the forest canopy cannot reach the floor, but this does not mean the floor is in almost total darkness or that no warmth reaches it from above. If no more than 8 percent of the warmth of the sunshine reached the floor of a forest where 92 percent was absorbed in the canopy, it is unlikely that the forest could survive at all because the ground at its base would be permanently frozen. Absorption and penetration vary according to the wavelength of the radiation, and a relatively high proportion of the absorbed radiation is at short wavelengths. More ultraviolet radiation and blue light are absorbed than red light and heat. In addition, solar radiation reaches the surface most intensely from the direction of the Sun only when the sky is cloudless. Clouds scatter incoming radiation, so it reaches the surface from all angles. It casts no sharp shadows because objects are illuminated evenly on all sides. In the forest, a proportion of the radiation, arriving almost horizontally, avoids absorption by foliage. Light intensity on the forest floor is greater on cloudy days than on sunny days. This may seem paradoxical, but on a sunny day the brightly lit patches of ground are surrounded by very deep shadow, whereas on cloudy days the entire area is illuminated by diffuse light.

Some types of forest allow more radiation to penetrate than do others. In a forest dominated by birch and beech, for example, about 50 percent of the incoming radiation may reach the floor in summer and 75 percent in winter. Pine forests allow between 20 percent and 40 percent of the incoming radiation to reach the floor, and the much denser fir and spruce forests allow between 10 percent and 25 percent. One consequence of this is that there are fewer hours of daylight inside the forest. This means that there is a shorter time for the ground to warm during the day than there is outside the forest and that the shading of the ground also reduces the extent of warming. Daytime air and ground temperatures are lower, so there is less radiative cooling at night, and the diurnal range of temperature is smaller inside the forest than it is outside.

In temperate latitudes, not all the heat that is absorbed by the ground during the day in summer is radiated away on summer nights because the nights are short, and before the balance can be restored, the Sun has risen once more. By the end of summer, the ground just below the surface is warmer than it was in spring, and it completes its cooling during the fall. Just as the diurnal temperature range is lower inside a forest than it is outside, so is the annual temperature range. In open country the ground reaches a higher temperature by day and over the summer than it does inside a forest.

Altering the Amount of Moisture in the Soil and in the Air

The warmer the ground, the more readily water will evaporate from it. This is an immediate and obvious effect that is easily verified. Scrape away the dead leaves or needles lying on the surface, and except in very dry weather, the ground inside a forest is usually wetter than the ground outside the forest. The forest floor is more shaded and cooler, so water is slower to evaporate from it. In a pine forest in Arizona, evaporation from the ground in summer is about 70 percent of that on open ground outside the forest.

That is only part of the effect forest clearance has on atmospheric humidity. All plants take water from the ground, transport it upward, and lose it by transpiration through their leaf stomata (see "Transpiration" on pages 48–50). The rate of transpiration varies from one species to another, depending on the efficiency with which they conserve moisture by preventing its loss. Pines transpire much less water than fir trees, for example, and trees transpire much less water in winter than in summer because leaf temperature is lower in winter and, therefore, so is the rate of evaporation from open stomata. Transpiration ceases in winter in deciduous trees, of course. In early June the relative humidity inside a Northern Hemisphere mixed deciduous forest of birch, beech, and maple might be about 1 percent higher than it is outside. By the middle of July the difference might have increased to about 5 percent, and in some forests it can reach about 11 percent. The difference in relative humidity inside and outside the forest is due to water vapor that is returned to the air by transpiration.

Trees also intercept falling rain, a proportion of which evaporates from leaf and bark surfaces without ever reaching the ground. Together, evapotranspiration and the evaporation of intercepted rain can vaporize a considerable amount of water. In the Harz Mountains of Germany, it has been calculated that Norway spruce (*Picea abies*) forests return to the atmosphere annually about 13.4 inches (340 mm) of precipitation by evapotranspiration and about 9.5 inches (240 mm) by the evaporation of intercepted rain. This makes a total of about 23 inches (580 mm).

Changing Precipitation

Altering the amount of water vapor entering the air affects the hydrological cycle by an amount that varies according to whether the ground is level or sloping and the direction in which sloping ground faces. The table on page 268 shows

Fate of the Annual Precipitation Falling on a European Oak Forest

	PERCENTAGE OF TOTAL
Transpired by trees; returned to the air	30.5
Penetrates ground; joins groundwater	18.75
Runs off ground vegetation	16.75
Intercepted by canopy; evaporates	13
Evapotranspired by ground vegetation	12.25
Runs off through the soil	8.75

what happens, on average, to all the precipitation that falls onto a European oak forest through the year.

It might seem that if all the trees were cleared from the Harz Mountains, the annual precipitation there would increase by about 23 inches (580 mm) a year, or that removing an oak forest would mean that the 43.5 percent of precipitation returned to the air from trees would reach the ground. Unfortunately, it is not so straightforward.

Transpiration returns to the air water that has entered plants through their roots. It is soil water. In other words, a proportion of the moisture present in the air above forests has been placed there by the trees themselves. Water evaporates because it is warmed, and, in evaporating, it absorbs the latent heat of vaporization from the surface from which it evaporates. The absorption of latent heat by water molecules cools the surface and is the principal mechanism by which plants avoid overheating to an extent that inhibits photosynthesis. Water vapor transpired by trees enters warm air and rises. On warm days, it is carried upward in rapid thermal upcurrents. As the air rises, it cools, and if it is moist enough, its water vapor will start to condense. It is not uncommon for clouds to form above forests. To some extent, therefore, the forest itself generates the precipitation that falls on it.

A forest can also force approaching air to rise in much the same way that a mountain does, though on a smaller scale. If the air is already unstable (see "Stable and Unstable air" on pages 64–65), and especially if the forest is on a hillside where the air has already started to rise, the additional forcing can destabilize the air further and cause clouds of the cumulus type to form. Stratified clouds may develop in stable air. Both cloud types can produce rain or snow, either as showers or in the form of lighter but more prolonged precipitation.

Remove the trees, therefore, and the air will be drier above the ground where the forest once stood. This does not necessarily mean precipitation will decrease in an area from which the forest has been cleared, however. Much

depends on the scale of the forest clearance because air moves horizontally, transporting its water vapor and clouds, and it may travel some distance before those clouds start to produce rain or snow. It does suggest that clearing forest from a large area is likely to reduce the average annual precipitation over that area and in regions adjacent to it.

Reducing the amount of water returned to the air by evaporation from tree surfaces means that a higher proportion of the precipitation will reach the ground. Even if the overall amount of precipitation is smaller, the actual amount reaching the ground may not be. At the same time, reducing the amount of water being transpired may allow water to accumulate in the ground. The most probable overall effect is that clearing a forest from well-drained land will make the ground drier, but on poorly drained land, the ground will become wetter.

Clearing temperate forests has no measurable effect on the climate of the world as a whole, but it can have a substantial local and regional effect. The climate is likely to become windier and the air dustier. Sea fogs may penetrate farther inland. There is likely to be an increase in the difference between maximum and minimum temperatures, daily and seasonally. Where the ground freezes in winter, it will do so earlier and to a greater depth than it did formerly. Precipitation is likely to decrease, and as the local climate becomes drier, in places this may increase the risk of dust storms.

■ FORESTS AND THE GREENHOUSE EFFECT

Sunshine warms the ground and the surface of the oceans. Warmed, these surfaces radiate heat. During the course of a year, the amount of heat that the surface of the Earth radiates into space is equal to the amount it receives from the Sun. If this were not so, the Earth would grow steadily warmer or cooler. Although incoming and outgoing radiation balance, the atmosphere retains some of the warmth. This is the greenhouse effect, and it is entirely natural.

Incoming radiation spans a wide wave band from ultraviolet at the shortwave end to infrared and heat at the long-wave end. It is most intense in the wave band of visible light between 0.39 micrometers (μm) at the violet end of the spectrum and 0.74 μm at the red end (1 μm = 10^{-6}m = 0.000394 inch). About 9 percent of the Sun's radiation is in the ultraviolet wavelengths down to about 0.30 μm, about 45 percent is visible light, and 46 percent is infrared and heat at wavelengths of more than 0.74 μm. The Earth emits radiation at wavelengths from about 4.0 μm to 100.0 μm, with a peak intensity at about 10.0 μm.

Trapping Radiation

The atmosphere is transparent to visible light. Shortwave ultraviolet radiation is absorbed in the stratosphere by oxygen and ozone, and some infrared and heat radiation is absorbed by water vapor and carbon dioxide, but almost all of the solar radiation that is not reflected by clouds reaches the surface. The surface, however, radiates at very much longer wavelengths to which the atmosphere is partially opaque. Water vapor absorbs radiation at about 1.0–4.0 µm, 6.0–9.0 µm, and 25.0–60.0 µm. Carbon dioxide absorbs at about 5.0 µm and 19.0–20.0 µm. Ozone, methane, and various other gases also absorb radiation, each at different wavelengths. There is a window at about 10.0 µm where no radiation is absorbed and heat from below can escape directly into space, but most radiation at wavelengths outside the window is trapped.

This complicates the radiation balance. Incoming radiation passes through the atmosphere and is absorbed by the surface of land and water. Most outgoing radiation is absorbed in the atmosphere. This warms the air, which then reradiates its heat in all directions, some of it to the sides and some downward to warm the air and the surface still further. Wherever the radiation is absorbed, it is reradiated, so radiation is moving in all directions. The overall effect is to trap heat near the surface. The energy balance is preserved because only the half of the Earth facing the Sun receives incoming radiation (it arrives only by day), but outgoing radiation leaves constantly from the whole of the surface (by night as well as by day).

The amount of solar radiation reaching the top of the atmosphere is called the solar constant, and it has been measured (about 128 watts per square foot; 1,380 W/m^2). The amount reaching the surface and the amount reradiated from the surface can be calculated, and from that, so can the surface temperature. These calculations indicate that the average temperature at the surface should be -0.4°F (-18°C). In fact it is 59°F (+15°C). The difference of 59.4°F (33°C) is due to the absorption of outgoing radiation by atmospheric water vapor, carbon dioxide, and certain other gases—the greenhouse effect. Were it not for the greenhouse effect, life on Earth would be exceedingly difficult.

Greenhouse Gases

Around the middle of the last century, as manufacturing industries expanded throughout Europe and North America and cities grew to accommodate the factories and their workers, coal production started to rise. It was the most widely used fuel, joined during the 20th century by oil and natural gas. All of these fuels are based on carbon. Their energy is released through the oxidation of their carbon to carbon dioxide, and the carbon dioxide is released into the air. Before the great industrial expansion, the atmospheric concentration of carbon dioxide was about 280 parts per million (ppm). It is now about 365 ppm, an increase of 30 percent, and the concentration is still rising. Carbon dioxide is a very minor constituent of the air, amounting to only 0.0365 percent of the total, but because it absorbs radiation at infrared wavelengths, a continuing increase in its concentration will lead to a general climatic warming. This is called an enhanced greenhouse effect, and carbon dioxide is known as a greenhouse gas. In this context, the amount in the atmosphere is usually reported as tonnes of carbon (1 tonne = 1.1 tons), and since the 1980s, carbon has been accumulating in the air at a fairly constant rate of 3.52 billion tons (3.2 billion tonnes) a year. Of all the carbon dioxide being released into the atmosphere each year as a result of human activities, about 77 percent is from burning fossil fuels, 20 percent from changes in land use, mainly in the Tropics, and 2 percent from cement manufacture.

Carbon dioxide is not the only greenhouse gas. Methane (CH_4), nitrous oxide (N_2O), ozone (O_3), and chlorofluorocarbon compounds (CFCs) are also greenhouse gases. These are much less abundant than carbon dioxide, however, so for convenience their climatic effect is counted as a multiple of that of carbon dioxide and called the *global warming potential* (GWP) for that gas. Methane has a GWP of 11, meaning that it causes 11 times more warming than carbon dioxide, and nitrous oxide has a GWP of 270.

Forests Absorb Carbon

All living organisms, including trees, are composed mainly of carbon. When organisms die, their remains decompose in the soil. The soil therefore contains carbon in the form of decaying organic material and carbon dioxide in the air trapped between soil particles. Scientists have estimated the amount of carbon held in this way by different types of vegetation, including forests. The estimated amounts vary, depending on the method by which they were calculated, but in the case of forests they are huge. Together, temperate broad-leaved and coniferous forests contain approximately 756–790 billion tons (687–718 billion tonnes) of carbon, 20 percent of it in the plants themselves and 80 percent in the soil. That is much more carbon than the 471–608 billion tons (428–553 billion tonnes) held by tropical forests.

Since about 1850, the total area of the world's forests has decreased by about 20 percent, due mainly to deforestation in the Tropics. This change, mainly from forest to agriculture, accounts for about 90 percent of the carbon dioxide entering the air due to changes in land use. Newly planted trees grow fast, and as they grow they absorb carbon. Their rate of absorption slows as they approach their full size, but planting forests, combined with slowing the rate at which tropical forests are being cleared, can increase the amount of carbon that the world's forests hold. Each acre of mature natural forest in temperate regions contains 250–253 tons of carbon (561–569 t/ha). This is carbon that the trees have removed from the air. Mature replanted forest would contain rather less carbon than this, but planting forests can clearly reduce the rate at which carbon dioxide accumulates in the atmosphere.

Plants could use more carbon dioxide than they can obtain from the air at present. Lack of carbon dioxide inhibits photosynthesis. Consequently, increasing the atmospheric concentration of carbon dioxide benefits plants because photosynthesis accelerates as the concentration of carbon dioxide increases. This is called the carbon fertilization effect, and many commercial horticulturists deliberately add carbon dioxide to the air in their greenhouses to take advantage of it. There will be a limit beyond which a shortage of water or an essential mineral nutrient imposes another constraint, but photosynthesis should stabilize at a higher rate than most plants have achieved in the recent past. If trees grow and mature faster due to carbon fertilization, planting forests may absorb more carbon than the calculations suggest.

Once a forest has matured, it will absorb no more carbon from the air, but land can be used to remove much more atmospheric carbon than this. To do so, the trees must be harvested and made into durable objects such as buildings, furniture—or books that remain for many years on the shelves of homes, schools, and libraries. The felled trees must then be replanted to absorb more carbon as they grow. Alternatively, trees can be grown as a crop for fuel. Burning them will release carbon dioxide, of course, but it will be carbon dioxide that the trees absorbed while they were growing. Burning biofuels adds no carbon dioxide to the atmosphere.

Global Warming

Eventually, from all these sources, the atmospheric concentration of carbon dioxide will reach about 560 ppm, which is double its preindustrial level, possibly by about the end of the 21st century. This would exert a warming influence, but one that is partly offset by emissions of sulfate and dust particles. These tend to reflect sunlight, and they also act as cloud condensation nuclei (see "How Cloud Forms" on page 46), increasing the formation of cloud. Recent warming has resulted mainly from increased nighttime cloudiness in winter, due to increased humidity with an abundance of condensation nuclei. Low-level clouds reflect and absorb long-wave radiation, which reduces the rate at which the surface cools at night, but the clouds also shade and cool the surface by day.

Allowing for the effect of particles, scientists calculate from computer models that a doubling of the concentration of carbon dioxide would cause the average global temperature to rise between 1.8°F (1°C) and 6.3°F (3.5°C), with the most likely increase about 2.7°F–3.6°F (1.5°C–2°C). There would be a small rise in sea levels of eight to 34 inches (0.20–0.86 m), due mainly to the expansion of seawater as it warmed, but there would be no major melting of the polar icecaps.

During the 20th century, the average global temperature increased by between 0.7°F and 1.4°F (0.4°C–0.8°C). The increase was not a steady one. There was a period of strong warming between about 1910 and 1945, followed by a stabilization, and then a very slight decrease in temperature—leading to fears of a new little ice age (see "Cycles of Climate" on pages 40–41). In 1976 a change in the sea-surface temperature of the North Pacific Ocean triggered a resumption of the temperature rise. Since then the average temperature has been increasing at a steady rate of 0.22°F–0.34°F (0.12°C–0.19°C) per decade, as recorded by surface measurements and calculated from satellite data. Almost all of the increase has occurred to the north of latitude 30°N, and it has been strongest in the Arctic, especially in Alaska and northeastern Siberia, although average temperatures appear to be falling in central Greenland and southern Iceland. It is most likely that this rate of warming will continue. Based on observation rather than computer models, this means that the global average temperature in 2100 will be about 2.7°F (1.5°C) higher than it was in 2000.

Calculating the Consequences

Should the average temperature continue to rise, the long-term consequences are very difficult to predict. It is not necessarily the case that the weather will remain much as it is now, only a little warmer, although this might happen. The predicted warming will produce temperatures in Europe that are significantly higher than those of medieval times, between about 800 C.E. and 1300 C.E. Average summer temperatures in Britain during that period appear to have been 1.3°F–1.8°F (0.7°C–1.0°C) higher than they are today, and in central Europe they were 1.8°F–2.5°F (1.0°C–1.4°C) higher. Fields were made and cultivated at higher elevations and in higher latitudes than has been possible since, with cereals, probably barley, being grown in Norway at 69.5°N, and vineyards prospered as far north as the English Midlands. Comparing the predicted temperature with that of the Medieval Warm Period (see "Cycles of Climate" on pages 40–41) is misleading, however. Although the likely temperature rise by 2100 is about 2.7°F (1.5°C), this is a global average, and so far, warming has occurred mainly in high latitudes. By 2100 Europe and North America may be markedly warmer than they were during the Middle Ages.

A rise in temperature means that more water will evaporate. This has a cooling effect because the latent heat of vaporization is absorbed from the surroundings. As the water vapor is carried aloft, it will be cooled and condense. This will release latent heat, warming the surrounding air. The water droplets will form clouds. If these are at a low or medium height, they will reflect incoming radiation, cooling the surface below them. If they form very high, on the other hand, the ice crystals from which they are made will absorb long-wave radiation faster than they reflect incoming radiation, warming the air. Plants will respond to an increase in the carbon-dioxide concentration, partly by growing faster but also by opening their stomata for shorter periods. This will also reduce the rate at which they lose water by transpiration, so they will use water more efficiently.

It may happen that, in effect, the climatic belts of the world are shifted toward the Poles. Scientists anticipate that such warming as occurs will be most marked in high latitudes, equatorial temperatures will change little or not at all, and temperatures will rise more over land than over the oceans. Precipitation will increase in high latitudes and in the monsoon regions, and in winter will increase in middle latitudes. In some parts of continental interiors, soils will become drier in summer.

This suggests a slight expansion of the humid tropical belt into what are now desert regions, including the Sahel zone, and an expansion of the tropical and subtropical arid belt into a higher latitude. Southern Europe could become desert, or at least semiarid, and the area with a Mediterranean climate could extend much farther north.

Migrating Forests

If the Mediterranean climate belt expands northward, the area in the Northern Hemisphere covered by temperate forests would remain unchanged, but it would be shifted northward. The North American and European boreal forest and Russian taiga would occupy most of what is now tundra, and because there is no land farther north into which the tundra could migrate, its extent would be much reduced. Mixed and broad-leaved deciduous forests would expand northward in Canada, Scandinavia, and Siberia, and sclerophyllous forest might appear in Europe, perhaps as far north as Paris, and in North America away from coasts about to the latitude of the Great Lakes. The southern continents barely extend into latitudes where the climate supports temperate forest, so this type of forest might disappear altogether from the Southern Hemisphere.

In the centuries following the end of the last ice age, plants migrated steadily northward. A few years ago many biologists were expressing the fear that this type of migration cannot be repeated, partly because the warming being anticipated will be much more rapid that that which occurred then. Probably this fear is misplaced. The amount of anticipated warming has been scaled down a little in recent years, and there is evidence that at times the postglacial warming was very rapid indeed. There is a much more serious risk, however, that migrating plants will find their way blocked. Unless societies are prepared to allow their farmlands to be colonized by wild plants migrating northward, the cultivated fields of Europe and North America will present a formidable obstacle. Farmers tend to remove plants that compete with their crops, and migrating species, including trees, may well be treated as weeds. Ordinarily, a climatic warming would be unlikely to cause the extinction of any plant or animal species, but with migration routes blocked, this becomes a risk.

Ocean Currents Transport Heat

Some scientists have suggested a very different kind of development. Warnings of the effects of a general warming have been based mainly on calculations of the way the atmosphere would respond, and they are at best approximate. They include many assumptions about the location and types of cloud that would form, and they have to be adjusted to make them correspond to the climatic conditions actually observed. More important, until recently they took little account of what would happen to ocean currents.

Ocean currents transport heat from low to high latitudes and return cool water to equatorial waters. The warm water of these currents warms the air crossing them and this has a significant effect on climates. The most important of these ocean systems is the one operating in the Atlantic and sometimes called the Great Conveyor (see the sidebar "The Great Conveyor" on page 36).

When seawater freezes, the salt is removed, so the ice is made of freshwater but the adjacent seawater becomes saltier. The added salt increases its density. At the same time, the water is at just above freezing temperature, and seawater is at its densest at 32°F (0°C). Because of this, in the north, near the edge of the sea ice, there is water that is denser than the water adjacent to it. The dense water sinks to the ocean floor and flows slowly south, as the North Atlantic Deep Water (NADW), all the way to Antarctica. Its place is taken by warmer water flowing northward at the surface, then cooling, becoming saltier, and sinking in its turn. The mechanism generating the formation of NADW is called the thermohaline circulation.

When the water of the Great Conveyor returns to the surface, the current is driven by the prevailing winds. The warm water flows through the Atlantic from the equator, through the Caribbean, along the southeastern coast of the United States, then westward across the ocean as the Gulf Stream. The Gulf Stream turns south in the latitude of Portugal and Spain, returning to the equator, but a branch breaks away, as the North Atlantic Drift (or Current). This heads northeastward, past the coast of Britain to Norway.

A warming of the atmosphere will increase evaporation. More cloud will form, and there will be more precipitation. Much of this will fall over the oceans because of the large area they cover, and river flows will increase. More freshwater will reach the ocean surface than reaches it today, and the freshwater will float on the surface because it is less dense than salt water. Freshwater freezes at a higher temperature than salt water but with less effect on the density of adjacent water because no salt is removed. Scientists calculate that a large increase in the amount of freshwater at the surface of the North Atlantic might alter the thermohaline circulation. NADW formation would weaken, and the Great Conveyor would slow, possibly causing the North Atlantic Drift to cease flowing, with a weaker Gulf Stream turning south in the latitude of southern Europe. Were this to happen, west European climates would cool.

Another consequence of increased precipitation would be a growth of ice sheets in Greenland and northeastern Canada, which would chill the air flowing across them. Some scientific studies suggest that this could trigger the rapid onset of an ice age. Far from temperate forests extending their range northward, climatic change of this kind implies that the tundra would extend into lower latitudes and the forests would retreat toward the equator.

The circulation of water in the Atlantic has not changed during about the last 10,000 years, but scientists now believe that this period of stability is unusual, that the circulation has been much less stable in the past, and that it might be destabilized rather easily. The danger arises not from the amount of carbon dioxide in the atmosphere but from the rate at which it accumulates. The present current regime would survive a slow accumulation and slow increase in precipitation, but a rapid change might disrupt it.

The influence of the present system of currents can be seen in temperatures. At St. John's, Newfoundland, at 47.57°N, for instance, the average summer temperature between May and September is 24.8°C (76.6°F). In Plymouth, England, at 50.35°N, it is 29.2°C (84.6°F). The difference in winter temperatures between October and April is even more marked. The average for St. John's is -10.0°C (14°F), and for Plymouth it is 16.1°C (61.0°F). Plymouth has a significantly warmer climate, despite being farther north than St. John's. The difference is due partly to the fact that in middle latitudes the prevailing winds carry weather systems from west to east. Eastern North America receives weather that has crossed the continent, while the weather reaching western Europe has crossed the ocean. Maritime climates are less extreme than continental climates (see "Continental and Maritime Air" on page 52)—they are milder in winter and cooler in summer. This accounts for some of the difference, but it leaves the difference in summer temperatures, where Plymouth is warmer than St. John's. A total loss of the thermohaline circulation would certainly bring cooler climates to western Europe. At present this seems unlikely.

What Is Most Likely to Happen?

There is no longer any doubt that average global temperatures are rising, nor is there any doubt that the accumulation of greenhouse gases is contributing to this rise. The extent of that contribution is uncertain, however. Some scientists maintain that the present warming is due almost wholly to the release of greenhouse gases. In their scenario, the warming results directly from human behavior and is potentially extremely grave. This is the view held by most environmentalist campaigners, and many climate scientists support it. Other climate scientists believe that the computer models that are used to calculate the long-term consequences of the warming are too crude to be reliable and that the influence of greenhouse gases has been overestimated. They believe that the present warming is mainly natural and is probably due to an increase in the output of solar energy, although they accept that greenhouse gases intensify the natural warming. Some scientists refer to the present rise in temperature as the Modern Warm Period, comparing it to the Medieval Warm Period and the Roman Warm Period that preceded it.

It is impossible to predict what the world will be like a century from now and unwise to trust forecasts that are made simply by extrapolating trends. If the present rate of warming continues, however, world temperatures will be a little higher in 2100 than they were in 2000. Most of that increase will take place in the temperate and arctic regions of the Northern Hemisphere, where forests are likely to respond by expanding along their northern boundaries and possibly by retreating along their southern borders. In other words, the temperate forests will shift northward.

■ FOREST CLEARANCE AND SOIL EROSION

Forests make the soil on which they stand. Thick tree roots run through the ground, and the roots of trees, shrubs, and herbs form an intricate network of fibers a little way below the ground surface. When roots die, they leave tunnels through the soil, ventilating it and facilitating the movement of animals and the growth of more roots. Falling trees overturn the soil (see "How Trees 'Plow' the Forest Soil" on page 28), and decomposing leaves, branches, and other plant material form a deep layer in which complex organic compounds are converted into plant nutrients, composed of molecules small enough to enter the root hairs of living plants.

When the trees are cleared, the soil quickly begins to change in ways that may be thought desirable. Historically, the purpose of forest clearance has usually been to convert forest into farmland by making space for cultivation while at the same time exploiting the accumulated fertility of the forest soil. Trees are felled and removed, an operation that may involve dragging them some distance across the ground; then the land is plowed and harrowed in preparation for sowing. It may also be, however, that clear-felling is the method of tree harvesting, with no intention of changing the use of the land. Once the crop has been removed, the land will be prepared for more trees.

Waterlogging

Farming or a second tree crop will usually succeed after an area of forest has been cleared, but there are dangers. In temperate regions, where most of the low-lying, level ground was converted to agriculture long ago, natural and plantation forests are often located on hillsides or on level ground at high elevations, places where the rainfall is often heavy. In these conditions, ground that is left bare may erode; soil may simply wash away down the slope. On fine-textured, usually peaty soil and soils that are strongly compacted in the subsoil, or B horizons (see "Digging Deep to Reveal the Soil Layers" on pages 19–20), there is also a risk of waterlogging.

Trees move large volumes of water from the ground and release it into the air by transpiration. When a stand of Norway spruce, transpiring the equivalent of 13 inches (330 mm) of rain a year (see "Transpiration" on pages 48–50), is felled that amount of water is no longer being removed from the ground. Fine texture or the compaction of subsurface layers inhibit the horizontal and vertical movement of water. If water enters the soil faster than it can be removed, the water table may rise. Before long it may be high enough to harm the following tree crop by producing cold, wet, airless soil conditions at a depth tree roots seek to penetrate.

Surplus water can be removed by installing land drains. These are set from about 66 feet (20 m) apart on peat soils with a shallow slope up to 164 feet (50 m) apart on coarser soil with a steeper slope, taking care to ensure that the drainage system does not allow water to collect in natural hollows. Where the subsoil is strongly compacted, deep plowing before the first tree crop is planted will break up the compacted layer, improving the movement of water and also root penetration.

Improving the drainage will prevent waterlogging, but it does not reduce the risk of erosion, the actual loss of soil. Vulnerability to erosion varies widely, but it can be predicted. There is a universal soil-loss equation that allows soil scientists to take account of all the factors affecting erosion, such as the intensity of rainfall, soil type, angle of slope, and the distance from the top to the bottom of the slope. The equation is very complicated to use, but it gives a value, in tons per acre (or tonnes per hectare) for the amount of soil likely to be lost by erosion each year. Wind erosion is rarely a problem on forest soils in temperate regions, which usually receive a high rainfall distributed evenly through the year, so they are unlikely to dry out sufficiently to be blown.

How Fast Does a Raindrop Fall?

Not all of the rain falling onto the canopy of a forest reaches the ground, and most or all of the rain that does reach the forest floor will have had its fall broken by leaves or branches. Breaking the fall of the raindrops slows them, so it seems obvious that they will strike the forest floor with less energy than the raindrops that fall on open ground. In fact, though, they strike the ground harder.

The drops form in a different way. Even in the middle of summer, in temperate latitudes almost all raindrops are snowflakes that have drifted gently down to a level, in their cloud or beneath it, where the temperature is several degrees above freezing. The snowflakes melt, and if the resulting drop of water is too heavy to be carried aloft by air currents, it falls to the ground. This is not what happens to water that has been intercepted on its way to the ground. As the rain runs off leaves and branches, tiny rivulets merge and the drops that fall are rather bigger than ordinary raindrops. They fall, not because they are too heavy to be carried aloft, but because they are flowing down a gradient and reach a point where they are no longer moving over a supporting surface.

Like falling bodies of any kind, once they start to fall they accelerate. If they drip from a height of about 33 feet (10 m), they will accelerate for long enough to reach their terminal velocity, which is the speed at which the gravitational force accelerating them and air friction slowing them, balance one another. When a falling body reaches its terminal velocity, that is the speed at which it continues to fall all the way to the ground. A drop of water more than about 0.2 inch (0.55 cm) in diameter is unstable and will quickly break into two or more smaller drops, but a drop of that size has a terminal velocity of about 20 MPH (about 32 km/h). The total energy of drops striking the ground is greater inside a forest dominated by certain tree species than it is on open ground outside, after allowing for the amount of rain that reaches the forest floor as small streams flowing down tree trunks and the amount that evaporates before reaching the floor. The rain does not impact on bare soil, of course, but on vegetation close to ground level and, below that, on the surface litter, a layer mainly of leaves or needles.

Splash Erosion

On bare ground, however, the situation is very different. Raindrops striking the soil knock particles free from the lumps to which they were attached. Very heavy rain can throw soil particles 2 feet (60 cm) into the air. This is called splash erosion, and it liberates very small particles that can be washed and beaten by further raindrops into the small pores and crevices in the soil surface. In extreme cases this can destroy the structure of the soil by filling all its pores with fine particles. Even without removing soil, this greatly reduces the ability of the soil to sustain plants. Strictly speaking, it is not erosion since no soil is lost, but the process is called puddle erosion because it is most likely to occur beneath large puddles.

Splash erosion partly seals the soil beneath an impermeable cap, reducing the amount of water that is able to drain vertically downward and increasing the amount flowing across the surface. Fine particles that are not held in crevices are then carried over the surface by the flowing water, eventually being removed from the area. If the subsoil is compacted and has not been broken up by plowing, in time erosion may remove the overlying material and expose the surface of the compacted layer. As it does so, the topsoil becomes progressively thinner and less able to support deep-rooting plants. Once the compacted layer is exposed, it will be almost impossible for roots to penetrate.

Years ago, this kind of surface damage was called sheet erosion, suggesting an even flow over the surface like a very wide, shallow river, but the image is misleading. In the first place, it is the splashing of raindrops that has by far the biggest effect. The kinetic energy of falling rain is more than 200 times greater than that of water running off the soil surface. Kinetic energy is the energy of motion, related to the mass and the speed of the moving body, and the more energy there is available, the greater the number of particles that can be transported.

It is easy to see why this is so. Kinetic energy (KE) is calculated by $KE = 1/2mV^2$, where m is the mass and V the velocity. Suppose a mass of rainwater W falling at 9 meters per second (m/s), and $KE = 1/2 \times W \times 9 \times 9 = 81W/2 = 40.5W$. If one-third of the water runs off at, say, 1 m/s (a realistic value), then the mass of runoff water is $W/3$ and $KE = 1/2 \times W/3 \times 1 \times 1 = W/6$. Comparing the two values for KE, as $40.5W$ divided by $W/6$, shows that in this example the falling rain has 243 times more kinetic energy than the runoff water.

Using real amounts of rainfall and measured values for rainfall intensity, calculations of kinetic energy can be used to determine the erosive power of rain and the results plotted on a map, with lines joining places of equal erosivity. These show, not surprisingly, that the regions of the United States most vulnerable to splash erosion are inland from the Gulf and Atlantic coasts from about San Antonio, Texas, to Jacksonville, Florida, which is the area that most often experiences rainstorms of tropical intensity.

Rills and Gullies

There is another reason why "sheet erosion" is a misleading term: Water does not flow across the surface as a smooth sheet. Bare soil is not smooth, like a parking lot. There are small depressions where it gathers and then overflows, cutting small channels as it does so. Some of these channels grow larger, with side channels feeding into them, and they may become established as rills. At this stage they are small enough to be removed easily by ordinary cultivation meth-

Gully erosion on a hillside in Georgia. The gully forms an ugly gash on the hillside, and water flowing along it carries soil washed from the surrounding land. *(Jeff Vanuga, Natural Resources Conservation Service)*

ods. Left untreated on bare ground, however, the rills may grow deeper and wider, large ones capturing the flow from smaller ones, until they become gullies. Hillsides that are crossed by gullies look severely damaged, but appearances can be deceptive. In deserts, where rain falls rarely but occurs as intense downpours when it happens at all, gullies are known as wadis (*wadi* is sometimes spelled *ouadi*) or arroyos. Between downpours these appear as dried riverbeds, but it is not practicable to cultivate the soil in deserts, so the land has no value for agriculture or forestry and, therefore, any erosion it suffers causes no harm. Similarly, where there are gullies in temperate latitudes, in most cases they have been allowed to develop because the land is of too little value for it to be worth the high cost of removing them.

This is not to say that gullies are unimportant, only that they often do little economic harm to the land on which they occur. They do indicate soil erosion, and this can be serious away from the eroded land. Gullies form when there is a large increase in the volume of water draining across a sloping surface at times of very heavy rain. Usually, the surface runoff finds its way to the small streams and rivers that tumble down the sides of many steep hills. Sometimes these streams carry very little water, and at other times their channels are full, but they are always where the water drains and the route by which it is removed. The system fails when the amount of water greatly exceeds the capacity of the stream channels. Water overflows, and new channels form that develop into gullies. After that, water has a network of channels through which to flow, and the channels comprising the network may grow into gullies.

Landslides and Mudslides

Soil often erodes by the removal of fine particles, but there are more dramatic alternatives on the steep slopes that are often exposed when forests are cleared. Water flowing below the surface may detach the overlying material. Destabilized, this then begins to move down the slope, sliding on the layer of mud at its base, the mud acting as a lubricant. The result is a landslide that carries away soil, rocks, and plants—including any remaining trees. If the ground is wet enough, the slide may be of mud, and if the lubricating mud layer is very close to the surface, the flow may occur as an avalanche of surface debris. Less spectacularly, on land left bare for any length of time, soil that is lubricated from beneath may creep down the slope a little at a time but as a mass.

Geologists described all these forms of erosion as mass wasting. Soil creep is not as dangerous as landslides, avalanches, and mudflows are, but it is a more serious form of soil loss. With each rainstorm, the topsoil shifts a short distance and then stops when the ground dries, only to move again with the next rain. Left unchecked, creep erosion can strip an entire hillside down to bare rock.

Where Does the Soil Go?

Soil that has been eroded is washed downhill, and much of it eventually enters rivers. After a few days of heavy rain, rivers are usually brown with the load of soil they are carrying. Brown water is also contaminated water because, in addition to soil particles, it carries organic matter that alters the chemical composition of the river. The waters flow rapidly, driven by pressure from the increased amount draining into them farther upstream, and their energy allows them to transport large quantities of soil particles and stones. Precisely how much they carry depends on the volume of water and its rate of flow and on the size of the particles.

When the river flow slows, the water has less energy to carry soil particles, and these begin to settle, the heaviest first. Rivers slow and often become wider when they cross a shallower gradient such as a plain. They also slow where they flow into the sea and where they enter a reservoir held behind a dam. After each storm, the rivers carry more soil, and downstream, at the coast, or in a reservoir, the soil is deposited. Layers of sediment form and grow thicker. Where a slow-moving river crosses a plain, the increasing depth of sediment on its bed reduces the depth of water. The channel becomes shallower but wider. This increases the risk of flooding because the river is now less able to carry the volume of additional water released by storms or the melting of snow near its source. Navigation is also restricted as the river depth decreases.

The problem may be even more serious in harbors. As the river meets the sea and the movement of sea currents opposes its flow, it can lose energy rapidly. At the same time, chemical reactions between saltwater and the fine particles carried by the freshwater cause the particles to adhere to one another (the technical term is *flocculate*) and sink. That is how mud banks form and eddies can carry the sinking sediment into harbors that then have to be dredged to maintain a sufficient depth of water for the vessels using them.

Dams are constructed to hold rivers back while reservoirs fill and to regulate river flow farther downstream. They halt river flow, so an inflowing river loses all its energy and deposits its entire load of soil particles on the bed of the reservoir. This progressively reduces the amount of freshwater that can be stored and shortens the life of the dam.

Bare Ground and Accelerated Erosion

There are many ways to increase the flow of surface water inadvertently. Most changes of land use will do so under certain circumstances, and one change very likely to do so is the clearance of forest. Interception and transpiration cease over the cleared area, so a larger proportion of the falling rain reaches the surface and a smaller proportion is removed from the ground. In addition, the ground surface has been laid bare, leaving it vulnerable to erosion. There have been measurements of the rate of erosion before and after the clearance of forest on steep slopes with sandy soil in England. These show that there was no erosion where the hillside was forested. Grass afforded less protection than trees, and on grass slopes, about one ton of soil was lost from each acre (2.4 t/ha) every year. On bare soil, erosion carried away about eight tons per acre (17.7 t/ha) per year.

Fire can also increase erosion because that, too, exposes the ground. In a 10-year period in Oklahoma during which the amount of rainfall was identical on two forest sites—one protected from fire and the other burned—10 times more soil was lost by erosion from the burned site than from the protected one. Results from a similar comparison in North Carolina were still more dramatic. In nine years there, the burned site lost more than 150 times more soil than the protected site.

Forest roads are a major cause of soil erosion. In plantations that will remain permanently forested, roads are often built in much the same way as public roads and are surfaced, but this is expensive, and until the tree crop is harvested, the roads carry only light traffic. Many forest roads, especially the minor ones, are little more than dirt tracks, made as cheaply as possible. They are often on steep slopes in regions of high rainfall, and very heavy vehicles use them at harvest time. Unless such roads are well maintained with drains across them at intervals and plants encouraged to grow on them, dirt roads on steep gradients turn into gullies, and their erosion can continue for years after the harvest operations that caused the initial damage.

Reducing the Risk

Erosion is caused by the splashing of raindrops onto bare ground, often causing fine soil particles to seal the surface, followed by the flow of water carrying soil down the slope. To prevent erosion, foresters minimize the area of ground that is laid bare by harvesting. Then they seek to prevent water flowing across the surface. Trenches cut at right angles to the direction of slope often help. The trenches themselves are level because it is not intended that water should flow along them. They capture water and retain it long enough for it to soak into the ground.

As soon after harvesting as is practicable, the ground is prepared for the next tree crop. Preparation nowadays often involves using machines to break up the soil around the place where a new tree is to be planted. The operation is called *scarification,* and it chops up and scatters the remains of small branches and foliage (called brash) from the previously harvested crop. Scarification leaves an uneven surface through which water penetrates more easily than it does through an unbroken surface. Except on very well drained soils, the young tree is planted on a mound made by exca-

vating soil from a hollow. This also breaks up the surface, aiding the vertical penetration of water and helping to prevent surface flow down the slope.

Erosion is a natural phenomenon, resulting from the same weathering processes that release the small mineral particles that form the basis of soil. Such geological erosion is unavoidable, and it is acceptable because soil forms at about the same rate as natural erosion removes it. Accelerated erosion, on the other hand, removes soil faster than it can be replaced, and this is unacceptable. Forest clearance is one of the two principal causes of soil erosion, the other being agriculture, and in both cases better land management can minimize it.

■ WARM-TEMPERATE AND MEDITERRANEAN-TYPE FORESTS

Rain forest is not confined to the Tropics. It may also develop in higher latitudes where temperatures are mild and the rainfall is abundant. Possibly the largest temperate rain forest is in the United States, extending for about 12 miles (19 km) along the valleys of the rivers Bogachiel, Hoh, Queets, and Quinalt in the Olympic National Park in the state of Washington. The park was established in 1938 to protect the Olympic Mountains, and in 1953 the boundaries were extended to include a 50-mile (80-km) stretch of coastline. The Olympic Mountains and the park lie on the Olympic Peninsula to the west of Seattle and are bordered to the north by Vancouver Island, Canada, across the Strait of Juan de Fuca. The whole park occupies an area of about 897,000 acres (363,000 ha).

This coast is washed by a branch of the Kuroshio Current. Part of the North Pacific gyre (see the sidebar "Gyres and Boundary Currents" on page 87), the Kuroshio Current begins where the North Equatorial Current, flowing from east to west, meets the Asian coast and turns northward. The current passes along the coast of the Japanese island of Honshu and then turns eastward at about latitude 45°N to flow directly across the Pacific. It divides at about longitude 150°W, roughly due south of Anchorage, one branch flowing southward as the cool California Current, the other flowing northward and then westward as the Oyashio Current that crosses the Pacific along the southern coast of Alaska to rejoin the Kuroshio Current. Warm water from one arm of the Kuroshio Current flows parallel to the coast of Washington and British Columbia. Air above the current is warmed, but the adjacent sea and air are cool, and the rate of condensation is high. In summer, sea fogs are frequent in the Olympic Peninsula, and the annual rainfall on west-facing slopes exceeds 140 inches (3,550 mm). Such conditions encourage the growth of mosses, lichens, and club mosses, and the trees are festooned with epiphytes. Every acre of the forest is said to contain about 6,000 pounds of these plants (6,700 kg/ha). The principal tree species are Sitka spruce (*Picea sitchensis*), western hemlock (*Tsuga heterophylla*), and groups of bigleaf maple (*Acer macrophyllum*). Curiously, considering the climate, this forest is maintained naturally by fire. Spells of dry weather are quite common in summer, and lightning occasionally ignites flammable material on the valley sides. The shapes of the valleys can funnel air, fanning the flames so that the fire spreads to the valley bottom. Red alder (*Alnus rubra*) grows among the ashes, followed by spruce and then hemlock. It is from spruces and western hemlocks taken from this forest that modern British conifer plantations were developed.

Part of the Olympic Forest was clear-felled in the early years of this century. Second-growth forest developed, matured, and was then logged, the area being replanted to Douglas fir (*Pseudotsuga menziesii*), although some of the old-growth forest remains. Elsewhere, the original forest survives, with spruces, some 300 or more years old, up to 280 feet (85 m) tall. Where old trees fall, they decay slowly, and young seedlings grow in rows along the decaying trunks. Because of the thickness of these mighty stems, the seedlings are several feet above ground level, and they develop long stilt roots that remain after the old trunks beneath them have disappeared. Other trees have buttress roots reaching to 13 feet (4 m) or more up the trunks. Because in many places the forest is fairly open, the ground vegetation is lush. The forest is rich in animal life. There are several species of deer, black bears, cougars, and about 140 species of birds. Grazing by large numbers of Roosevelt elk (a local variety of the wapiti, *Cervus canadensis*) controls the regeneration of trees in some of the valleys.

Rain forests also occur farther south in California, dominated by the Sierra redwoods (*Sequoiadendron giganteum*) of the Yosemite National Park. Elsewhere, such forests are more common near the eastern coasts of continents.

Southern Hemisphere Temperate Rain Forests

In Australia, the coastal belt of Victoria and New South Wales supports temperate rain forests dominated by various species of southern beeches (*Nothofagus*) and coachwood (*Ceratopetalum*), merging farther north in Queensland with bunya bunya (*Araucaria bidwillii*) and kauri pines (*Agathis* species). Still farther north, the temperate rain forest merges almost imperceptibly with tropical rain forest.

The natural vegetation over much of New Zealand is also temperate rain forest, dominated by *Nothofagus*, red pines (*Dacrydium* species), and podocarps (*Podocarpus* species),

with kauri pines in North Island. Maori people once built their war canoes from kauri pines. The brown kiwi (*Apteryx australis*), the emblem of New Zealand, forages for berries, worms, and insects on the floor of the New Zealand rain forest and the tui (*Prosthemadera novaeseelandiae*), a bird resembling a starling, hunts in the canopy.

There are much smaller areas of rain forest in South Africa, along the coast and in the Drakensburg Mountains. In South America, the Valdivian rain forest, west of the Andes in Chile, is famous for the Chile pine, or the monkey-puzzle tree (*Araucaria araucana*), and Patagonian cypress (*Fitzroya cupressoides*).

Kyushu, the southernmost island of Japan, and sheltered parts of Honshu to its north also support temperate rain forest as the natural climax vegetation. This forest is similar in composition to the rain forest of southern China. Both are dominated by a variety of oaks (*Quercus* species) and beeches (*Fagus* species), with some coniferous trees. A mild, wet climate favors rice-growing as well as rain forest, and much of this part of eastern Asia is forest no longer but is agricultural.

Himalayas and Caucasus

Temperate rain forests occur in tropical latitudes in the Himalayas at elevations of about 4,000 feet (1,200 m). In Nepal, for example, these forests include large stands of bamboos up to 50 feet (15 m) tall. In Nepal, as elsewhere in Asia, forests of all kinds have been seriously depleted to provide fuel (see "Wood as Fuel" on pages 246–247) as well as land for farming. Clearance has led to soil erosion, and now steps are being taken to protect such natural forest as remains and to increase the forest area.

A temperate rain forest that has been protected since 1924 occupies part of the Caucasus Biosphere Nature Reserve in Russia. The reserve covers 1,017 square miles (2,635 km²), and the rain forest is on the western side, which receives the highest rainfall in Russia, about 197 inches (5,000 mm) a year. Rhododendron, laurel, holly, and oak dominate the rain forest.

Mild temperatures and abundant rainfall mean that temperate rain forests, like their tropical counterparts, tend to produce giant trees. These are of great value for timber, and there is a clear commercial pressure to exploit them. The forests are ecologically fragile, however, and even where their appearance is restored by secondary growth, some of the species they support may be lost. Obviously, even more are lost when plantations replace the natural forests. The risk to temperate rain forests is recognized, and many are now protected. Those of North America are contained within the boundaries of national parks, and conservationists closely guard the integrity of those in Australia and New Zealand.

Mediterranean Forest

Around the Mediterranean, the original forest, dominated mostly by holm oak (*Quercus ilex*) and Aleppo pine (*Pinus halepensis*), has disappeared almost completely. Cleared in classical times to provide farmland, its soils subsequently eroded and now support the scrub vegetation known as maquis or, when more open, as garrigue. Maquis is a type of chaparral, the best known example being in southern California on the western side of the Sierra Nevada. There the composition varies according to the aridity, but in many parts of the mixed chaparral of intermediate rainfall, the dominant plants are evergreen oaks such as California scrub oak (*Quercus dumosa*), tall shrubs such as California lilac (*Ceanothus* species), and chamise (*Adenostoma fasciculatum*). Chaparral also grows in Arizona, where the dominant tree is shrub live oak (*Quercus turbinella*). A somewhat similar Mediterranean-type vegetation, known as fynbos, grows in Cape Province, South Africa. Fynbos differs from maquis and chaparral in having very few trees, although otherwise it is rich in plant species. In Chile, the matorral is another variety of chaparral.

In southern and western Australia, the climate produces plant communities known as mallee scrub. This is dominated by *Eucalyptus* species, silk oaks (*Grevillea* species), *Banksia,* and *Protea* small trees or shrubs, all broad-leaved evergreens, as well as small trees belonging to the family Epacridaceae. Most of these are adapted to dry conditions, and many have very thick trunks. The Epacridaceae family includes the Southern Hemisphere equivalents of the heather family (Ericaceae) of the Northern Hemisphere. In places, silver wattle (*Acacia dealbata*), a tree that grows to 100 feet (30 m) tall and is known to florists as mimosa, grows as an understory species where fire has removed competition. (True mimosa is an unrelated genus [*Mimosa*] of about 400 tropical and warm-temperate species.)

Marginal Forests

Where sclerophyll forest and plant communities of the chaparral type grow near coasts, they are at risk. The hot, dry summers and mild, wet winters that produce this vegetation are very attractive. Understandably, people who are able to do so choose to move to this benign climate, but in clearing space for the homes, roads, and services they need, the natural vegetation is likely to be destroyed. The comparative rarity of this vegetation type and the risk it faces are now recognized in most countries. The American chaparral is protected, but nowhere is protection perfect, and forests have never fared well in any competition over land use.

Temperate rain forests and sclerophyllous forests occupy a much smaller area than the commoner broad-leaved deciduous and boreal forests. They have always inhabited the vulnerable margins of the forests. Rain forests are found in foggy, wet places where people are not eager to build homes, but they produce giant trees that are attractive to logging companies. Sclerophyllous forest produces some useful timber and many more useful herbs, including lavenders, rosemary, thymes, myrrh (*Cistus creticus*), and many others that are now cultivated. Its attraction arises not so much from its plants, however, as from its climate, and where this coincides with spectacular scenery, the combination is powerfully destructive. What remains of these marginal forests is in especial need of protection.

■ DECIDUOUS FORESTS

Broad-leaved deciduous forests develop naturally where moderate rainfall is distributed fairly evenly through the year, winters are not too long although temperatures may remain a few degrees below freezing for several weeks, and summers are warm. This is the kind of climate found at low elevations throughout the temperate regions. The climate that suits these forests so well also attracts farmers, with the fertile soils that form beneath the forests as an added bonus. In Britain and many other parts of Europe, for example, farmland made more than 2,000 years ago by clearing the natural forest (see "The First European Farmers" on page 223) is as productive today as it has ever been. So far as anyone can tell, with good farming methods these soils can remain in cultivation forever. Not surprisingly, therefore, of all the types of forest in the world, it is the broad-leaved deciduous forests of the Northern Hemisphere that have suffered most from the ax and saw. For the most part, these are the forests that have been felled to provide the land that feeds and clothes the people of Europe and North America.

Despite the need for cultivable land, however, quite large tracts of forest survived. There were several reasons. Forests were themselves an important source of food—as game and as sport (in the procurement of it). Hunting was an amusement reserved for the rich and powerful, and it was they who protected areas of forest. Timber and small wood were also needed, and in the days before plantation forestry based on conifers, broad-leaved forests were the only source of these materials in regions where this was the natural climax vegetation. So some forests survive, and today in the world as a whole, broad-leaved deciduous and mixed broad-leaved and coniferous forests occupy about 8.4 million square miles (22 million km^2). This is not all ancient or old-growth forest: The total includes secondary growth and broad-leaved plantations. Nevertheless, the area is substantial. It is about two-thirds the area of the forests of the humid Tropics.

North American Old-Growth Forest

North American old-growth forests survived as patches separated by fields, as areas the farmers never needed for crops, and in remote places that the farming families never reached. These forests are found mainly in the mountains of the East from the Adirondacks down to the southern Appalachians and on the floodplains farther south. With so vast an area available to them, settlers could afford to pick and choose, and farmers preferred the areas with the best climate. Mesophytic forests, typical of moderate climates, were the ones most likely to be cleared for farming. These were the forests dominated by oak, hickory, sugar maple, tulip tree, and beech, as well as chestnut until this was destroyed by disease (see "Tree Predators and Parasites" on pages 173–180).

Forests also provided lumber and fuel. Many areas, not destined to become farmland, were cleared for their timber, pitch, and resin and then abandoned, the land being required for no other use. Forest returned to these areas as secondary growth.

Comparisons between the histories of North American and European forests can be misleading. There is a temptation to assume that only Europeans were involved in both cases, that the people who had created the agricultural landscapes of Europe crossed the Atlantic and reproduced the process by clearing the previously untouched American forest. This implies that the colonists migrated to a continent with no human inhabitants or with human inhabitants who had not altered the natural forest in any substantial way. That is not so. Native Americans had made major modifications during the thousands of years they occupied the land before the settlers began to arrive.

The Native American inhabitants of the forested regions lived in villages, owning land as tribes rather than as individuals, and grew plant crops by a type of slash-and-burn cultivation. Around the village they cleared the forest, burned it, and grew their crops on the exposed land. They also used fire to make paths through the forest, to drive game, and to keep away dangerous animals and those that damaged their crops. They used the forest as a source of fuel, gradually cutting it back farther and farther from the village until, after a number of years, the entire village moved to a new location. The old site was abandoned and a fresh area was cleared. No new trees were planted, and forest clearance encouraged the growth of grass, herbs, and tree seedlings on which herbivores such as deer grazed and browsed. Herbivore populations increased and strongly affected the regeneration of the forest.

In some places, therefore, what appears to be old-growth forest may be secondary growth, but no record has survived of the original clearance. This presettlement treatment is not confined to the east, of course, but extended to all forests. There are oakwoods in California that are known

Tree squirrels, instantly recognizable by their long, bushy tails, can damage trees and are often considered pests in plantations. Ground squirrels live in grassland. This Belding's ground squirrel (*Spermophilus beldingi*) lives in alpine meadows above the tree line in the western United States. *(John and Karen Hollingsworth, The U.S. Fish and Wildlife Service)*

to have been burned and converted into fields long before the arrival of Europeans.

If this casts doubt on the status of North American old-growth forest, there is no question about the condition of secondary growth. It has been expanding rapidly throughout this century. Second-growth forest now covers between 65 percent and 85 percent of upland New England, for example. Ecologists participating in the Harvard Long Term Ecological Research program report that in many ways the landscape of New England appears more natural today than it has at any time since the 18th century.

Białowieża Forest

Fragments of the original primeval forest also survive in Europe. Some were hunting forests, protected by their status for long enough to survive the expansion of agriculture. Others are on land of a quality too poor to be farmed economically.

Probably the largest single forest is the one that covers about 502 square miles (1,300 km²) in Poland and Belarus and is known by its Polish name of Białowieża. Until about 200 years ago, it was a hunting forest, but since then it has had other uses, not to mention disturbance from the warring armies that have passed its way. Cattle have been grazed in it, leaf litter removed for horticultural use, and parts may have been cleared and cultivated. It suffered a major fire early in the 19th century, from 1895 to 1915 the deer population was deliberately increased to such a level that in places the forest came close to disappearing, and then the deer were removed and the last herd of European bison with them. Both animals have now returned. The forest itself is divided into numbered square blocks by a grid of wide, straight paths called rides.

The history of the forest is complex, as is its ecology, but there is no doubt of its value. At least in places, it is a remnant of the types of lowland forest that were once widespread in Europe from France to Russia. The most valuable area, of 18.32 square miles (47.47 km²) lying at the center, has been fully protected since 1921 as the Białowieża National Park. This area contains broad-leaved deciduous forest dominated by oak, lime, and hornbeam on higher ground, alder and ash near the rivers, and ash and elm elsewhere, with spruce and pine on the very wet or acid soils. In all, the forest contains about 3,000 species of plants, some of them rare, and 8,500 of insects and other animals. The large mammals include the wolf (*Canis lupus*), red deer (*Cervus elephas*), wild boar (*Sus scrofa*), beaver (*Castor fiber*), lynx (*Felis lynx*), and the last herd of European bison or wisent (*Bison bonasus*) living freely (rather than in zoos and parks). They were reintroduced in 1929.

In 1991, the governments of Poland and Belarus agreed to plans to establish the forest as a Biosphere Reserve, under the auspices of the UN Educational, Scientific and Cultural Organization (UNESCO). In August 1996, however, the Polish government merely doubled the size of the national park. Outside this securely protected area, most of the forest is used for recreation and some for logging, although a moratorium on logging was declared in 1995. Conservationists are campaigning to extend national-park status to the entire forest.

European Primeval Forest

In many parts of central and eastern Europe, the conversion of forests to farms was slow, and hunting forests survived for a long time. Prior to medieval times, settlements grew along river valleys, where land was cleared for cultivation and the remaining forests were managed to provide timber, small wood, and fuel. Once developed, this pattern of land use changed little for centuries. Many areas in the mountains were remote and largely unoccupied. Major changes in land use began early in the 19th century and accelerated as roads and railroads improved communications and European economies industrialized. Factories were built where there was access to the raw materials they needed, and the forests provided fuel. Clearances accelerated, and natural forest was being logged regularly in the early 1930s in Ukraine and as late as the 1980s on the border between Croatia and Slovenia. Even more forest might have been lost had the establishment of forest reserves not helped to check the process. Landowners were the first to appreciate the value of their forests, and it was they who began to create reserves. The very first reserve to protect virgin forest was established in 1838 at Zofin, Czech Republic, and others followed.

Today, patches of natural forest survive and are protected throughout Europe. Most are to be found in the southeast, in Austria, Slovenia, Croatia, Bosnia and Herzegovina, and especially in the Czech Republic and Slovakia. There are also some in Spain and France.

No more than about 6.5 percent of the total area of the temperate forests of the world is protected by having been formally designated a national park, wilderness area, or allotted to one of the other internationally recognized conservation categories. Conservationists are campaigning to extend protection more widely, especially to old-growth and primeval forests. Such designation seldom confers absolute security, but it does acknowledge the importance attributed to an area. It may be, however, that the value of forests and especially of natural forest is now widely appreciated and guarded by the vigilance of those who live in and near them. That is the most reliable protection of all.

■ BOREAL FORESTS

Canada occupies a total area of 3.85 million square miles (9.98 million km²), of which 3.56 million square miles (9.22 million km²) is land and the remainder freshwater. Forests cover 1.61 million square miles (4.18 million km²). In other words, almost half the land area of Canada is forested, and Canada is the second largest country in the world (after Russia). The Canadian forest is the third largest in the world, after those of Eurasia and the South American tropics, and it accounts for one-tenth of the entire forested area of the world.

There is broad-leaved deciduous forest in the south and temperate rain forest along the western coast, but most of the Canadian forest is boreal. The boreal forest lies in a broad belt extending from the Atlantic to the Alaskan border and from northern Newfoundland, Québec, and the Yukon at about 60°N to the Great Lakes, with a tongue stretching southward along the eastern slopes of the Rockies. Along its northern edge, the forest becomes more open, like parkland, merging with the tundra, and it merges with grasslands in northern Manitoba, Saskatchewan, and Alberta, with the prairie farms to the south.

Boreal forests grow in a cool, wet climate where summers are short and winters long. Boreas, the name from which the word *boreal* is derived, was the Greek god of the north wind. The climate is unsuitable for farming, so only about 6 percent of the Canadian forest has been cleared to provide farmland, but the forest is of immense economic importance. Forestry and industries immediately associated with it provide about 347,000 jobs—about 2 percent of all employment—but when the all the jobs these industries generate indirectly are included, the total comes to about 1.2 million, or 7 percent of the national workforce. There are 348 communities that depend wholly on the forestry industries. Forest products account for about 16 percent of Canada's total exports. Obviously, the forest is managed. About 907,000 square miles (2.35 million km²) of forest, amounting to 58 percent of the forest area, produces timber commercially of which about 23.23 million acres (9.4 million hectares), or 0.4 percent, is harvested each year.

Managing Canadian Forest

The composition of the boreal forest is less diverse than that of forests in lower latitudes. In eastern Canada, black spruce (*Picea mariana*), white spruce (*P. glauca*), and balsam fir (*Abies balsamea*) are the only trees in large areas. In central Canada there are jack pine (*Pinus banksiana*), lodgepole pine (*P. contorta*), and tamarack (*Larix laricina*), and in the west Sitka spruce (*Picea sitchensis*). To the south, these merge with some broad-leaved species, including paper birch (*Betula papyrifera*), balsam poplar (*Populus balsamifera*), and quaking aspen (*P. tremuloides*). This may make the forest seem monotonous, but these are only the dominant species. In all, the Canadian forest contains 131 species of trees and about 3,000 other species of plants, not to mention 200 species of mammals, almost 550 of birds, and almost 90 of reptiles and amphibians, as well as about 100,000 invertebrate species.

Management of this forest is controversial, partly because of its vast size. The 10 provincial governments are responsible for the administration of about 80 percent of the overall forest, about 11 percent is administered by the federal government, and about 9 percent is owned by private individuals and corporations. Usually, the provincial authorities license private companies to harvest timber on the land they control. This authorization specifies the harvesting methods that may be employed and includes a requirement for reforestation.

Old-Growth Forest

Conservationists classify about 30 percent of the area producing timber, amounting to 175 million acres (71 mil-

lion ha) as old-growth forest that should not be altered. This area includes stands of Sitka spruce in British Columbia that are known to be 500 years old and areas of pine forest in central Canada that are about 140 years old. These and similar areas are clearly old-growth forest, but the status of some of the forest included in the conservationist estimate is more controversial. Even within old-growth forest, companies are permitted to remove certain old, very large trees that are close to dying and liable to become highly flammable or are prone to pest infestation or diseases that could harm the trees around them. Old-growth forest contains trees of great age, but this fact alone may be insufficient reason for preserving them in a country where the forest is so vast and so much of it has never been exploited. Old trees are large, and this makes them valuable. Harvesting them yields profits that can be invested in the management of protected areas of greater scientific or historical importance.

Patches of forest that exemplify particular forest types or are of especial ecological interest are being identified and protected. These include old-growth forest, of course, but also secondary-growth forest that supports a wide range of species or is important for some other reason. Varying levels of protection are provided by a hierarchy of designations, including natural areas, ecological reserves, wilderness areas, wildlife parks, provincial parks, and national parks. There are more than 3,000 protected areas, and when it is completed, the network of 34 national parks will represent all 39 of the natural types of plant and animal community found in the country. In 2006 about 8 percent of the total Canadian forest area was completely protected. This amounted to 334,890 square miles (865,000 km²)—more than the combined areas of Texas and Louisiana. It is the largest area of protected forest within the borders of any country in the world.

Clear-Cutting

When trees are harvested, the method used may cause unacceptable damage, and conservationists throughout North America are especially worried about the effects of clear-cutting. This is the harvesting method in which all the trees are removed from a tract of land, leaving it bare. There is no doubt that clear-cutting is a visually intrusive technique. The bare ground, often incongruously rectangular with sharply defined edges set amid intact forest, can be seen from far away, and it is ugly. As the years pass, of course, new growth restores the appearance, but there are other concerns. Clear-cutting destroys habitat, and although this returns, the recreated habitat may be inferior to the one that was lost. It is possible that some patches of the clear-cut area may remain bare. During the time the ground is bare, erosion might damage it (see "Forest Clearance and Soil Erosion" on pages 272–276), and soil washed from the land may harm fish in nearby rivers. Erosion can also cause distress to local people who live by hunting.

The dangers are recognized, and clear-cutting is only one of the harvest techniques that are used, but sometimes it can replicate the natural events experienced in a forest where from time to time storms and fires destroy all the trees in particular areas. In Canada, about 0.5 percent of the commercial forest is damaged each year by fire and insect attack. This is larger than the area harvested each year. Where many of the trees are very old or all of much the same age, clear-cutting may be the best way to open up the forest, allowing light to penetrate and young saplings to grow. Clear-cutting policy is under review in Canada, however, and henceforth it is likely to be used with greater sensitivity, for example by reducing the size of cutover areas.

Replanting and Regrowth

Under Canadian provincial law, every area of harvested forest must be regenerated promptly, and it is not enough for companies simply to leave the site and hope for the best. They are required by law to monitor the area and ensure that regeneration succeeds. If it does not, the company must replant the area and manage it until the forest is fully restored. Eventually, plantations and the natural regeneration of secondary-growth forests should supply most timber and forest products, but until then logging of the primary forest will continue. Each year about one billion young trees are planted in Canada. This is roughly double the number that are felled each year, so the Canadian forest is expanding. In the United States, 1.5 billion tree seedlings are planted every year, and the forest area increased by nearly 15,500 square miles (40,000 km²) between 1991 and 2001.

Natural regeneration is the preferred method for restocking, and this is what is encouraged on more than half the harvested area. The reason is partly ecological and partly commercial. Regeneration restocks the area with those trees that thrive best, and because it is a natural process, the resulting forest closely approximates that which is best adapted to the site. Trees that are well adapted to their site are likely to grow more vigorously than introduced species and to be healthier, so they will yield a more valuable crop more quickly. Not all areas regenerate easily, however, so some planting is necessary. Replanting is used on less than half the harvested land, and the seedlings used are chosen to maximize the genetic diversity.

Scandinavian Forest

Across the Atlantic, the boreal forest continues in Scandinavia. The table shows the extent of the forest and the proportion of the land area it occupies in Finland, Norway, and Sweden. Of the total forest area, boreal forest accounts

Forest Area and Proportion of Total Land Area in Scandinavia			
COUNTRY	FOREST AREA		PERCENTAGE OF LAND AREA
	miles²	km²	
Finland	86,850	225,000	73.9
Norway	36,233	93,870	30.7
Sweden	106,258	275,280	66.9
Total	229,342	594,150	53.42

for about 66 percent in Finland, 57 percent in Sweden, and 23 percent in Norway.

Ownership is more complex than in Canada, being shared between private individuals and families, forestry companies, communities, and the state. In Norway, private individuals or companies own about 86 percent of the forest, in Sweden about 80 percent, and in Finland 72 percent. There are about 125,000 privately owned productive forests in Norway, and in Finland nearly 440,000 private forests of 2.5 acres (1 ha) or more in area.

This division of ownership also complicates management, and there have been worries that old-growth forest was being lost and replaced by monoculture plantations. Along the Finnish border with Karelia (Russia), for example, old-growth forest has been clear-cut in patches up to 1,235 acres (500 ha) in area. These are densely forested countries, however, with long traditions of living in, with, and from the forest. One Finnish family in five owns a forest, even if it is only a small one. Livestock are grazed in them and game hunted, and for centuries roundwood, timber, and tar have been important exports. The ecological value of certain types of forest has been recognized only recently.

The importance of maintaining biodiversity is recognized now and, if rather belatedly, the Scandinavian countries are protecting their forests. The total area of forest enjoying formal protection remains small. In Finland 2,000 square miles (5,200 km²), or 2.6 percent of the productive forest is inside the boundaries of reserves. This is expected to increase to 3.5 percent of the forest, or 2,700 square miles (7,000 km²). Most of the protected forest is located in the north of Lapland. Norway protects 240 square miles (620 km²) of forest, which is about 1.06 percent of the total productive forest. These are old-growth forests located in the mountains. In Sweden 3.6 percent of the productive forest is protected, amounting to 3,200 square miles (8,320 km²). Most of the protected forest lies in the northern mountains, where the protected area amounts to 46 percent of the total forest. The overall area of Scandinavian boreal forest is maintained by replanting, a practice the Norwegian government has been subsidizing since 1863.

There is no possibility of the boreal forest disappearing, and it is very unlikely that its area will diminish in the forseeable future. Indeed, the trend is for the area to increase. Its composition may change if natural forest is replaced by plantation, but the consequences of such a change are now widely appreciated. Not only would the ecological value of the forests be diminished, in the accessible areas so would their commercial value for tourism and recreation. In Canada and Scandinavia, considerable importance is now attached to the conservation of natural forest.

THE TAIGA

Across northern Eurasia from Finland to the Pacific and from the Arctic Circle south to about latitude 52°N, the latitude of Irkutsk and the southern tip of Lake Baikal, there stretches the taiga. It is a vast ocean of trees, more open in the north, where it merges gradually into tundra, and in the south, where it gives way to the steppe grasslands. In all, forests cover nearly 47.9 percent of the land area of the Russian Federation, and the taiga accounts for almost all of this. Each year Russia produces 5.6 billion cubic feet (158.1 million m³) of timber, almost all of it from the taiga.

The taiga is the second largest forest in the world. Only the tropical forest of South America is slightly bigger. The Russian taiga covers about 3 million square miles (7.8 million km²), almost all of it in Russia. Larch (*Larix* species) occupies about 1.03 million square miles (2.65 million km²), Siberian stone pine (*Pinus sibirica*) 154,400 square miles (400,000 km²), other pine species 444,000 square miles (1.15 million km²), and spruces (*Picea* species) 297,000 square miles (770,000 km²).

Forest Wildlife

As might be expected in so vast an area with so sparse a human population, the taiga supports much wildlife, although the harsh winters and lack of variety among habitats limit the range of species. The Siberian spruce grouse (*Falcipennis falcipennis*) is one of several species that are found nowhere else. One of Europe's rarest mammals, the Russian desman (*Desmana moschata*) occurs in the taiga, although it is also found elsewhere in eastern Europe. It is a member of the mole family (Talpidae) that has abandoned the subterranean life and taken to the water, acquiring a waterproof coat, a flattened tail it uses as a rudder and for swimming, and webbed feet it uses as paddles. It is fully protected in Russia. The wisent or European bison (*Bison bonasus*) became extinct in 1925 but has been reintroduced.

There are eight or nine subspecies of wolf (*Canis lupus*), an animal still thriving in the taiga but now extinct in most parts of Europe. The wolverine (*Gulo gulo*), a member of the weasel family strong enough to kill a reindeer, grows in the North American and Scandinavian boreal forest as well as the taiga. In Russia it is hunted for its fur, as are the Siberian weasel or kolinsky (*Mustela sibirica*) and sable (*Martes zibellina*). Moose (*Alces alces*), known in Europe as the elk, feed on water plants in the swampy parts of the forest and by rivers and streams and on grass and young shoots in forest clearings. Moose are abundant in some areas, and their numbers are controlled by hunting, as they are in North America.

Conservation and Forest Resources

Despite the size of their forests, the Russians have long recognized the need to conserve them. Conservation laws were enacted in 1888 with the aim of protecting all forests (see "Russian Forestry" on page 243). The Barguzinski reserve, established in 1916 near Lake Baikal, was Russia's first scientific nature reserve. Today there are 99 of these special-purpose preserves (zapovedniki) with a total area of more than 1.1 million square miles (3 million km²) or 1.4 percent of the land area of Russia. The zapovedniki enjoy the highest level of protection and are used as outdoor laboratories for ecological research and to protect the nation's biodiversity. Activities within the reserves, including recreation, are limited and strictly controlled. Together the zapovedniki protect 49 percent of Russia's vascular plant species, 36 percent of the bryophytes, 44 percent of the fungi, and 86 percent of the lichens. They protect 515 species of birds, 168 species of mammals, 40 species of reptiles, and 26 species of amphibians. The zapovedniki are not distributed evenly, but experts in Russian conservation are working to identify gaps. A few zapovedniki belong to universities or the Russian Academy of Science, but most are the responsibility of the Department of Zapovedniki within the Russian Department of Nature Protection.

In addition, there are 34 national parks and similarly protected areas. These attract many tourists and are established mainly for recreational use. The first national parks were designated in 1983, and they are managed by the Federal Forest Service. There are also many designated areas (zakazniki) in which certain economic activities, such as logging, mining, grazing, or hunting, are forbidden. In addition, small specific sites such as bird colonies, rock formations, and scenic landscapes are protected as nature monuments. More than 16 percent of Russian forests are specially protected under the Forest Code of Russia. Clear-cutting is forbidden in these areas, strict regulations govern logging, and all forestry operations are controlled.

A herd of bison and the Russian desman live in the Priokso–Terrasny Biosphere Reserve, a mixed-forest reserve of 19 square miles (49 km²) not far from Moscow.

There are also Russian desmans in the Oka Reserve, a little to its east, and the reserve contains 800 species of flowering plants and 230 species of birds. Since 1956, the Oka Reserve contains the site of the central ornithological station for the Russian Federation. Established in 1935, the reserve covers 88 square miles (229 km²) of which 75 square miles (194 km²) are taiga.

About 745 miles (1,200 km) to the northeast of Moscow, the Pechoro–Ilyich Reserve, created in 1930 on the River Pechora in the western foothills of the Ural Mountains, is much bigger. Its total area is 2,784 square miles (7,213 km²), of which 2,411 square miles (6,246 km²) is taiga, with 204 species of birds and 43 of mammals. Taiga comprises about 90 percent of the Pinezh Reserve, about 466 miles (750 km) northwest of the Pechoro–Ilyich Reserve in the Archangelsk region. It was created in 1975 and has an area of 159 square miles (412 km²). In the mountains bordering Mongolia in the south of central Siberia to the southeast of Novosibirsk, the Sayano–Shushen Reserve, established in 1976, has an area of 1,504 square miles (3,896 km²), almost 60 percent of which is taiga forest. The Baikal State Nature Reserve, established in 1979 on the shores of Lake Baikal, includes 452 square miles (1,172 km²) of taiga; the Sokhondo Reserve, east of Irkutsk near the Mongolian border, protects 814 square miles (2,110 km²); and the Stolby Reserve, by the Yenisei River to the east of Tomsk, contains 175 square miles (454 km²) of forest.

Uncontrolled Logging

The constitutional changes that followed the collapse of the Soviet Union led to the privatization of many state enterprises, but the Russian Federation is reluctant to privatize forests (as are Ukraine and Belarus). Consequently, Russian forests are almost all owned by the state, and their finances are the property of the federal government. The Basic Forest Law of 1993, replaced in 1997 by the Forest Code, introduced the leasing of areas of forest and auctions of standing timber. Any organization may now lease an area of forest, and the regional authorities issue felling licences by negotiation, tender, or auction. Most leaseholders are former state-owned logging companies. There are internal paradoxes, such as that by which the Federal Forest Service, which is responsible for forest conservation, is financed by revenues from its own logging and is one of the country's largest logging organizations, producing up to 20 percent of all harvested timber. Conservationists fear that combining the functions of management, conservation, and commercial exploitation in a single body has created opportunities for corruption and uncontrolled logging.

The system of reserves, national parks, and nature monuments affords excellent protection, but the rules are not always enforced. Where the rules are applied, logging is permitted only in exceptional circumstances in zapovedniki. Gradual and selective logging—not clear-cutting—is permitted in other protected forests. Clear-cutting is the predominant logging method in the remaining forests. There has been serious overharvesting in Siberia where the forests are close to major markets in Asia. Today, however, two-thirds of the logging takes place in the European part of the Russian taiga which comprises only 20 percent of the total forest area. The logging industry aims to increase its logging, mainly in the European forests, to 1 billion cubic feet (300 million m³) a year by about 2010. That is almost double the present output, but it is uncertain whether the goal is attainable.

Karelia

For several years, increased logging earned Russia foreign currency, which it badly needed, but it did so by allowing bribed officials to sell timber at low prices. This is what nearly happened in Karelia. The Republic of Karelia (in Russian Kareliya, in Finnish Karjala) is part of the Russian Federation bordered by Finland. The eastern part of Karelia has been Russian since the 1323. Peter the Great acquired the western part from Sweden in 1721, and in the 19th century when Russia administered all of Finland, Karelia became Finnish once more. In 1920 Karelia was divided, Russia taking the eastern part and Finland the western part, but then Russia took all of it in the Russo-Finnish war of 1939–40.

Enso, a major Finnish timber company, had been prepared to start felling, licensed by the Republic of Karelia, but in October 1996 the company was persuaded to declare a moratorium. With warm support from regional conservation organizations that had pressed for it, Enso, the Ministry of Environment of Finland, and the Karelian government announced that they would establish a working group to prepare a conservation plan for the most valuable ecological areas of Karelian forest. These would include areas of old-growth forest adjacent to those in northern Finland.

As well as defining areas in need of protection from all forestry activities, the working group would devise improved forestry practices for the commercially managed forests. This would lead to the international certification of Karelian forests by the Forest Stewardship Council (FSC), an organization representing the international forestry and timber industries and conservation bodies. With FSC certification, companies and individuals purchasing Karelian timber or products made from it would know it came from well-managed forests.

The arrangement was ideal, but less than a year later it was in trouble. Enso was abiding by the agreement, but it looked as though requests that other companies respect it were being ignored. Members of the Finnish Forest Industries Federation are required to observe certain standards of practice that include observation of the Karelian agreement, but not all Finnish companies belong to the federation. One of the smaller and, in the view of conservationists, less scrupulous companies obtained a logging permit from the Karelian authorities to clear 5,000 acres (2,024 ha) in an area that conservationists were seeking to have designated as the Kalevala National Park. There was no forest reserve in that part of Karelia, so the arrangement was perfectly lawful. Near Lake Onega to the south of Karelia, the Kivach Nature Reserve protects 41 square miles (105 km²) of the Karelian taiga, but this is the only area with formal protection.

Although it was alleged that Karelian timber was being sold at far below the proper price, the authorities were in a difficult position. Iron-ore mining and the metallurgical industry have provided income and employment since the reign of Peter the Great in the 17th century. The name of the capital, Petrozavodsk, means "Peter's factory." Despite this, the forest is Karelia's principal natural resource, and forestry its most important industry. The authorities claimed that the economy could not survive without the sale of some timber from old-growth forests. Environmental groups publicized the issue, but the fate of the Karelian old-growth forest remains uncertain. Only 5.3 percent of the total area of Karelia is protected, amounting to 3,686 square miles (9,550 km²), of which designated areas (zakazniki) account for 60 percent, natural parks for 25 percent, and zapovedniki for 6 percent.

Long-Term Risks

Much of the taiga is remote and inaccessible. Harvesting its timber is impracticable without roads or railroads to transport the logs, so there is a check to the rate at which logging can expand. Nor is the taiga subject to the pressures causing deforestation in the tropical lowlands. There is no urgent demand for agricultural land to draw migrants in the wake of the logging companies, clearing more forest to either side of the forestry roads. Russia is not short of farmland, and in any case, the taiga cannot be farmed for climatic reasons.

The risk is not that the taiga will diminish in size significantly but that its most ecologically valuable areas may be damaged or even cleared. This would reverse some of the achievements made by conservationists over many decades.

■ THE NEED FOR CONSERVATION

Throughout the temperate regions of the world, people strongly oppose the clearance of forests. In the Republic of Karelia, it was environmental groups opposed to clear-cutting that protested against the timber company that ignored the agreement to halt logging in the old-growth forest. This local and regional insistence on preserving the integrity of particular forests and parts of forests is mirrored at the international level by attempts to regulate the activities of logging companies. The UN Conference on Environment and Development, or Earth Summit, held in Rio de Janeiro in 1992, failed to agree on a plan to protect forests, but the desirability of doing so was not in doubt.

Five years later, on June 26, 1997, forest protection was debated again at a special session of the UN General Assembly held in New York and known as the Earth Summit + 5, although little progress was made. The European Union, supported by the governments of Canada and Malaysia and some environmental groups, tried to push through a legally binding convention that would limit logging throughout the world. This was opposed by other environmental groups and governments with strong forestry interests, including those of the United States, Brazil, and India. They argued that the convention likely to be agreed upon would be a weak compromise that did more harm than good. An alternative idea, devised by the World Wide Fund for Nature (WWF) and supported by the World Bank and 20 countries, including China, was to protect areas equal to one-tenth of each of the types of forest in the world with a network of reserves that would be in place by 2000.

So far, none of these schemes has attracted sufficient support to be implemented, but the search will continue until some workable program has been devised. The main threat is to tropical forests, but a different threat to temperate forests is also acknowledged. These are unlikely to disappear, but in some places they are being converted from natural and old-growth forest to plantation forest.

Plantation Forests

Conservationists seek the protection of old-growth forest and are not impressed by demonstrations of an increase in the total number of trees or area of forest if this is achieved by commercial afforestation. Although plantation forestry may be inferior to the gentle management of natural forest, however, it has real environmental advantages. Temperate forests are now recognized as a huge sponge, absorbing carbon dioxide (see "Forests Absorb Carbon" on pages 269–270), and plantations are likely to be the best way to maximize carbon dioxide absorption. Not all tree species and the soils they produce are equally efficacious, so there could be advantages in growing the essentially monocultural stands that conservationists find objectionable. Plantations do contain wildlife, of course, and proper management can improve the quality of the habitats they provide. In Britain the expansion of conifer plantations has allowed the wildcat (*Felis sylvestris*), which had become extremely rare, to expand its range into regions from which it had long been excluded.

Conservationists seek to protect natural habitat for the species it supports, and it is natural forests that provide the widest variety of habitats. Once adequate safeguards are in place, habitats can be improved. Northern Rocky Mountain wolves, a subspecies of the gray wolf (*Canis lupus*), are native to Yellowstone National Park, but so many were killed to reduce their numbers that by 1940 packs of wolves were rarely seen. In 1970 scientists found no trace of resident wolves in the park, although individuals might occasionally stray in from adjacent areas. After several years of careful planning, U.S. and Canadian wildlife experts captured a number of wolves near Dawson Creek, British Columbia, and Hinton, Alberta, and released them into Yellowstone. They released 14 wolves in 1995 and 17 in 1996. Today the Yellowstone wolf population is increasing. Large predators, such as wolves, help regulate the populations of prey species. They complete the ecosystem, which improves the habitat. It has been suggested that wolves be reintroduced in remote parts of the highlands of Scotland, although there is strong opposition from farmers and many local people. There are also plans to reintroduce the European beaver (*Castor fiber*) to Scotland, where it has been extinct since the Middle Ages. The reintroduction of European bison into Polish and Russian forests is also a form of habitat enrichment.

Habitat Conservation

Wolves and bison need large areas of forest and freedom from disturbance. Without these, they cannot survive. These animals are not unique, of course, but merely familiar examples of a general principle. Every plant and animal species has needs that must be supplied if it is to survive. This fact leads to the concept of the habitat conservation plan (HCP), devised in the United States and based on recommendations from a team of scientists working for a number of agencies, including the U.S. Fish and Wildlife Service (FWS) and the National Marine Fisheries Service. HCPs allow commercial logging and the conservation of endangered species to prosper side by side.

In Britain it is not an offense to injure or kill a member of an endangered species if this was the accidental consequence of a lawful activity. The U.S. Endangered Species Act makes no such allowance. Any landowner wishing to conduct an activity that might unintentionally harm (the technical term is *take*) any species listed as endangered or threatened must first obtain an incidental take permit from the Fish and Wildlife Service. In order to obtain a permit, the landowner must produce an HCP showing the steps that will be taken to minimize the harmful effect. The FWS offers help in developing the HCP, and once it has been approved and the permit issued, the activity can proceed. An HCP lasts for a specified period.

On January 30, 1997, Kathleen A. McGinty, chair of the White House Council on Environmental Quality, Bruce Babbitt, Secretary of the Interior, and Jennifer Belcher, Commissioner of Public Lands for the state of Washington, announced in Seattle the largest program to date on forested land. Covering 1.6 million acres (650,000 ha) and lasting between 70 and 100 years depending on the particular areas protected, the HCP aimed to protect more than 285 species across the state. It used as its emblem the northern spotted owl (*Strix occidentalis*), an endangered species found in the old-growth forests of the northwest over which foresters and conservationists had been arguing bitterly for years. The marbled murrelet (*Brachyramphus marmoratus*) was another bird species the HCP would protect. It is a member of the auk family (Alcidae) that feeds at sea but flies inland to breed in forests and mountains.

By the summer of 2006, more than 350 HCPs had been approved, covering more than 30 million acres (12 million ha) of land. It is only relatively recently that the importance of temperate forests has come to be fully recognized. Many of the plants and animals native to the temperate regions occur in or near the edges of forests. Only forest, and especially old-growth forest, may provide the habitats they need to survive. In addition, temperate forests absorb and store carbon in amounts that are significant in terms of the global climate. The desirability of retaining the natural forests that still exist, and increasing the overall area of forest is not disputed. Today, the arguments revolve around the best ways this may be achieved.

■ SUSTAINABLE FOREST MANAGEMENT

Everyone now accepts that if forests are to be managed, sustainability should be the most important aim of that management. Sustainability is a concept that was introduced some years ago. It may have appeared first in the *World Conservation Strategy: Living Resource Conservation for Sustainable Development*. This was published in 1980 by the International Union for Conservation of Nature and Natural Resources (IUCN), in collaboration with the World Wide Fund for Nature (WWF), United Nations Environment Programme (UNEP), the Food and Agriculture Organization of the UN (FAO), and the UN Educational, Scientific and Cultural Organization (UNESCO). In 1987 the concept was defined in *Our Common Future,* the report of The World Commission on Environment and Development, chaired by Gro Harlem Brundtland, who was then the prime minister of Norway. "Sustainable development," the report stated, "is development that meets the needs of the present without compromising the ability of future generations to meet their own needs." In the case of forests, the commission suggested that policies should begin with "an analysis of the capacity of forests and the land under them to perform various functions." Following proper analysis, some forests might be cleared for arable or livestock farming, some managed for increased timber production, and some left intact to protect the way water drains through the land, to conserve living species, or to provide recreational amenities. Whatever plan is implemented, the commission urged that it be made in collaboration with local people.

At one level, the sustainable management of a temperate forest would define the forest as an area of land producing timber and other forest products and would aim to ensure that it continued to do so indefinitely. This management aim is hardly original, however. In Europe it has been applied since medieval times (see "Managing the Forests" on pages 235–236). Nowadays, throughout the temperate regions, forest plantations provide a regular harvest of products grown by methods that safeguard the capacity of the site to sustain a desired level of production. Forestry differs from farming only in the much longer time needed for each crop to mature. It aims to be sustainable indefinitely.

Sustaining Biodiversity

Sustainability can also have another meaning based on a different definition of a forest. This describes a forest not simply as a large stand of trees but also as a community of living organisms, the integrity of the whole being dependent on the diversity of its component species. The more species there are, the more stable the system will be. It follows, therefore, that the sustainability of a forest depends on maintaining its biodiversity. Whether biodiversity is linked to ecological stability is uncertain. Boreal forests cover vast areas with stands that contain very few species, yet these are no less stable than the much more diverse tropical forests. Stability seems to depend on the characteristics of the species comprising an ecosystem rather than their variety.

Conservation or Forestry?

Conserving biodiversity as a means of achieving sustainability often leads to conflict between conservationists and foresters. Holding the condition of the primeval forest prior to the arrival of humans as an ideal and regarding humans as standing outside nature and opposed to its interests, conservationists are inherently suspicious of human intervention. Many would prefer people to leave the forests strictly alone, a view that conflicts with their wish to popularize forests and encourage people to visit them. Foresters, in contrast, grow and harvest trees and consider it desirable to manage forests in ways that promote tree growth and health.

Plantations resolve the conflict, at least in principle. There, foresters can plant, tend, and harvest tree crops. Forest plantations need not consist wholly of a limited number of species exotic to the area. British conservationists criticized the Forestry Commission for planting large upland tracts with introduced conifers, but the commission was compelled to do so in order to create a national forest that could provide a strategic reserve of timber. Once that objective had been achieved, the "coniferization" policy was relaxed. As conifer stands mature and are cleared, mixtures of species are being planted. The transformation of British forests is slow because the growth and maturation of trees takes decades, but as the years pass, changes in the composition of the state forests will become increasingly evident.

It matters, of course, that plantations do not expand at the expense of natural forest. In Britain, where most of the original forest was lost many centuries ago, there is sufficient room for plantations outside ancient forests, but in other countries this is not always so. In North America, Scandinavia, and the Russian Federation, exploiting natural forest and replacing some of it with plantation forest is unavoidable because the natural forest is so vast.

Commercial Forestry and Conservation

Commercial forestry is not necessarily environmentally disruptive and does not preclude conservation. Responsible logging can increase biodiversity and reduce the risk of catastrophic fires. In Britain, forests on public land are now managed as recreational amenities as well as sources of timber, and so far as possible the wildlife value of forest habitat is enhanced. Some areas within state forests are now managed primarily for conservation, and visitors are informed of the species present in the forest.

Similar efforts to accommodate conservation needs are being made by other countries. Canadian forests are encouraged to regenerate naturally wherever possible, and where replanting of cutover areas is necessary, this tries to mimic the distribution patterns of species in the surrounding forest. This is a conservation measure that is supported by sound economic arguments (see "Replanting and Natural Regrowth" on page 257), and a planned tree-genetics center will underpin both the selection of trees for replanting and the establishment of a network of ecological forest reserves. In Finland, the Ministry of the Environment is identifying and mapping the location of old-growth forest. Already those areas on public land that deserve protection have been identified, and attention has turned to areas on private land. In Norway and Sweden, commercial forestry now takes account of conservation needs.

Several schemes are designed to promote the sustainability of North American forests. The Sustainable Forest Initiative sets standards for forestry methods and certifies forests that meet them according to independent inspectors. Approximately 54.5 million acres (22 million ha) of Canadian forests and 38.5 million acres (16 million ha) of U.S. forests have been certified. The Forest Stewardship Council operates in more than 60 countries, certifying forest products made from wood that has been grown sustainably. The scheme has certified about 15 million acres (6 million ha) of U.S. forests. The American Tree Farm System began in 1941. It certifies 12 million acres (5 million ha) of forests in 46 states. Green Tag Forestry certifies woodland that is managed in ways that maximize biodiversity and sustainability. It has certified about 66,000 acres (26,710 ha) in 12 states.

10

Management of Temperate Forests

At one time, most people lived in villages surrounded by the fields that produced their food and forests that supplied them with timber, small wood, fuel, and food in the form of nuts, berries, and game. These forests, adjacent to human habitations, were resources that people used and on which they depended. Beyond those forests, there lay large tracts of forest that people seldom visited. Where possible, travelers sought treeless high ground, following tracks that gave them a better view of the surrounding landscape and better warning of approaching dangers. People avoided the forests, where there were no landmarks for guidance and where dangerous animals might attack. People made no use of those forests. They were not a resource.

Today the situation is very different. All forests are valued. Many provide essential materials. Others provide opportunities for recreation. Forests sustain wildlife, including many species that could not survive at all outside a forest habitat.

If forests are to continue providing materials and facilities that people need, then they must be managed, and throughout history, forests that are utilized have been managed. Techniques such as coppicing were devised to provide small wood, and when trees were felled, more were planted. Trees were planted to create park woodlands where deer, cattle, and sheep could find food and shade. Timber production became more like farming, with trees that were known to yield wood of the desired quality being grown in plantations, like field crops. Ways were found to reduce the impact of tree diseases and attacks from insect pests.

All of these forms of management related to the sustainable production of raw materials. In modern times, people have found new ways to appreciate their forests. Forests are now valued as havens for wildlife and as places to visit for relaxation.

■ TRADITIONAL MANAGEMENT

"The well-wooded and humid hills are turned into ridges of dry rock, which encumbers the low grounds and chokes the watercourses with its debris. . . . There are parts of Asia Minor, of Northern Africa, of Greece, and even of Alpine Europe, where the operation of causes set in action by man has brought the face of the earth to a desolation almost as complete as that of the moon."

Should we fail to manage forests in a sustainable fashion, more is at stake than a shortage of timber some time in the future. George Perkins Marsh (1801–82) saw this clearly more than a century ago. The passage quoted here is taken from his book *Man and Nature,* published in 1864, and it is meant as a warning to Americans. It attracted a great deal of attention and led to the establishment of federal forest reserves in the United States and a range of conservation measures in other countries.

Marsh was a remarkable man. A lawyer, diplomat, and politician who spoke 20 languages, from 1862 until his death, he was U.S. ambassador to Italy. It was there, among impoverished communities scratching a living from the eroded lands around the shores of the Mediterranean, that he saw the consequences of forest clearance and there that he wrote his impassioned plea for greater care in the management of the natural environment. "Man has too long forgotten that the earth was given to him for usufruct alone," he wrote, "not for consumption, still less for profligate waste."

Americans were on the move in the middle of the 19th century, settling ever farther west. Railroads were improving communications, and new towns were springing up across the continent, towns in which many of the buildings were made from wood. Wood was also needed for railroad ties, telegraph poles, wagons, and, of course, for fuel to drive the locomotives and furnaces as well as to heat homes and cook food. That wood was taken from the seemingly endless forests. Marsh was not the only person to recognize that the forests were not really endless and that they were being plundered. His book expressed in words a concern amounting to fear that was felt by many, which is why it struck so clear a chord.

Sustainability

Reserves, like those that were established in response to Marsh's book, protect forests from exploitation, at least in principle, but protection is not management in an economic sense. Economic management permits the forest to supply products and, therefore, to sustain industries, but in a controlled way.

This does not necessarily imply planting replacements for trees that are felled. In boreal forests and the taiga, the preferred method nowadays is to encourage natural regeneration and to plant only where this is impracticable. A stand of just one species of coniferous trees may be harvested by clear-cutting, but in mixed and broad-leaved forests, it may be just certain trees that are commercially valuable. Perhaps, therefore, the best way to manage the forest is to remove the more valuable trees and leave the rest.

Techniques for sustainable forest management are now being tested in the Tropics, but they are controversial and the arguments surrounding them may be relevant to temperate forests. On one side are those arguing for sustainability. By this they mean that the most valuable trees in an area should be cut, seedlings of commercially desirable species planted to replace them, and the seedlings protected while they grow by weeding and the prevention of damage by animals. In this way gaps made in the forest by harvesting favored species will be filled, and the forest will remain productive indefinitely.

Those who disagree maintain that the technique does not work very well. In removing trees and protecting the replacement seedlings, so much ground has to be cleared that the forest is seriously damaged. What is more, the method is difficult and in remote areas often impossible to police. There is no way to guarantee that quotas for particular species are observed. This type of forest management is a good deal less sustainable than its supporters suppose. The alternative, they suggest, is to allow forestry companies to take as many as they wish of the valuable tree species. Once those have gone, the economic value of that area of forest will plummet. The area can then be purchased cheaply by a government agency or voluntary group and made into a protected reserve.

Both sides agree that forests should yield a reliable harvest of timber. They disagree over definitions of sustainability. The conventional approach to sustainability is concerned with yield, but this would allow natural forest to be converted to a plantation, provided the timber yield from the plantation is sustainable. A sustainable yield does not mean that the forest is sustained when it is regarded as a complex community.

Combining Different Activities on the Same Land

A third style of management would integrate forestry with other types of land use. One version of this is now widespread in temperate regions, where managed forests and plantations also provide recreational and educational facilities for the general public. Another version is also common in Europe, where livestock are raised in very open forests. Many of these parklike woodlands were planted deliberately by landscape designers to produce a kind of arcadian effect (see "Park Woodland" on pages 292–294).

In parts of Germany, a system of management was once practiced that seems to have had no equivalent elsewhere. An area of forest would be clear-cut, and once all the useful timber had been removed, the remaining vegetation would be burned. Then a crop of rye would be grown in the ashes between the stumps of the felled trees. By the time the cereal was ready to harvest, tree seedlings would be appearing naturally, and these would be left to grow.

In Britain, the consequences of uncontrolled deforestation became evident many centuries ago. Vast areas of the original forest were cleared during the Roman occupation (see "Roman Britain: Forest to Factory" on pages 232–233). By medieval times most of the countryside in lowland England consisted of islands of forest in an ocean of farmed fields. The woodland itself was classified according to its use. Those areas producing mainly small wood were distinguished from those producing timber. Most forests were owned privately but with local people having the right to use them for specified purposes. Wasteland was land owned and used by the community, rather than by a landowner.

Whether the ownership was private or communal, the trees were protected. Fences, or more substantial boundaries made from a ditch and bank, kept out livestock that might trample or eat seedlings or damage bark. Beneath the rough, outer bark of a tree, there lies a very thin layer, the cork cambium, and beneath that the phloem through which nutrients are transported around the plant. This is renewed from the vascular cambium, an underlying layer usually one cell thick. Together, these layers comprise the outer and inner bark. Should the inner bark be damaged, the tree will be left with a wound through which infection can enter and which will leave a scar when it heals. Cutting through the inner bark all the way round the tree is called *girdling* or ring-barking, and it kills the tree because it severs the channels transporting essential nutrients. Animals that might nibble or gnaw through the bark must be excluded from well-managed forest.

Occasionally, animals were allowed to graze in certain parts of a forest during a stage in its cycle of growth when they were unlikely to cause much damage. Such grazing was known as agistment, and it was more commonly used in park woodland than in closed-canopy forest, as was pannage—allowing pigs to forage for acorns, beech mast, and fallen fruit. People were sometimes granted the right of herbage, entitling them to cut and remove grass and other herbs. Bracken was gathered for use as bedding for people as well as livestock. Bast, the inner part of the bark (phloem) from lime trees, was used as a source of fiber. Small branches and sometimes even the leaves were also used. Bark itself was used in tanning, and certain types of galls were used to make ink. Once a tree had been felled, nothing was wasted.

Coppicing

Although girdling a tree will certainly kill it, felling the tree may not. Girdling makes it impossible for nutrients to reach those parts of the tree above the injury, but a felled tree has no parts above the injury, and in many species, although not all, new growth will sprout from the stump, called a stool. Cutting a tree close to the ground in order to allow the stool to sprout is called coppicing, and the technique was discovered in prehistoric times. The remains of coppiced trees have been found in peat in Somerset, England, and dated as 6,000 years old. Coppicing was being practiced widely in Britain by the time of the Norman conquest in 1066.

Coppicing does not produce a new, single trunk, of course, because a tree trunk grows outward and the center is composed of dead wood (see "Primary and Secondary Growth" on page 113). Instead, the new growth occurs around the edges of the cut stool as a circle of thin shoots. Coppice growth is rapid. Often within about five years, depending on the species, it produces a dense clump of poles. These are harvested, cutting down to the original stool, and some years later there is a fresh crop of poles to be cut. The traditional rotation varied in length from about four to about 30 years, with an average interval of about 12–15 years between cuts. A stool will continue producing coppice poles for many years. Indeed, coppicing seems to increase the longevity of most trees.

Uses for Underwood

Obviously, coppicing does not produce timber, only thin poles known as underwood, but this is what was needed. Small wood was used to make furniture, tool handles, and other articles of everyday use (see "Small Wood" on pages 250–254), but such items were durable. In those days society was not founded on the principle of consuming goods by frequently replacing them and throwing away the old ones. Items were handed down from one generation to the next until they were so worn as to be useless. As a result, the demand for wood for these uses was modest. Builders exerted a rather bigger demand for small wood. They needed laths known as wattle to make ceilings and walls that they would then daub with mud or plaster in the type of construction known as wattle-and-daub that was widely used. Wattle-and-daub sounds flimsy and reminiscent of mud huts. Done well, however, it lasts a very long time. There are houses with

wattle-and-daub walls in England that are still standing and occupied up to five centuries after they were built.

There was a much more vigorous demand for fencing. Metal wire was far too costly to be used to confine or exclude livestock. Instead, hedges made from growing plants, most commonly hawthorn (*Crataegus monogyna*), were used for permanent enclosures. Gaps in hedges were sealed by driving rows of stakes firmly into the ground and weaving thin, flexible rods between them. The stakes and rods were coppice products. Hurdles were used as temporary fences to corral animals and to make the most efficient use of pasture by controlling their grazing. These consisted of flexible rods woven horizontally between vertical poles and made as sections that could be joined as necessary. Finally, wood was the most widely used fuel, either directly or as charcoal, and small wood makes the best fire.

How Coppicing Spread

The practice of coppicing spread farther and farther northward through Britain, eventually into the highlands of Scotland, as demand for iron and steel grew in the early decades of the Industrial Revolution. The ironmasters who owned the foundries contracted with landowners for a regular, reliable supply of the charcoal that they needed, and it was only by coppicing that sufficient charcoal could be produced. By the time it reached its peak in the late 18th century, coppicing was by far the most common system of forest management. It was being practiced in most of lowland Britain, and most broad-leaved tree species were being treated in this way. Coniferous trees die when they are felled, so they cannot be coppiced.

It mattered little which tree species were used for coppicing. Most types of wood could be used as fuel and to make charcoal. Hazel (*Corylus avellana*) was popular, especially for making hurdles and fences, sweet chestnut (*Castanea sativa*) was also good for fencing, and willow (*Salix* species) was used for basket-making. Oak (*Quercus* species), alder (*Alnus* species), ash (*Fraxinus excelsior*), field maple (*Acer campestre*), hornbeam (*Carpinus betulus*), and several other species were also coppiced. Coppicing developed as a method for managing natural forest, but from about 1800, landowners began to improve their coppiced forests by clear-cutting and replanting with the species they preferred. Areas of coppiced chestnut and hazel in the south and east and oak in the north and west progressively replaced the remaining natural forests.

In medieval times the most valuable coppice land contained no standard (full-size) trees. These shade the ground, creating an area around themselves in which coppice shoots grow only slowly, if at all, so they reduce the output of the most important crop. Large timber is also needed, however, for such purposes as building timber-frame houses, bridges, and ships. It was usual, therefore, for coppiced forest to include a certain number of trees that were allowed to grow to their full, standard size. The method was then known as coppice-with-standards. Standard did not mean quite what it means today. The selected trees were allowed to grow until the trunks stood about 16–33 feet (5–10 m) tall with no branches and was surmounted by a crown that spread quite widely. Oak provided the most popular constructional timber, and it was the tree most often grown as a standard.

Had coppicing been the method of forest management favored in North America, George Perkins Marsh would not have thought it necessary to utter his warning. As it was, European settlers obtained most of the forest products they needed simply by logging, sometimes followed by burning. From the middle of the 17th century, some of the charcoal needed for smelting iron was obtained from coppice underwood, but the technique was never used extensively. At its greatest extent, in the last century, it occupied only about 494,000 acres (200,000 ha), and coppicing was already ceasing to be fashionable.

By the middle of the 20th century, coppicing was little more than a memory. No longer practiced in America or Britain, the remaining areas of coppiced forest became overgrown thickets. Many were cleared to provide farmland. Then interest in them was revived as their significance for conservation came to be recognized. For centuries, coppices had supplied local needs. Underwood and timber had been taken from them, not to mention nuts, fruit, and a wide variety of other forest products, and nothing had been returned to the soil. Gradually the fertility of the soil had declined. This had no effect on the trees, but it did affect the shallow-rooting herbs by denying any advantage to the more aggressive species with a high demand for nutrients that might otherwise have crowded out their competitors. Ground plants benefited from the intensity of light reaching the forest floor and by the shelter from wind afforded by the trees. Areas of old coppice were found to support a wide range of plant species, and coppicing was seen as a simple technique for producing great ecological diversity.

■ PARK WOODLAND

A forest, even a natural one, is not necessarily the best place to look for the biggest, oldest trees. The really majestic trees, hundreds of years old, more often live in parks. There is a reason for this. A tree does not die when it reaches a certain age. Provided it has leaves and can draw water and nutrients from the ground, each year it will grow another layer of cells, and it will continue to do so indefinitely. Eventually it dies, of course, because insects will bore into its wood, allowing fungi to infect it, so the tree sickens, but the process of dying may take a century or more. Foresters do not

allow trees to reach this stage of their lives. They fell trees when they reach maturity, that is, when the tree has reached its full height but long before it starts to suffer serious fungal attack. Most natural forests are managed, which makes it unlikely that they contain many ancient giants of trees. Parks are different.

Nowadays, a park is a place for recreation. It is where families walk on weekends, where people exercise their dogs, where children play. The park is a green space in an urban area, a patch of simulated countryside surrounded by streets and buildings, usually with well-made paths, flower beds, toilets, and notices to tell visitors what they may and may not do. Alternatively, in some European countries, a park is a large, formal garden where the careful control of nature is flaunted. Here, the visitor may delight in natural things, but they are highly ordered. Lawns have sharply defined edges, flowers grow in weed-free beds, and trees grow where they will guide the eye toward splendid vistas or into secluded corners where there are scenes to surprise and delight. Such gardens celebrate human mastery of the environment, reflecting the order imposed on nature by the straight furrow lines of a plowed field. Town parks of these kinds have existed at least since the 17th century, and the less formal type of park, with more open space, more trees, and fewer flower beds, incorporates three earlier traditions.

In-Fields and Out-Fields

Early farmers in northern Europe employed an in-field and out-field system of cultivation. The in-fields were small, irregular in shape and formed an approximate circle around the dwellings and farmyard. They were cultivated all the time. Beyond them there lay a circle of larger fields that were cultivated in a rotation, periodically being left fallow. These were the out-fields. They were enclosed by walls or hedges, and each was called "park." A park was an out-field—an enclosed area of land. This use still survives in some place and field names.

Later, in an extension of this meaning of the word, wealthy landowners attached the name *park* to the uncultivated land a little way from their houses. The park was managed, but increasingly its value was aesthetic rather than commercial. It was where the wealthy could ride or rest in the shade of a tree. That meant the ground had to be fairly open, with the trees scattered.

Deer Parks

A park was also an area of enclosed land in which animals were kept, and the animals most often found there were fallow deer (*Dama dama*), sometimes known as park deer. Originally a native of southern Europe and Asia Minor, the fallow deer was introduced to Britain probably in medieval times, perhaps by the Normans, and kept for hunting in parks and royal forests. The deer were a source of food, at least for the rich, and raising them was a type of land use, rather like the ranching of cattle. Other animals shared the parks with them. In various places there were wild pigs (*Sus scrofa*), red deer (wapiti, *Cervus elaphus*), and white cattle, sometimes called park cattle.

Hunting forests were not securely enclosed, but deer parks were. Agile as goats, fallow deer are good at escaping, and keeping them captive was costly. The park had to be surrounded by a wall too high for the deer to jump or climb, or by a tall, dense hedge, or by a strong fence made from chestnut stakes called a pale (hence the expression *beyond the pale* meaning "in the outer wilderness").

In time, the deer altered the landscape. Deer eat young tree seedlings, and grass grows on ground that is not shaded by trees, so unless deer are controlled, in time they will convert forest to pasture. One way to deal with the problem is to allow much of the park to develop as pasture, while retaining the bigger, stronger trees as small groups or isolated individuals.

Pollarding

Isolated trees could be protected by pollarding, a management technique that is still applied to trees lining city streets. Pollarding involves cutting the tree about 6.5 feet (2 m) above ground level. What remains of the trunk is called a *bolling*. At first a bolling is no more than a stump, but soon thin poles sprout from around its edge. When they grow to a suitable size, these can be cut and used. Pollarding differs from coppicing only in the height at which the cut is made. Like coppicing, pollarding increases the lifespan of the tree. Its advantage in a park is that the young shoots are well out of reach of deer, and deer cannot harm the tough lower trunk. Trees and deer can live together. Not all the park trees were pollarded. Some were allowed to continue growing, like the standards in a coppice-with-standards system, but usually to a larger size. These big trees supplied timber.

Some landowners practiced a form of management with deer but without pollarding. This involved dividing the land into areas, some retained as closed forest and others as either open woodland or grassland. When trees were felled, deer would be excluded from that area until replacement trees had grown too big to be harmed by them. The more open areas provided grazing and space in which the deer could be hunted.

Hunting is all very well as an amusement for the wealthy, but during the centuries, improvements in livestock husbandry made it a relatively inefficient way of procuring meat. A landowner might well conclude that the sensible way to manage the park was to remove the deer and other half-wild animals. Then, depending on its quality, the land could be brought into cultivation or developed as coppice. Parks might well have disappeared, but a third force was at work.

Trees for Shade and Shelter

In winter, trees provide shelter from the wind and from wind-driven rain and snow, and in summer they provide shade. To farmers and estate owners with homes in exposed locations, these were important attributes. Houses were drafty at the best of times, and English rain can fall in horizontal sheets, while in summer it was thought pleasant to sit outdoors shaded from the full intensity of the sunlight. So trees were preserved or planted to provide shelter and shade around houses and also around churches.

It was not only landowners who planted them. By the 17th century, trees had been planted around cottages and in villages where they served an additional purpose: People dried their laundry on them. Londoners also grew them. A visitor noted in 1748 that trees had been planted in the garden of almost every house and square and that trees lined both sides of the roads outside the city center.

As trees came to be appreciated for their utility, their beauty was also recognized. Writers in the 17th century bemoaned the absence of groves on bare hillsides, and there was stiff opposition when attempts were made to sell off patches of woodland so that the land could be converted to agricultural use. People had learned to love trees, and a visually attractive countryside was held to be one where trees, as individuals or small groves, were scattered among the pastures and cultivated fields. This produced a much richer landscape, full of detail to catch and hold the eye, yet a landscape that was clearly under human control. It was a benign landscape that fed and sheltered its human inhabitants, and its trees were tamed. These were not the trees of the dark, dangerous, primeval forest.

Landscape Architecture

The concept of an attractive landscape having been defined, by the 18th century, wealthy landowners were paying to have existing landscapes modified so they conformed more closely to the ideal. A tradition began of what came to be known as landscape architecture. Lancelot ("Capability") Brown (1715–83) is the most famous name associated with this movement. Having started his working life as a gardener's boy, Brown obtained a post at Stowe, Buckinghamshire, which was one of the most renowned gardens of its day, where his job included showing visitors around the garden. He earned his nickname from his habit of saying that a site "had capabilities." Thousands upon thousands of ash, oak, elm, and beech were planted under Brown's direction, most of them in small groves. Critics thought these "clumps" unnatural, as indeed they are, but many of them remain to this day.

By Brown's time, tree-planting had become fashionable and a recognized way to improve an estate. Not all landowners could afford to embark on large-scale landscaping, but they could plant trees, and they did. The result was a major expansion of the area of what was, in effect, park woodland. It looked informal, cattle and sheep grazed among the trees, and some of these parks even contained fallow deer but now for ornament rather than for food.

Over much of Europe, what had once been closed-canopy forest was replaced by open woodland intermingled with farms and pasture. Cultivated fields are the most important component of lowland landscapes, but it is on farmland, in scattered groves, on field boundaries, and on pastureland where they are surrounded by grass that a significant proportion of European trees is now to be seen. It is there, too, in the parklands, that the biggest and oldest trees are to be found.

■ PLANTATION FORESTRY

Many present-day forests are plantations and as different from natural forest as an arable farm is from the original prairie. Plantation trees grow in rows, and each tree is much the same size as those around it. Forest workers spend their time planting, spraying, and harvesting. It looks simple. It is not. A plantation does not just happen as the result of some workers going out to plant trees. It must be planned, and the planning begins with collecting information.

At the start, the forester knows only the area of land available for afforestation and its location. The first step, therefore, is to identify clearly the boundaries of the plantation site, its altitude, the location of any roads, railroads, ponds, rivers, streams, or power lines that cross it, the gradients of its hillsides and the directions they face, the soils, and the existing vegetation cover. Work begins, therefore, with a good map, aerial photographs, preferably stereoscopic ones, measuring, and a lot of walking.

With these important features clearly marked, the next step is to divide the site into compartments and draw their boundaries on the map. The compartments are the fairly large areas bounded by forest roads, rivers, or other natural features. They will vary in size, most being up to about 50 acres (20 ha), and they are permanent. Within each compartment, there will be several individual stands of trees in subcompartments with boundaries that may change over the years as circumstances require.

Protecting Against Wind and Fire

This initial survey will also involve learning as much as possible about the local climate. Average summer and winter temperatures and the amount and seasonal distribution of precipitation obviously affect all growing plants, but in the case of trees, so does the wind. It is important to identify those parts of the site where trees will be exposed to winds strong enough to blow them down. Often, cotton flags are the tools used to measure the wind strength. Leave a flag flapping in the wind, and little by little, it will tear at its outer edge until it is very tattered. How long it takes to reach this sorry state depends on the strength of the wind, so the rate at which cotton flags become tattered is a good indication of places where trees may be blown down, or "windthrown." When the places at risk of windthrow have been identified, the shape of the forest edge can be modified to reduce the likely damage.

As the details are drawn on the map, safety features must also be included. Should part of the forest catch fire, the fire must be contained if at all possible to prevent it from spreading to adjacent blocks. Firebreaks are strips of ground separating blocks of trees. They must be at least 33 feet (10 m) wide and contain nothing that will burn readily. Forest roads and rides, which are paths wide enough for horse riders, give little protection against fire. Some tree species, such as larches, are difficult to ignite and do not burn well. Surrounding a subcompartment with a belt of these trees can help slow the spread of a fire.

Predicting the Yield

Armed with the results of the survey, forest planners can decide which tree species are likely to be most suitable for the conditions. Once that is decided, it is possible to estimate how many trees the site will support, the eventual annual sustainable yield of timber and small wood, and how soon that yield will be obtained. This is obviously important. Managers must know the size of the financial investment needed to secure a particular yield and the rate of return they may expect on that investment. They must also know how many workers they will need, and when and what machinery. If the first harvest of full-grown trees will not take place for 60 years, there is no point in buying heavy harvesting machines just to have them sit idle for more than a half-century, quietly rusting and growing obsolete.

Future yield will depend on the species grown, and it is estimated from the "yield class." As a stand of trees grows, each year the total volume of useful wood increases. This increase includes the volume of dead trees and trees removed from a stand by thinning. The increase, measured in cubic meters of timber per hectare of land, is known as the mean annual increment (MAI). At first the MAI increases rapidly, as young trees grow to their full size, but beyond a certain age, the rate of increase declines. Plotted on a graph, it appears as a curve. The peak of the curve, where the MAI reaches a maximum, gives the yield class for that particular stand. This is an abstract number equal to the maximum MAI that the stand can achieve. A stand of some coniferous trees, for example, may attain a peak MAI of 30 cubic meters

British Forests

	CONIFERS		BROAD-LEAVES		TOTAL	
	Miles²	Km²	Miles²	Km²	Miles²	Km²
Forestry Commission						
England	583	1,510	205	530	787	2,040
Wales	367	950	46	120	417	1,080
Scotland	1,671	4,330	104	270	1,776	4,600
Total	2,621	6,790	355	920	2,980	7,720
Nonforestry Commission						
England	838	2,170	2,706	7,010	3,543	9,180
Wales	247	640	436	1,130	687	1,780
Scotland	2,374	6,150	1,007	2,610	3,381	8,760
Total	3,462	8,970	4,150	10,750	7,612	19,720
Total						
England	1,421	3,680	2,911	7,015	4,330	11,220
Wales	614	1,590	482	1,250	1,104	2,860
Scotland	4,045	10,480	1,111	2,880	5,157	13,360
Total	6,083	15,760	4,505	11,670	10,592	27,440

(Source: Forestry Commission National Statistics, June 15, 2006.)

per hectare (429 feet³/acre), giving that stand a yield class of 30. Many broad-leaved trees have very much lower yield classes, some as low as four.

So far, not a single tree has been planted, so there is nothing to measure. In a general way, yield classes are known for each tree species, so they can be looked up in published tables, but once a stand is growing, it can be calculated by a rule of thumb from the age of the stand and the height of its trees. This is possible because the age and height of a tree are closely related to the volume of timber it will produce. Knowing the age of a stand of a given species, the forester need measure only the average height of the trees. A stand of Sitka spruce that is 50 years old and with trees 30 feet (9 m) tall, for example, has a yield class of about 17 (which is not very good for Sitka). This is an estimate known as a general yield class (GYC). When the plantation has been established long enough for there to be actual production figures from particular parts of it, these will make it possible to calculate the yield class much more precisely, as a local yield class (LYC).

Planning for Conservation and Recreation

Plantation forestry is a commercial operation, an industry that grows trees in order to sell timber and wood. At one time that might have been thought a sufficient objective, but other factors must also be considered nowadays. A temperate forest is not merely a large number of trees. It is a community of plants and animals (see "Modern Forestry" on pages 304–306), and account must be taken of their needs. Forests serve an important conservation role, and they must be planned with this in mind.

Wildlife conservation may have a very high priority on sites of particular value. Elsewhere, it may be enough to keep forest trees away from the banks of streams so that these can be colonized naturally, and to manage roadsides and open spaces sensitively to maximize their quality as habitat. Wildlife will fare best in a forest composed of tree species that grow naturally in the area, so most modern plantations include stands of native species even where the important timber species are exotic. This is especially important in Britain, where Scotch pine (*Pinus sylvestris*) is the only native conifer. In 2006, conifers occupied about 6,083 square miles (15,760 km²) in Great Britain, and native broad-leaved trees occupied 4,505 square miles (11,670 km²). For years, British conservationists were opposed to afforestation, not because they disliked forests as such, but because plantations consisted of just a few species of imported conifers. As the figures in the table above show, that situation is changing. As plantations complete the first cycle of their rotation, more broad-leaved species are being included in the replanting, and in new plantations they are grown from the beginning.

Forests near large population centers are also public amenities. People demand to be allowed to visit them to walk, ride bikes or horses, picnic, and generally enjoy themselves in an informal atmosphere. Again, at one time such free public access was thought to be incompatible with the needs of forestry, but the public won. At least in publicly owned forests whose managers are accountable to elected politicians, people are now welcomed, and basic facilities such as toilets, maps, and information about the trees and wildlife are provided. As with conservation, recreational requirements must be allowed for in the initial planning. Most of the time, forest blocks take care of themselves, and tree-growing can be combined with recreational uses of the land. During forestry operations, the public must be excluded for safety reasons.

Preparing the Ground

So far, people living nearby will have seen very little sign of activity on the land they have been told is due to become a plantation forest. They may have noticed a few individuals walking around. There were some surveyors with their striped poles, measuring tapes, and theodolites. One day a helicopter seemed to spend rather a long time flying back and forth, probably taking photographs, but that is about all. The local residents might conclude that the scheme had been abandoned or at least postponed. In fact, though, the activity has been intense, but almost all of it has taken place indoors.

The surveys complete, it is time to prepare the site. Now the neighbors will see some action—and they may not like what they see. Wet ground may have to be drained because trees will not grow well or in some cases not at all where the water table is higher than the depth to which their roots extend. Installing drains means that large, plow-like machines will cut deep gashes across the land, clearing away the existing vegetation as they go and leaving the area looking as though tanks have fought a battle across it. Some roads may have to be realigned, which means destroying an existing road and making another.

These operations will prepare the whole site, but only a part of the site will be planted in the first year. The eventual aim is to produce an annual yield, which means that once the rotation is established, a certain number of trees must be ready to harvest each year. Unless a change in policy calls for it to be removed, an established plantation becomes a permanent feature of the landscape. Each year some of its subcompartments will be clear-cut, but these are usually quite small compared to the size of the plantation as a whole, and a few years later, with a new crop of trees, those subcompartments merge once more with the surrounding blocks. The species chosen may change from one rotation to the next. Little by little this will alter the appearance of the forest, but the forest will still be a plantation.

Cultivation

It follows, therefore, that the first trees to be cut must also be the first to be planted. The site will be planted a few areas at a time, and it will be several years before the whole of it is forested. At first, therefore, it is only in some blocks that the ground needs to be made ready for planting.

Preparation means cultivation. A large machine, depending on the soil type either a scarifier or a disk trencher, breaks up the ground surface. Then an earthmoving machine scoops up topsoil to make mounds into which the young trees will be planted. It is now standard practice to plant trees on mounds. Despite the large machines, the aim is to cause the least disturbance possible, and the ground is prepared only in those patches where the trees will be located.

On some sites, the mounds together with areas about three feet (1 m) in diameter around them are sprayed with herbicide once cultivation is completed. This is to kill weeds as they are germinating and to suppress them for long enough to allow the trees to become established. Once the trees are too tall to be shaded, further weed control is unnecessary.

Tree Nurseries and Orchards

Now, at long last, planting can begin. It may be feasible to sow tree seeds directly into the prepared mounds, but in most cases, and especially on a new site, it is not seeds that are planted but tree seedlings. These are produced by tree nurseries, which are run as part of the forestry operation. Young trees are grown from seed in the nurseries, and a major part of the nursery workers' task is selecting the best seed for the purpose.

Trees are individuals, just as animals are. Even within a species, no two individuals are completely identical. The differences distinguishing them may not be visible, but they may be very important for all that because they result from the ways trees adapt to a particular site for many generations. The trees all look the same, but one individual may be better suited than another to the soils or climate at the new plantation or more resistant to the local pests and diseases. The only way to find out is to keep meticulous records of all the trees and the places from which they were originally taken and of the source of the seeds.

Seeds taken from trees that have been chosen for their desirable qualities are germinated to test that they germinate well and then grown in orchards. Pollination is carefully controlled so that the parents of each seedling are known. Especially suitable trees are often cloned by taking cuttings from them or, increasingly nowadays, by growing them from tissue samples, a technique called micropropagation. Cloning produces a stand of trees that are genetically identical to one another. Trees grown in this way provide seed of a quality that is known thoroughly and can be guaranteed. A great deal of forestry research takes place in the seed nurseries and orchards.

Planting and Protection

Depending on the species, by the time they are delivered for planting, the seedlings will have spent one or two years in seedbeds and a similar length of time in an orchard. They will be 8–16 inches (20–40 cm) tall, with well developed roots, and they will be planted 6.5–10 feet (2–3 m) apart.

The young trees will need protection. If the subcompartment adjoins farmland, livestock might enter and damage the trees. It is the responsibility of the farmer to prevent stock from straying, but if the field has no stockproof fence, the foresters may have to erect one. They may also need to put up a fence to keep out wild deer. Alternatively, trees can be protected by enclosing them in cylinders of wire or plastic mesh. This method is very widely used because not only does the guard protect the tree from animals but it also shelters it, stimulating it to grow, and makes it very easy to distinguish small trees from weeds. Where herbicides are used to kill weeds, the guard also protects the trees from being sprayed accidentally.

Until they become established, some tree species need more substantial shelter from the weather. The usual way to provide this type of protection is to plant a "nurse" tree beside the more delicate tree. The nurse belongs to a species that grows well on the site and that grows faster than the tree it shelters. By the time the nurse is large enough to be felled, the tree it shelters will be tall enough to survive alone. In Britain, lodgepole pine (*Pinus contorta*) is used on peat soils to nurse Sitka spruce (*Picea sitchensis*), and Scotch pine (*Pinus sylvestris*) is widely used as a nurse species.

Beating Up and Thinning

Inevitably, some young trees will die within the first couple of years. If the loss leaves unacceptably large gaps, these will have to be filled, but by now the surviving trees will be much bigger than the seedlings ordinarily supplied by the nursery. The replacements will have a better chance of competing for light and nutrients if they are about the same age as the rest of the stand or if they belong to a different, faster-growing species. Replacing trees lost in this way is called *beating up*.

After a few more years, unwanted woody undergrowth is cleared away, and some branches may be pruned from the trees to improve the quality of timber they will produce. Unless people need to move freely among the trees, foresters no longer remove the lower branches. This operation, called *brashing*, involves removing all the branches below a height of about 6.5 feet (2 m), but it is time-consuming and therefore costly.

Thinning is also time-consuming, but it contributes to its own cost. The aim is to remove the thinner, weaker trees to make more room for the stronger ones, which will then grow thicker trunks. Thinning makes the strong trees more valuable, but the thinnings can also be sold as small wood or for fuel (see "Forests as Sources of Fuel" on pages 246–248 and "Small Wood" on pages 250–254).

Harvesting

Finally, after 50 years or more, harvest time arrives. Harvesting is the most dangerous forestry operation. Trees are felled and their branches removed using chain saws, and when a tree is cut, it falls. Fallen trees are then removed from the site and loaded onto vehicles. Workers engaged in harvesting must be highly trained in the use and maintenance of their equipment. They must also know how to work safely and how to deal with emergencies. Under no circumstances must an untrained person be allowed to use a chain saw, which is an extremely dangerous tool.

A forestry worker using a chain saw to fell a tree. He is working safely. He wears a "hard hat" helmet, a visor covering his face, ear protectors, thick gloves, heavy boots, and a brightly colored vest to make him clearly visible from afar. Tree-felling is potentially a very dangerous job. *(The Natural Resources Conservation Service)*

Bound together into rafts, the logs are now floating down the Big River, California. Note the log float carrying gear in the foreground. This scene is from the 1870s. *(Ira C. Perry, reproduced courtesy of KRIS Big River Project, Mendocino Historical Society)*

Like all forestry tasks, harvesting is planned carefully and performed methodically. First the routes, called racks, along which timber will be extracted must be identified and marked. The size and species of trees determines the type of chain saw that is used. Lightweight chain saws are usually used to cut conifers and small broad-leaved trees. The block to be cleared is then divided into sections, and one or two workers are assigned to fell the trees in each section. Felling is planned so that a distance not less than twice the height of the tallest tree always separates workers. The first few trees to be cut all fall in the same direction, and then the remainder are cut so that they fall at right angles, across the first trees. That arrangement makes it much easier to remove their branches (an operation called *snedding*) and lift the trunks.

Broad-leaved trees are especially difficult. If they fall awkwardly or their branches are removed clumsily, the wood may split. If that happens, the tree may be ruined. Broad-leaved trees are also dangerous. There may be dead branches that fall when the tree starts to move, not only from the tree being felled but also from adjacent trees. When the tree is on the ground, it may roll, and branches may spring back suddenly.

It is no easy matter to transport a tree trunk that may be more than 65 feet (20 m) long, and foresters use heavy machinery for this task. A forwarder is a large tractor fitted with a crane. It lifts the timber clear of the ground and loads it onto a trailer. Trailers can carry 6.5–11 tons (6–10 t) of timber. Alternatively, the harvesting team may find a skidder more suitable for the site. This is also a tractor, but with either a rear-mounted winch or a grapple with which the machine can seize and lift one end of a tree trunk and drag it from the site. There are different types of forwarders and skidders, but all of them have a safety cab and much stronger guarding than a farm tractor. Conditions at the site determine which machine to use, and sometimes it is necessary to use cranes. Because of the high cost of the machinery and the skilled workforce needed, specialist contractors often carry out the harvesting.

After felling, the logs are driven away to the sawmill, the chain saws and heavy machinery leave the area, and the subcompartment is left abandoned and bare. It is time

Floating logs to the mill may seem straightforward, but things can go wrong. This is a log jam, and clearing it will be difficult and time-consuming. *(U.S. Forest Service)*

Loading a logging truck. Newly felled trees are bulky and heavy, and large industrial machines are needed to transport them out of the forest.   *(The Natural Resources Conservation Service)*

then to think about preparing the ground for the next crop. Nowadays natural regeneration is usually the preferred method on an established site where a subcompartment has been cleared and is to be restocked. Mature trees surrounding the cleared block will supply the necessary seeds, and over several generations the trees will adapt to the site conditions. Where regeneration is not practicable or where a different species is to be grown, seed or seedlings will be needed and mounds made to receive them. The rotation is complete and the cycle begins again.

Where a large river is available, logs can sail to the sawmill. In this 1902 scene, workers are sitting on a pile of logs that are waiting to be rolled into the nearby Spanish River in Algoma District, Ontario. They will then drift downstream to the sawmill.   *(John Boyd, Archives of Ontario, Canada)*

■ PEST CONTROL

Elms were once an important feature of English lowland landscapes, but in the 1960s Dutch elm disease killed more than 80 percent of them (see "Dutch Elm Disease" on page 174). A fungus caused the disease, but the fungus was spread by bark beetles. Trees, like other plants, are susceptible to insect attack, and those attacks can be serious, occasionally even devastating.

Dutch elm disease was spectacular, highly visible, and widely publicized. So too, some 60 years earlier, was the outbreak of fungal disease that destroyed almost all American chestnut trees. Temperate forests in more remote regions frequently suffer devastation on a comparable scale. Each year, for example, the equivalent of nearly two-thirds of the Canadian timber harvest is lost to insect pests and fungal diseases. In Britain, the common or silver fir (*Abies alba*) cannot be grown commercially because it suffers so badly from attacks by a bug, *Adelges nordmannianae*. Spruce bark beetles, of which there are several species, cause extensive damage in many parts of the temperate regions. Rigorously enforced controls on imported timber helped exclude them from Britain for many years, but the great spruce bark beetle (*Dendroctonus micans*) managed to establish itself in the country in 1982. It breeds in large spaces that it excavates beneath the bark, causing severe damage to the bark and trunk and sometimes killing the tree. The larch bark beetle (*Ips cembrae*), a pest mainly of larch as its name suggests, is widespread throughout most of Europe and temperate Asia.

Trees are large and offer a variety of habitats, so it is not surprising to discover that they harbor many species of insects. There are opportunities for insect larvae and adults that feed on leaves, including the needles of conifers, either munching their way around the edges or through the inside, which is the method favored by leaf miners. There is nourishment for those that feed on the materials they find in the crevices of the bark and those that feed on wood beneath the bark. There are insects that feed on flowers and fruits, seedeaters, those that live belowground and eat roots, and many others. More than 170 insect species can be found on pine trees, about 90 on spruces, and there are even more on broad-leaved trees.

The great majority of these insects do no harm to the tree. Even those that feed on its leaves or bark do not eat sufficient material to cause any real damage, and the insect population includes predators as well as herbivores. To some extent, the insect population regulates itself at levels which are usually too low to cause serious harm. If the numbers of one species start to increase, so do those of the predators that feed on it. Many of these are insects, but birds and bats also eat insects, and they are voracious. All of the bats that live in temperate regions feed exclusively on insects. There are also other arthropods with an appetite for insects. On a morning when dew makes them clearly visible, spider webs can be seen festooning every shrub and the spaces between shrubs. There are more spiders on the ground below the webs, ones that chase their prey rather than trapping it. In addition there are centipedes and other carnivores. Small mammals will also eat any insects they come across in the course of their general foraging, nor is the curbing of insect numbers a job only for predators. Insects suffer from parasites as do all animals, as well as from diseases caused by viruses.

Why Plantations Are Healthier than Natural Forest

Self-regulation by the insect population and all the other insectivores is very effective, but it is not the whole story. Time also plays a part. A seedling, standing barely taller than the herbs around it, harbors few insects, and most of those are just alighting to rest, drink, or feed before moving on. They do not live on the young tree. As the seedling grows into a sapling, some resident insects will establish themselves, and more will join them as the years pass. Trees do not start life with a full complement of insects. They acquire them gradually. For this reason, there are fewer species of insects among stands of young trees than there are among stands of mature trees. Young trees are also better able to resist insect attacks than are old trees, because they are healthier and more vigorous.

Healthy trees are less vulnerable to pest attack than unhealthy trees. An unhealthy tree will have dead and rotting branches or twigs, open wounds where branches have been torn away or bark stripped, or cracks and fissures providing access to the xylem and phloem. Xylem and phloem are the channels through which water, nutrients, and infections are transported to all parts of the plant (see "Xylem" and "Phloem" on pages 104–105). Insects find ample food among this dead or damaged tissue, and they either carry fungal spores with them or make it easier for fungal spores to enter and start growing.

In this respect, a tree is little different from any other organism. Wounds can become infected, infections can spread, and an individual who is injured or already sick finds it more difficult to fight off new infections. Regardless of the size and composition of the insect population, therefore, a healthy tree is less likely to be harmed than a sickly one.

Plantation forests comprise blocks of trees that are all the same age, and the trees are harvested as soon as they reach marketable size. The trees are not permitted to reach old age, unlike the trees in a natural forest. The youth of plantation trees greatly reduces their susceptibility to diseases transmitted by insects. Plantations are healthier than natural forests, in the sense that they contain less disease.

Is Monoculture a Bad Thing?

Many people suppose that monoculture, which is the growing of one or a very few species of plants over a large area, encourages pest infestations. That is because herbivorous insects tend to specialize on particular plants, so a large monocultural stand represents a vast food supply to the insects specializing in those species, and their numbers are almost certain to increase. Eventually, so the argument goes, such a large number of specialist insects will be feeding that they will do real harm to the plants. Where these specialists find their preferred plants scattered among other species they find unpalatable, not only is there less food for them overall but it is also more difficult for them to migrate from one plant to another. Food plants are harder to find, they are likely to be occupied when they are found, and there is nothing for the traveling insects to eat while on the move. Fungal diseases are more likely to attack monocultures for the same reason.

This is partly true for agricultural and horticultural crops, although the dynamics of insect populations is fairly complex and monocultural agricultural and horticultural systems can be ecologically stable. It is not at all true for forests. Monocultural plantation forests are far less susceptible to pest damage and diseases than are more diverse natural forests. This is probably because plantation trees are felled before they bear wounds that could be infected, and the plantations themselves have not been in existence long enough for the full complement of insects and fungi to have accumulated. Natural boreal forest contains large areas of just one, or a very few, tree species, so it resembles a monoculture, but it enjoys a different type of protection: Insects are immobilized during its long, very cold winters.

Old-Growth Forest

An old tree will invariably have dead and decaying tissue and bear the scars of old wounds. In themselves, these do little or no harm. Any old-growth broad-leaved forest contains hollow trees in which so much of the center of the trunk has rotted away that what remains is barely sufficient to prevent the tree from falling. Around it on the ground, there may be old, dead branches that the tree has shed. No animal could possibly survive for long in such an apparently decrepit condition, yet the tree produces leaves each spring and may well continue to do so for another century or more, provided it does not succumb to a major infection or insect infestation. The tree is alive but barely so. It is old and sick—and vulnerable. Once the tree becomes infected, disease can spread from it to its neighbors and from there throughout a large area of the forest.

Old-growth forests contain old and sick trees growing alongside young and healthy ones, so these are the forests most likely to suffer from major pest attacks and outbreaks of disease. This seems paradoxical. An old-growth broad-leaved forest contains several tree species. Here and there one of the component species may form a pure stand, but it will not be very extensive. The old-growth forest is fairly diverse, and species diversity is popularly supposed to confer a degree of protection. That protection extends only to pests and diseases that attack only one or two species, however, and its benefits are far outweighed by the capacity of old, sick trees to act as pest and disease reservoirs.

The Cost of Pests and Diseases

When a forest suffers a serious infestation, the effects can be catastrophic. There are moths and sawflies that produce larvae capable of defoliating and killing entire trees and large areas of forest. Bark beetles excavate tunnels and galleries beneath bark, greatly reducing the value of the timber and introducing fungal infections, such as Dutch elm disease. Bark beetles, loopers, bud moths (with larvae known as budworms), and various weevils can destroy entire forests—and sometimes they do.

Countries that rely heavily for their timber on logging in old-growth forests cannot afford to accept the level of damage pests and diseases are capable of inflicting. Every year, for example, Canada loses the equivalent of almost two-thirds of its timber harvest to insect pests and fungal diseases simply because most Canadian forests are mature or contain a high proportion of very old trees. In the United States, insects and fungal disease destroy several times more trees than are lost by fire. It is reasonable to suppose that similar losses occur in the European boreal forest and the taiga. An obvious consequence of this is that achieving a target yield of timber involves clearing a substantially larger area of forest than would be needed if the pest and disease losses were smaller. Logging causes inevitable damage to wildlife, so an obvious way to reduce its adverse environmental effect is to reduce outbreaks of pests and diseases.

Weed control may also be necessary. Natural regeneration or planting will repopulate an area of old-growth forest that has been clear-cut, but the natural process may be slowed by competition from other plants, growing rapidly where they suddenly find themselves exposed to full sunlight. Some of these are likely to be trees, useless as sources of timber but able to grow faster than the more desirable species. Control of weeds may involve cultivating an area that is to be planted. This kills weeds that have emerged and allows planted seedlings to establish themselves. Cultivation cannot be used in areas where trees are regenerating naturally, because plowing would destroy tree seedlings, so it may be necessary to use a herbicide that is toxic to the most vigorous weeds but harmless to the trees. Unfortunately, herbicides are not very specific in the plants they affect, so unless the crop plant has a natural immunity to the herbicide, weed control must take place before the crop appears (see "Cultivation" on page 296). Genes for herbicide resistance have been inserted into some agricultural field crops, and in years to come, commercially valuable plantation trees may also carry them, but, of course, existing mature trees in old-growth forests cannot benefit from them.

Defining Pests and Weeds

Herbivorous insects can experience sudden vast increases in their populations, so they kill all the trees of certain species over a large area simply by feeding on them, and the seedlings that will replace the lost trees may be choked by competing plants. Despite this, pests and weeds are never found in natural forests. Such destruction is entirely natural, unavoidable, and it is not at all serious. All types of forest experience it, including temperate forests, and in time they recover. It is quite wrong to suppose that a natural forest remains unaltered throughout eternity or even for a mere thousand years. It changes constantly, and sometimes the change is dramatic but temporarily so. Pests and weeds can be defined economically, but ecologically the terms have no meaning. There are no pests or weeds in the natural forest as long as the forest remains unexploited. They exist only in those forests from which the landowner needs to obtain timber, and they exist then because they reduce the value of the forests.

Deciding whether the plants are weeds that should be suppressed and whether insect numbers have increased to such an extent as to constitute an infestation requiring control is a matter of economics, of costs and benefits. The question is very simple: Will the value of the timber lost to the weeds or pests be greater or smaller than the cost of treatment? Answering the question is rather more difficult because it involves estimates of how the situation will develop under varying circumstances. Nevertheless, finding the right answer is extremely important, both environmentally and economically. Many of the problems associated with pesticide use in the 1940s and 1950s arose from incorrect diagnoses, leading to the spraying of crops that would have recovered had they remained unsprayed. Trees are more likely to recover than farm crops because they live for much longer and are physically large. They can tolerate a surprising amount of injury with little or no long-term ill effect.

Insecticides and Herbicides

Once a pest infestation is causing widespread damage in a mature forest, it can probably be dealt with only by spraying an insecticide, and because the damage is occurring in the crowns, spraying has to be by aircraft. Aerial spraying is an expensive last resort, and it must be done with great care to ensure that the insecticide reaches its target without drifting outside the affected area. Quite apart from the cost, insecticide spraying always carries some risk of harm to other species, although modern insecticides are much safer than those used in the past.

A pesticide is any substance that is used to kill insects (insecticide), plants (herbicide), or fungi (fungicide), and a chemical company seeking to market a new pesticide nowadays has to satisfy the regulatory authorities of its safety in the environment. This requires a scientifically rigorous testing process conducted through several years and covering the effects of the product in every kind of environment it might reach. The relevant legislation differs in detail from one country to another, but a manufacturer seeking to export a new product must study all the national regulations and satisfy the most stringent.

There are also strict rules governing the use of pesticides, and in most countries these are legally enforceable. Not only must the pesticide itself be approved, so must the purposes for which it is used and the way it is applied. In many cases these regulations are very specific and restricted.

Foresters use insecticides to deal with infestations in commercial forests that can be controlled in no other way. Infestations may occur at any stage in the growth of the forest. Herbicides are used only to protect young seedlings. Once trees are established, weeds can no longer harm them.

Control by Bacteria

No one supposes that insecticides are popular, and alternatives are constantly being sought, but at least the modern chemicals sprayed in forests are fairly safe. Organophosphate compounds are often used. These do not persist in the environment, breaking down into harmless compounds, and they are only mildly poisonous to wildlife, but they are highly toxic to the insects against which they are used. There are also a few biological insecticides, the best known of which is based on *Bacillus thuringiensis*, a species of bacteria. A dusting powder containing *B. thuringiensis* is a very safe and effective insecticide, used widely in Canadian forests.

Bacillus and some other genera of bacteria survive adverse conditions by turning into spores, and as they form spores, some species produce crystals of protein. Protein crystals produced by *B. thuringiensis* are extremely poisonous if ingested by the caterpillars of many species of moths, but they are harmless to other animals and to plants. Any spores and crystals that are not eaten by caterpillars break down quite quickly. The bacterial spores can be cultured and dried on an industrial scale, and different bacterial strains have been developed for use against particular pests. Other bacteria are used in the same way, and so are viruses that cause fatal diseases in pest insects.

Genetic Engineering

Herbicides must be used with care because they may harm the tree they are meant to protect as well as the weeds competing with it. Depending on the species, a seedling needs to be protected against weeds for the first three to seven years if the tree is to grow rapidly and the plantation is to be highly productive. In recent years, the ability to modify the genetic constitution of the crop species has made herbicides more effective.

Trees can be genetically modified to make them resistant to broad-spectrum herbicides—herbicides that will kill most plants. This means that the herbicide can be applied at the time and in a dose that is most likely to kill competing plants with no risk of damaging the young trees. Weed control is made more effective, so the seedlings need to be sprayed less often. Fewer sprayings mean less herbicide is used, reducing costs and adverse effects on the environment. There are dangers, however. Among the weeds that compete with the crop, there are also woodland plants that provide food and shelter for a variety of animals. More effective use of herbicides might reduce biodiversity, and if shrubs and understory trees that compete with the crop also acquire resistance to the herbicide, the weed problem could become worse.

Plants known as Bt varieties have also been genetically modified to produce *Bacillus thuringiensis* toxin. The technique was first applied to farm crops, but Bt varieties of trees are now being developed. Bt plants are immune to attack from a number of serious pests. They do not need to be sprayed, and the amount of BT toxin they produce is very much smaller than the amount that would be applied by spraying. Inserting a gene to produce the insecticidal Bt protein is the first application of genetic engineering to the protection of trees, and there are others. For example, an elm tree has been engineered to render it resistant to Dutch elm disease.

Control by Management

Pest and disease problems can sometimes be reduced or even eliminated by good management. Modern pest control begins by acquiring a detailed knowledge of the life cycle of the pest. This reveals the stages at which a pest species is most likely to be vulnerable to attack, and sometimes it suggests a simple preventive method that involves no spraying at all. For example, the fungus *Cronartium ribicola* causes a serious disease called white pine blister on the branches of several commercially important pines. The vulnerable species include eastern white or Weymouth pine (*Pinus strobus*), western white pine (*P. monticola*), sugar pine (*P. lambertiana*), whitebark pine (*P. albicaulis*), limber pine (*P. flexilis*), and southwestern white pine (*P. strobiformis*). The fungus often attacks trees that have been damaged by a stem-feeding insect, *Pineus strobi,* although the insect causes little harm in itself. Although the fungus remains permanently in infected pine trees, the infection cannot spread directly from one pine tree to another. The only species *C. ribicola* can infect are currant and gooseberry bushes (*Ribes* species), and the infection spreads to pines from *Ribes*. So if *Ribes* species are never grown anywhere near pine trees, the disease is prevented. Scotch pine (*P. sylvestris*) suffers from a similar blister disease caused by the fungus *Coleosporium senecionis*. In this case the secondary hosts are groundsels (*Senecio* species), so efficient weeding prevents fungal infection.

Biological Control

Biological pest control is similar to control by management, but it involves manipulating the behavior of the pests rather than modifying the way the crop is grown. Females of many species of flying insects use chemical attractants, called pheromones, to attract males for mating, and sexually mature male bark beetles use a pheromone to attract females. Insect pheromones can form the basis for one method of biological pest control. If the pheromone to which a pest species responds can be identified and synthesized, it can be used to attract males or females into traps where they can be killed. Females of many insect species mate only once, and this offers another opportunity to manipulate behavior. Males are separated from colonies of pest insects bred for the purpose, sterilized, and then released. They can be relied on to follow the pheromone trail, but they mate unproductively.

With some species, pest numbers can also be held in check by encouraging their natural predators. The great spruce bark beetle (*Dendroctonus micans*) causes serious damage to all species of spruce (*Picea*), but there is another beetle, *Rhizophagus grandis,* that feeds on *D. micans*. It is bred for release into spruce forests as a biological control agent.

Plantation forests suffer fewer pest attacks and outbreaks of disease than do mature and old-growth forests, so it makes good sense to obtain timber and other forest products from plantations. This would allow the old-growth forests to be conserved, responding in their own way to the insects and fungi they harbor, while at the same time reducing the need for pest control. Forest nurseries, on the other hand, are treated differently from the plantations themselves. Nurseries are much more like horticultural enterprises or fruit orchards. So far as possible, seedlings must be kept free of pests and parasites to ensure that they carry no infection into a plantation.

Pesticides will continue to be used for some years to come, but the scale of their use should not be exaggerated. Forests are sprayed much less than farms. Of all the chemical pesticides used in Canada, only 2 percent are sprayed over forests, and this proportion, which is probably typical of most temperate regions, is unlikely to increase. The expansion of plantation forestry will reduce pest and disease outbreaks, and the development of resistant varieties and new pest control methods will reduce reliance on pesticides.

MODERN FORESTRY

Forestry is the commercial production of timber, small wood, and other materials from a forest, either by logging or by raising trees for the purpose. Silviculture is the care of every aspect of the forest, not only its trees, regardless of whether or not any part of the forest is being exploited commercially. Clearly, the two are different. For many years forestry was the primary concern of the owners of forested land in the temperate regions of the world. Today, although many forests remain commercial enterprises, the emphasis is moving strongly in the direction of silviculture. Attitudes change over the years. For a long time, the shift is gradual and passes unnoticed by people who are not directly involved. Then, apparently all of a

sudden, the new ideas are everywhere, and it is as though this is what everyone thought and believed all along. Usually, and certainly in the case of the transition from forestry to silviculture, several factors drive the process of change.

Forestry was born out of the realization that, first, all societies have always needed wood, people use it in countless ways, and there is no reason to suppose a time will ever come when wood is totally obsolete. Second, societies obtain their wood by felling trees. Third, if communities continue to fell trees when they need wood and that is all that they do, then a time is bound to come when the forests have been cleared from the land and there are too few trees left to meet the demand. This need not be so. Plant trees to replace those that are felled, or assist cleared areas of forest to regenerate naturally and establish plantations to satisfy as much as possible of the demand, then there will always be wood. Conservation measures aimed at ensuring a continuing supply of forest products were introduced in Europe in the Middle Ages (see "Managing the Forests" on pages 235–236). The application of resource management to forests is not a novel idea, although plantation forests are a much more recent invention. So far, however, the purpose remains economic. Resource management is an economic activity, and the trees themselves are evaluated according to the type, quality, and volume of wood they represent.

Difficulties arose when it came to implementing this type of management. Forests occupy land, and in many cases that land can be put to more profitable use. Farms are more profitable than forests, so forests have tended to be squeezed onto land that is unsuitable for agriculture, mainly at high elevations on exposed hillsides. Trees are plants, however, with physical requirements not much different from those of other plants. Conditions that are harsh for farm crops are also harsh for trees. It proved possible to grow trees in the hills, of course, but the enterprise was never highly profitable. In fact, the return on the capital invested was usually rather lower than the return offered by most investments. A sensible economist with no interest in fancy ideas about resources would have handed the better land to farmers and abandoned upland afforestation schemes, investing the money in a much sounder industrial undertaking.

Wildlife Conservation

Some other reason was needed for supporting the forests. Happily, one was already available, and it was supplied by the naturalists, comprising in Europe and North America literally millions of members of the societies, associations, clubs, and groups devoted to the study and protection of plants and animals. These folk enjoyed traveling into the countryside to observe wildlife, but they often found commercial forests closed to them and clearly being managed only for monetary profit. Naturally they complained, long and loud,

Forests provide many opportunities for peaceful recreation in beautiful surroundings. This boat carries visitors along the Grand River, Michigan, past forests resplendent in their fall colors.   *(Environmental Protection Agency)*

and eventually they were heard. When the new reason was needed, that was it. Forests were seen to exist for wildlife conservation as well as commercial production. From there it was but a short step to recognizing the amenity value of forests and opening them to the public for recreational use.

Management styles were modified, and most forest workers themselves welcomed the change. They live and work in the forests, after all, and it is not surprising that they are familiar with and care about the plants and animals that share the land with them. Some animals have to be excluded or otherwise controlled because they damage trees, but in general, forestry and conservation can function side by side. Silviculture is possible.

A few elementary changes can make a disproportionately big difference. Widening rides, which are the wide paths suitable for horse riding, and then leaving the edges to develop as long strips of habitat for wild plants has a major effect. This change can be made when subcompartments are thinned. If

thinning is done early and vigorously, sunlight will penetrate to patches on the forest floor long enough for many plants to establish themselves securely before the canopy closes and shades them. Nesting boxes for bats and certain bird species, strategically placed on trees, provide accommodation that may not be available in the trees themselves.

Naturalists used to point out that conifer plantations provide poorer wildlife habitat than broad-leaved stands. This is usually true, but the difference can be narrowed. Pine and larch have fairly open canopies, and other species, including spruces, can be thinned so that they allow light to penetrate. With good light penetration, by the time these trees mature, the wildlife around them is as diverse as that in many broad-leaved areas.

Apart from general modifications in management style, conservation of forest wildlife now begins at the initial planning stage. Among all the other information they yield, site surveys must also identify sites of particular importance not only for wildlife but also for sites of geological, archaeological, or other interest. Marked on the site maps, wildlife areas will be managed differently from the rest of the forest, perhaps with different mixtures of tree species growing to different ages. Operations near places that are important for other reasons will take care not to damage or, in some cases, obscure valuable features.

The Silvicultural Forest

Silviculture creates a rather different forest. Plantations no longer consist of endless straight rows of trees all of the same species and the same size, standing above the deep shade of a silent, bare floor. There is more light, more variety of tree species to provide visual interest, and a great many more flowers, birds, mammals, butterflies, and other wildlife.

Obviously, it is the forests within reach of population centers that attract visitors, and those are the forests in which facilities for visitors must be provided. Where there are few accessible forests, there are often plans to plant new ones. In Britain, which has a smaller forest area than many countries, such plans are well advanced. The new community forests will be very extensive and will contain only native species. They will not really be simulations of the original primeval forest and, even less, of the medieval forest. Rather, they will be forests of our time, existing for the enjoyment of people and for the benefit of wild plants and animals. They will be managed, of course, but the management will be silvicultural. England has 12 community forests, each inside or within easy access from a major urban area. Since 1990, approximately 25,000 acres (10,000 ha) of forest have been newly planted, and 67,000 acres (27,000 ha) of existing forest has been converted to community forest.

Elsewhere, in the more remote places, forests see few visitors. Recreational facilities are not needed in them, but they, too, are being affected by the change in management style, and those who work in them are as aware as anyone else of the desirability of conserving habitats. Wildlife conservation is at least as important there. Indeed, it may be more so, because the absence of people means less disturbance to natural habitats, and forests in remote regions often occupy a larger area than those closer to the cities, so they can accommodate animals that need large ranges in which to seek food.

Forests and Cultural Heritage

Throughout the temperate regions of the world, forests are now regarded as having great cultural importance. This arises from their role as communities of plants and animals. These are not just any plants and animals, however, but the ones that are featured in the folktales and fairy tales, the myths and legends of every nation. As well as being havens for wildlife, forests provide the settings for stories that are told and retold down the generations.

Obviously, the change in attitude is not yet complete. There are still regions in which old-growth broad-leaved and primeval coniferous forests are being cleared and not replaced, but overall the area of temperate forest is increasing (see "How Much Forest Is There?" on pages 255–256), and schemes involving permanent clearances arouse fierce opposition among local people.

A large proportion of the European forest was cleared centuries ago, and, more recently, major inroads were made into the North American forest. Europeans and Americans saw there was a risk of losing all of their forests. Britain actually paid for its loss in the form of the high cost of importing almost all its timber. The lesson was learned, and the losses began to be reversed. Provided we remember the lesson and continue to plant more trees than we fell, the temperate forests will continue to expand and prosper.

Conclusion

At one time, long ago, forest covered most of lowland Europe and large parts of North America away from the dry continental interior, as well as parts of Australia and New Zealand. Where the climate was mild and moist, the forest consisted wholly or mainly of broad-leaved species of trees. In the south, where summers are hot and dry, many of the broad-leaved trees were evergreen, and coniferous trees grew among them. Elsewhere the trees were deciduous, shedding their leaves in winter. The northern forests were mainly coniferous.

The forests developed and spread into higher latitudes as the climate grew warmer and wetter, starting at the end of the last ice age. When the first humans arrived in the temperate regions, they found vast, apparently endless forests, and that is where those people lived, hunting game, fishing in the rivers and lakes, and gathering wild fruits and roots. In their free time, the people told each other stories. Everyone enjoys listening to stories. It is part of what makes us human. Stories explain mysteries, and those ancient tales peopled the forest with spirits that eventually became fairies, witches, and other supernatural beings, some in the form of familiar animals but with the power of speech that had the power to intervene in people's lives for good or for ill. So there are two forests, one the natural forest inhabited by real trees, herbs, and animals and alongside it the mythical forest, a place of adventure, romance, danger, and magic. Both are extremely important.

This book has concentrated mainly on the real forest. It has described how a real forest works, how its climates bring rain and sunshine, warmth and cold. It has explained how trees grow, find nourishment from the air, water, and soil, and how they reproduce. It has provided brief descriptions of the trees most often found in temperate forests. Many plants and animals inhabit the forests. Ecology is the scientific study of the relationships between different species and between those species and their physical and chemical surroundings. The book has described the ecology of forests and also the way ecological ideas have developed over the years.

As soon as people moved into the forests, their history became interwoven with that of the plants and animals around them. People began to clear the forest, and, as farming replaced hunting and the gathering of wild food, the rate of clearance accelerated. The book gives a brief outline of this process. Finally, the book has described the present state of the world's temperate forests.

Modern forests are valued as places where people can find peace and take healthy exercise in safety. *(Fogstock.com)*

Agriculture is no longer expanding in the temperate regions. Modern American, European, Australian, and New Zealand farms produce as much food and fiber as world markets need. There is space once more for the forests, and the forests are expanding. Forests supply the world with timber, small wood, and firewood, of course, but they provide much more than raw materials and fuel. Today they are valued for the wildlife they sustain and for the opportunities they provide for quiet recreation and relaxation, a temporary escape from the noise and rush of city life, and as city dwellers find tranquillity in the forests, so artists are beginning to decorate the forests with sculptures. The forests are once again becoming places of stories and wonders.

The temperate forests will not disappear, but neither will greed. Developers will be tempted by the profits to be made from clearing the trees and building on land within reach of urban centers. Country clubs and golf courses will always generate more income than forests that are open to everyone with free access. The forests can continue to expand, but they will do so only for as long as people use them, appreciate them, and vigorously oppose any attempt to remove them.

Tables

Proportion of Land Area Forested in Selected Countries

COUNTRY	PERCENT OF LAND AREA	COUNTRY	PERCENT OF LAND AREA
Finland	73.9	Italy	33.9
Japan	68.2	Canada	33.6
Sweden	66.9	Lithuania	33.5
Slovenia	62.8	United States	33.1
Estonia	53.9	Germany	31.7
Russian Federation	47.9	Norway	30.7
Latvia	47.4	Poland	30.0
Austria	46.7	Greece	29.1
Bosnia and Herzegovina	43.1	France	28.3
Portugal	41.3	Belgium	22.0
Slovakia	40.1	United Kingdom	11.8
Georgia	39.7	Netherlands	10.8
Spain	35.9	Ireland	9.7

Source: Food and Agriculture Organization (FAO), *Global Forest Resources Assessment 2005*.

Area of Northern Hemisphere Temperate Forest

	SQUARE MILES	SQUARE KILOMETERS
North America	2,615,011	6,774,640
Europe	3,903,981	10,113,940
Australia and New Zealand	663,870	1,719,870
Total temperate forest	7,182,862	18,608,450
Total world forest (all types)	39,509,974	102,357,447

Source: Food and Agriculture Organization (FAO), *Global Forest Resources Assessment 2005*.

Roundwood Production from Temperate Forests 2000

	TOTAL (million ft³; million m³)	FUELWOOD AND CHARCOAL (million ft³; million m³)	INDUSTRIAL ROUNDWOOD (million ft³; million m³)
Europe	18,530.0; 524.8	2,841.3; 80.5	11,881.4; 336.5
Russia	5,582.5; 158.1	1,847.8; 52.3	3,734.7; 105.8
North America	24,225.9; 686.1	2,715.0; 76.9	21,511.0; 609.2
Australia and New Zealand	144.4; 40.9	95.6; 2.7	1,348.3; 38.2
Total	48,482.8; 1,409.9	7,499.7; 212.4	38.475.4; 1,089.7

Source: Food and Agriculture Organization (FAO), *Yearbook of Forest Products* and Encyclopaedia Britannica, *2006 Book of the Year.*

SI Units and Conversions

UNIT	QUANTITY	SYMBOL	CONVERSION
Base Units			
meter	length	m	1 m = 3.2808 feet
kilogram	mass	kg	1 kg = 2.205 pounds
second	time	s	
ampere	electric current	A	
kelvin	thermodynamic temperature	K	1 K = 1°C = 1.8°F
candela	luminous intensity	cd	
mole	amount of substance	mol	
Supplementary Units			
radian	plane angle	rad	$\pi/2$ rad = 90°
steradian	solid angle	sr	
Derived Units			
coulomb	quantity of electricity	C	
cubic meter	volume	m³	1 m³ = 1.308 yards³
farad	capacitance	F	
henry	inductance	H	
hertz	frequency	Hz	
joule	energy	J	1 J = 0.2389 calories
kilogram per cubic meter	density	kg/m³	1 kg/m³ = 0.0624 lb/ft³
lumen	luminous flux	lm	
lux	illuminance	lx	
meter per second	speed	m/s	1 m/s = 3.281 ft/s
meter per second squared	acceleration	m/s²	
mole per cubic meter	concentration	mol/m³	
newton	force	N	1 N = 7.218 lb force
ohm	electric resistance	Ω	
pascal	pressure	Pa	1 Pa = 0.145 lb/in²
radian per second	angular velocity	rad/s	
radian per second squared	angular acceleration	rad/s²	

(table continues)

SI Units and Conversions (continued)

UNIT	QUANTITY	SYMBOL	CONVERSION
Derived Units			
square meter	area	m^2	1 m^2 = 1.196 yards^2
tesla	magnetic flux density	T	
volt	electromotive force	V	
watt	power	W	1W = 3.412 Btu/h
weber	magnetic flux	Wb	

Prefixes attached to SI units alter their value.

Prefixes Used with SI Units

PREFIX	SYMBOL	VALUE
atto	a	$\times 10^{-18}$
femto	f	$\times 10^{-15}$
pico	p	$\times 10^{-12}$
nano	n	$\times 10^{-9}$
micro	μ	$\times 10^{-6}$
milli	m	$\times 10^{-3}$
centi	c	$\times 10^{-2}$
deci	d	$\times 10^{-1}$
deca	da	$\times 10$
hecto	h	$\times 10^{2}$
kilo	k	$\times 10^{3}$
mega	M	$\times 10^{6}$
giga	G	$\times 10^{9}$
tera	T	$\times 10^{12}$

Glossary

abscission   the rejection by a plant of a branch or leaves

abscission layer   a layer of cells at the end of the leafstalk of a deciduous tree or shrub where the leaf is detached and shed

acicular   needle-shaped

acid   a substance that liberates hydrogen ions and accepts electrons; it has a pH of less than 7

adherence   the attachment of water molecules to the sides of a container

adiabatic   describes a change in temperature that involves no exchange of heat with an outside source.

albedo   a measure of the amount of light a surface reflects

alga   a single-celled or multicelled organism (protist) that performs PHOTOSYNTHESIS; seaweeds are algae

alkali   a substance that accepts hydrogen ions and releases electrons; it has a pH greater than 7

alternation of generations   the two types of individual that occur in the life cycle of a plant; the DIPLOID generation is called a SPOROPHYTE, and the HAPLOID generation a GAMETOPHYTE

angiosperm   a flowering plant

anther   the structure in a male flower, located at the tip of a FILAMENT, where POLLEN is produced

aquifer   an underground body of permeable material (e.g., sand or gravel) lying above a layer of impermeable material (e.g., rock or clay) that is capable of storing water and through which GROUNDWATER flows

archegonium   a female sex organ found in liverworts, mosses, ferns, and most GYMNOSPERMS

base   a substance that will react with an acid to form a SALT

beating up   replacing trees that have died

biodiversity   biological diversity at all levels, but especially diversity of species

biogeography   the study of the distribution of plants

biomass   the total mass of all the living organisms present in a specified area

bolling   the trunk of a pollarded tree

boreal   pertaining to the north

boundary current   an ocean current that flows parallel to the eastern or western coast of a continent

bract   a modified leaf that forms part of a flower

brash   *see* BRASHING

brashing   removing small branches and leaves (brash) from a tree trunk before the tree is removed from the forest

buffering   the resistance to a change in pH following the addition of an ACID or ALKALI that occurs in a solution (a buffer) of a weak acid and its weak BASE

callus   tissue that forms over a plant wound or that develops in a tissue culture that is dividing rapidly

calyx   the collective name for the SEPALS

cambium   a layer of cells in the stem, branches, and roots of woody plants, located between the XYLEM and PHLOEM, that divide to produce new xylem and phloem; cell division continues throughout the life of the plant

canopy   the forest cover, more or less shading the floor, formed by the touching and overlapping branches and foliage of adjacent trees

capillarity (capillary attraction)   the movement of water against gravity through a fine tube or narrow passageway

capillary attraction   *see* CAPILLARITY

capillary fringe   the region immediately above the WATER TABLE into which water rises by CAPILLARITY

carpel   the female reproductive organs of a plant, comprising the STIGMA, STYLE, and OVARY

catena   a sequence of soils of the same age and usually on the same PARENT MATERIAL that is repeated across a section of landscape

chaparral   SCLEROPHYLLOUS scrub found in western California

chemical weathering   WEATHERING that results from chemical reactions between rocks and substances carried in the soil solution

chitin   a strong, lightweight material that is a major component of the cell walls of fungi (and of the exoskeleton of insects)

chlorophyll   the pigment present in the leaves and sometimes stems of green plants that gives them their green color; chlorophyll molecules trap light, thus supplying the energy for PHOTOSYNTHESIS

chloroplast   the structure in plant cells that contains CHLOROPHYLL and in which PHOTOSYNTHESIS takes place

chromosome   one of the threads of DNA found in the nucleus of a cell. Each chromosome carries some of the organism's genes, and the full complement of chromosomes (a number that varies according to species) contains the full set of genetic instructions (the genome) for that organism

climax   the final stage in a plant SUCCESSION in which the plant community reaches a stable equilibrium with the environment

clone   a group of cells or individuals that are genetically identical

codominant   a tree that is shorter than the DOMINANTS but belongs to the same species; should a dominant fall, a codominant will grow up to take its place

coherence   the attraction between water molecules that draws water up the side of a container

cold lightning   lightning that releases heat beneath the bark of a tree but does not sustain the heat long enough to set the tree on fire

cold pole   one of the places that experiences the lowest mean temperatures on Earth

cold wave   a sudden and large fall in temperature; in most of the United States, it is a fall in temperature of at least 20°F (11°C) within a period of not more than 24 hours that reduces the temperature to 0°F (-18°C) or lower, and in California, Florida, and the Gulf Coast states, it is a fall in temperature of at least 16°F (9°C) to a temperature of 32°F (0°C) or lower

colpus   an aperture or groove on the surface of a POLLEN grain

compensation point   the intensity of sunlight at which the amount of carbon dioxide a plant absorbs in PHOTOSYNTHESIS is precisely equal to the amount it releases in RESPIRATION

compound leaf   a leaf consisting of two or more parts

cone   the INFLORESCENCE of a member of the Coniferophyta

coniferous   bearing cones

conservative plate margin   a boundary between two tectonic plates at which the plates are moving past each other

constructive plate margin   a boundary between two tectonic plates at which the plates are moving apart and new rock is solidifying between them

continentality   the extent to which a climate resembles the most extreme type of continental climate (dry, hot in summer, cold in winter)

continental drift   the movement of continents across the Earth's surface

continental shelf   the gently sloping part of a continent that lies below sea level

convection current   a vertical movement of a fluid caused by heating from below

coppice   the practice of cutting trees close to ground level in order to stimulate the growth of poles; an area of woodland that has been managed by coppicing

CorF   see CORIOLIS EFFECT

Coriolis effect (CorF)   the deflection due to the Earth's rotation experienced by bodies moving in relation to the Earth's surface; bodies are deflected to the right in the Northern Hemisphere and to the left in the Southern Hemisphere

cork (phellem)   A layer of protective tissue, comprising dead cells, that forms immediately beneath the outer surface of the trunk and branches of a woody plant

cotyledon   a seed leaf; the leaf that emerges from a germinating seed

crust   the solid, uppermost layer of the Earth

cuticle   a thin, waxy, outer coat that protects the leaves and stem of a plant

cyanobacterium   a bacterium that possesses CHLOROPHYLL and that practices PHOTOSYNTHESIS

deciduous   describes parts of a plant or animal that are shed at the same season each year

dendrochronology   the scientific dating of material by counting annual growth rings in trees

dendroclimatology   the reconstruction of past climates through the study of annual growth rings in trees

dendroecology   the study of the relationships between patterns in the growth rings of trees and the ecological factors that may have caused them

desert   a region in which, for most of the time, less precipitation falls that could evaporate during the same period;

the amount of precipitation needed for a desert to develop varies according to the temperature, but a desert is likely anywhere that the annual precipitation is less than 10 inches (250 mm) a year and is highly erratic

destructive plate margin   a boundary between two tectonic plates where the plates are colliding

diagnostic horizon   a SOIL HORIZON that is used to identify the type of soil

dioecious   possessing male and female reproductive organs on separate plants

diploid   describes a cell containing two sets of CHROMOSOMES, or an organism made up of such cells

disjunct distribution   the occurrence of related species in places separated by major geographical barriers, such as an ocean or a mountain range

dominant   the species comprising the largest plants in a community, or the species with the most influence of the character of the community

ecology   the scientific study of the relationships among organisms inhabiting a specified area and between the organisms and the physical and chemical conditions in which they live

ecosystem   a clearly defined area or unit within which living organisms and their physical and chemical surroundings interact to form a stable system

ecotone   a narrow, well-defined transition zone between two ECOSYSTEMS

effective precipitation   the amount of precipitation after deducting the amount that evaporates; it is the amount of precipitation that is available to plants

embryo   a young plant contained within a plant seed, or a young animal contained within a fertilized egg or other reproductive structure. (In humans an embryo is called a fetus after the first eight weeks of pregnancy)

emergent   a forest tree that stands taller than those around it

endosperm   a store of food found in the SEEDS of many ANGIOSPERMS

endotherm   a animal that maintains a fairly constant internal body temperature by physiological means (e.g., shivering, sweating)

epiphyte   a plant that grows on the surface of another plant, using that plant only for physical support

equinox   March 20–21 and September 22–23 when the noonday Sun is directly overhead at the equator and day and night are of equal length everywhere in the world

evapotranspiration   evaporation and   TRANSPIRATION considered together

evergreen   describes a plant that bears leaves at all times of year; although it sheds its leaves, it does not shed all of them at the same time

exine   the tough outer coat of a POLLEN grain

exotherm   *see* POIKILOTHERM

exotic   describes a species that has been introduced; a species not native to the area in which it is found

fastigiate   with a conical or tapering shape

Ferrel cell   part of the general circulation of the atmosphere in which air rises in about latitude 60° in both hemispheres, moves toward the equator at high altitude, subsides in about latitude 30°, and flows away from the equator at low level

fertilization   the union of male and female GAMETES to produce an EMBRYO

field capacity   the water that remains in the soil after excess moisture has drained freely from that soil

filament   the stalk of the STAMEN of a flower, bearing an ANTHER at its tip

floodplain   the part of a river valley that is periodically flooded and is covered by loose sediment carried by the river

floret   an individual flower forming part of an INFLORESCENCE

floristics   *see* PHYTOGEOGRAPHY

forb   a herbaceous plant other than a grass

forest   a plant formation composed of trees with crowns that touch to form a closed CANOPY; the trees on an area of land that is forested

fungus   an organism belonging to the kingdom Fungi, comprising nonphotosynthesizing organisms that feed by absorbing organic substances from their surroundings and reproduce by SPORES

gamete   a sex cell, that is, a spermatozoon or an ovum

gametophyte   the HAPLOID stage in the life cycle of a plant. In simple plants, such as mosses, the gametophyte is the visible plant; in GYMNOSPERMS and ANGIOSPERMS, the gametophyte is inconspicuous

geostrophic wind   the wind that blows approximately parallel to the ISOBARS above the layer where air movement is influenced by surface features

girdling (ring-barking)   making a cut all the way around the trunk of a tree that is deep enough to penetrate the bark and sever the VASCULAR TISSUE, thereby killing the tree by preventing the transport of water and nutrients

glacial   a period when polar ice sheets advance; an ice age

glaciation   an ice age

glacier   a large mass of ice that is attached to the surface but that flows

global warming potential   the capacity of a greenhouse gas to absorb infrared radiation compared with the capacity of carbon dioxide, which is given a value of 1

Gondwana   the former SUPERCONTINENT of the Southern Hemisphere that comprised South America, Africa, Madagascar, India, Sri Lanka, Australia, New Zealand, and Antarctica

gravitational water   water that moves downward through the soil under the influence of gravity

greenhouse effect   the absorption and reradiation of long-wave radiation emitted by the Earth's surface by molecules of water vapor, carbon dioxide, ozone, and several other greenhouse gases, warming the air

greenhouse gas   *see* GREENHOUSE EFFECT

groundwater   underground water that flows through an AQUIFER

guyot   a flat-topped submarine mountain

gymnosperm   a seed plant in which the OVULES are carried naked on the scales of a CONE; coniferous (i.e., cone-bearing) trees are the most abundant gymnosperms

gyre   a circular or spiral system of ocean currents

habitat   the living place of a species or community

Hadley cell   the tropical part of the general circulation of the atmosphere; air rises over the equator, moves away from the equator at high altitude, subsides over the subtropics, and flows toward the equator at low altitude

haploid   describes a cell nucleus that contains only one set of CHROMOSOMES

hardwood   the wood from broad-leaved trees

heartwood   the central part of a tree trunk, made from dead cells

heliosis   *see* SOLARIZATION

herb   a small, nonwoody plant in which all the parts above ground die back at the end of each growing season

hot lightning   lightning that sustains its current long enough to generate sufficient heat to ignite material that it strikes

humus   decomposed plant and animal material in the soil

hydrogen bond   the attraction that links molecules in which hydrogen is bonded to nitrogen, oxygen, or fluorine

hypha   one of the minute threads that form the main part of a fungus

ice storm   wind-driven rain that falls through air and onto surfaces below freezing temperature where it freezes on contact, forming thick layers of ice

inflorescence   a mass of small but complete flowers (called florets) growing together and giving the appearance of a single flower; sunflower and grass flowers are inflorescences

integuments   the coats of the OVULE in ANGIOSPERMS

interglacial   a period of warmer weather between two GLACIALS

interstadial   a time of warmer weather lasting 1,000–2,000 years during a GLACIAL

isobar   a line drawn on a weather map to join points on a surface (not necessarily the ground surface) of equal air pressure

iteroparity   having more than one reproductive cycle in the course of a lifetime

jet stream   a winding ribbon of strong wind about 5–10 miles (8–16 km) above the surface. Jet streams are typically thousands of miles long, hundreds of miles wide, and several miles deep

krummholz   stunted, gnarled, small trees that grow on mountainsides between the upper limit for forest and the TREE LINE

K-selection   natural selection favoring species living in a stable environment; they produce few offspring and devote much care to their young; most of the young survive

lapse rate   the rate at which the air temperature decreases (lapses) with increasing altitude; in unsaturated air the dry ADIABATIC lapse rate is 5.38°F per thousand feet (9.8°C/1,000 m); in saturated air the saturated adiabatic lapse rate varies but averages 2.75°F per thousand feet (5°C/1,000 m)

latent heat   the heat energy that is absorbed or released when a substance changes phase between solid and liquid, liquid and gas, and solid and gas; for water at 32°F (0°C), the latent heat of melting and freezing is 80 cal./g (334 J/g); of vaporization and condensation, 600 cal./g (2,501 J/g); and for sublimation and deposition, 680 cal./g (2,835 J/g)

Laurasia   the former SUPERCONTINENT that comprised North America, Greenland, Europe, and Asia

leaching   the removal of soil materials in solution

legume   a plant belonging to the bean family (Fabaceae), most of which have nodules attached to their roots containing colonies of bacteria that fix atmospheric nitrogen

liana   a free-hanging climbing plant

lichen   a composite organism comprising a fungus and an ALGA or CYANOBACTERIUM

lichen forest   open coniferous woodland where the ground is mostly covered by lichen

lignification   the process by which LIGNIN accumulates in the cells of woody plants after the cells die

lignin   a hard substance that forms in the cells of woody plants, binding other cell components together and remaining in position after the cells have died

living fossil   a member of a species that is almost identical to a species known from the fossil record, implying that it has change little in a very long time

loess   wind-deposited sediment mainly comprising quartz particles 0.0006–0.002 inch (0.0015–0.005 cm) in size

lumen   a space or opening enclosed by cell walls

mantle   that part of the Earth's interior lying between the outer edge of the inner core and the underside of the crust

maquis   drought-resistant Mediterranean scrub

megaphyll   a large leaf with branched VEINS

megaspore   the first cell of the female GAMETOPHYTE generation

megasporocyte   the DIPLOID cell in an OVULE that divides to form HAPLOID cells, one of which becomes the MEGASPORE

meiosis   a form of cell division that occurs in sexually-reproducing organisms, in which the cell divides twice, producing four HAPLOID daughter cells

meristem   plant tissue composed of cells that are capable of dividing indefinitely

mesophyll   the tissue lying just below the surface of a leaf, where PHOTOSYNTHESIS takes place

micropyle   a canal in the covering of the NUCELLUS through which the POLLEN tube passes during FERTILIZATION

mitosis   cell division in which the cell divides to produce two identical daughter cells, both of which are DIPLOID

monoecious   having separate male and female organs on the same individual plant

mutualism   a close relationship between members of two species that benefits both

mycelium   the mass of hyphae (*see* HYPHA) that make up the main part of a fungus

mycorrhiza   a close physical association between a fungus and the roots of a plant from which both organisms benefit

natural   describes a community of plants that occur without human intervention (i.e., they have not been introduced)

nectar   a liquid secreted by a nectary that is up to 60 percent sugar

niche   the function an organism performs in its environment

nucellus   the tissue in a plant OVULE that contains the EMBRYO sac

oceanicity   the extent to which a climate resembles the most extreme kind of maritime climate

old-growth forest   a forest that existed prior to European settlement and over which humans are believed to have exerted little or no influence.

osmosis   the movement of water or another solvent from a region of low concentration to a region of higher concentration across a membrane that permits the passage of solvent molecule but not molecules of the solute.

ovary   the female reproductive organs of a flower

ovule   the structure in ANGIOSPERMS and GYMNOSPERMS that develops into the seed following FERTILIZATION

paleobotany   the study of fossil plants

paleoecology   the study of fossils and sedimentary deposits using ecological principles and concepts in order to determine past environmental conditions

palynology   the study of living and fossil POLLEN grains, SPORES, and certain other microfossils

Pangaea   the SUPERCONTINENT that came into existence about 260 million years ago and began to break apart about 220 million years ago

Panthalassa   the world ocean that surrounded PANGAEA

parasite   an organism that lives on or inside the body of a host organism (which is usually larger), from which it obtains food, shelter, or some other necessity

parenchyma   plant tissue composed of unspecialized cells

parent material   the original material from which a soil has developed

partial pressure   in a mixture of gases, the proportion of the total pressure that can be attributed to one of the constituent gases

pedogenesis   the natural process of soil formation

pedology   the scientific study of soils

peduncle   the stalk attaching an INFLORESCENCE to a plant

permafrost   permanently frozen ground; to become permafrost, the ground must remain frozen throughout a minimum of two winters and the summer between

permeability   the ability of a material to allow water to flow through it

petiole   the stalk that attaches a leaf to the stem of a plant

pH   a value between 0 and 14 that gives a measure of the acidity or alkalinity of a substance; a neutral medium has a pH of 7, a pH lower than 7 indicates acidity and a pH higher than 7 indicates alkalinity

phellem   *see* CORK

phloem   tissue through which the products of PHOTOSYNTHESIS and hormones are transported from the leaves to all parts of a vascular plant

photorespiration   a light-activated process that occurs in the CHLOROPLASTS of many plants in which oxygen is absorbed and carbon dioxide released, but without providing the plant with energy

photosynthesis   the sequence of chemical reactions in which green plants and CYANOBACTERIA use sunlight as a source of energy for the manufacture (synthesis) of sugars from hydrogen and carbon, obtained from water and carbon dioxide respectively; the reactions can be summarized as: $6CO_2 + 6H_2O + light \rightarrow C_6H_{12}O_6 + 6O_2\uparrow$, the upward arrow indicating that oxygen is released into the air; $C_6H_{12}O_6$ is glucose, a simple sugar

physical weathering   WEATHERING due to the action of wind, rain, and ice

phytogeography (floristics)   the study of the geographic distribution of plants

pinna   one of the leaflets of a PINNATE leaf

pinnate   describes a COMPOUND LEAF comprising leaflets arranged on either side of a RACHIS

pinnule   one of the smallest divisions of a fern frond

plate tectonics   the theory holding that the Earth's crust comprises a number of rigid sections, or plates, that move in relation to each other

poikilotherm (exotherm)   an animal whose body temperature varies with that of its surroundings (e.g., a fish)

polar cell   part of the general circulation of the atmosphere in which air subsides over the North and South Poles, moves away from the poles at low level, rises in about latitude 60°, and flows back toward the poles at high altitude

polar molecule   a molecule in which the electrons are shared unequally among the nuclei, so that although the molecule carries no net electric charge, one side of the molecule carries a positive charge and the opposite side carries a negative charge; water molecules are polar

pollarding   cutting off the top of a tree about 6 feet (1.8 m) above ground level to produce a crop of poles that emerge too high for browsing animals to reach

pollen   the grains containing male sex cells that are produced in the ANTHERS of flowers

pore space   the total interconnected space between the mineral particles in a soil

porosity   the percentage of the total volume of a material that consists of spaces between particles

potential temperature   the temperature air (or any other fluid) would have if it were subjected to sea-level pressure and its temperature changed ADIABATICALLY as the pressure adjusted

predator   an organism that obtains food by consuming and usually killing another organism

primeval forest   forest that has never been subjected to human influence

rachis   the axis that bears the flower of a plant or the leaflets of a COMPOUND LEAF

radicle   the part of a plant EMBRYO that will grow into a root

receptacle   the part of a flower stalk from which all parts of the flower arise

refugium   an isolated area in which plants and animals survive major climatic changes taking place elsewhere

regolith   unconsolidated, weathered material, including rock fragments and mineral grains, that lies on top of unaltered, solid rock

relict   an organism that has survived while related species became extinct

respiration   the sequence of chemical reactions in which carbon in sugar is oxidized with the release of energy; the opposite of PHOTOSYNTHESIS; the reactions can be summarized as: $C_6H_{12}O_6 + 6O_2 \rightarrow 6CO_2 + 6H_2O + energy$; $C_6H_{12}O_6$ is glucose, a simple sugar

rhizosphere   the area of soil immediately surrounding plant roots

ribosome   a granule composed of RNA and protein that is the site of protein synthesis in all types of cells; all cells contain many ribosomes

ring-barking   *see* GIRDLING

root plate   the roots of a tree, together with the soil attached to them, that are exposed as a flat, circular, platelike structure when a tree is blown down by wind

Rossby wave   a wave with a wavelength of 2,485–3,728 miles (4,000–6,000 km) that develops in moving air in the middle and upper TROPOSPHERE

r-selection   natural selection that maximizes the rate of increase of a population inhabiting an unstable environment; r-species produce many offspring and devote little attention to them; most of the offspring die, but enough survive to continue the species, and should conditions improve, many more will survive and the population will increase

salt   the product of the chemical reaction, in water, between an ACID and a BASE

sap   the substance that exudes from ruptured VASCULAR TISSUE or PARENCHYMA

saprobe   an organism that feeds on dead plant or animal matter

sapwood the active, living part of the trunk or branch of a woody plant, lying immediately beneath the bark

scarification breaking up the soil and BRASH and scattering the brash in preparation for planting the next tree crop in a plantation

sclerophyllous applied to evergreens with small leaves that are often hard, leathery, or stiff, and sometimes have prickles around the edges, thick bark, and well-protected buds

seafloor spreading the theory that the ocean floor is created at ridges where MANTLE material rises to the surface and the crustal rocks move away from the ridges on either side, causing the ocean basin to widen as the seafloor spreads

seamount an isolated submarine mountain that rises more than 3,300 feet (1,000 m) above the ocean floor

sedimentary rock rock formed from particles that have settled to the seabed and later been compressed

seed the body, formed from a fertilized OVULE, from which a young plant emerges

seismic pertaining to earthquakes

selective logging harvesting individual trees

semelparity having only one reproductive cycle in the course of a lifetime

sepal one of the modified leaves attached to the RECEPTACLE that enclose the flower bud and surround the petals of a flower after it opens

serotiny the retention of seeds by a plant until conditions favorable for their germination triggers their release

shrub a perennial woody plant less than 33 feet (10 m) tall with several main stems arising at or close to ground level, but with no clearly identifiable trunk

sieve element one of the long, slender, tapering cells, terminating in a perforated region called a SIEVE PLATE, that join end to end in a sieve tube, forming part of the PHLOEM tissue

sieve plate the perforated region at the end, and occasionally the side, of a sieve cell

snag the standing trunk of a tree that has snapped off above ground level

snedding removing the branches from felled trees

softwood the wood from coniferous trees

soil horizon a horizontal layer in a SOIL PROFILE that differs in its mineral or organic composition from the layers above and below it, and from which it can be clearly distinguished visually

soil moisture tension the upward force acting on water held in a container

soil profile a vertical section cut through a soil from the surface to the underlying rock

solarization (heliosis) the inhibition of PHOTOSYNTHESIS when the intensity of light is very high

solstice one of the two dates each year when the noonday Sun is directly overhead at one or other of the Tropics and the difference in length between the hours of daylight and darkness is at its most extreme; the solstices occur on June 21–22 and December 22–23

sorus a fruiting body that consists of a mass of SPORES or sporangia (*see* SPORANGIUM)

specific heat capacity the amount of heat that must be applied to a substance in order to raise its temperature by one degree, it is measured in calories per gram per degree Celsius (cal/g/°C) or in the scientific units of joules per gram per kelvin (J/g/K; 1K = 1°C = 1.8°F)

sporangium a saclike structure in which fungal SPORES are formed

spore a reproductive unit, usually consisting of a single cell, that can develop into a new organism without fusing with another cell

sphorophyll a leaf that bears sporangia (*see* SPORANGIUM); sporophylls occur in clubmosses, *Equisetum,* and ferns

sporophyte the reproductive stage in the life cycle of a plant; in GYMNOSPERMS and ANGIOSPERMS, this is the dominant stage, comprising the visible plant; in mosses, liverworts, and hornworts, the sporophyte is small and inconspicuous

stadial a prolonged period of cold weather that is shorter and milder than a GLACIAL

stamen the male reproductive organ of a flower, comprising the FILAMENT and ANTHER

stigma part of the female reproductive structure in a flower; it has a sticky surface that holds POLLEN grains

stomata (sing. stoma) small openings, or pores, on the surface of a plant leaf through which the plant cells exchange gases with the outside air; stomata can be opened or closed by the expansion or contraction of two guard cells surrounding each stoma

stratosphere the region of the atmosphere that extends from the TROPOPAUSE to an altitude of about 31 miles (50 km)

style the part of the female reproductive structure in a flower that connects the STIGMA to the OVARY

subdominant trees belonging to the same species as the DOMINANTS and CODOMINANTS, but smaller than either

subduction   the sinking of one crustal plate beneath another at a DESTRUCTIVE PLATE MARGIN

succession   a sequence of changes in the composition of a plant and animal community occupying a site that continues until a stable equilibrium, the CLIMAX, is attained

supercontinent   a land mass formed by the merging of previously separate continents as a result of CONTINENTAL DRIFT; PANGAEA was a supercontinent comprising all the present-day continents

superorganism   a group or community of organisms that possesses some of the features of a single organism; the most extreme example is found in some interpretations of the Gaia hypothesis, which hold that all the living organisms on Earth comprise a single superorganism maintaining environmental conditions favorable to life as a whole

suture   a line of deformed rocks that are believed to mark the boundary where two continents have collided

swidden farming   a farming method in which farmers clear forest from an area of land and grow crops there for several seasons before abandoning the site and moving to another, allowing the forest to regenerate

taiga   the coniferous forest forming a belt across northern America and Eurasia

taxonomy   the scientific classification of organisms and, by extension, of soils and clouds

thermal equator   the region, between latitudes 23°N and 10–15°S, where the surface temperature is highest; its mean position is about 5°N

timberline   the boundary below which there is forest and above which there are no trees at all

tracheid   a long, cylindrical cell with a tapering, perforated end; tracheids join end to end to form the XYLEM tissue in GYMNOSPERMS

trade winds   the winds that blow toward the equator in equatorial regions, from the northeast in the Northern Hemisphere and from the southeast in the Southern Hemisphere

tramp species   species that have spread around the world as a result of human commerce

transform fault   a type of fault that occurs in rocks on the ocean floor where two adjacent crustal plates meet at a CONSERVATIVE PLATE MARGIN; compared with similar faults on land, the direction of movement is reversed, or "transformed"

transpiration   the evaporation of water through leaf STOMATA when these are open for the exchange of gases

tree   a perennial, woody plant that is more than 33 feet (10 m) tall and has one or more than one clearly identifiable trunk

tree line   the elevation or latitude beyond which the climate is too severe for trees to grow

trophic   pertaining to food or feeding

Tropics   two lines of latitude at 23.5°N (tropic of Cancer) and 23.5°S (tropic of Capricorn) where the Sun is directly overhead at noon at one of the SOLSTICES; that part of the Earth lying between the Tropics

tropopause   the boundary separating the TROPOSPHERE from the STRATOSPHERE; it occurs at a height of about 10 miles (16 km) over the equator, 7 miles (11 km) in middle latitudes, and 5 miles (8 km) over the North and South Poles

troposphere   the layer of the atmosphere that extends from the surface to the TROPOPAUSE; it is the region where all weather phenomena occur

tundra   a treeless plain in the Arctic or Antarctic where the vegetation is dominated by grasses, sedges, rushes, and wood rushes, together with dwarf shrubs, LICHENS, and mosses

turgor   rigidity of plant tissues due to water held under pressure in the cells

vascular tissue   plant tissue through which water and nutrients are transported

vein   VASCULAR TISSUE in a leaf in the form of one or more vascular bundles of PHLOEM and XYLEM tissue lying parallel to each other and very close together

vessel element   one of the cells forming the XYLEM tissue in ANGIOSPERMS

vorticity   the tendency of a mass of fluid that is moving in relation to the Earth's surface to rotate about a vertical axis

water pressure   pressure exerted below the surface of water by the weight of overlying water

water stress   the physiological effect on a plant that occurs when it is obtaining insufficient water from the soil or water is evaporating from its leaves faster than the vascular system can replace it; the wilting of leaves is the first symptom in broad-leaved trees

water table   the upper margin of the GROUNDWATER; soil is fully saturated below the water table but unsaturated above it

weathering   the breaking down of rocks at or near the Earth's surface as a result of physical and chemical processes; these include the action of wind and water, and the expansion of water as it freezes, and chemical reactions between rock constituents and substances dissolved in rainwater or released by living organisms

xeromorphic resistant to drought

xylem plant tissue through which water entering at the roots is transported to all parts of the plant

zonal index the strength of the winds blowing from west to east between latitudes 33° and 55° in both hemispheres

expressed either as the horizontal pressure gradient or as the corresponding GEOSTROPHIC WIND

zygote the fertilized ovum (egg) of a plant or animal at the stage where it is DIPLOID, but before it has begun to divide

Bibliography and Further Reading

Allaby, Michael. *A Change in the Weather.* New York: Facts On File, 2004.

———. *Fog, Smog, and Poisoned Rain.* New York: Facts On File, 2003.

———. *Basics of Environmental Science.* 2nd ed. New York: Routledge, 2000.

Ashman, M. R., and G. Puri. *Essential Soil Science: A Clear and Concise Introduction to Soil Science.* Cambridge, Mass.: Blackwell Science, 2002.

Brewer, Richard. *The Science of Ecology.* 2nd ed. Fort Worth, Tex.: Saunders College Publishing, 1998.

Burroughs, William James. *Climate Change: A Multidisciplinary Approach.* New York: Cambridge University Press, 2001.

Emiliani, Cesare. *Planet Earth: Cosmology, Geology, and the Evolution of Life and Environment.* New York: Cambridge University Press, 1995.

Foth, H. D. *Fundamentals of Soil Science.* 8th ed. New York: John Wiley, 1991.

Gaston, Kevin J., ed. *Biodiversity: A Biology of Numbers and Difference.* Cambridge Mass.: Blackwell Science, 1996.

Heywood, V. H., R. K. Brummitt, A. Culham, and O. Seberg. *Flowering Plant Families of the World.* Richmond, U.K.: Royal Botanic Gardens, Kew, 2007.

Hibberd, B. G. *Forestry Practice.* 11th ed. Forestry Commission Handbook 6. London: Stationery Office, 1991.

Hora, Bayard, consultant editor. *The Oxford Encyclopedia of Trees of the World.* New York: Oxford University Press, 1981.

Lowe, J. J., and M. J. C. Walker. *Reconstructing Quaternary Environments.* 2nd ed. Harlow: Pearson Educational, 1997.

Peterken, George F. *Natural Woodland.* New York: Cambridge University Press, 1996.

Rackham, Oliver. *Trees and Woodland in the British Landscape.* 2nd ed. London: Weidenfeld and Nicolson, 2001.

Roberts, Neil. *The Holocene: An Environmental History.* New York: Basil Blackwell, 1989.

Tansley, A. G. *Practical Plant Ecology.* London: George Allen and Unwin, 1923.

———. *Introduction to Plant Ecology.* London: George Allen and Unwin, 1946.

Williams, Joseph H., and William E. Friedman. "Identification of Diploid Endosperm in an Early Angiosperm Lineage." *Nature* 415 (Jan. 31, 2002): 522–526.

The World Commission on Environment and Development. *Our Common Future.* New York: Oxford University Press, 1987.

Web Sites

Borealforest.org. *World's Boreal Forests: Forest Management in Russia.* Borealforest.org. Available online. URL: www.borealforest.org/ world/rus_mgmt.htm. Accessed June 30, 2006.

Kandler, O. "The Air Pollution/Forest Decline Connection: The 'Waldsterben' Theory Refuted." FAO. *Unasylva* No. 174. Available online. URL: www.fao.org/docrep/ v0290e/v0290e07.htm. Accessed June 26, 2006.

Natural Resources Canada. *Genetic Engineering for Forestry in Canada: A Canadian Forest Service Discussion Paper 2002.* Available online. URL: www.nrcan.gc.ca/ biotechnology/english/d_implic.html. Last updated March 20, 2002. Accessed July 24, 2006.

Royal Society of New Zealand. *Genetic Engineering—An Overview.* Available online. URL: www.rsnz.org/topics/ biol/gmover/3.php. Accessed July 24, 2006.

Russian Conservation News. *Types and Forms of Russian Protected Areas.* Available online. URL: www.russian-conservation.org/opttypes.html. Accessed July 3, 2006.

Segelken, Roger. "Cyclosporin Mold's 'Sexual State' Found in New York Forest: Cornell Students' Discovery Could Target Additional Sources of Nature-Based Pharmaceuticals." *Science News.* Cornell University, September 16, 1996. Available online. URL: www.news.cornell. edu/releases/Sept96/cyclosporine.hrs.html. Accessed May 18, 2006.

United Nations Environment Program. *Geo-2000.* Chapter 2: The State of the Environment: Europe and Central Asia. UNEP. Available online. URL: www.unep.org/ geo2000/english/0076.htm. Accessed June 26, 2006.

World Energy Council. *Survey of Energy Resources: Wood (Including Charcoal).* Available online. URL: www. worldenergy.org/wec-geis/publications/reports/ser/ wood/wood.asp. Accessed June 21, 2006.

Worrall, Jim. *White Pine Blister Rust.* Forestpathology.org. Available online. URL: wwwforestpathology.org/dis_ wpbr.html. Last modified January 27, 2006. Accessed July 25, 2006.

Index

Note: *Italic* page numbers indicate illustrations; *m* refers to maps

rain forests, temperate 7, 165–166, *166*
rauli 218
receptacle 109
red alder 201
red ash 202
red buckeye 206
red fir 190
red maple 212–213
red silver fir 190
"red sky at night" 57
red spruce 199, 263–264
red willow 220
redwoods 7, 50, 94, 145, 153, 165, *165*, 198
refugia 95–98, 96*m*
regolith 19
Regosols 26
relative humidity (RH) 44
relative pollen frequency (RPF) 230
rendzinas 22
renewable resources 244–246
replanting 257, 282
reproduction strategies 139
reserves 245
respiration 167, 168
RH (relative humidity) 44
rhizomorphs 178
rhizosphere 104
rhododendron 217
Rhododendron ponticum 8, 9, 217
rhodora rhododendron 217
ribosomes 105
rills 274–275
Rio Summit (Earth Summit) 183
river birch 205
Robin Hood 239
roble de Chiloe 218
roble de maule 218
roble pellin 218
rock maple 212
rock oak 215
Roman Empire 224–226, 225*m*, 232–233
Roman mythology 241
root plate 28
roots 49–50, 103–104
rosebay rhododendron 217
Rossby, Carl-Gustav Arvid 59
Rossby waves 60
rowan 165, 217, 241

RPF (relative pollen frequency) 230
r-selection 139
ruil 218
Russia 243
Russian folklore 238

S

Saffir–Simpson Hurricane Scale 68
Sakhalin fir 191
Sakhalin spruce 200
sallow willow 221
salmon population 40–41
SALR. *See* saturated adiabatic lapse rate
salts 27
sand 21–22
sand pear 217
sap 50
saprobes 131
sapwood 115, 178
Sargent spruce 200
Sargent's rowan 217
saturated adiabatic lapse rate (SALR) 46, *64*, 65
saturation 44
savin 193
sawflies 177
Scandinavia 282–283
scarification 276
scarlet oak 215
Schimper, Andreas Franz Wilhelm 82
Schrenk's spruce 200
Schurz, Carl 243
sclerophyllous forest 6, 8
Scotch elm 207–208
Scotch pine 80, *80*, 88, 95, 165, 177, 181, 197
Scotland 234*m*
Scott, Michael 185
seafloor spreading 13
seamounts 13
sea owler 218
seasons 36, *36*
sedimentary rock 13
seed drill 226, *227*
seeds 100–102, 110–112, 141–142
seismic techniques 13
selective logging 9

sepals 109
September elm 208
sequoia 198
Sernander, J. Rutger 128
serotiny 111
service tree 218
shade 158, *158*, 267, 293
shade-tolerant trees 157–158
shagbark hickory 210
Shakespeare, William 238
sheep 167, 223
shingle oak 215
Shore pine 194
shortleaf pine 196
shrubs 160
Shumard oak 215
Siberia 5
Siberian crab apple 207
Siberian elm 208
Siberian fir 191
Siberian larch 193
Siberian spruce 200
Siberian stone pine 197
Siebold's beech 205
Siebold's walnut 220
sierozem 22
Sierra Nevada 6
Sierra redwood 198
sieve element 105
sieve plates 105
Sikkim larch 193
Sikkim spruce 200
silt 21
silver birch 205
silver fir 191
silver lime 204
silver maple 213
silver pendant lime 204
silviculture 306
single-leaf ash 202
SI system xv
Sitka alder 201
Sitka spruce 9, 80, *80*, 187, 198
slash pine 196
sleet 48
slippery elm 208
small-leaved lime 204
smelting 247–248
Smith, Adam 242
Smith, Robert Angus 257–258
smoothbark hickory 210